Aquatic and Terrestrial Humic Materials

Aquatic and Terrestrial Humic Materials

Edited by

Russell F. Christman
Egil T. Gjessing

ANN ARBOR SCIENCE
THE BUTTERWORTH GROUP

230 Collingwood, P.O. Box 1425, Ann Arbor, Michigan 48106

Library of Congress Catalog Card Number 82-71526
ISBN 0-250-40550-4

Manufactured in the United States of America

Butterworths, Ltd., Borough Green, Sevenoaks
Kent TN15 8PH, England

PREFACE

This volume is based on selected manuscripts that were presented at the Symposium on Terrestrial and Aquatic Humic Materials in Chapel Hill, North Carolina, on November 4 and 5, 1981. The symposium attracted substantial interest from the scientific community and consisted of plenary addresses, formal presentations and poster sessions concerning analytical methods for structure determination, reactions of humic materials and effects of these materials on living systems. The organization and delivery of the symposium was partially supported by grants from the Norwegian Institute for Water Research and the U.S. Environmental Protection Agency, Office of Drinking Water. The symposium concept was derived, in part, from associations developed during the North Atlantic Treaty Organization Committee on Challenges to Modern Society (NATO/CCMS) Drinking Water Pilot Project, which examined a broad spectrum of drinking water quality and health-related issues.

An international steering committee worked closely with us in selecting both the presentations for the symposium and the manuscripts for this publication. We wish to express our gratitude for this effort to Joseph A. Cotruvo, Christopher S. Martens, John W. Parsons, Johannes J. Rook, Alan A. Stevens and E. Michael Thurman. Special acknowledgment is due to Myrna L. Bower and Margaret C. Schimert for their roles in designing trouble-free arrangements.

Opportunities for scientists engaged in humic research to compare experimental methodologies, debate the significance of recent results and plan future experiments have been rare. In this context, the symposium was important given the recent public health emphasis attached to humic research. Our objective in publishing this work is to provide a record of most of the scientific deliberations shared at this symposium. We cannot capture the atmosphere of amiable cooperation and mutual regard that was evident at the symposium, nor can we measure the stimulation that resulted from the almost continuous interaction between presenters and attendees. To all who participated in this symposium, we respectfully dedicate this volume.

Russell F. Christman
Egil T. Gjessing

Christman Gjessing

Russell F. Christman received his PhD in Chemistry from the University of Florida in 1962 for research on the chemical structures of natural product organic materials in water. He joined the faculty of the University of Washington in Seattle in December 1962, and in 1970 was appointed Assistant to the Provost and Director of the Division of Environmental Affairs. In September 1973 Dr. Christman joined the faculty of the University of North Carolina at Chapel Hill, where he is the administrator of the Department of Environmental Sciences and Engineering.

Dr. Christman was one of the original members of the National Drinking Water Advisory Council and has served as a consultant to the U.S. Environmental Protection Agency and the World Health Organization. He has been a contributing member to the several committees and panels of the National Academy of Sciences that authored the 1974 edition of the Federal Water Quality Criteria. Since 1974 he has served as the editor of *Environmental Science and Technology*, a journal of the American Chemical Society. He is the author of numerous research papers and book contributions in the field of environmental studies. Dr. Christman's main research interest has been aquatic organic chemistry, particularly the structural identification and properties of aquatic humic substances and their reactions with oxidizing agents such as $KMnO_4$, Cl_2, ClO_2 and O_3.

Egil T. Gjessing has been affiliated with the Norwegian Institute for Water Research (NIVA) since 1962, where he is currently Assistant Manager of the Environmental Engineering Division. He is a graduate of the University of Bergen, Norway, in Chemistry, and worked with the Norwegian Fisheries Research Laboratory for a two-year period following graduation. As a research scientist with NIVA, Dr. Gjessing has directed major projects focusing on the properties of natural organic matter in water and is the author of some 30 papers on the subject. He has also participated in other NIVA projects, including a four-year study of the effects of polluted precipitation on Norwegian surface waters, a major Norwegian interdisciplinary study on acid precipitation and its effect on forest and fish, and a two-year study on toxic pollutants in the aquatic environment. In recent years he has focused on drinking water problems. In 1965 he was on leave from NIVA on a U.S. postdoctorate fellowship at the Water Chemistry Laboratory, University of Wisconsin, working with organic matter in water. Dr. Gjessing is the author of *Physical and Chemical Characteristics of Aquatic Humus*, published by Ann Arbor Science.

CONTENTS

CHAPTER 1

STRUCTURAL STUDY OF HUMIC SUBSTANCES: NEW APPROACHES AND METHODS

E. M. Thurman and R. L. Malcolm

U.S. Geological Survey
Denver Federal Center
Denver, Colorado

ABSTRACT

This report proposes a threefold approach to study the structure of humic substances. This approach uses: (1) separation and purification of humic substances by liquid chromatography; (2) degradation by oxidative procedures and methylation of humic substances by ^{13}C reagents; and (3) identification of products and structural components by capillary gas chromatography-mass spectroscopy and ^{13}C nuclear magnetic resonance (^{13}C-NMR). Using this approach, we fractionated and characterized fulvic acid from an aquatic source—the Suwannee River in southeastern Georgia. This example of fractionation and characterization shows that adsorption liquid chromatography separates fulvic acid from inorganic substances, carbohydrates and fatty acids. Standard methods of characterization, elemental analysis and functional-group titration are compared against the new method of ^{13}C-NMR for the Suwannee River sample. The corroboration of methods shows that ^{13}C-NMR is a useful technique for the structural analysis of humic substances. Finally, based on the preceding characterization, fulvic acid in the Suwannee River is thought to consist of a mixture of plant degradation products, including lignin, terpenoids, flavonoids and hydrolyzable tannins.

INTRODUCTION

Humic and fulvic acids are, by the simplest definition, the sodium hydroxide extract of soil. The analytical problem of humic substances is complicated by the simplicity of this definition because: (1) soil organic matter is a complex mixture of many organic substances; (2) sodium hydroxide extracts many of these substances, including simple organic compounds, inorganic compounds, and complex mixtures of polysaccharides and humic substances; and (3) humic substances themselves are complex mixtures that represent products of microbial degradation, chemical polymerization and oxidation.

Where does one begin on this complex analytical problem? Over the past 50 years, others have proposed analytical schemes for humic substances; today, our research objective is to evaluate the most recent analytical methods for a new insight on an old problem. The subject of this chapter is how to use the new separation procedures, degradation schemes and analytical methods to characterize, and someday structurally identify, the nature of humic substances.

The three key words of this chapter are separate, degrade and identify. This is shown diagramatically in Figure 1. Here we review the new analytical

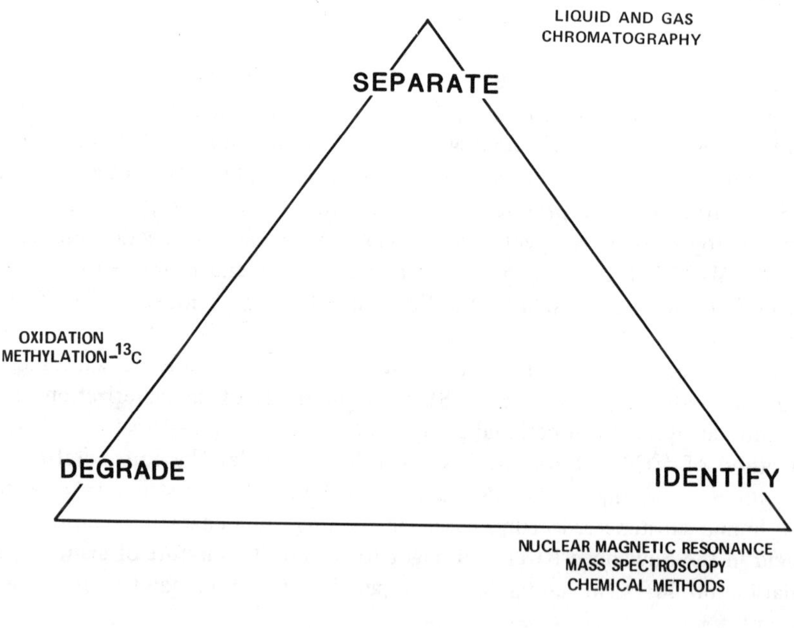

Figure 1. Three-step approach for structural analysis of humic substances.

advances that have been applied to humic substances over the past few years. These analytical advances, beginning at the apex of the triangle in Figure 1, include separation methods, such as capillary gas chromatography and liquid chromatography; methods of identification, such as mass spectroscopy and carbon-13 nuclear magnetic resonance spectroscopy (^{13}C-NMR); and methods of degradation, such as chlorination and ^{13}C-methylation. This is not an inclusive list, but rather a representative one we have subjectively chosen that contains some excellent techniques to apply to humic substances.

A review of these methods comprises the first part of this report. The second part describes the application of these techniques to a sample of fulvic acid from the Suwannee River in southeastern Georgia. The combination of the three step approach shown in Figure 1 gives us new insight into the nature of aquatic humic substances from the Suwannee River. Surely the application of these analytical methods will prove equally useful on humic substances from both water and soil.

EXPERIMENTAL

Sample Location and Processing

The Suwannee River was sampled in January 1977; the fulvic acid was removed from the water by the procedure of Thurman and Malcolm [1].

Characterization Methods

The amino acid analysis was done by Beckman* amino acid analyzer after an 8-hr digestion in refluxing 6 *N* hydrochloric acid. Carbohydrate analysis was by the phenol-sulfuric acid method [2]. Liquid ^{13}C-NMR was done by the method of Mikita and others [3].

Titration of phenolic and carboxylic acids was done by titration with standard base to an endpoint of pH 8 for carboxylic acids and pH 10 for phenolic groups. Phenolic groups were assumed to be 50% titrated at this pH, and were corrected by multiplying the titrant volume by 2. Because the acid-dissociation constant of phenol is a pH of 10, this is a minimum estimate of phenolic content.

Elemental analyses were carried out by Huffman Laboratories, Wheatridge, Colorado. Molecular weight was determined by X-ray scattering [4]. The Quaternary Research Center at the University of Washington performed age determination. A 2.5-g sample was used. Delta ^{13}C was determined by the

---

*Any use of trade names is for descriptive purposes only and does not imply endorsement by the U.S. Geological Survey.

same laboratories. Solid-state ^{13}C-NMR spectra were obtained from a home-built spectrometer, located at Colorado State University, at a field strength of 1.4 T.

DISCUSSION

Separation Methods

Humic substances are structurally complex, polyelectrolytic, dark-colored organic acids that are found in soils, sediments and natural waters. These substances are fractionated into three fractions, including: (1) humic acid, which is soluble in dilute base, but precipitates in dilute acid; (2) fulvic acid, which is soluble in both dilute base and dilute acid; and (3) humin, which is a humic-like polymer that will not alkaline-extract from soil or sediment.

Delineation of these three fractions is the first step in separation methods. In fact, most studies address only humic and fulvic acids. Humin, because it does not alkaline-extract, is more difficult to study. In natural waters, humic and fulvic acids are the major two types of humic substances. Humic substances account for 40 to 80% of the dissolved organic matter in water; they are an important and ubiquitous fraction.

The number of separation methods that may be applied to humic substances are many: acid precipitation, metal precipitation, electrophoresis, dialysis, liquid extraction, ion-exchange and liquid chromatography. Of these methods, we would like to address liquid chromatography, because many advances in technology and mechanisms of separation have been made. It is also a separation technique that has only been applied recently to humic substances from soils, sediments and water. The fact that it works well on polar organic acids makes it a suitable separation method for humic substances.

There are many types of liquid chromatography that function or separate by different mechanisms. Table I lists the types of liquid chromatography

Table I. Types of Liquid Chromatography Applied to Solutions of Humic Substances

Type of Chromatography	Mechanism of Separation
1. Gel filtration	Size or molecular weights
2. Adsorption	Hydrophobicity
3. pH gradient adsorption	pKa (acid dissociation constant)
4. Weak anion exchange	Hydrogen bonding or anion exchange
5. Cation exchange	Charge exclusion

that are applicable to solutions of humic substances. They are listed in order of their use for humic substances, with gel filtration the most popular.

Gel Filtration

Size separation has several problems that limit its use:

1. Sephadex gels are used most frequently to fractionate humic substances. These resins may adsorb humic substances onto the gel [5]; this retards the elution of the material and results in an underestimate of molecular weight.
2. A slight negative charge present on the resin repulses the molecule, which also has negative charge. Thus, the molecule travels the column more quickly than a noncharged solute of similar size, and an error in molecular size is made: the molecule appears larger than its actual size [6].

These problems, coupled with the fact that major size differences occur between humic acid and fulvic acid, but not among fulvic acids, limits the use of this type of chromatography. In spite of the above arguments against gel chromatography, it is widely used. A better approach in size separation may be through the humic/fulvic acid split and ultrafiltration of the fractions, using PM-membranes (a brand name of Amicon, Lexington, Mass.). We concluded on the basis of another study [4] that size separation of fulvic acid is difficult because of the small range in molecular weight of aquatic fulvic acid (from 1000 to 2000). Other researchers also have shown that fulvic acid from soil is in that same range [7].

Adsorption Chromatography

The second type of chromatography listed in Table I, adsorption chromatography, has also been widely used, but only on aquatic substances [6,8,9], although Aiken et al. [10] recommended its use on soil extracts. This type of chromatography works by the hydrophobic effect: the nonpolar carbon skeleton binds or adsorbs to the resin. The polar groups (hydroxyl and carboxylic groups) orient into the aqueous phase and work toward desorbing the molecule. Adsorption and desorption may be controlled by balancing the polar and nonpolar interactions. This is done by adjusting the pH to 2.0, which makes the humic substances nonionic, and they adsorb onto the resin. At pH 5.0 or greater, humic substances are ionic and desorb, because of the ionic character of the carboxyl group.

The use of adsorption chromatography to separate various nonhumic substances from fulvic acid is shown in Figure 2. Simple carbohydrates, uronic acids and hydroxy acids pass through the XAD resin and do not adsorb, because they are hydrophilic molecules. The fulvic acid adsorbs onto the resin and is separated from these substances. Thus adsorption chromatography is a simple way to separate nonassociated carbohydrates from fulvic acid.

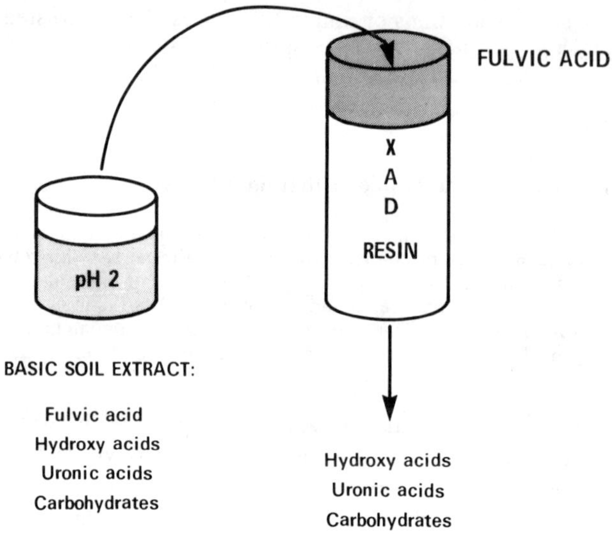

Figure 2. Adsorption chromatography of basic soil extract.

In fact, aquatic and soil fulvic acid differ considerably in their carbohydrate content. Aquatic humic substances have considerably less carbohydrate, because they are processed by adsorption chromatography. Table II shows major differences in soil and aquatic fulvic acid because of resin separation. Carbon content is increased, oxygen content is lowered and the colorimetric test for carbohydrates shows much less for aquatic material. We have taken soil fulvic acid and separated the free carbohydrates with resin adsorption. Because carbohydrates are molecules about the size of fulvic acid, they are not separated by classical fractionation, such as dialysis. However, soil fulvic acids processed by resin adsorption dropped 75% in carbohydrate content after resin processing (Table III), and approached the levels of carbohydrate in aquatic fulvic acid.

Table II. Differences Between Soil and Aquatic Fulvic Acid from Carbohydrate Content

	Percent by Weight		
	Carbon	Oxygen	Carbohydrate
Soil fulvic acid	46	45	10-15
Aquatic fulvic acid	52	39	2-5

Table III. Change in Carbohydrate Content by Adsorption Chromatography on XAD Resins

	Percent Carbohydrate	
	Before Adsorption	After Adsorption
North Carolina fulvic acid	7.5	2.0
Prairie soil fulvic acid	20.0	5.0

pH Gradient Adsorption Chromatography

The third type of chromatography, pH adsorption, separates humic substances by adsorbing them at two pH regions. First, they are adsorbed at pH 7 in a sodium bicarbonate buffer. At this pH, the weak acids are nonionic and adsorb onto the resin; the stronger carboxylic acids are ionic and pass through the resin. Humic acids generally adsorb at pH 7, whereas fulvic acids do not adsorb.

Within the humic acid fraction, the material may be separated into fractions containing more and less carboxylic acids. For example, Table IV shows an Irish peat humic acid fractionated by adsorption chromatography at pH 7. The fraction that adsorbed at pH 7 contained 4.2 mmol/g carboxylic acids, and the fraction that did not adsorb contained 5.2 mmol/g. A third fraction, eluted in methanol, was less than 4.0 mmol/g carboxylic acids. Thus, the fractionation gives materials of varying amounts of acidity. Note that the phenolic content increased as carboxyl content decreased. This is typical for humic substances separated by pH gradient chromatography [11].

Weak Anion Exchange

Adsorption chromatography removes the more hydrophobic organic matter from water by Van der Waals interactions of the carbon skeleton with the matrix of the neutral adsorbent [8,12,13]. Solutes that contain many

Table IV. Humic Acid Fractionated by pH Gradient Adsorption

Fraction Irish Peat Humic Acid	Carboxyl (mmol/g)	Phenol (mmol/g)
Adsorbed at pH 2	5.2	3.2
Adsorbed at pH 7	4.2	5.1

functional groups (one functional group for every one to three carbon atoms) are commonly too hydrophilic (water-soluble) to adsorb on the XAD adsorbents. These organic acids are called hydrophilic acids [14].

Humic substances isolated from soil and water are usually removed by XAD adsorbents, but some hydrophilic acids are not retained and pass through the resin. This hydrophilic fraction may be adsorbed and recovered quantitatively by weak-base anion exchange. Degradation studies of humic substances generate hydrophilic organic acids, which are a second example of hydrophilic acids that may be quantitatively recovered by weak-base anion exchange. This method follows the procedure of Leenheer [15].

First, the sample is hydrogen-saturated with cation exchange resin (Figure 3), and then it flows through the weak-base resin in its free base form. Therefore, the organic acids coming out of the cation exchange resin elute as R-COO^-H^+, and major anions such as H^+Cl^-, $H_2^+SO_4^=$ and so forth. These salts and organic acids are retained by the resin as the hydrogen ion converts the resin to an anion exchange resin; then the accompanying anion is retained. Thus, the sample flowing out of the anion exchange resin is now deionized water and contains only hydrophilic neutral organic solutes that are present in the sample.

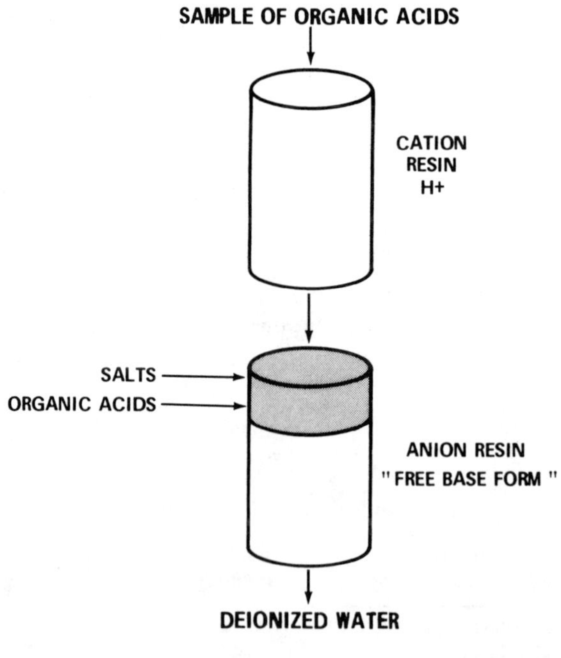

Figure 3. Adsorption of organic acids by weak-base ion exchange [15].

The weak-base resin is then eluted with 3 N NH_4OH (ammonium hydroxide) to remove the organic acids. The inorganic salts coelute with the organic acids. This concentrate is then ready for analysis by conventional techniques.

The weak-base resins may also be used for chromatography of humic substances. This method has been useful for separating a phenolic-rich fraction from aquatic fulvic acids. This type of chromatography uses the hydrogen bonding effect of phenolic groups for the nitrogen groups present on the weak-base resin.

The sample is passed through the column at pH 7 with a sodium bicarbonate buffer. The phenolic groups are nonionic at this pH and adsorb by hydrogen bonding to the amine groups present on the resin. The amines also are nonionic at this pH. This separation mechanism is shown in Figure 4. Adsorption occurs when the phenolic functional group interacts with the resin, while the hydrophilic part of the molecule interacts with water. If the hydrophilic interaction is not as energetic as the adsorption of the functional group, the molecule remains on the resin. However, if there are sufficient carboxylic acid groups or other hydrophilic groups, hydroxyl or carbohydrate, the molecule does not adsorb at pH 7. The difference in functional group content between the fraction that adsorbs at pH 7 and the fraction that does not adsorb is shown in Table V. The fraction that absorbs at pH 7 (elutes at pH 13) has three times or more phenolic content than the fraction that passes through the resin at pH 7. Thus, weak-base ion exchange is a good method to separate a phenolic-rich fraction from fulvic acids in soil and water.

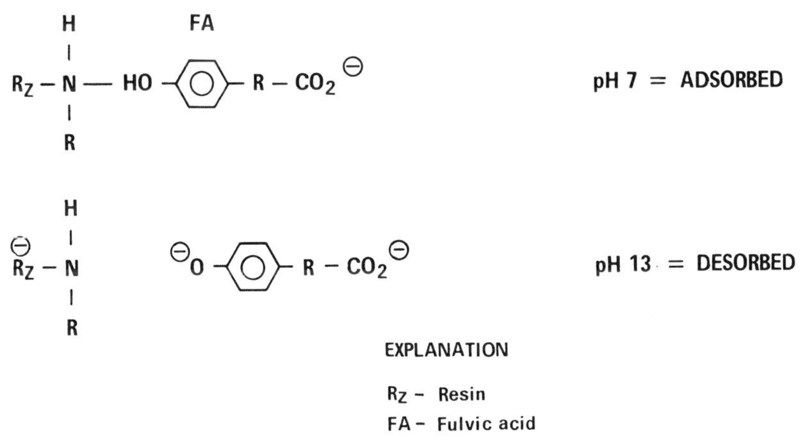

Figure 4. Phenolic interaction on a weak-base resin.

Cation Exchange

Cation exchange is used commonly to hydrogen-saturate humic substances and convert them from sodium salts to a free acid form. It also removes trace metals and other cations associated with the humic material.

Any amines or amino acids associated with the humic substances are retained on the cation exchange resin. If humic substances contain free amines, they would be removed on the cation exchange resin [16]; these are the usual applications of cation exchange resins.

There are chromatographic uses for 10-μ cation exchange resin. These resins are used in high-performance liquid chromatography (HPLC) to separate hydrophilic acids [17]. Although there are no published results on humic substances with these resins, they are effective at separating complex mixtures of acids and should work well on the fulvic acid fraction.

Degradation Techniques

Degradation of humic substances traditionally involves harsh oxidative conditions using alkaline permanganate and copper oxide [7,18,19]. The products of these conditions are monocarboxylic acids, dicarboxylic acids and benzene carboxylic acids. These products are the skeletal remains of oxidation, because the powerful oxidative agents take one-third of the molecules to form carbon dioxide [19]. These products have given us an idea of the structural core of humic substances–that is, an aromatic core with aliphatic chains.

Milder oxidative conditions, such as alkaline hydrolysis, have been tried [18,19]. Phenols and phenolic acids are found, as well as aromatic and

Table V. Functional Group Content of Two Fractions from a Weak-Base Resin (ES-771-Duolite Diamond Shamrock)

Fraction	Carboxylic Acid (mmol/g)	Phenolic Hydroxyl (mmol/g)
Biscayne fulvic acid		
pH 7	7.3	1.1
pH 13	3.7	5.4
Colorado River fulvic acid		
pH 7	5.8	0.1
pH 13	3.4	3.8

aliphatic acids. These data suggest a core of aromatic content with ester bridges of aliphatic content [19].

Permanganate oxidation of methylated humic substances has been used; this protects phenolic OH groups against attack by permanganate. Phenolic acids have been a major product of this procedure [18].

Nitric acid oxidation produces nitrophenolic acids. Thus, the conclusion from the oxidative procedures is a humic substance that contains a core, both aromatic and phenolic. One must be careful that the low yields of these procedures do not mislead us.

Reductive procedures, such as zinc dust distillation and Na amalgam reduction, have not been productive, because of the large oxygen content of the humic material; yields are only 1% or less [18]. Sodium amalgam reduction has produced compounds, such as vanillic, syringic and p-hydroxybenzoic acids, and indicates that lignin may be a source material for humic acid [18].

Because of the interest in chlorination of water, chlorination of aquatic humic substances has been used as an oxidative degradation procedure [20,21]. These data are relatively new to the humic-substances area, so the high points of this work will be reviewed. Also, several new types of degradation studies that modify humic substances so that ^{13}C-NMR may be used to elucidate structure will be discussed.

New Procedures

Chlorination: Oxidative Procedure. The oxidative procedure follows that of Christman et al. [20] and uses sodium hypochlorite, NaOCl, in alkaline conditions of pH 12. The sample is methylated and products determined by gas chromatography/mass spectroscopy [20]. These authors determined mass balance of carbon in their systems; they identified 8% of the humic acid and 17% of the fulvic acid.

The principal products from humic acid were both aliphatic and aromatic di-, tri- and tetracarboxylic acids. Phenolic and methoxy structures were absent, which suggests chlorine attack on aromatic rings and their cleavage and destruction. Two of the most abundant compounds in the ether extract were dichloro- and trichloroacetic acid. These have been shown to be products of the oxidation of resorcinol and 3-methoxy, 4-hydroxy cinnamic acid [20].

The principal products from chlorination of fulvic acid were similar to the humic acid sample. In total, there were 41 identified products from humic acid and 71 products from fulvic acid.

The important results of their work show that three classes of compounds were identified, and all were acids: (1) nonchlorinated substituted aromatic acids; (2) nonchlorinated straight-chain acids; and (3) chlorinated straight-chain acids. One interesting finding was phthalonic acid, which results from the oxidation of naphthalene, and constitutes evidence for fused-ring structures in humic acid.

It was concluded on the basis of chlorination studies that 17% of the carbon was identified; if this 17% is representative, then the approximate ratio is 67:19:2.5 for aromatic to dibasic to monobasic compounds for aquatic humic acid. They also find that aromatic esters are important. The carbon linkage is generally 3 to 4 carbon-carbon bonds per aromatic ring to account for tri- and tetra-substituted acids. Milder chlorine oxidation procedures produce more aliphatic than aromatic products [21].

Methylation with ^{13}C Reagents. This method, developed by Wershaw and applied by Mikita, Steelink, and Wershaw [3] involves the methylation of hydroxyl groups present in the humic substances with ^{13}C-enriched reagents. The enriched material is then analyzed by ^{13}C-NMR. An example is shown in Figure 5 and is discussed in the section on NMR spectroscopy.

Identification

In the past 10 years, two methods of identification have been developed that are especially useful in the study of humic substances: ^{13}C-nuclear magnetic resonance (^{13}C-NMR) and gas chromatography/mass spectroscopy (GC/MS). In this section, we will compare these methods and their application to humic substances, with the older standby methods of analyses, such as elemental analysis, functional group titrations, molecular weight and wet chemical methods. Finally, we will present confirmation of results between the wet chemical methods and ^{13}C-NMR.

^{13}C-Nuclear Magnetic Resonance (^{13}C-NMR)

Although proton NMR (^{1}H-NMR) became commercially available in 1953 and in the mid-1950s it was realized that ^{13}C spectra could be obtained, it was about 1970 before the introduction of commercial Fourier transform (FT) spectrometers [22] for ^{13}C. Because the natural abundance of ^{13}C is approximately 1%, there is considerably less sensitivity of this nucleus compared to the proton. Thus, pulsed FT is needed to enhance the signal-to-noise ratio of the NMR; the result is a "good" ^{13}C-NMR spectrum.

The first reports of NMR spectroscopy of humic substances by ^{1}H- and ^{13}C-NMR appeared in the late 1960s by Schnitzer and Skinner [23] and Ishiwatari [24]. Vila et al. [25] had the first definitive report on ^{13}C-NMR. They showed that Fourier transform ^{13}C-NMR gave direct measurement of the carbon atom in humic substances, so detailed structure was now possible.

They pointed out that the chemical shifts in ^{13}C-NMR are an order of magnitude or more than ^{1}H-NMR. Their spectra show aromatic and aliphatic carbons, but poor signal-to-noise ratio hindered resolution of other structure. Stuermer and Payne [26] also used ^{13}C-NMR to indicate differences in aliphatic character between marine and terrestrial fulvic acids.

Useful ^{1}H-NMR spectra of humic substances were obtained by Lentz et al. [27] in NaOD solutions, because both aromatic and aliphatic protons are present. Wilson et al. [28] present ^{1}H-NMR and ^{13}C-NMR for humic acids. Their spectra show the presence of both aromatic and aliphatic carbons and protons; however, signal-to-noise ratio is poor.

Recently both ^{1}H-NMR and ^{13}C-NMR spectra were obtained by Hatcher et al. [29,30] that give meaningful structural analysis of humic substances. The ^{13}C-NMR spectra, obtained in NaOD solutions, clearly show aliphatic carbons, methoxyl carbons, carbohydrate carbons and aromatic carbons. The ^{1}H-NMR spectra presented reinforced the interpretations of the ^{13}C-NMR spectra.

Hatcher et al. [29,30] calculate areas for both aliphatic and aromatic carbon, pointing out that aquatic sediments contain humic substances that are 10 to 15% aromatic, while terrestrially derived humic acids are 20 to 35% aromatic. Hatcher et al. warn that relative areas may be incorrect because of the Nuclear Overhauser Enhancement, a coupling of a proton and ^{13}C nuclei that enhances the size of the ^{13}C peak.

Hatcher et al. [29] also present ^{13}C-NMR spectrum in the solid state and compare the spectra with the liquid ^{13}C-NMR in NaOD. The comparisons are very encouraging for structural analysis of humic substances. The solid-state ^{13}C-NMR uses both magic angle spinning (54.7°) and cross polarization to sharpen and define the spectra. (See Miknis et al. [31] for an explanation of these techniques on coals and organic polymers.) This is state-of-the-art use of NMR on humic substances; results are very encouraging.

In their second report, Hatcher et al. [30] report solid-state ^{13}C-NMR of humic acids and humins from recent marine sediments. Cross polarization and magic angle spinning (CP/MAS) is used for the ^{13}C-NMR. The spectra show the differences in humin from terrestrial and aquatic environments; terrestrial are more aromatic and less aliphatic in character.

The ^{13}C-NMR spectra being obtained open new areas of research in humic substances. For example, Mikita et al. [3] have pioneered an enrichment of ^{13}C using methylating agents, as previously discussed. These enriched and methylated humic substances give very good spectra of the functional groups present in humic substances. A spectrum obtained by the methylation procedure of Wershaw (described in the preceding section) is shown in Figure 5.

The peak at 52 ppm is the carboxyl peak; 55 ppm is the phenolic hydroxyl; and 57 ppm is the hydroxyl region, including carbohydrate. These spectra are now compared with the solid-state ^{13}C-NMR in CP/MAS mode for the same sample (Figure 6). Note that the region compared in solid-state ^{13}C-NMR is from 0 to 200 ppm, and in the methylation procedure from 50 to 60 ppm. The solid-state ^{13}C-NMR is for natural abundance ^{13}C. The peak at 30 ppm is aliphatic carbon; at 75 ppm it is carbon bonded to hydroxyl; at 125 ppm it is aromatic carbon; at 140 ppm it is phenolic hydroxyl; at 160 ppm it is carboxylic carbon; and at 200 ppm it is carbonyl carbon. In both spectra, the area under the peaks is integrated to determine the percentage of each functional group.

We integrated these samples and compared data for both Suwannee River

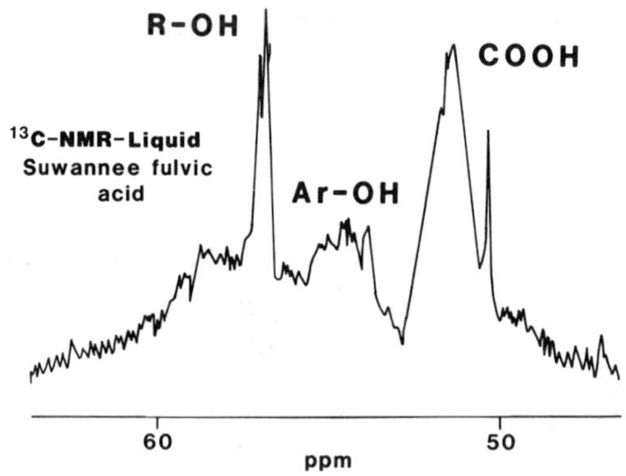

Figure 5. Liquid-state ^{13}C-NMR for Suwannee River fulvic acid.

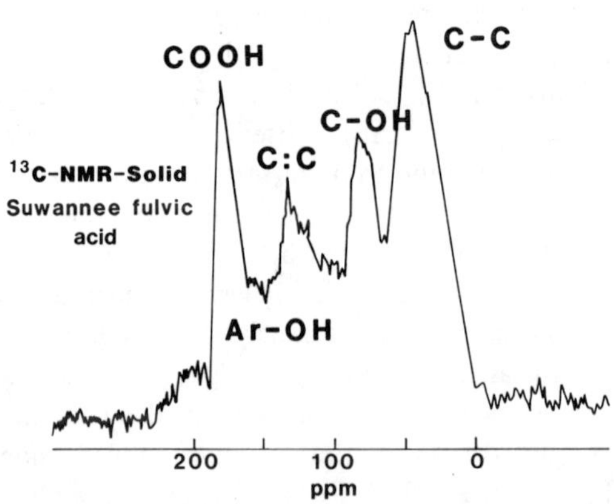

Figure 6. Solid-state ^{13}C-NMR for Suwannee River fulvic acid at a field strength of 1.4 T [31].

humic and fulvic acids and for the Biscayne humic and fulvic acids. The Biscayne is a groundwater humic substance isolated from the Biscayne aquifer near Miami, Florida. Functional group comparisons by the NMR methods and a titration method are shown in Table VI.

It is important that these different NMR methods and the titration method are similar in their results, because this means that further enhancement of spectra will better define the functional group content of humic substances. This, coupled with methods of separation and degradation, opens a new research area of humic acid chemistry. Elemental analysis can be used to check solid-state ^{13}C-NMR (Table VII). Again, this shows that integration methods for solid-state ^{13}C-NMR are accurate. Elemental analyses are valuable to cross-check ^{13}C-NMR analysis of humic substances.

Table VI. Comparison of Functional Groups of Fulvic Acids from Suwannee River and Biscayne Aquifer

Method	Carboxyl (mmol/g)	Phenol (mmol/g)	Hydroxyl (mmol/g)
		Biscayne Fulvic Acid	
Titration	6.3	1.8	
Solid-state ^{13}C-NMR	7.2	1.3	7.5
Methylation ^{13}C reagents NMR	6.3	1.9	7.9
		Suwannee Fulvic Acid	
Titration	6.0	2.1	
Solid State ^{13}C-NMR	6.2	1.7	8.6
Methylation ^{13}C reagents NMR	6.0	3.6	5.4

Table VII. Comparison of Methods for Humic Substances: Solid-State ^{13}C NMR and Elemental Analyses

Sample	Elemental Analysis for Oxygen (wt%)	^{13}C[a]-NMR Oxygen Analysis (wt%)
Suwannee fulvic acid	42.9	39.1
Suwannee humic acid	40.9	39.7
Biscayne fulvic acid	35.4	37.1
Biscayne humic acid	30.1	34.4

[a]Calculated from the number of C-O bonds.

Gas Chromatography/Mass Spectroscopy (GC/MS)

Refinements in capillary gas chromatography and high-resolution mass spectroscopy have revolutionized the identification of organic compounds. For example, in the work of Christman et al. [20] on degradation products of humic substances, capillary GC and high-resolution mass spectroscopy separate and identify over one hundred compounds. The combination of retention time on the GC and accurate mass measurements (to ±10 ppm) gives high assurance of compound identification, even when standard compounds are not available.

Improvements in electromagnets, so that fast scanning is available [less than 1 sec for 1 decade (100 to 1000 amu)], allow the high resolution of the electromagnets, and yet give the fast scanning times of the quadrapole mass spectrometers. Generally, the quadrapole mass spectrometers have a resolution in parts per thousand; the higher resolution of the electromagnet is necessary for research on humic substances, where standard compounds do not exist.

New probes have been developed; for instance, a fast atom bombardment (FAB) is available using chemical ionization that gives spectra on compounds with molecular weights of 500 or more [32,33]. This tool gives promise for fulvic acid fractions, especially those isolated from water, where molecular weights range from 500 to 2000. If methylation procedures are used prior to MS, this may increase volatility so that solid probe high-resolution mass spectroscopy will be an important tool for humic structural studies. Although still in the development stage, the use of FAB/MS/MS may allow structural identification of humic substances as complex mixtures. In this method the first MS acts as a chromatographic separator to separate molecular ions, and the second MS identifies daughter products of the molecular ions [34].

RESULTS

Combining Separation; Degradation, and Identification

In this section, we apply separation, degradation and identification methods on humic substances from the Suwannee River in southeastern Georgia (Figure 7). Step 1, separation, concentrated the sample from water and separated it from inorganic salts, free amino acids and free carbohydrates. Water-soluble organic acids, such as uronic acids and hydroxy acids, are also separated. This separation was done by adsorption onto XAD-8 resin. This same procedure works very well on humic extracts from soil. If this simple procedure were done on the fulvic acid extract, the free saccharides, salts

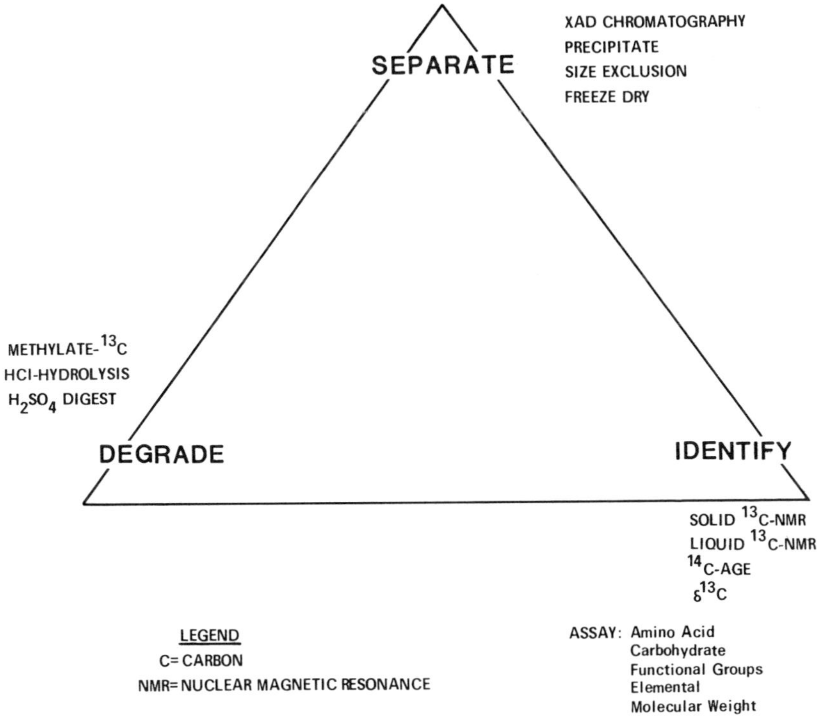

Figure 7. Three-step approach applied to the Suwannee River fulvic acid.

and many free organic acids would be removed. Preparation of a hydrogen-saturated low-ash fulvic acid is essential for thorough characterization and identification. After the precipitation of humic acid, the fulvic acid is freeze-dried. This step further purifies the sample by lyophilizing any fatty acids or phenols that may be present at trace levels. A polymer of fulvic acid remains, ready for degradation and identification studies.

The Suwannee sample was degraded by: 6 *N* HCl hydrolysis and amino-acid assay, sulfuric-acid digestion for carbohydrates and methylation with ^{13}C reagents for ^{13}C-NMR.

The Suwannee River fulvic acid was then identified by the 9 methods listed in Figure 7. The results of these tests give us the following information on the Swannee River fulvic acid (summarized in Table VIII).

The Suwannee River originates in the Okefenokee Swamp in southeastern Georgia. The $^{13}C/^{12}C$ (ratio) of −31.4 suggests a terrestrial origin [35] and the ^{14}C age shows that the organic matter contains large amounts of "bomb"

Table VIII. Characterization of Suwannee River Fulvic Acid

1. Elemental analysis			
C = 51.3	N = 0.56	ash = <0.05	
H = 4.32	S = <0.2		
O = 42.9	P = <0.1		
2. Functional groups (mmol/g)			
$-CO_2H$ (titration)	= 6.0	phenolic –OH	= 2.1
$-CO_2H$ (NMR-solid)	= 6.2		= 1.7
$-CO_2H$ (NMR-liquid)	= 6.0		= 3.6
hydroxyl (NMR solid)	= 8.6	carbonyl (NMR) solid	= 1.7
hydroxyl (NMR liquid)	= 5.4		
3. Carbon distribution (%)			
aromatic	20.8		
aliphatic	36.7		
C-OH	20.1		
CO_2H	14.6		
phenolic	3.9		
carbonyl	4.0		
4. Amino acids = 36 nmol/mg			
% N (AA) = 10%			
5. % carbohydrate = 4%			
6. Molecular weight = 1200 ± 200			
7. ^{14}C age = less than 30 yr			
8. $^{13}C = ^{12}C = -31.4$			

^{14}C. This means that the fulvic acid is very recent material, probably plant organic matter of recent age, as well as organic matter degrading over the past 25 years, not old organic matter from accumulating peat in the swamp.

The molecular weight, as shown by X-ray scattering, vapor pressure osmometry, and ultrafiltration [3], is 1000 to 1700. The humic acid, which constitutes only 5% of the organic matter has a molecular weight of from 3000 to 5000.

The major functional groups are hydroxyl and carboxyl. The carboxyl content is 6.0 mmol/g, or approximately 1 carboxyl group per 7 carbon atoms. On the average, this is one carboxyl group per aromatic ring. This assumes that carboxyl groups are equally distributed, which may not be a valid assumption. Hydroxyl content is approximately 8.6 mmol/g, or approximately 1 hydroxyl for every 5 carbon atoms. Phenolic hydroxyl content is approximately 2.5 mmol/g (average of 3 methods) or 1 hydroxyl for every 17 carbon atoms. Carbonyl content is 1.7 mmol/g, or 1 carbonyl for every 25 carbon atoms.

Summarizing for all functional groups, we find the fulvic acid contains

18.8 mmol/g functional groups, or approximately 4 functional groups for every 9 carbon atoms. Thus, the polymer is highly substituted by oxygen-containing functional groups.

From the elemental analysis and the molecular weight, we deduce that the molecular formula is approximately $C_{74}H_{72}O_{46}N_{0.7}$, with a molecular weight of approximately 1700; this is a rough estimate for the sake of structural interpretation of the molecule. Using solid ^{13}C-NMR, we find that of the 74 carbons, 20.8% (i.e., 16.0 carbons) are aromatic carbons. This could be interpreted as between 2 and 3 aromatic rings present in the molecule. From the molecular formula and the characterization data, the accounting of functional groups and aromatic and aliphatic content of the fulvic acid molecule is compiled in Table IX.

Let us now use this characterization data to construct structures that are consistent with the data. First, liquid chromatographic data suggest that the fulvic acid fraction in water is a mixture of compounds containing phenolic groups, carboxylic acid and hydroxyl groups. The weak-base anion exchange chromatography suggests that a more highly phenolic fraction exists where phenolic groups exceed carboxylic acid groups, and that it may account for 15 to 30% of the organic matter in some samples.

Therefore, the structures we will present are that of a mixture; the components of the mixture are major higher plant products that have undergone oxidation and perhaps some phenol polymerization; polymerization is probably a less important process than oxidation. It must be noted that these structures are only proposed as possible structures.

Let us examine the hypothetical components in Figure 8. First are lignin products. This molecule is highly substituted and is soluble after oxidation. Only the structure of spruce lignin is well known, but other lignins are known

Table IX. Carbon Assignments from Molecular Formula and Structural Information

Assignment	Number of Carbon Atoms	Number of Hydrogen Atoms	Number of Oxygen Atoms
Aromatic	16	?	
Carbohydrate	6	12	6
Carboxyl	10	10	20
Phenolic	4	4	4
Carbonyl	3	0	3
Hydroxyl	8	8	8
Aliphatic	25	?	
Ether	?	0	5
Total	72	34	46

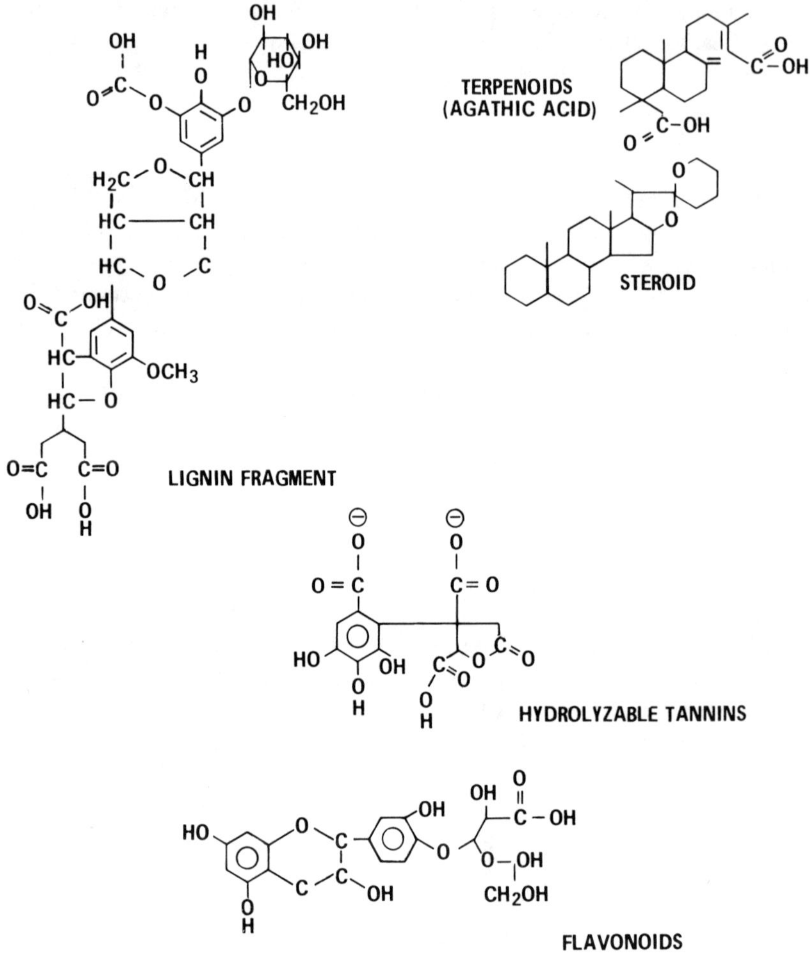

Figure 8. Possible structural components in fulvic acid from Suwannee River.

to contain glucosides that contribute solubility and carbohydrate character to the lignin. Oxidation and opening of aromatic rings would increase solubility and carboxyl content. A structure like this may be more likely for the humic acid in water, because of the greater aromatic character. However, the aliphatic character of lignin does fit the solid-state ^{13}C-NMR data for the aliphatic region, i.e., highly substituted aliphatic carbons and a low H/C ratio.

The second possible component in the mixture is the hydrolyzable tannins. These are esters of gallic acid connected to a central glucose unit. Esters have been reported to be possible structures in aquatic fulvic acid, and are good candidates, because of solubility and widespread occurrence. These molecules emphasize a core that is aromatic and carbohydrate-rich. Microbial breakdown of these products would most likely yield derivatives with gallic acid. The aliphatic carbons are not abundant, but those that are present are greatly substituted.

Compounds that would increase aliphatic character of the mixture are terpenoids and steroids. Generally these are insoluble oils, but acids of these compounds do occur, as well as glycosides, which would greatly enhance solubility. Note that these are highly substituted aliphatic carbons that keep the H/C ratio low and would give NMR signals consistent with the spectra we found. The terpenoid acid is agathic acid.

The final component to consider is the flavonoids. These are aromatic, colored, water-soluble compounds. They are widespread and usually occur as glycosides. In flavonoids, we have considered the possibility of partial oxidation of glycosidic side-chains that are consistent with NMR signals that show a large amount of hydroxy groups; however, the carbohydrate tests show low carbohydrate content. Partial use of carbohydrates by bacteria may yield these types of products.

SUMMARY

In summary, we emphasize that structural analysis of humic substances is best accomplished by a combination of three methods: separation, degradation and modification, and identification. The prudent use of these techniques, especially detailed preparative separations, will improve understanding of the structure of humic substances.

Methylation procedures with ^{13}C reagents and liquid and solid ^{13}C-NMR give new insights on structure of both carbon atoms and functional groups.

Finally, gentle degradation procedures and use of capillary gas chromatography and high-resolution mass spectroscopy increase the knowledge of the structure of humic substances. A simultaneous effort, applying these three methods, is a valuable research approach for analysis of the structure of humic substances.

REFERENCES

1. Thurman, E. M., and R. L. Malcolm. "Preparative Isolation of Aquatic Humic Substances," *Environ. Sci. Tech.* 15:463-466 (1981).

2. Kedeti, G., and W. H. Lederer. *Handbook of Micro-Methods for the Biological Sciences* (New York: Van Nostrand Reinhold, 1974) p. 51.
3. Mikita, M. A., C. Steelink and R. L. Wershaw. "Carbon-13 Enriched Nuclear Magnetic Resonance Method for the Determination of Hydroxyl Functionality in Humic Substances," *Anal. Chem.* 53:1715-1717 (1981).
4. Thurman, E. M., R. L. Wershaw, R. L. Malcolm and D. J. Pinckney. "Molecular Size of Aquatic Humic Substances," *Org. Geochem.* 4:27-35 (1982).
5. Brook, A. J. W., and K. C. Munday. "The Interaction of Phenols, Anilines, and Benzoic Acids with Sephadex Gels," *J. Chromatog.* 47:1-8 (1970).
6. Thurman, E. M., and R. L. Malcolm. "Concentration and Fractionation of Hydrophobic Organic Acid Constituents from Natural Waters by Liquid Chromatography," U.S. Geol. Survey Water-Supply Paper 1817-G, 16, Washington, DC: U.S. Government Printing Office (1979).
7. Schnitzer, M., and S. U. Khan. *Soil Organic Matter* (Amsterdam: Elsevier Scientific Publishing Company, 1978), p. 319.
8. Mantoura, R. F. C., and J. P. Riley. "The Analytical Concentration of Humic Substances from Natural Waters," *Anal. Chem. Acta* 76:97-106 (1975).
9. Aiken, G. R., E. M. Thurman, R. L. Malcolm and H. F. Walton. "Comparison of XAD Macroporous Resins for the Concentration of Fulvic Acid from Aqueous Solution," *Anal. Chem.* 11:1799-1803 (1979).
10. Aiken, G. R., R. L. Malcolm and E. M. Thurman. "Isolation of Humic Substances from Water and Soil Extracts Using Amberlite XAD Resins," Abstracts of 1977 Soil Science Society of America Meeting, Los Angeles, California (1977), p. 140.
11. MacCarthy, P., M. J. Peterson, R. L. Malcolm and E. M. Thurman, "Separation of Humic Substances by pH Gradient Desorption from a Hydrophobic Resin," *Anal. Chem.* 51:2041-2043 (1979).
12. Gustafson, R. L., and J. Paleos. Faust, S. J., Ed. *Organic Compounds in Aquatic Environments* (New York: Marcel Dekker Inc., 1971), p. 213-237.
13. Thurman, E. M., R. L. Malcolm and G. R. Aiken. "Prediction of Capacity Factors for Aqueous Organic Solutes Adsorbed on a Porous Acrylic Resin," *Anal. Chem.* 50:775-779 (1978).
14. Leenheer, J. A., and E. W. D. Huffman, Jr. "Classification of Organic Solutes in Water by Using Macroreticular Resins," *J. Research, U.S. Geol. Survey,* 4:737-751 (1976).
15. Leenheer, J. A. "Comprehensive Approach to Preparative Isolation and Fractionation of Dissolved Organic Carbon from Natural Waters and Wastewaters," *Environ. Sci. Technol.* 15:578-587 (1981).
16. MacCarthy, P. and S. O'Cinneide, "Fulvic Acid: I. Partial Fractionation," *J. Soil Science* 25:420-428 (1974).
17. Turkelson, V. T. and U. Richards, "Separation of the Citric Acid Cycle Acids by Liquid Chromatography," *Analytical Chemistry* 50:1420-1423 (1978).
18. Schnitzer, M., and S. V. Khan. *Humic Substances in the Environment,* (New York: Marcel Dekker, 1972), p. 327.
19. Christman, R. F. "Chlorination of Humic Substances," U.S. Environmental Protection Agency, Report No. R8044430 (Washington, DC: U.S. Government Printing Office, 1979).

20. Christman, R. F., J. D. Johnson, F. K. Pfaender, D. L. Norwood, M. R. Webb, J. R. Haas and M. J. Babenrieth. "Chemical Identification of Aquatic Humic Chlorination," *Water Chlorination Environ. Impact and Health Effects Vol. 3,* Jolley, Brungs and Cumming, Eds., (Ann Arbor, MI: Ann Arbor Science Publishers, Inc., 1980) pp. 75-84.
21. Havlicek, S. C., J. H. Reuter and M. Ghosal. "Identification of Major and Minor Classes of Natural Organic Substances Found in Drinking Water," U.S. Environmental Protection Agency, Report No. 68-10-4480 (Washington, DC: U.S. Government Printing Office, 1979).
22. Abraham, R. J., and P. Loftus. *Proton and Carbon-13 NMR Spectroscopy* (London: Heyden, 1978) p. 230.
23. Schnitzer, M., and S. I. M. Skinner. "Gel Filtration of Fulvic Acid, a Soil Humic Compound," *Isotopes and Radiation in Soil Organic Matter Studies,* Vienna: Int. Atomic Energy Agency, 1968) pp. 41-55.
24. Ishiwatari, R. "Fractionation and Characterization of Humic Acid from a Lake Sediment," *J. Geochem.* 2:175-184 (1969).
25. Vila, F. J. G., H. Lentz and H. D. Ludemann. "FT-^{13}C Nuclear Magnetic Resonance Spectra of Natural Humic Substances," *Biochem. Biophys. Res. Comm.* 72:1063-1069 (1976).
26. Stuermer, D. H., and J. R. Payne. "Investigations of Seawater and Terrestrial Humic Substances with Carbon-13 and Proton Nuclear Magnetic Resonance," *Geochim. Cosmochim. Acta* 40:1109-1114 (1976).
27. Lentz, H., H. D. Ludemann and W. Ziechmann. "Proton Resonance Spectra of Humic Acids from the Solum of a Podzol," *Geoderma* 18:325-328 (1977).
28. Wilson, M. A., A. J. Jones and B. Williamson. "Nuclear Magnetic Resonance Spectroscopy of Humic Materials," *Nature* 276:487-489 (1978).
29. Hatcher, P. G., R. Rowan and M. A. Mattingly. "^{1}H and ^{13}C NMR of Marine Humic Acids," *Organ. Geochem.* 2:77-85 (1980a).
30. Hatcher, P. G., D. L. VanderHart and W. L. Earl. "Use of Solid-State ^{13}C NMR in Structural Studies of Humic Acids and Humin from Holocene Sediments," *Organ. Geochem.* 2:87-92 (1980b).
31. Miknis, F. P., G. E. Maciel and V. J. Bartuska. "Cross Polarization Magic Angle Spinning ^{13}C NMR Spectra of Soil Shales," *Organ. Geochem.* 1:169-176 (1979).
32. Taylor, L. C. E. "Fast Atoms Allow MS Study of Large Molecules," *Ind. Res. Development* 23:124-128 (1981).
33. Barber, M., R. S. Bordoli, R. D. Sedgwick and A. N. Tyler. "Fast Atom Bombardment of Solids as an Ion Source in Mass Spectrometry," *Nature* 293:270-275 (1981).
34. Cooks, R. G., and G. L. Glish, "Mass Spectrometry/Mass Spectrometry," *Chem. Eng. News* 59 (48):40-52 (1981).
35. Stuermer, D. H., E. E. Peters and I. R. Kaplan. "Source Indicators of Humic Substances and Proto-Kerogen. Stable Isotope Ratios, Elemental Compositions and Electron Spin Resonance Spectra," *Geochim. Cosmochim. Acta* 42:989-997 (1978).

CHAPTER 2

APPLICATION OF ^{13}C-NMR SPECTROSCOPY TO THE STRUCTURAL ELUCIDATION OF DISSOLVED MARINE HUMIC SUBSTANCES AND THEIR PHYTOPLANKTONIC PRECURSORS

A. H. Gillam

Department of Oceanography
The University of British Columbia
Vancouver, B.C., Canada

M. A. Wilson

CSIRO Division of Fossil Fuels
North Ryde, N.S.W., Australia

ABSTRACT

In the past five years, nuclear magnetic resonance (NMR) spectroscopy has been used for structural group analysis of humic substances, including determinations of the aromaticities of humic substances. Recent studies of aquatic (freshwater) humic substances have shown that differences in soil organic materials of the surrounding watershed areas are a major source of variations in the aromatic content of aquatic humic substances.

This chapter presents the results of the first NMR study of planktonic precursors of dissolved marine humic substances. Extracellular humic precursors from a large-scale culture of a marine diatom were analyzed using a novel ^{13}C-NMR technique. The results obtained are discussed in terms of the aliphatic, oxygenated aliphatic, aromatic and carboxylic carbon contents of the humic fractions. The use of ^{13}C-NMR estimates to differentiate between the allochthonous and autochthonous components of aquatic humic substances is critically evaluated.

INTRODUCTION

^{13}C nuclear magnetic resonance spectroscopy (^{13}C-NMR) has been shown to be a powerful tool in the elucidation of the structure of humic materials [1–8]. Provided due care is taken in establishing quantitative conditions, ^{13}C-NMR can be used to measure the amounts of aromatic, alcoholic, polymethylene and carboxyl carbon in humic acids [7,9]. Nevertheless, very little work has been reported on marine humic materials. Most of the published work to date has been confined to the analysis of soils [8].

The first ^{13}C-NMR spectra of humic materials isolated from a marine environment were recorded by Stuermer and Payne [10], but the interpretation of the spectra was severely hampered by the poor signal-to-noise ratio. Greater success has been had by recent workers using more advanced spectrometers which have better signal-to-noise capabilities. The cross polarization technique with magic angle spinning (CP/MAS) [11] has enabled spectra to be obtained within a few hours, rather than the several days previously required [6,7,9]. There are real possibilities that CP/MAS will become a routine technique for the analysis of humic substances, but at present considerable care is needed in establishing the optimum experimental conditions. Some questions still need to be answered about the quantitative nature of the technique [12].

The first ^{13}C-NMR spectra of humic materials isolated from marine sediments were recorded by Hatcher et al. [6]. They were able to show that aliphatic structures are the dominant components of marine humic acids from sediments collected from a marine sapropelic lake and from the New York Bight. The values obtained for the aromatic fraction (f_a) of the humic acids were lower than those obtained for terrestrial humic acids. It was suggested that differences in f_a are largely due to the different lignin contents of the terrestrial and marine plants that are precursors of the humic acids.

A number of factors may be important in determining the structure of the organic matter in soils, marine sediments and aquatic environments. First, the nature of the parent material may be important, because if the decomposing plant life contains a larger proportion of one class of chemical compounds than another, then the organic matter will contain different degradation products. Secondly, it is clear that geographical and geochemical factors can also be important. The freshwater humic substances concentrated from Lake Celyn (N. Wales) were shown [9] to be more aromatic than terrestrial humic substances. Wilson et al. [9] suggested that the major portion of humic material in the lake arises from water-transportable material from the surrounding peat.

Another important factor is the degree of microbiological activity. This influences the rate of transformation of plant and animal debris, which, in turn, will mean that the contribution of weakly and strongly modified organic material to the total will differ in environments with different microbial populations.

Climate too will influence both the amount and type of vegetation and microbiological activity in a particular environment [13]. Furthermore, the electrophoretic mobility of humic acids has been reported [14] to be dependent on the climatic conditions prevailing on sampling. A recent NMR study has shown that the aromaticity of whole soils can vary considerably from sample to sample [15]. Thus, it is clear that the nature of soil organic matter can vary from soil to soil.

Although microbiological activity can be much lower in marine environments than in terrestrial ones, microbial life is abundant in most coastal areas of the ocean. The photosynthetic phytoplankton, a mixture of diatoms and phytoflagellates, are, directly or indirectly, the major source of dissolved organic carbon (DOC) in marine waters. Bacterioplankton are suspended bacteria which utilize the DOC released by metabolizing or decaying microorganisms. However, very little is known of their contribution to the formation of marine organic matter [16].

As part of a study to examine the production of possible allelopathic metabolites by *Phaeodactylum tricornutum* Bohlin, large-scale cultures of this marine diatom were prepared. Both intracellular and extracellular organic extracts were then isolated from the cells and filtrate of this species. Marine humic substances were isolated from a large-volume (860-liter) sample of coastal seawater to enable a comparison to be made of the CP/MAS ^{13}C-NMR spectra of the naturally occurring dissolved organic matter and that isolated from *P. tricornutum.* Since the relative contributions of aromatic, aliphatic and carbohydrate carbon to the chemical composition of several terrestrial and freshwater humic substances have been estimated [8,9], the results of this study should assist in elucidating the chemical transformations taking place during the humification process.

EXPERIMENTAL

Coastal Seawater Humic Substances

The large-volume (860-liter) sample of coastal seawater was collected from the submerged inlet of the seawater pumping system of the West Vancouver Laboratory, Department of Fisheries and Oceans, Canada, and was stored in 200-liter polyethylene barrels at 18 °C prior to filtration. Details of the analytical scheme used for the concentration of marine humic substances are similar to those given by Wilson et al. [9]. Briefly, the seawater was filtered (GF/C), acidified to pH 2.0 and passed through columns (30 $\times$ 2 cm) of acetone-extracted Amberlite XAD-2 resin (BDH Chemicals, Vancouver, Canada). After the passage of the sample, the column was rinsed with cold distilled water until the eluate was chloride-free. The humic substances were then eluted by the passage of >5-bed volumes of a 1:1 (v/v) mixture of methanol and 2 *M* ammonium hydroxide. The solution was evaporated to dryness in a rotary evaporator at <35 °C. The dried humic extract was then redissolved in 100 ml of distilled water and exhaustively extracted with chloroform to remove lipoidal material. The aqueous solution was filtered (Whatman #1) and then rotary-evaporated at <35 °C. The dried humic extract was stored in vacuo over $Mg(ClO_4)_2$. The chloroform-soluble fraction was dried over Na_2SO_4 and rotary-evaporated before storing in vacuo over $Mg(ClO_4)_2$.

Organic Extracts of *P. tricornutum*

A unialgal, axenic culture of *Phaeodactylum tricornutum* Bohlin was obtained from the Pacific Northeast Culture Collection at the University of British Columbia. Initially 1-liter axenic cultures of *P. tricornutum* were grown in charcoal-treated seawater, which had been autoclaved prior to the addition of sterile ES nutrient enrichment [17]. Cells were grown to late exponential phase on a 16:8 light:dark cycle at 18 °C under 60 $\mu E\ m^{-2}s^{-1}$ illumination. A large-scale, unialgal culture was subsequently grown in a 200-liter polyethylene drum using charcoal-treated, filtered (0.45 μ) seawater containing ES enrichment, minus EDTA and vitamins. Cells were grown to late stationary phase and then removed via continuous centrifugation (Sharples TL). The filtrate was acidified (pH 2.0) and extracted with XAD-2 resin in a similar manner as that described for the coastal seawater sample. Both the water- and chloroform-soluble materials obtained from the extracellular products of *P. tricornutum* were dried and stored in vacuo over $Mg(ClO_4)_2$.

The cellular material (84.6 g wet weight) was placed in a precleaned (Soxhlet extraction, $CHCl_3$, 24 h) Pyrex extraction thimble and extracted continuously with $CHCl_3$ (7 days, with fresh solvent each day). The chloroform-soluble material was filtered (Whatman #1), dried over Na_2SO_4 and rotary-evaporated at <35 °C to an oily residue. The material was stored in vacuo over $Mg(ClO_4)_2$. The chloroform-extracted residue was dried to remove residual chloroform and then extracted for 48 h under nitrogen, with 500 ml of 0.5 *M* NaOH. The cellular debris was removed (Whatman #1) and the alkaline solution was passed over a 15- × 1-cm column of Dowex 50W-X8 (H^+ form). The neutral extract was acidified to pH 2.0 (no precipitation occurred) and passed (5 times) over a 15- × 1-cm column of precleaned XAD-2 resin. The column was rinsed with cold distilled water and the adsorbed organic material was eluted with 300 ml of 1:1 (v/v) mixture of

MeOH/2 *M* NH_4OH. The humic extract was rotary-evaporated at <35 °C and the dried residue was stored in vacuo over $Mg(ClO_4)_2$.

NMR Spectroscopy

NMR spectra were obtained on a Bruker CXP100 instrument using the cross polarization technique with magic angle spinning. A boron nitride rotor was used with a kel-F base. Spinning speed was ~3.5 kHz. Carbon spectral frequency was 22.63 MHz. Contact times were from 0.25 to 2 msec for cross polarization. Contact times of >1 msec were necessary for maximum cross polarization. Recycle times were 0.3 to 1 sec. Signal-to-noise in the spectra was >20:1 after 5000 scans.

RESULTS AND DISCUSSION

Origin of Marine Humic Substances

The source of the majority of the dissolved organic carbon in seawater would appear to be marine organisms [16,18], with a smaller contribution from allochthonous sources.

Several types of structure have been suggested for aquatic humic materials [19]. Phenols, quinones and the breakdown products of proteins and carbohydrates have all been considered as possible starting materials for their synthesis. Condensation, rearrangement, cyclization and decarboxylation processes involving sugars and amino acids have been thought to lead to the formation of yellow to brown polymeric melanoidins [20]. Degens [21] has suggested that marine organic matter is composed largely of nitrogen-containing compounds. A clathrate-type structure, involving both polar and neutral organics surrounding a metal ion or clay template, was suggested for these heteropolycondensates. Kalle [22] has discussed the role of melanoidin formation in the synthesis of aquatic humic compounds and has distinguished between these "carbohydrate humic acids" and aquatic humic compounds originating from phenolic or polyphenolic material. Melanoidin formation is believed to be favored by the mildly alkaline conditions prevailing in seawater and may be one of the dominant mechanisms in the formation of marine humic substances.

Assignments for Marine Humic Substances

The CP/MAS ^{13}C-NMR spectrum of the coastal seawater sample of dissolved marine humic substances is shown in Figure 1(a). The presence of a broad range of aliphatic components is indicated by the resonance in the

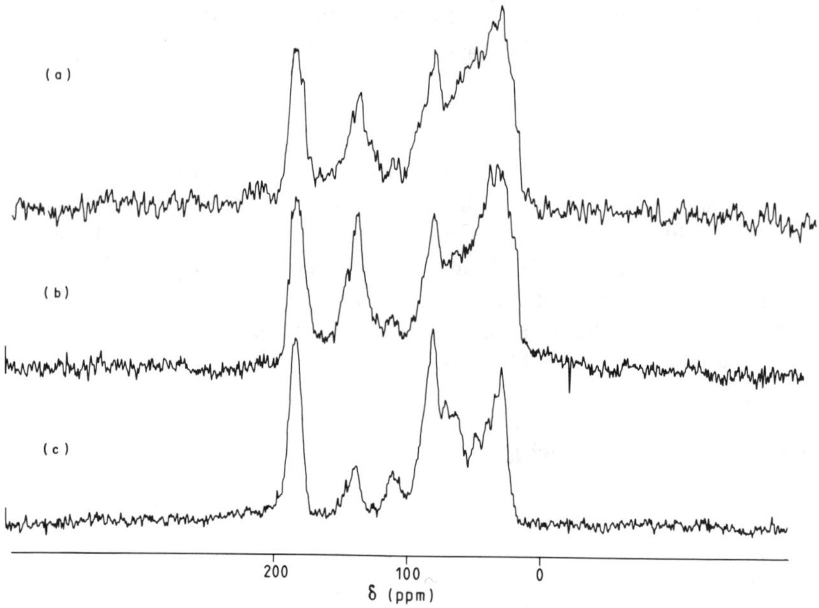

Figure 1. CP/MAS spectra of dissolved marine humic substance and phytoplanktonic precursors: (a) humic material, (b) extracellular extract of *P. tricornutum,* and (c) intracellular extract of *P. tricornutum.*

10 to 50 ppm range. The signal shows maxima at 30 and 47 ppm. The large peak at 30 ppm is due to the presence of a large number of repeating polymethylene units in the humic macromolecule. The broad resonance, centered at ~78 ppm, is due to alcohol and ether carbons, and has been partially attributed to the oxygenated carbons in carbohydrates.

The resonance, ranging from 125 to 150 ppm and centered at 135 ppm, is due to protonated aromatic carbon, bridgehead and alkyl-substituted carbon and hetero-substituted (O,N) aromatic carbon. Olefinic carbons also resonate in this region; their contribution to the structure of marine humic substances, however, remains unknown. The large resonance at 183 ppm arises from carboxyl, amide and ester carbons. The presence of substantial amounts of carboxylic groups in marine humic substances has been described by Gillam [23].

Nature of the Extracellular Material from *P. tricornutum*

Phaeodactylum tricornutum Bohlin is a marine diatom of the class Bacillariophyceae and the division Chrysophyta. It is a major constituent of the diet of mussels and oysters. The nutritional and environmental requirements of *P. tricornutum* have been widely studied [24–26]. Large-scale culturing of this species was undertaken in this laboratory to study the possible release of allelopathic metabolites which could account for the dominance of this diatom in mixed populations during large-scale mass culturing [26]. Subsequently, a large-volume (200-liter) culture of *P. tricornutum* was made available for the work described here. Although *P. tricornutum* is not a dominant species in coastal waters, it is reasonable to assume that other diatoms and phytoflagellates would excrete a broadly similar range of organic materials. It is hoped to extend the approach described in this chapter to a wide range of "ecologically" important algal species.

Marine phytoplankton produce a wide range of extracellular organic materials, but the extent of and the reasons causing this excretion are not clear [27–29]. Hellebust [30] found that glycolic acid (hydroxyacetic acid) was excreted to a small extent by several diatoms, including *P. tricornutum.* Glycerol and mannitol were also excreted by several species, also in relatively minor amounts. In a study of two diatoms (*Chaetoceros curvisetus* and *C. affinis*), Myklestad et al. [31,32] have shown that up to 40 mg/liter of extracellular polysaccharide material was excreted. Very few extracellular metabolites from marine phytoplankton have been characterized [33]. Most of the reports of extracellular phytoplankton material have suggested that organic acids, amino acids and carbohydrates are the major components excreted [34].

Assignments for the Extracellular Material from *P. tricornutum*

The CP/MAS spectrum of the extracellular material from the marine diatom *P. tricornutum* is shown in Figure 1(b). The general features of the spectrum are similar to those from the spectrum of the coastal seawater humic substances. There is a broad resonance extending from 10 to 50 ppm; this is typical of methyl, methylene and methine carbons in a wide range of sterically hindered or highly branched environments. A considerable amount of fine structure in the 17 to 37 ppm region can be seen. The extracellular material from the marine diatom also shows a resonance, centered at 77 ppm, due to the presence of oxygenated carbons, such as those in carbohydrates. A further resonance is seen at ~110 ppm, due to anomeric carbons in the polysaccharides. Like the seawater sample, the extracellular diatom extract shows a signal in the aromatic region. There is some fine structure at 136 and 138 ppm. This may be indicative of the presence of alkyl-substituted, rather than hetero-substituted, aromatic systems; the latter would show resonances in the 150 ppm region. Carboxyl, amide and ester carbons resonate at 180 ppm in this extract.

Nature of the Cellular Material from *P. tricornutum*

There are very few reports of the cellular organic components of *P. tricornutum.* A study of the cellular carbohydrates of this diatom [35] revealed a carbohydrate content of ~20%, which was thought to comprise glucose and mannose, with smaller amounts of xylose, rhamnose and ribose.

Assignments for the Cellular Material from *P. tricornutum*

The CP/MAS spectrum of the cellular material extracted by NaOH from *P. tricornutum* is shown in Figure 1(c). Clearly, the general features of the spectrum are similar to those shown in Figures 1(a) and 1(b). Resonances can be seen at 30, 77, 110, 135 and 180 ppm. The carbohydrate content, as indicated by the resonances at 77 and 110 ppm, is much greater in this NaOH extract of the diatom than in the extracellular, water-soluble material. Also, there are fewer substituted aromatic compounds in this fraction, as indicated by the resonance centered at 135 ppm, when compared to the extracellular material.

Significance of Results

The CP/MAS ^{13}C-NMR spectra of the three extracts of this study all show the same general features. The results suggest, therefore, that a major contributor to marine humic materials may be material from marine diatoms. Quite clearly, there are no gross differences in the chemical compositions of the seawater organic material and that derived from the marine diatom.

Close examination of the spectra (Figure 1) reveals that the spectrum of the extracellular material from the diatom [Figure 1(b)] more closely resembles the spectrum of the seawater material [Figure 1(a)] than does the intracellular material [Figure 1(c)]. This suggests that extracellular material is a more important contributor to the humic material than intracellular material.

The results clearly show that dissolved marine humic materials are highly aliphatic, but they also show that aromatic materials can be significant components. The presence of aromatic precursors of marine humic substances has been suggested [36]. Several phaeophyceae were shown to produce extracellular UV-absorbing materials. Preparative isolation of the exudate of *Fucus vesiculosus* indicated that these extracellular compounds contained polyphenolic components. The UV spectrum of the extracellular material from *F. vesiculosus* was similar to that reported for water from the Irish Sea [37]. Thus, it appears that there is a possible source of aromatic precursors in macrophytic and other marine algae which could account for the presence of aromatic moieties in marine humic substances. It seems unlikely that aromatic material is metabolized or diminished in concentration during humification. Although unsaturated, nonaromatic carbons are known to be abundant in the lipids and carotenoids of marine algae, these should be removed to a great extent by the solvent extraction procedure used during the isolation of the organic materials of this study. Regretfully, the contribution of olefinic carbon to the chemical structure of marine humic substances and their phytoplanktonic precursors remains unclear, and future studies should attempt to clarify this issue.

CONCLUSIONS

This chapter presents the results of the first NMR study of planktonic precursors of dissolved marine humic substances. The results show clearly that the extractable material from the marine diatom *P. tricornutum* has a chemical composition similar to that of the dissolved marine humic material. These results suggest that marine diatoms are important contributors to marine humic materials. Despite this limited study, it appears that the factors that can influence the structure of terrestrial humic substances, namely climate, microbiological activity and geography, do not make an important contribution to the structure of seawater humic substances. Thus, the organic material present in the coastal seawater of this study resembles that released from phytoplanktonic precursors.

ACKNOWLEDGMENTS

The CP/MAS ^{13}C-NMR spectra of this study were obtained with the aid of a rotor kindly provided by Drs. Pugmire, Grant and Zilm, Chemistry Department, University of Utah. Culture facilities for this work were provided by Dr. R. J. Andersen, Department of Oceanography, The University of British Columbia, Canada. The comments of C. G. Trick, Department of Oceanography, The University of British Columbia were most helpful.

REFERENCES

1. Vila, F., H. Lentz and H. Lüdemann. "F.T.-C13 Nuclear Magnetic Resonance Spectra of Natural Humic Substances," *Biochem. Biophys. Res. Comm.* 72:1063-1069 (1976).
2. Wilson, M. A., and K. M. Goh. "Proton Decoupled Pulse Fourier-Transform ^{13}C Magnetic Resonance of Soil Organic Matter," *J. Soil Sci.* 28:645-652 (1977).
3. Grant, D. "Chemical Structure of Humic Substances," *Nature* 270:709-710 (1977).

4. Wilson, M. A., A. J. Jones and B. Williamson. "N.M.R. Spectroscopy of Humic Materials," *Nature* 276:487-489 (1978).
5. Ogner, G. "The ^{13}C Nuclear Magnetic Resonance Spectrum of a Methylated Humic Acid," *Soil Biol. Biochem.* 11:105-108 (1979).
6. Hatcher, P. G., R. Rowan and M. A. Mattingly, "^{1}H and ^{13}C NMR of Marine Humic Acids," *Org. Geochem.* 2:77-85 (1980).
7. Newman, R. H., K. R. Tate, P. F. Barron and M. A. Wilson. "Towards a Direct Non-Destructive Method of Characterising Soil Humic Substances Using ^{13}C N.M.R.," *J. Soil Sci.* 31:623-631 (1980).
8. Wilson, M. A. "Applications of Nuclear Magnetic Resonance Spectroscopy to the Study of the Structure of Soil Organic Matter," *J. Soil Sci.* 32:167-186 (1981).
9. Wilson, M. A., P. F. Barron and A. H. Gillam. "The Structure of Fresh Water Humic Substances as Revealed by ^{13}C-n.m.r. Spectroscopy," *Geochim. Cosmochim. Acta* 45:1743-1750 (1981).
10. Stuermer, D. H., and J. R. Payne. "Investigation of Seawater and Terrestrial Humic Substances with Carbon-13 and Proton Nuclear Magnetic Resonance," *Geochim. Cosmochim. Acta* 40:1109-1114 (1976).
11. Schaefer, J., and E. O. Stejskal. "Carbon-13 Nuclear Magnetic Resonance of Polymers Spinning at the Magic Angle," *J. Amer. Chem. Soc.* 98:1031-1032 (1976).
12. Hagaman, E. W., and M. C. Woody. "Statistical Coal Structure Analysis by CP/MASS-^{13}C-NMR Spectroscopy: Quantitative Evaluation of Aromaticity Parameters." In: *Proc. Int. Conf. Coal Sci.* (Dusseldorf, 1981), pp. 807-809.
13. Goh, K. M. "Dynamics and Stability of Organic Matter." In: *Soils of Variable Charge,* B. K. G. Theng, Ed. (New Zealand: *N.Z. Soil Sci. Soc.,* 1980).
14. Turenne, J. F. "Molecular Weights of Humic Acids in Podzol and Ferrallitic Soils of the Savannas of French Guyana and Their Evolution Related to Soil Moisture," *Tropic. Agric.* (Trinidad) 51:133-144 (1974).
15. Barron, P. F., M. A. Wilson, J. F. Stephens, B. A. Cornell and K. R. Tate. "Cross-Polarization ^{13}C NMR Spectroscopy of Whole Soils," *Nature* 286:585-587 (1980).
16. Sieburth, J. McN. *Sea Microbes* (New York: Oxford University Press, 1979), pp. 24-27.
17. Provasoli, L., in *Cultures and Collection of Algae,* A. Watanabe and A. Hatton, Eds. (Proc. U.S.-Japan Conf., Hakone, Sept. 1966: Jap. Soc. Plant Physiol.), pp. 63-75.
18. Menzel, D. W., in *The Sea,* Vol. 5, E. D. Goldberg, Ed. (New York: John Wiley and Sons, Inc., 1974), pp. 659-678.
19. Duursma, E. K., in *Chemical Oceanography,* Vol. 1, J. P. Riley and G. Skirrow, Eds. (London: Academic Press, 1965), pp. 433-475.
20. Hedges, J. I. "The Formation and Clay Mineral Reactions of Melanoidins," *Geochim. Cosmochim. Acta* 42:69-76 (1978).
21. Degens, E. T., in *Symposium on Organic Matter in Natural Waters,* D. W. Hood, Ed. (Institute of Marine Science, University of Alaska, Occasional Publication No. 1, 1970), pp. 77-106.
22. Kalle, K., in *Oceanogr. Mar. Biol. Rev., Vol. 4,* H. Barnes, Ed. (London: George Allen and Unwin Ltd., 1966), pp. 91-104.

23. Gillam, A. H. "The Chemical and Physical Characteristics of Aquatic Humic Substances," PhD Thesis, The University of Liverpool, Liverpool, England (1979).
24. Spencer, C. P. "Studies on the Culture of a Marine Diatom," *J. Mar. Biol. Assn. U.K.* 33:265-290 (1954).
25. Raymont, J. E. G., and M. N. E. Adams. "Studies on the Mass Culture of *Phaeodactylum*," *Limnol. Oceanogr.* 3:119-136 (1954).
26. Goldman, J. C., and J. H. Ryther. "Temperature-Influenced Species Competition in Mass Cultures of Marine Phytoplankton," *Biotech. Bioeng.* 18:1125-1144 (1976).
27. Smith, D. F., and W. J. Wiebe. "Constant Release of Photosynthate from Marine Phytoplankton," *Appl. Environ. Microbiol.* 32:75-79 (1976).
28. Sharp, J. H. "Excretion of Organic Matter by Marine Phytoplankton: Do Healthy Cells Do It?," *Limnol. Oceanogr.* 22:381-399 (1977).
29. Mague, T. H., E. Friberg, D. J. Hughes and I. Morris. "Extracellular Release of Carbon by Marine Phytoplankton; A Physiological Approach," *Limnol. Oceanogr.* 25:262-279 (1980).
30. Hellebust, J. A. "Excretion of Some Organic Compounds by Marine Phytoplankton," *Limnol. Oceanogr.* 10:192-206 (1965).
31. Myklestad, S., and A. Haug. "Production of Carbohydrates by the Marine Diatom *Chaetoceros affinis* var. *willei* (Gran) Hustedt. I.," *J. Exp. Mar. Biol. Ecol.* 9:125-136 (1972).
32. Myklestad, S., A. Haug and B. Larsen. "Production of Carbohydrates by the Marine Diatom *Chaetoceros affinis* var. *willei* (Gran) Hustedt. II.," *J. Exp. Mar. Biol. Ecol.* 9:137-144 (1972).
33. Trick, C. G., P. J. Harrison and R. J. Andersen. "Extracellular Secondary Metabolite Production by the Marine Dinoflagellate *Prorocentrum minimum* in Culture," *Can. J. Fish. Aq. Sci.* 38:864-867 (1981).
34. Hellebust, J. A., in *Algal Physiology and Biochemistry*, W. D. P. Stewart, Ed. (Berkeley, CA: University of California Press, 1974), pp. 838-863.
35. Ford, C. W., and E. Percival. "The Carbohydrates of *Phaeodactylum tricornutum*. Part 1. Preliminary Examination of the Organism, and Characterisation of Low Molecular Weight Material and of a Glucan," *J. Chem. Soc.* 7035-7041 (1965).
36. Craigie, J. S., and J. MacLachlan. "Excretion of Coloured Ultraviolet-Absorbing Substances by Marine Algae," *Can. J. Bot.* 42:23-33 (1964).
37. Armstrong, F. A. J., and G. T. Boalch. "The Ultra-Violet Absorption of Sea Water," *J. Mar. Biol. Assn. U.K.* 41:591-597 (1961).

CHAPTER 3

SOLID-STATE ^{13}C-NMR OF SEDIMENTARY HUMIC SUBSTANCES: NEW REVELATIONS ON THEIR CHEMICAL COMPOSITION

Patrick G. Hatcher and Irving A. Breger

U.S. Geological Survey
Reston, Virginia

Larry W. Dennis* and Gary E. Maciel

Colorado State University
Fort Collins, Colorado

ABSTRACT

Humic acids, fulvic acids and humins from samples of a variety of depositional environments have been analyzed by solid-state ^{13}C nuclear magnetic resonance (NMR) using the cross polarization and magic angle spinning (CP/MAS) technique. This technique yields quantitative signals for aromatic and paraffinic structures; carboxyic, carbonyl and ether functional groups; and carbohydrates. Aromaticities of humic acids and humins generally range from 20 to 70%, whereas fulvic acid isolates tend to be less aromatic. Paraffinic carbons are major components of most humic isolates. Paraffinic structures predominate in submerged sediments, such as peats, algal sapropels and marine sediments, as well as in neutral or poorly drained soils. These structures, thought to be derived from algal or microbial contributions, have not previously been detected in soil humic substances in such large proportions by conventional degradative methods of structural analysis. Consequently, prior to the development of NMR methods, they were not recognized as major contributors to the composition of terrestrial humic substances.

*Current address: Exxon Research and Engineering Co., Baytown, Texas.

Comparisons among terrestrial humic acids, fulvic acids and humins reveal that the humins contain a high proportion of paraffinic structures, whereas humic acids generally contain more aromatic structures. The aromatic structures are thought to be largely derived from lignin in vascular plants. Fulvic acids, in contrast to humic acids and humins, contain significant amounts of polysaccharide-like substances. Soils examined in this study generally contain few phenolic carbons; this suggests that phenolic structures are not significant components.

INTRODUCTION

The major organic constituents of terrestrial and aquatic plants are carbohydrates, proteins (amino acids) and lipids. Terrestrial plants additionally contain a macromolecular substance called lignin. When plant residues are incorporated into sedimentary deposits, most of the organic components are altered within a relatively short period of time. They may decompose completely to CO_2, H_2O, NH_3 and CH_4, or they may be transformed to new compounds having different chemical properties; under certain conditions, they may even remain unaltered. These processes, collectively called humification, result in the formation of complex macromolecular compounds known as humic substances.

Studies show that the chemical structures of humic substances derived from terrestrial plants differ from those of humic substances derived from aquatic plants [1–4]. Terrestrial humic substances are largely aromatic and contain methoxyl, hydroxyl, carboxyl and ether functional groups [5,6]. These groups are, in part, derived from lignin residues, which are predominantly aromatic in structure and occur only in terrestrial vascular plants. Humic substances derived from aquatic flora, such as algae and microorganisms, are very different [1,2,7,8]; their main distinguishing feature is their aliphatic structure [3,9]. This difference is probably related to the fact that lignin-like structures are not significant components of aquatic plants. Because of these differences in structure, terrestrially and aquatically derived humic substances can be chemically distinguished from one another even though physical and petrographic distinctions may not be possible.

Because humic substances are macromolecular and are largely heteropolymeric, their structural composition has been virtually impossible to define in detail. In the past, structural studies were limited to determining the C, H, O, and N percentages and functional-group contents. Atomic H/C ratios and H contents were used extensively to determine whether humic substances were of significant aliphatic or aromatic character. This technique lacked specificity but, in lieu of better methods, it provided a rough insight into the structure. With the advent of sophisticated instrumentation for infrared (IR) absorption analysis, electron-spin resonance

(ESR), NMR and gas chromatography/mass spectrometry (GC/MS), these new techniques, both destructive and nondestructive, were applied in an effort to unravel in more detail the structural features of humic substances [5].

Because GC/MS requires volatile compounds, humic substances often have been subjected to pyrolytic, oxidative or reductive chemical degradation to break down the macromolecules to fragments amenable to study by GC/MS. When identified, these fragments have been used to infer the presence of structures in the original macromolecules [5,10]. Although much has been learned about possible structures present in humic substances by these techniques, it is generally agreed that much structural information is lost in the degradation process inasmuch as molecular moieties can be chemically and thermally destroyed [8]. Obviously, nondestructive analytical techniques are preferred.

Unfortunately, nondestructive methods for structural elucidation have lacked resolution and sensitivity. IR spectrometry has provided valuable measurements of amounts of different functional groups present; however, very little information has been obtained for the carbon structure. ESR has yielded evidence that free radicals are stabilized in humic substances, probably via quinone or hydroquinone structures. Aside from this finding, ESR has contributed little structural information. NMR has been used as an analytical tool for structural elucidation; however, poor resolution, low sensitivity and, most important, the need for complete solubility have limited the usefulness of this technique for humic substances until recently. With new advances in the design of NMR spectrometers and techniques, both ^{1}H and ^{13}C nuclear environments can now be observed for humic substances in both liquids and solids [11]. Increased sensitivities and higher resolution have made NMR an attractive tool for structural studies of humic substances [3,4,11-21].

NMR previously was applied to the study of humic and fulvic acids in solution, but some early attempts to obtain usable spectra were disappointing [10]. Recent applications of improved NMR methods for the study of humic substances dissolved in aqueous solutions have been more successful [12,15,17]. In fact, high resolution ^{1}H- and ^{13}C-NMR spectra have recently been published to demonstrate the potential of solution NMR for structural studies of humic and fulvic acids [3,18–21].

NMR techniques become increasingly complex for studies of humin, as spectra must be obtained from solid samples. Because of C-H dipolar broadening and chemical-shift anisotropy, ^{13}C-NMR of solids has, in the past, provided very little structural information. Although dipolar broadening was eliminated with the advent of high-power ^{1}H decoupling and cross polarization (CP) [22,23], the chemical shift anisotropy still hindered ^{13}C resolution, and only marginal resolution between aromatic and aliphatic resonances was possible [24]. Cross polarization has recently been combined with magic angle spinning (MAS) (54.7° to the magnetic field) making it possible to achieve ^{13}C resolution approaching that of NMR in solution [25–29]. With such resolution, we are now able to obtain useful structural information for solids such as humins, as well as for anhydrous humic and fulvic acids. As a result, CP/MAS techniques have been applied to coals [29–34] and to kerogens [35–37] and more recently, to humic acids, fulvic acids and humin [3,11,19]. In all these substances, aromatic and aliphatic structures can be clearly identified and directly measured, representing a significant advancement over previously available methods.

The CP/MAS ^{13}C-NMR technique thus offers great promise for structural characterization of insoluble macromolecular substances. For this reason, CP/MAS ^{13}C-NMR was used as a primary tool for the structural characterization of humic substances in this study in an attempt to improve our understanding of their origin, chemical structure and diagenesis. Other chemical analyses, such as carbohydrate or sugar analyses, elemental analyses, infrared spectroscopy and ^{1}H- and ^{13}C-NMR analyses in solution, were also used to substantiate results of the CP/MAS ^{13}C-NMR analyses.

MATERIALS AND METHODS

Sample Origins

Aquatic Sediments

Samples were collected from a number of aquatic sedimentary deposits whose locations are listed in Table I. For each sample locality appropriate references are given for information on their geologic settings and readers are referred to these articles for such information [2,38–48].

Table I. Sample Localities for Humic Substances from Aquatic Sedimentary Deposits and Peat

Sample Name	Location	Reference
K6-77-0	Hudson-Raritan estuary, New York.	38
K6-77-4	Continental shelf, New York Bight, 10 naut mi from Rockaway Point, New York.	38
MN-4	Continental Shelf, New York Bight, 33 naut mi from Rockaway Point, New York.	38

Table I, continued

Sample Name	Location	Reference
HuG"R"	Hudson Canyon, New York Bight, 266 m water depth, 100 naut mi from Rockaway Point, New York.	
Guaymas Basin	Guaymas Basin, Gulf of California, 1968 m water depth, from a core taken during the Scripps Institution of Oceanography Hypogene Experiment (site 30G). Sample obtained from B. R. T. Simoneit (Oregon State University).	39
Walvis Bay, 8-0	Continental Shelf, Walvis Bay, Namibia (West Coast of Africa), 846 m water depth, from a surficial sediment sample (0 to 6 cm) taken on a cruise (AII-93-3) of the *Atlantis II* in 1975 by J. W. Farrington (Woods Hole Oceanographic Institution).	2
Walvis Bay, 12-0	Same as above but in 124 m water depth.	2
Black Sea	Sample collected in the Black Sea in 1969 on a cruise (AII-49) of the *Atlantis II* at site 1461K in 1788 m water depth and a 0 to 40 cm sediment depth.	40,41
Cariaco Trench	Sample collected in the Cariaco Trench (840 m water depth) on the continental shelf of Venezuela on a cruise (AII-60) of the *Atlantis II* at site 1798G. Sample obtained from B. R. T. Simoneit.	42
Mangrove Lake, Bermuda	Samples were obtained from a 16-m core containing a gelatinous sapropel underlain by a fibrous peat. Sample horizons are denoted by the following code: ML71-2-X where X is a number that refers to a specific horizon in the core.	43,44
Potomac River	Sample of humic acids obtained from G. Diachenko (U.S. Department of Agriculture). Sediment was collected at Point of Rocks, Maryland, in the Potomac River.	
Minnesota peats: Cedar Creek	Sample obtained from a sedge peat in Cedar Creek Lake, Anoka County, Minnesota.	45
Rice Lake	Sample obtained from a sedge peat in Rice Lake, Minnesota, by P. Zubovic (U.S. Geological Survey)	
Okefenokee peat	Sample obtained by C. B. Cecil (U.S. Geological Survey) at Billy's Island site, Georgia, in a *Taxodium*-dominated peat	46,47
Everglades peat	A 1-m core of sawgrass peat was collected by Z. S. Altschuler (U.S. Geological Survey) in Conservation District 1A in the Everglades, Florida.	46,48

Soil Samples

Humic and fulvic acid isolates from a variety of soils were obtained from M. Schnitzer (Agriculture Canada, Ottawa) and R. Malcolm (U.S. Geological Survey, Denver, Colorado). These soils vary widely in geographical extent and in pH regime, and have been described in a number of studies [5,49–54]. In general, all soil samples can be grouped into the following three major categories: (1) cool, temperate-climate acidic soils; (2) subtropical neutral soils; and (3) temperate-climate neutral soils. Humic and fulvic acid isolates prepared by Schnitzer and Khan [5] and Malcolm [51] are directly comparable to those prepared in this study because the extraction and purification procedures were similar. The value of obtaining such samples is that they have been studied extensively, and a large amount of information exists concerning their structural composition.

Peat Deposits

Samples were collected from a number of peat deposits listed in Table I. The samples include two sedge peats from Minnesota; two sawgrass peats, one a buried peat from Mangrove Lake, Bermuda, and the other from The Everglades, Florida; and a *Taxodium* (cypress-dominated) peat from the Okefenokee Swamp, Georgia.

Methods

Extraction of Humic Substances

Samples were treated with water to remove free and low-molecular-weight carbohydrates, amino acids and other low-molecular-weight compounds, with a 1:1 (v/v) mixture of benzene and methanol to remove lipids, and with 0.01 *N* HC1 to dissolve carbonate minerals and some low-molecular-weight organic compounds. Residues were then treated with an aqueous 0.5 *N* NaOH solution overnight, with a better than 20:1 solution-to-sediment ratio and under nitrogen. Soluble humic substances (humic and fulvic acids) were separated from the insoluble residue (humin) by centrifugation. The NaOH solution was filtered through glass fiber filters (0.5-μ pore size) to remove colloidal clays. The NaOH extraction of the sediment was repeated for some samples but only small amounts of additional soluble humic materials were obtained. The method of extraction used here is similar to that used by Schnitzer and Khan [5] and is the most commonly used method, having a high extraction efficiency.

The filtered solution of soluble humic substances was treated to separate humic and fulvic acids using a modification of the method reported by Malcolm [51], namely, ion-exchange separation. In general, the 0.5 *N*

NaOH solution is passed through a column of Dowex* 50W X8 resin in the H^+ form to exchange the Na^+ for H^+; the humic and fulvic acids are thereby protonated. In this manner, protonated and soluble humic and fulvic acids were both eluted simultaneously in neutral solution that was then acidified to pH 2 by addition of 6 *N* HC1 in order to precipitate humic acids. These precipitates were separated from fulvic acids by centrifugation and washed with 0.1 *N* HC1. Humic acids were then freeze-dried.

Fulvic-acid solutions obtained in the centrifugate were usually free of salts. The solutions were evaporated to a small volume on a rotary evaporator and freeze-dried. For some samples, such as Continental Shelf sediments, fulvic acids were further purified by ultrafiltration through an Amicon PM10 membrane (10,000 molecular weight exclusion) as described by Kwak *et al.* [55] to remove low-molecular-weight saccharides.

Humins, recovered as residues following alkali extraction, were generally washed several times with distilled water, acidified to pH 6 by addition of concentrated HC1, and then washed repeatedly. In the case of peats, these residues were lyophilized and stored dry. The residues from marine sediments and the Mangrove Lake sapropel were treated with a 1:1 (v/v) mixture of concentrated HF and HC1 and heated on a steam bath. The HF/HC1 was repeatedly decanted, and fresh HF/HC1 solution was added. After several such treatments, only HC1 was added after decantation. The residue/HC1 mix was diluted to neutrality with deionized water and centrifuged to separate the humin, which was then freeze-dried.

Elemental Analyses

The amounts of carbon, hydrogen and nitrogen in the humic substances were determined by combustion in oxygen using a Perkin-Elmer 240 Elemental Analyzer. The values were corrected for moisture and ash contents measured by heating the samples at 105 °C and 960 °C, respectively.

Oxygen was usually measured by difference; however, some of the samples were analyzed for elemental oxygen by a direct pyrolysis method using a Carlo Erba model 1106 Elemental Analyzer. The ash contents of these samples were then determined by difference.

Infrared Spectroscopy

Dried samples (0.5 to 1.0 mg) were mixed in a vibrating ball mill with 300 mg of dry KBr, and the mixture was placed in an 11-mm pellet die.

*Any use of trade names in this chapter is for descriptive purposes only and does not constitute endorsement by the U.S. Geological Survey.

Approximately 15 tons of pressure was applied to produce a clear KBr pellet suitable for infrared spectroscopy using a Perkin-Elmer 621 spectrophotometer. The spectrum was scanned from 4000 cm^{-1} to 385 cm^{-1} against a blank KBr pellet in the reference beam.

NMR Spectroscopy

Solution NMR. Humic acids were placed in a fritted glass filter funnel and washed four times with a 0.01 *N* DCl in D_2O solution to exchange hydrogen atoms with deuterium atoms. The humic acids (100 mg) were dried between washings and then redissolved in 3 ml of a 0.5 *N* NaOD solution. ^{13}C spectra of humic acids were obtained at 25.2 MHz on a Varian Model XL-100 FT NMR having a Nicolet NT 440 multiobserved nuclei accessory for 12-mm tubes. Multiple transients (10,000 to 150,000) were acquired locking onto the deuterium signal. The chemical shift is reported relative to tetramethylsilane (TMS) as an external reference.

Solid-state CP/MAS ^{13}C NMR. Because solid samples are analyzed by this technique, humic acids, fulvic acids and humins were all dried prior to analysis. CP/MAS ^{13}C-NMR spectra were obtained on "homebuilt" spectrometers at the National Bureau of Standards and at the Colorado State University. Both spectrometers operate at a magnetic field of 1.4 T, which translates to nominal radio frequencies of 60 MHz for ^1H and 15.1 MHz for ^{13}C. The magic angle spinning probe is a single-coil, double-tuned design in which a spinner assembly with a polytrifluorochloroethylene rotor was used. The finely ground dry samples were pressed into the sample spinner. The probe operates at ambient temperature, approximately 28 °C.

Spectra were obtained using the spin-lock cross polarization technique [23] with the Hartman-Hahn condition [56] fulfilled for the 1 to 2 msec cross-polarization time with a radiofrequency field strength of 64 kHz on both the ^{13}C and the protons. Repetition times for the experiment ranged from 7.5 to 1.0 sec. Alternation of the phase of the initial 90° proton pulse by 180°, coupled with the addition and subtraction of consecutive scans, eliminated baseline artifacts [57]. Details of the mechanics involved in obtaining a signal from fossil fuels and polymers using this CP/MAS experiment have been published [25,27,28,29,35,58].

Quantitative Measurements by CP/MAS ^{13}C-NMR

Peak areas for all spectra were measured with an automatic digital integrator (Numonics Corp. Model 1224 Digitizer). Successive measurements of the absolute areas of well defined peaks provide deviations of (±)5%. Peaks that are not well resolved give relative deviations of ±10%. In calculations, we have assumed that the relative errors in peak areas range from 5 to 10% of the total area of the peak under discussion.

The question of signal-intensity distortions resulting from the cross polarization experiment has been addressed in several reports [24,33,34, 37,59]. The quantitative significance of ^{13}C intensities depends on the relationships among pertinent relaxation times and relevant experimental parameters. Values of the relaxation times are seldom known explicitly for the samples in question. Sullivan and Maciel [33] have shown that the relaxation properties of a Powhatan No. 5 coal are conducive to quantitative solid-state ^{13}C-NMR.

Intensities of peaks representing various carbons in CP/MAS spectra may, in principle, be distorted because certain structures (e.g., condensed aromatic structures) can isolate carbons from protons so that cross polarization may not be effective and full intensity may not be realized. Also, preferential concentration of free radicals and quadrupolar nuclei such as ^{14}N–which have a tendency to broaden the NMR signals–on one structural component may also lead to nonrepresentative signal intensities. VanderHart and Retcofsky [24] compared the CP/MAS spectra of some coals with spectra obtained by the conventional Fourier transform (FT) experiment (90° pulse followed by observation of the free-induction decay with a long waiting period between pulses). Aromaticities calculated by both methods were comparable, an indication that intensity distortions were minor or nonexistent.

Maciel et al. [59] have shown that the intensities measured in a CP/MAS ^{13}C-NMR experiment on coal are essentially the same as those obtained in a less efficient MAS experiment without cross polarization, further indication of the quantitative significance of CP/MAS ^{13}C-NMR intensities in coal.

Coals are more likely to contain condensed aromatic regions as opposed to humic substances that are not thought to contain an abundance of such structures. Because intensity distortions are not significant in spectra of coals, distortions due to carbons isolated from protons probably are not serious in humic substances.

Recently, Wind et al. [60] have demonstrated that unequal distributions of free radicals induce intensity distortions in CP/MAS spectra. Their studies were based on a coal that probably had a high content of free radicals, i.e., a spin density several orders of magnitude greater than that observed in humic substances. Their results, however, are not directly applicable to humic substances, and the question of signal distortion due to free radicals remains open to conjecture. In addition, we must point out that even in the coal examined by Wind et al. [60], the distortions, although measurable, were small. Considering the lower free radical spin density of humic substances [61], we conclude that signal distortions cannot be a serious problem.

Though free radicals and isolated carbons are not expected to cause signal distortions in CP/MAS ^{13}C-NMR spectra of humic substances, other factors such as contact times and repetition rates could also influence relative intensities. Studies of large numbers of coal samples and humic substances [31,62] have shown that contact times on the order of 1 to 2 msec and repetition rates of 1 to 4 sec have generally been adequate to obtain representative signals. In the present study, the effects of contact times and repetition rates were examined for several humic substances. Contact times of 1 to 2 msec and repetition rates of 1 to 4 sec were tested on a number of samples, and results indicated that a contact time of 1 msec and a repetition rate of 1.5 sec, and for some samples 1.0 sec, were adequate for quantitative results.

To obtain some additional satisfaction that representative and quantitative signals were being generated from CP/MAS ^{13}C-NMR spectra of humic substances, the Rice Lake (Minnesota peat) humic acid was examined by three NMR techniques. First, a conventional high-resolution ^{13}C-NMR spectrum was obtained in solution (0.5 *N* NaOD) using a 30° pulse and 0.6-sec acquisition time and repetition rate. Second, two normal CP/MAS ^{13}C-NMR spectra were recorded; one was obtained at Colorado State University using a 1-msec contact time and a 2-sec repetition rate, and the other was obtained at the National Bureau of Standards using a 2-msec contact time and a 7.5-sec repetition rate. Third, a long-term FT spectrum was obtained at Colorado State University using a 90° pulse and a 25-sec repetition rate. All spectra are shown in Figure 1. Although some differences can be observed in relative intensities, the values are well within 10%, the error expected in the measurement of areas. No serious distortions are apparent, except possibly in the solution spectrum.

Figure 1 indicates that the solid-state spectra closely approximate the solution spectrum; however, the resolution appears slightly better in solution. The most likely source of additional line-broadening in the solid samples is the conformationally induced dispersion of chemical shifts [25,27,28,58]. In solution, the humic acid molecules are free to undergo rapid intramolecular motions such that the observed chemical shift is the weighted average of the shifts of all the conformations available to the molecule. In solids, molecules may be held in particular conformations, each of which may have a different set of chemical shifts. This conformational broadening is a direct reflection of chemical information, not a shortcoming of the technique, and is not averaged by line-narrowing techniques. Although the lines are broader in the solid spectra than in the solution spectrum of humic acid, the relative areas of the corresponding component peaks in each spectrum are nearly the same. Thus, quantitative measurements can be made from either solution or solid spectra. At first

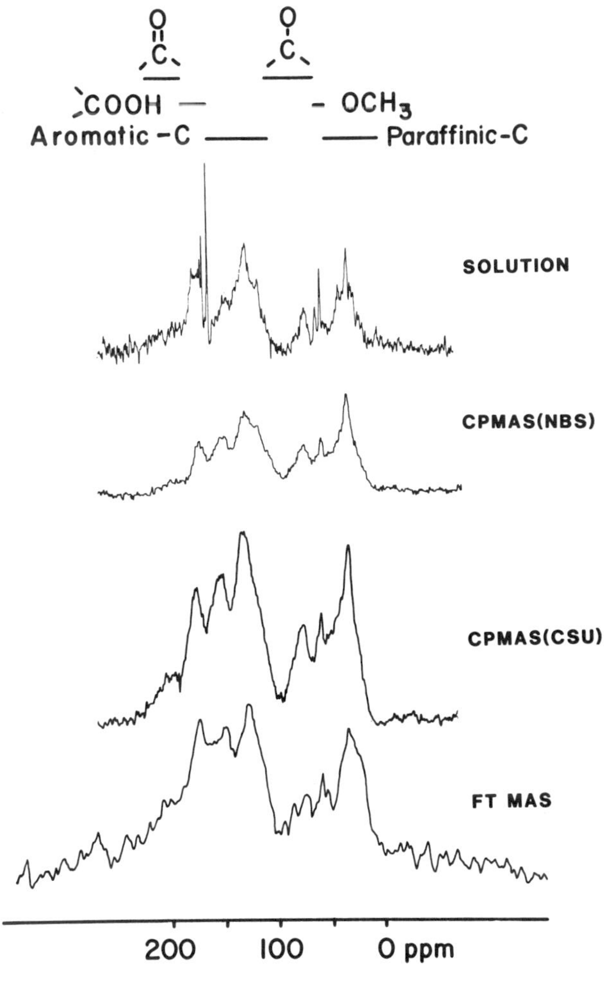

Figure 1. ^{13}C-NMR spectra of humic acids from the Rice Lake peat sample obtained in solution (0.5 *N* NaOD) and in the solid state by CP/MAS and by conventional Fourier transform (FT) with magic angle spinning. One of the CP/MAS spectra was obtained by D. VanderHart of the National Bureau of Standards (NBS) and the other was obtained at Colorado State University (CSU). Details of the NMR techniques are given in the text.

glance, the carboxyl region (160 to 190 ppm) of the solution spectrum might appear to be more pronounced than that of the CP/MAS spectra; however, by the techniques used for area measurement, the relative peak areas are very nearly the same (e.g., 24% for the carboxyl peak in the solution spectrum and 19% for same peak in the CP-MAS spectrum).

Because an error of 5 to 10% can be expected, these two measurements are not significantly different.

The results of these comparisons all point to the fact that signal distortions arising from free-electron interactions, condensed aromatic regions poor in hydrogen distribution, rapid retention rates and contact times of 1 msec are not serious impediments to obtaining quantitatively significant spectra.

RESULTS AND DISCUSSION

Aquatic Sediments

Humic substances were obtained from samples of aquatic sediments that included marine sediments, a river sediment, and an anoxic algal sapropel (Mangrove Lake, Bermuda). The sapropel yielded only humic acids and humin (the fulvic-acid content was so low that material obtained for analysis was insufficient). Aquatic fulvic acids are from samples of marine sediments of the Continental Shelf of the New York Bight.

Fulvic Acids

Fulvic acids isolated from continental shelf samples of the New York Bight were examined by ^{1}H- and ^{13}C-NMR in solutions of NaOD as reported by Hatcher et al. [19]. This work demonstrated that polysaccharides containing uronic acid groups (polyuronic acids) are the dominant components of these fulvic acids. An example of a CP/MAS ^{13}C-NMR spectrum for fulvic acids from aquatic sediments is shown in Figure 2. The CP/MAS ^{13}C-NMR spectrum is similar to the solution ^{13}C-NMR spectra [19], indicating the large contribution of polysaccharides (peaks at 72 and 105 ppm). Though low-molecular-weight carbohydrates can also contribute to the peaks at 72 and 105 ppm, the major contributors are polysaccharides because, by definition, fulvic acids should not contain appreciable quantities of low-molecular-weight carbohydrates. By integration, polysaccharide carbons account for approximately 52% of total carbons. Other peaks are observed in the spectrum, but their contributions to the total signal are small. Paraffinic carbons and associated protons have resonance bands in the region of 0 to 50 ppm; methyl carbons (15 ppm) are also significant contributors. Peaks at 18 to 40 ppm have been assigned to methylene carbons.

The presence of carboxyl or amide carbons is confirmed by the peak at 175 ppm in the ^{13}C spectrum; these carbons account for 15% of all the carbons. These carboxyl groups could be in part derived from uronic acid

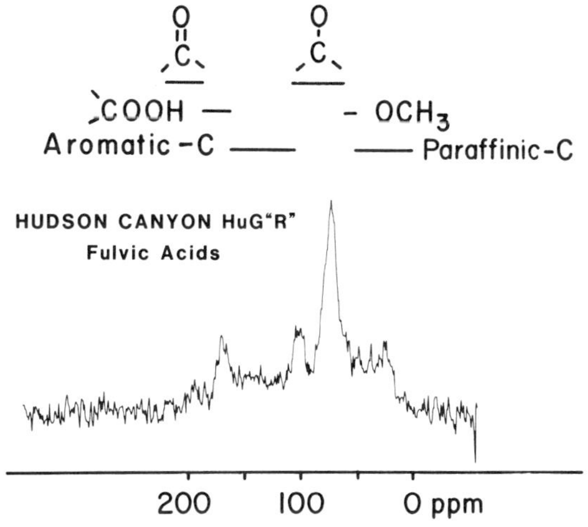

Figure 2. A representative CP/MAS ^{13}C-NMR spectrum of fulvic acids from marine sediments. The spectrum is obtained from Hatcher et al. [19].

moieties associated with polysaccharides (polyuronic acids). Infrared spectroscopic and colorimetric tests have confirmed the conclusion that polysaccharides containing uronic acids are dominant components of these marine sedimentary fulvic acids [19].

Of the studies of the nature of marine sedimentary fulvic acids, few have unambiguously identified the major structural components. Ishiwatari [63] observed that polysaccharides account for a large fraction of sedimentary fulvic acids. Steelink [64] and Rashid and King [65], suggested that polysaccharides or polyuronic acids could be isolated from sediments as fulvic acids or humic acids. Ernst [66] isolated uronic acids from fulvic acids in sediments of the North Sea. The evidence presented by Hatcher et al. [19] indicates that polysaccharides containing uronic acid moieties are by far the most abundant components of the marine sedimentary fulvic acids examined. The solubility of polysaccharides and polyuronides in weak acidic and basic solutions has long been known [67]; in most sedimentary deposits containing plant and microbial residues, these polysaccharides are partially extracted by alkali.

Humic Acids

Aquatic sedimentary humic acids, extracted from Continental Shelf and deep sea sediments and an algal sapropel from Mangrove Lake, Bermuda, have elemental compositions and IR spectra similar to those of aquatic sedimentary humic acids examined by Nissenbaum and Kaplan [68], Stuermer et al. [2], Huc and Durand [69], Debyser and Gadel [70], and Pelet and Debyser [71]. Elemental compositions, yields and stable isotopic compositions for aquatic humic acids examined in this study have been reported by Hatcher [62]. The most notable features of the data, when compared to those for soil humic acids [2,5], are the high nitrogen contents (values generally greater than 3%), the high hydrogen contents (values greater than 5%), and the isotopically heavy values for $\delta^{13}C$* (−20 ‰ to −25 ‰). Nissenbaum and Kaplan [68] have indicated that marine sedimentary humic acids having such nitrogen and carbon contents and $\delta^{13}C$ values are autochthonous.

The high hydrogen and nitrogen contents of marine sedimentary humic acids have been attributed to their predominantly aliphatic nature and to the contribution of nitrogen-rich proteinaceous materials [70]. Stuermer et al. [2] have suggested that the high nitrogen contents are indicative of the fact that melanoidin-type mechanisms are responsible for their formation. Huc and Durand [69], as well as Debyser and Gadel [70], showed, however, that most of this nitrogen could be eliminated by hydrolysis. This suggests that proteinaceous material can be associated with humic acids either via direct bonding or via weak hydrophobic attractions that are more or less related to the solubility behavior of proteins. Thus, proteins, because of their functional groups, may be soluble in weak alkali but insoluble in weak acid and, owing to the definition of humic acids, they can be extracted with and labeled humic acids.

Infrared spectra of aquatic sedimentary humic acids examined in this study resemble closely those published by Stevenson and Goh [72], as well as Huc and Durand [69] and Debyser and Gadel [70]. In general, the spectra indicate the presence of hydrogen-bonded OH and of paraffinic, carbonyl and carbohydrate structures. A band observed at 1520 cm^{-1} assigned to the amide II band disappeared on hydrolysis [62], providing partial evidence for the presence of proteinaceous substances.

Aquatic sedimentary humic acids examined by CP/MAS ^{13}C-NMR are shown in Figure 3. Solution NMR spectra of some of these samples (both ^{1}H- and ^{13}C-NMR) have been published by Hatcher et al. [3,4]. The ^{13}C-NMR spectra of the various aquatic sedimentary humic acids display signals principally in five major regions, whose assignments are listed in Table II. Region I (0 to 50 ppm) is assigned to paraffinic carbons or carbons that are chemically bonded only to other carbons. This is generally the most intense region in spectra of marine humic acids, indicating

*Stable isotope ratios of carbon are listed in the δ notation which is the per mil ‰) deviation of the $^{13}C/^{12}C$ ratio from that of the Chicago PDB standard.

Table II. ^{13}C-NMR Chemical-Shift Regions for Various Types of Carbons Typically Present in Humic Substances

Region I (0 to 50 ppm)	Region II (50 to 110 ppm)	Region III (110 to 160 ppm)	Region IV (160 to 190 ppm)	Region V (190 to 220 ppm)
paraffinic C:C–C*–C	alcohols C–OH	olefinic–C	carboxyl (C(=O)–OH)	aldehyde (C(=O)–H)
(C)₂C*–C	amines (C–NH_2, C–NHR, C–NR_2)	aromatic–C	ester (C(=O)–OR)	ketone (–C(=O)–)
C–C*(C)(C)–C	carbohydrates		amide (C(=O)–N)	
	ethers (C–O–C)			
C–CH_3	methoxyl (OCH_3)			
	acetals (O–C–O)			

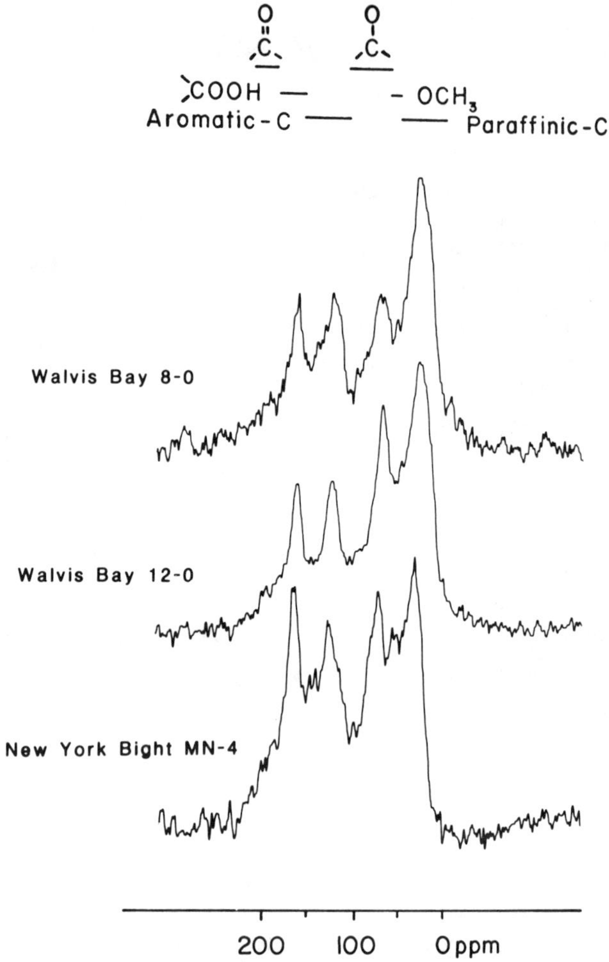

Figure 3. CP/MAS ^{13}C-NMR spectra

that paraffinic structures are major contributors. The major peak in this region is centered at 30 ppm, which is the chemical shift of methylene carbons in either long-chain paraffinic structures or cycloparaffins. The presence of long-chain *n*-paraffins is not likely based on the observation of Hatcher et al. [3] of significant numbers of methyl groups in the

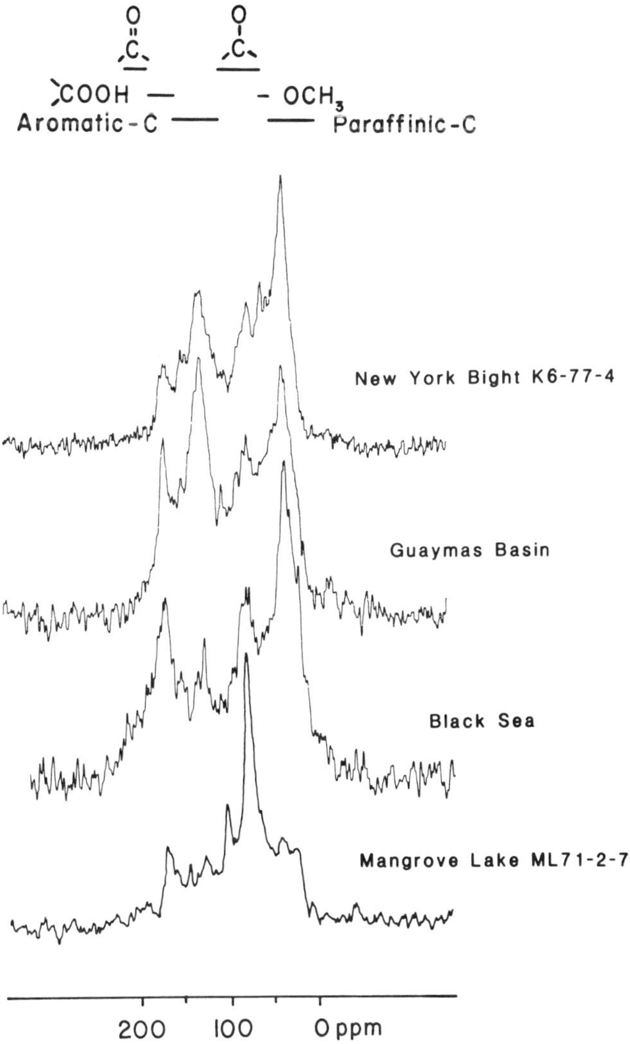

of some aquatic sedimentary humic acids.

^{1}H-NMR spectra. Thus, the peak at 30 ppm probably is derived mostly from cycloparaffinic carbons. Integration of areas indicates that paraffinic carbons contribute from 32 to 45% of the total carbons in most of the aquatic sedimentary humic acids.

Region II (50 to 110 ppm) is assigned to oxygen- and nitrogen-substituted carbons as in carbohydrates, ethers, alcohols and amines (Table II). Methoxyl carbons usually display resonances at approximately 55 ppm. Only the spectra of humic acids from New York Bight samples (Figure 3) and the Potomac River sample (not shown in Figure 3) show a distinct peak for methoxyl carbons. Because methoxyl groups are usually associated with lignin or terrestrial humic substances, these peaks may indicate the presence of some lignin residues in these particular humic acids. This is not totally unexpected considering the proximity of these samples to terrestrial sources.

Though the peak at 72 ppm can be assigned to either carbohydrates or aliphatic alcohols, it nonetheless is most likely to be derived from carbohydrates, considering the fact that carbohydrates have been identified as major components of some sedimentary humic acids [63]. A good clue that polysaccharides are present is the observation of a peak at 105 ppm in the ^{13}C-NMR spectra. This chemical shift is assigned to anomeric carbons of polysaccharides. Low-molecular-weight saccharides are excluded by definition. The ^{13}C-NMR spectrum of humic acids from Mangrove Lake (Figure 3) is characterized by a series of three large peaks in region II of the spectrum. These peaks predominantly reflect carbohydrate carbons as in polysaccharides [73]. The same three peaks are also observed in ^{13}C-NMR spectra of all other humic acids; however, the relative contributions are small, indicating that polysaccharides are not major constituents of those humic acids.

Region III, the 110 to 160 ppm region (Table II), reflects aromatic and olefinic carbon resonances. Because olefinic carbons are generally not major constituents of humic acids, aromatic carbons are expected to be the major contributors to peaks in this region. Therefore, by area measurements, the aromatic (and olefinic) carbon content of humic acids can be calculated. The amounts of aromatic carbons (Region III) vary greatly among aquatic sedimentary humic acids and range from 12 to 33%. Because aromatic carbons are thought to originate primarily from terrestrial sources, the humic acids containing significant signal intensities in this aromatic carbon region may, in fact, have some input from terrestrial sources, and the humic acids may not be entirely of an aquatic origin. The Walvis Bay samples are known to be derived almost entirely from aquatic plant remains, mostly diatoms [2]; thus, the aromaticity of these humic acids (16 and 18%) reflects the maximum aromaticity to be expected from humic acids that are strictly of aquatic origin.

Humic acids from the Potomac River sample (spectrum not shown in Figure 3), Guaymas Basin sample (Figure 3), and New York Bight sample K6-77-4 (Figure 3) are notably richer in aromatic carbon structures than humic acids that are strictly of aquatic origin. Because of their proximity to land, these three sediments are receiving significant terrestrial inputs. The New York Bight sample (K6-77-4) receives additional terrestrial inputs from the dumping of sewage sludge nearby [38]. Thus, it appears that the aromaticity provides a good clue as to the relative contribution of terrestrial input to these sediments.

A distinct peak is observed at a 175 ppm in all ^{13}C spectra of aquatic sedimentary humic acids. This peak in Region IV (160 to 190 ppm, Table II) represents carboxyl, amide and ester carbons, all of which have been identified as important components of humic acids [5]. This peak contributes approximately 15% of the total signal. Additional resonances are observed downfield of the carboxyl resonance in the region of approximately 200 ppm, Region V (190 to 220 ppm, Figure 3). These resonances are characteristic of aldehyde and ketone carbons. Although no distinct resolved peaks are observed, their presence is obvious, with contributions of about 5% of the total signal.

In summary, CP/MAS ^{13}C-NMR offers great potential for structural studies of marine and terrestrial humic acids, especially as the wealth of information available from the spectra can be obtained in a nondestructive fashion. The quantitative structural data available can provide information from which we can begin to draw molecular framework models of humic acids from aquatic environments. Examination of the spectra have borne out several interesting observations for marine humic acids.

The high carbohydrate content of the humic acid from the Mangrove Lake sapropel is unique. Humic acids of other samples have low but varying amounts of carbohydrates. What mechanism, then, is operative in the incorporation of carbohydrates in humic acids? Preliminary results suggest that the observed differences in the carbohydrate or polysaccharide content of these various humic acids are a function of the definition of humic acid. For instance, in the sapropel of Mangrove Lake, conditions are such that only carbohydrates are isolated as humic acids. Normally, carbohydrates are not expected to appear as humic acids, as the extraction procedure used is specifically designed to exclude neutral sugars and low-molecular-weight saccharides. Thus, carbohydrates recovered as humic acids may contain polar functional groups that are capable of imparting the solubility properties of humic acids to the carbohydrate polymers (i.e., carboxyl groups as in uronic acids). Hatcher et al. [19] have shown that fulvic acids of aquatic sediments could be composed of large amounts of polysaccharide-like substances. Their ^{13}C-NMR spectra are nearly identical to the spectrum observed for humic acids of the Mangrove Lake sapropel.

Interestingly, the yield of humic acids in the sapropel of Mangrove Lake is quite low, being less than 2% of the total organic matter. This finding is surprising, as aquatic sedimentary humic acids have been shown to constitute a much larger fraction of oceanic sedimentary organic matter [74,75], on the order of 10 to 60%. It is possible that aquatic sedimentary humic acids of the type shown to be present in other marine environments are not forming in the Mangrove Lake sapropel. However, because the sapropel is rich in carbohydrates, probably from an algal source, and some of those carbohydrates are probably uronic acids or polysaccharides containing some uronic acid monomers (alginic acid), the materials isolated in low yields as humic acids are most likely polyuronic acids. In fact, humic acids isolated from carbohydrate-rich marine sediments elsewhere probably contain small percentages of polyuronic acids. Although the spectra of humic acids of all other samples examined in this study suggest the presence of carbohydrates, they are probably present in relatively small quantities.

All other ^{13}C-NMR spectra examined indicate that paraffinic structures are the dominant components of aquatic sedimentary humic acids, although the spectra show significant differences in aromaticity of various aquatic sedimentary humic acids. The ^{1}H-NMR data of Hatcher et al. [3] and Dereppe et al. [21] suggest that these paraffinic structures are highly branched by virtue of the high proportion of methyl groups. Thus, a highly branched network of paraffinic structures having varying degrees of aromaticity forms the basic macromolecular framework for all aquatic sedimentary humic acids examined other than those from Mangrove Lake.

The results presented by Huc and Durand [69], Dereppe et al. [21] and Debyser and Gadel [70] indicate that proteinaceous materials may be associated with these aliphatic structures. The presence of nitrogen and the amide II band in infrared spectra of aquatic humic acids examined in this study suggest that proteinaceous substances may be components.

Humin

The insoluble macromolecular humic substances in aquatic sediments are collectively called humin, although some have referred to them as protokerogen [2]. In this study, humins from a number of different aquatic sediments have been examined; the humins were generally recovered from samples from which humic acids were isolated. Elemental compositions IR spectra and CP/MAS ^{13}C-NMR spectra (Figures 4 and 5) were obtained. The elemental data are generally similar to those of the aquatic humins published by Stuermer et al. [2], indicating that aquatic humins have a high proportion of nitrogen and hydrogen as compared to humin of soils and peats [5,76]. The nitrogen contents are generally greater than 4%, indicating that nitrogenous residues such as proteins and chlorophyll derivatives are probably being incorporated to a greater extent than in terrestrial humins, which generally have a lower nitrogen content. The high hydrogen contents (~7%) and atomic H/C ratios (~1.5) of humins derived strictly from aquatic sources are an indication that the structural components are more aliphatic than their terrestrial counterparts. Interestingly, some aquatic humins–samples HuG"R", K6-77-4 and MN-4–have low nitrogen and hydrogen contents as well as low atomic H/C ratios, indications that they consist of significant contributions from terrestrial

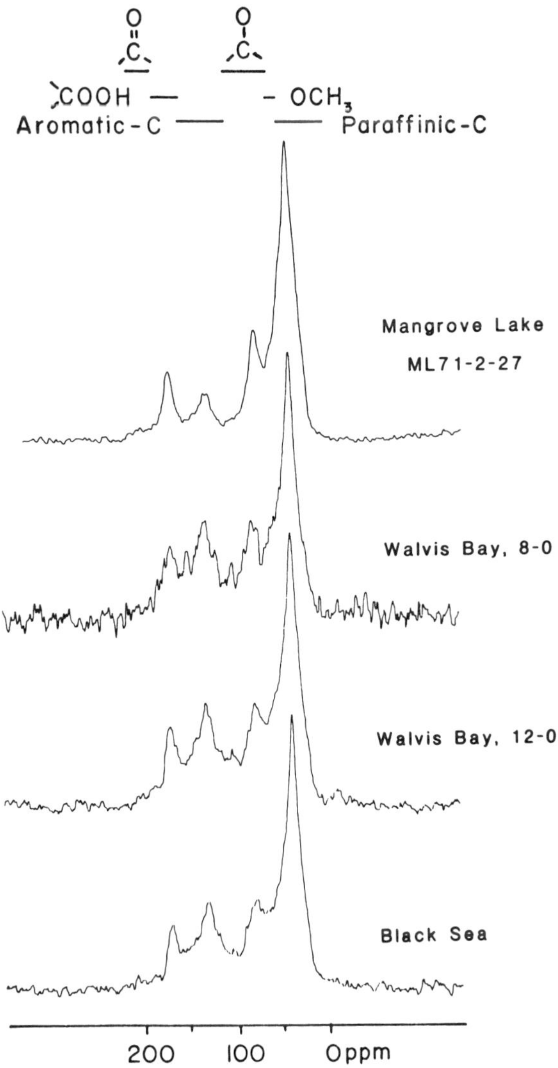

Figure 4. CP/MAS ^{13}C-NMR spectra of aquatic humins whose primary source is algal or microbial remains.

sources. Isotopic ratios (δ^{13}C) also indicate that these samples have significant terrestrial input by the fact that their δ^{13}C values are more like those of terrestrial sediments (approximately −25 ‰). Other aquatic humins have δ^{13}C values of approximately −20 ‰, similar to those of total organic carbon in marine sediments [77].

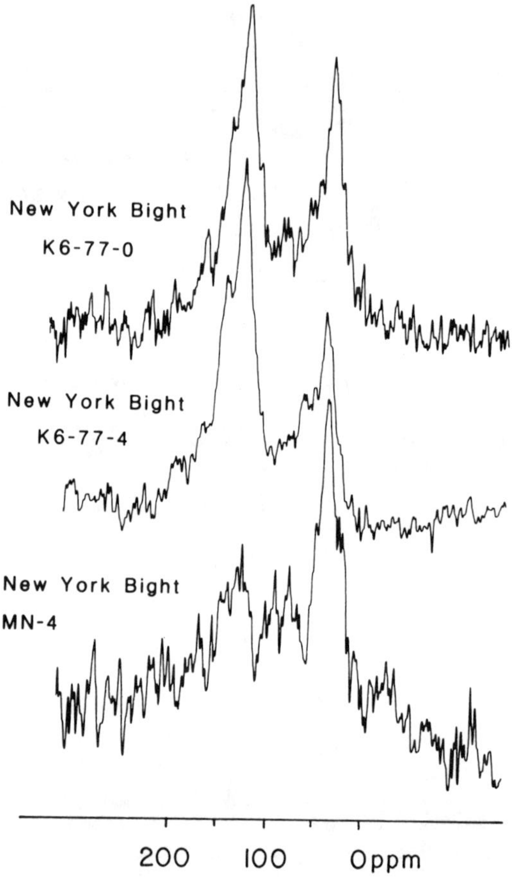

Figure 5. CP/MAS ^{13}C-NMR spectra of humic

The most notable features of IR spectra of aquatic humins are the strong aliphatic C-H stretching bands at 2860 cm^{-1} and 2920 cm^{-1} indicative of aliphatic structures, the C = 0 band of carboxyl and carbonyl groups at 1710 cm^{-1}, the amide II band at 1540 cm^{-1} characteristic of nitrogeneous substances, and the band at 1050 cm^{-1} characteristic of C-O stretching

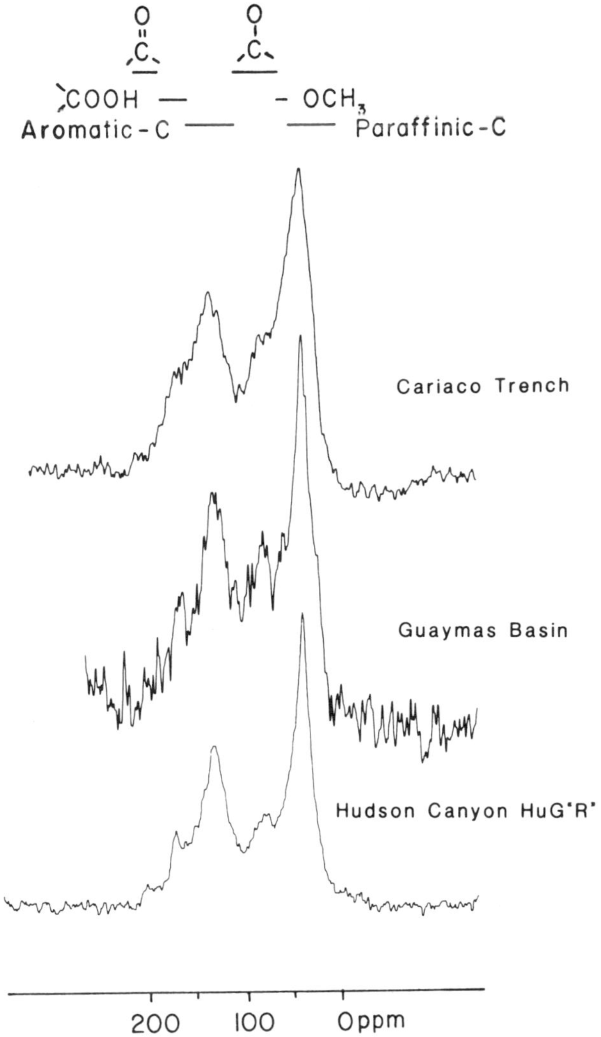

acids from various continental shelf sediments.

frequencies in ethers and carbohydrates. The strong resonances at 2860 cm^{-1} and 2920 cm^{-1} indicate that these humins are highly aliphatic. The strong amide II band is consistent with their relatively high nitrogen content.

CP/MAS ^{13}C-NMR spectra of aquatic humins are shown in Figures 4 and 5.

As was done earlier in the case of humic acids, the spectra are subdivided into five regions (Table II) whose relative areas were measured. The CP/MAS ^{13}C-NMR spectra confirm that aquatic humins derived strictly from aquatic sources (i.e., spectra in Figure 4) are aliphatic in structure. The largest peaks are present in the aliphatic carbon region, which spans the range of from 0 to 105 ppm, and these humins, therefore, appear to be predominantly aliphatic in nature. Oxygen-substituted carbons (Region II) are clearly resolved from paraffinic carbons (Region I), which resonate between 0 and 50 ppm where the major peak is at 30 ppm. The 30 ppm peak is dominant in all spectra (Figures 4 and 5) and contributes from 24 to 67% of the total area in aquatic humins examined in this study.

The peaks at 72 ppm and 105 ppm (Region II) are characteristic of carbohydrates, which are major components of Holocene sediments. Ether-linkage carbons (C-O-C) are known components of humic substances [5], and their resonances can also be detected in the chemical-shift region of carbohydrates (Table II). The peak at 72 ppm may also have a contribution from ether linkages. Considering the fact that these aquatic humins were treated with concentrated HF/HC1 to remove the mineral matter, significant contributions from carbohydrates are not expected because these carbohydrates have probably been hydrolyzed and removed by the acid treatment. Thus, the peak at 72 ppm probably represents mostly a contribution from etheric or alcoholic carbons rather than from carbohydrates. Regardless, this peak contributes only 6.8 to 30% of the total area in all aquatic humins examined. It is important to mention that the humin is operationally defined as the residue after HF/HC1 treatment. It is likely that macromolecular substances in the raw sediments contain a greater proportion of polysaccharide-like or proteinaceous material than represented by the spectra of humins.

Aromaticities of aquatic humins from sediments that are derived primarily from aquatic sources (e.g., Mangrove Lake and Walvis Bay humins, Figure 4), are low; generally, less than 20% of the carbons are aromatic. Humin from the Black Sea sediment also falls into this category, as its spectrum (Figure 4) is similar, if not identical, to spectra of Mangrove Lake and Walvis Bay humins; its aromaticity is 18%. Other aquatic humins whose spectra are shown in Figure 5 are definitely more aromatic, with aromaticities ranging from 23 to 52%. We have already suggested that high aromaticity is a characteristic of contributions from vascular plants; it is likely, therefore, that the humins with the higher aromaticities have large contributions from terrestrial sources. Thus, spectra of humins displaying the greatest aromaticities are those of samples in close proximity to land. As expected, the New York Bight samples K6-77-4 and K6-77-0 have higher aromaticities (52 and 50%, respectively) than other samples of aquatic humins. The samples from Hudson Canyon (HuG"R"), New York Bight (MN-4), Cariaco

Trench, and Guaymas Basin are all located on or adjacent to continental shelves, and some contributions from terrestrial materials are also expected.

Carboxyl or amide carbons (160 to 190 ppm, Region IV) are observed in the spectra of most humins. These peaks appear to be most prevalent in spectra of aquatic humins of Figure 4—those whose aquatic origin is unquestioned. Spectra of samples having larger contributions from terrestrial sources (i.e., greater aromaticity) do not display sizable carboxyl or amide peaks. This observation suggests that carboxyl or amide carbons may be partly associated with highly aliphatic structures of humin.

Finally, resonances to lower shielding than the carboxyl carbon region, from 190 to 220 ppm (Region V), assigned to carbonyl carbons of aldehyde and ketone functional groups, are discernible as shoulders on the carboxyl peak, suggesting that such functional groups are present. The relative areas are small (less than 5%), indicating that these functional groups are only minor components of aquatic humins.

In summary, aquatic humins appear to be composed of paraffinic structures having little aromaticity in sediments whose organic matter is thought to be derived entirely from aquatic plants. Humin from the sapropel of Mangrove Lake is representative of such humins. In sediments adjacent to land, such as continental shelf sediments, aromaticities increase in response to contributions from terrestrial vascular plants. These humins are probably physical mixtures resulting from dual inputs.

Terrrestrial Deposits

Terrestrial deposits such as soils and peat bogs are generally thought to receive their primary contribution of organic matter from vascular plant remains. Lignin, a relatively nonbiodegradable macromolecular substance composed of oxygen-substituted phenylpropane structures [78], is thought to be the primary contributor to humic substances of terrestrial sediments. Various lignin preparations have been examined [79,80].

CP/MAS ^{13}C-NMR spectra of two lignin preparations are shown in Figure 6 (see Sarkanen and Ludwig [78], for details of the methods used for lignin preparations). The upper spectrum, reproduced from Bartuska et al. [79], is of dioxane spruce lignin, and the lower spectrum is of periodate spruce lignin. Aside from being remarkably similar, the spectra confirm many published conclusions concerning the constitution of lignin, namely that oxygen-substituted phenylpropane structures form the basic molecular framework of these macromolecules [78].

The methoxyl carbon peak at 55 ppm, a major peak in the spectrum, confirms chemical studies that indicate the presence of 10 to 15% methoxyl by weight in lignin. No significant peaks are observed upfield of 50 ppm in the paraffinic carbon region, suggesting that most aliphatic

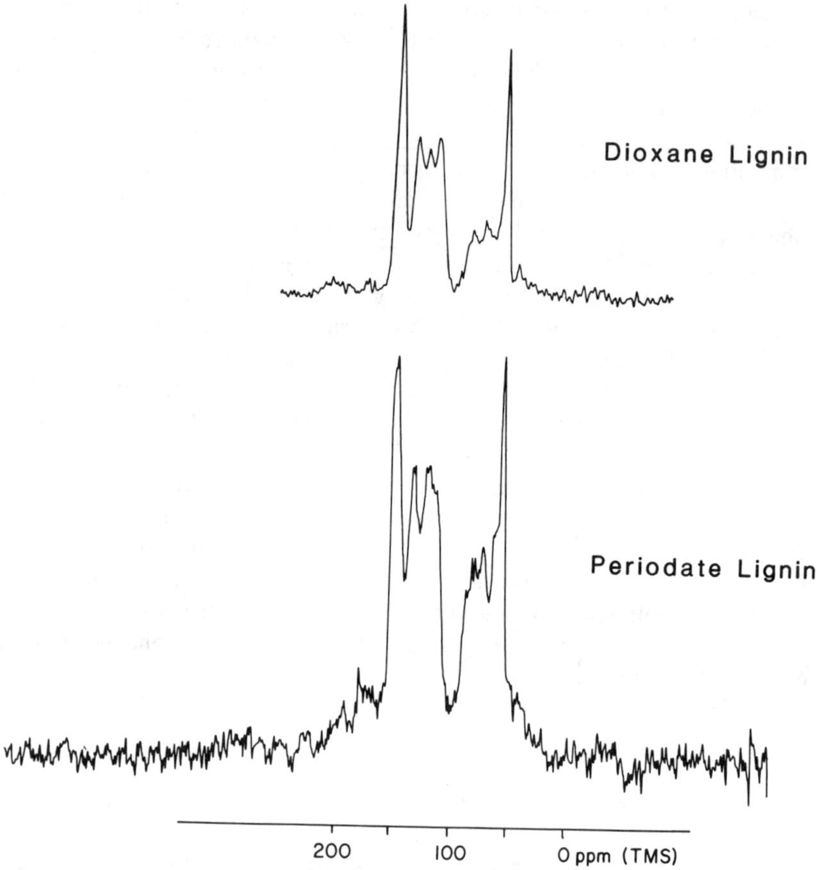

Figure 6. CP/MAS ^{13}C-NMR spectra of periodate and dioxane lignin. The spectrum of dioxane lignin was reproduced from Bartuska et al. [79].

carbons in lignin are substituted by oxygen or oxygen-containing functional groups. Resonances between 50 and 100 ppm typify such carbons.

The main feature of the CP/MAS spectra of lignin is the presence of intense signals for aromatic carbons (110 to 160 ppm). The dominant and clearly resolved aromatic signal at 150 ppm is that of oxygen-substituted aromatic carbons that are important constituents of lignin. Phenolic groups and methoxyl-substituted aromatic rings are the structures responsible for such signals. Aromatic carbons not substituted by oxygen have resonance bands upfield of this region between 110 and 135 ppm. The peaks farthest upfield, at approximately 120 ppm, are assigned to such aromatic carbons that are *ortho* or *para* to oxygen-substituted aromatic carbons. Other aromatic carbons, those not substituted by oxygen and those not *ortho* or *para* to the substituted carbons, have resonance bands centered at approximately 130 ppm. Distinct peaks can be observed for these three different aromatic carbon environments, indicating some definite regularity in structural composition of these aromatic systems.

If lignin is to be considered as a major source material for terrestrial humic substances, then highly aromatic structures should be observed as major components of terrestrial humic substances. In this study, humic substances from soils and peats were examined by CP/MAS ^{13}C-NMR to delineate their structural compositions in relation to their source materials.

Soil

Humic substances in soil have generally been regarded as being highly aromatic in structure. Elemental analyses, spectroscopic studies, and chemical degradation studies have indicated that fulvic acids, humic acids and humin are all composed of macromolecular structures with aromatic groups as building blocks [5,10].

CP/MAS ^{13}C-NMR spectra of humic and fulvic acids from various soil types have been published [6], and Figure 7 shows some representative spectra. Elemental analyses, functional group analyses, and aromaticities calculated by chemical degradations coupled to gas chromatography/mass spectrometry have been published by Schnitzer and co-workers [49,50,81,82] for most of the humic and fulvic acids examined in this study. In nearly all spectra of humic and fulvic acids in soils, the dominant signals are those of aromatic/olefinic carbons (Region III, 110 to 160 ppm) and carboxyl carbons (Region IV, 160 to 190 ppm). Other signals that are present to varying degrees are those of paraffinic carbons (Region I, 0 to 50 ppm), methoxyl carbons, ether and/or carbohydrate carbons (Region II, 50 to 110 ppm) and carbonyl carbons (190 to 220 ppm).

CP/MAS ^{13}C-NMR spectra of humic and fulvic acids vary widely among samples, and the greatest variations are observed for acidic soils [6]. Humic acids from an andosol and an inceptisol of Japan, as well as a North Carolina spodosol (Figure 7), have high aromaticities (46 to 73%). Most humic acids from neutral soils have aromaticities that range from 34 to 43%. Humic and fulvic acids of an Armadale soil, a spodosol from Prince Edward Island, Canada, have significantly lower aromaticities (21 to 29%); the largest signals are those of paraffinic carbons and carbohydrate carbons (Figure 7). The presence of carbohydrates in humic acids is not unexpected, as mentioned above in the discussion of aquatic sedimentary humic acids. Humic acids from the A_0 or uppermost horizon of the spodosol are likely to contain significant amounts of carbohydrates because

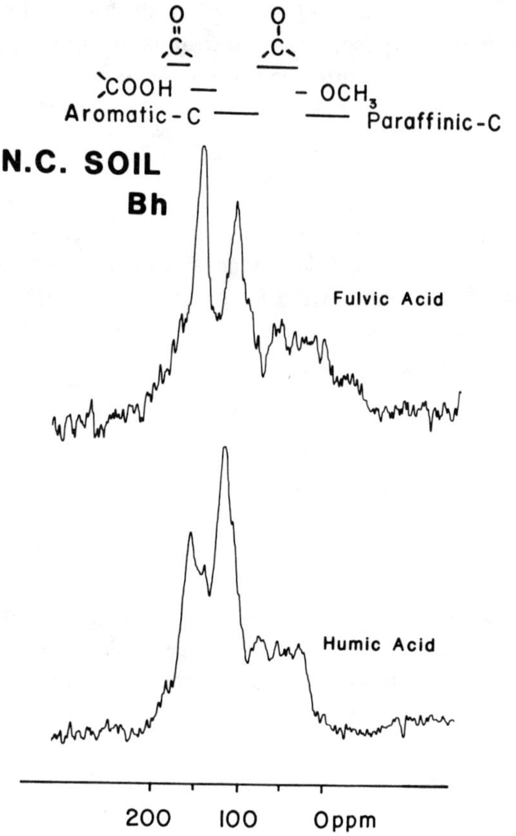

Figure 7. CP/MAS ^{13}C-NMR spectra of humic and fulvic acids from two acidic Armadale soil are shown.

the soil at this horizon is enriched in nondegraded plant litter and carbohydrates such as hemicelluloses extracted by the alkali.

An interesting observation made from the spectra of all soil humic acids examined is the presence of a significant peak for paraffinic carbons. The relative intensity for this peak varies from 7.1 to 36%, and probably

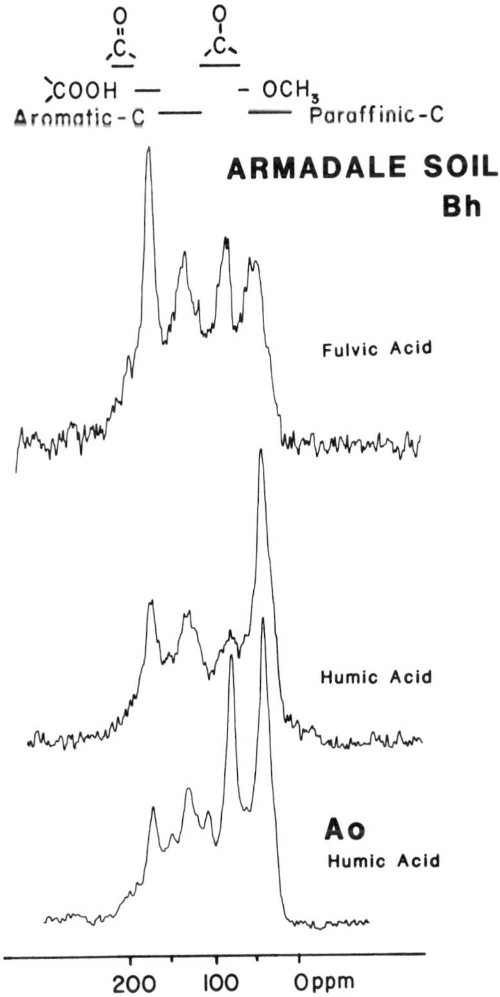

spodosols from temperature climates. Spectra of both the A_0 and B_h horizons of the

accounts for variations in calculated aromaticities for each of the humic acids. This peak has been largely attributed to microbial or algal components [4]. Spectra of humic acids from both the A_0 and B_h horizons of the Armadale soil display strong signals for paraffinic carbons that contribute 34 and 36% of the total area, respectively. Earlier it was shown that aquatic sedimentary humic acids give rise to such a signal at exactly the same chemical shift (30 ppm). Residues similar to those that contribute to aquatic sedimentary humic acids may also be important contributors to the humic acids of the Armadale spodosol. The spectra of humic acids from the A_0 and B_h horizons would be identical if signals from carbohydrates (72 ppm) in the sample from the A_0 horizon were removed. Degradation and loss of carbohydrates in the lower horizons of soils are well known to soil chemists [83] and is confirmed by the spectra.

Aromaticities calculated by chemical degradations range from 61 to 100% and are high for all the humic acids examined compared to aromaticities calculated by NMR [6]. In fact Hatcher et al. [6] have demonstrated that chemical degradation methods have overestimated the aromaticities of these humic acids. Matsuda and Schnitzer [49], in their analysis of oxidation products, alluded to the fact that greater quantities of aliphatic acids were produced from humic acids of the Armadale soil than from humic acids of other soils examined in this study. They concluded that aliphatic acids were probably derived from cleavage of aromatic rings. A more logical explanation, based on the results of the NMR study, is that aliphatic acids were produced from the degradation of paraffinic structures. Chemical techniques have apparently led to overestimated aromaticities in these instances, probably because aliphatic structures may have been completely oxidized to CO_2, H_2O, or other compounds not detected by the gas chromatographic methods. Obviously, one must be skeptical concerning the accuracy of chemical degradation techniques in calculating aromaticities. The nondestructive NMR techniques obviously lead to a far better estimate of aromaticity than do the chemical degradations.

As determined by NMR (Figure 7), fulvic acids appear to be significantly different from their respective humic acids. Fulvic acids have significantly more carboxyl or amide carbon (peak at 175 ppm) than do the humic acids, as is consistent with the chemical data [82]. The fulvic acids also have more carbohydrate/ether carbons and are less aromatic. Because of their solubility, carbohydrates will most likely be associated with the fulvic acid fraction [83–85] and the spectra are consistent with this observation. The most striking difference between the spectra of fulvic acids and those of humic acids appears in the paraffinic carbon region. The humic acids contain significantly more paraffinic carbon than do the fulvic acids. Additionally, the peaks are centered at different chemical shifts, i.e., at 30 ppm for the humic acids and at 50 ppm for fulvic acids. These differences indicate that the paraffinic structures in each are not similar. Because the chemical shift of 50 ppm typifies paraffinic carbons that are α to electron withdrawing groups such as phenyl, carboxyl, amino or amide groups it is likely that paraffinic structures of fulvic acids are highly substituted by such groups. If carboxyl groups which are so abundant in the fulvic acids were attached to aliphatic carbons, then the chemical shift of the substituted carbons would be expected to be at this 50-ppm position.

Phenolic carbons have generally been considered to be integral building blocks of soil humic substances [5,10], and the peaks representing oxygen-substituted carbons should be resolved adequately from peaks for other aromatic carbons, as is observed in CP/MAS spectra of lignin (Figure 6), which contains significant proportions of phenolic groups [79]. Indeed, a peak at 150 ppm is observed in the spectra of humic acids from the B_h and A_0 horizons of the Armadale spodosol (Figure 7) as well as in the spectrum of the North Carolina humic acid shown in the same figure. However, humic acids from other soil types have spectra that do not show a distinct peak at 150 ppm. Thus, the intensity of this peak in all spectra is significantly less than would be expected from the chemical data [82], which suggest that it should be as intense as the peak for carboxyl carbon at 175 ppm. In the spectra of the two fulvic acids (Armadale spodosol and North Carolina spodosol, Figure 7) no resolved peak for these oxygen-substituted aromatic carbons is observed at 150 ppm, in contrast to their reportedly high concentrations [82].

The apparently low intensity of this peak in spectra of soil humic and fulvic acids might be explained in several ways. First, phenolic carbons may in fact be overestimated by the chemical methods, as the amounts are usually measured by subtracting carboxyl contents from total acidities. Errors can easily be magnified by such procedures, especially when one considers that the carboxyl and total acidity determinations may not be accurate in macromolecular structures such as humic acids. Second, the NMR techniques may discriminate against phenolic carbons, especially if these carbons contain associated stable free radicals that might broaden the signals under the conditions of the CP/MAS experiment. The free-radical content of the Armadale fulvic acid (1.5×10^{17} spins/g [61]) is probably not high enough to account for total elimination of phenolic signals. Phenolic carbon signals are readily observable even in lignins that have free-radical concentrations similar to those of the fulvic acid (see Sarkanen and Ludwig [78]). It is not likely that the signals for phenolic carbons have been discriminated against as a result of free radical interactions.

In summary, humic and fulvic acids from nearly all soils appear to have highly aromatic structures, and this finding is in agreement with reports in the literature [82]. In some instances, however, aromaticities can be very low, and paraffinic or carbohydrate-like structures predominate. The NMR spectral features of paraffinic structures are very similar to those of aquatic sedimentary humic acids that were previously considered to be composed of highly branched and cross-linked macromolecular structures. Such amazing similarity may indicate that similar structures are involved and that they are derived from microbial or algal residues. Consistent with these observations, Mendez and Ogner [86] suggested that aliphatic structures in humic acids are derived from microbial exudates or possibly algal lipids. Consequently, the structure of humic acids from soils may not be composed of an association of large molecules that have both aliphatic and aromatic carbons loosely attached or bonded to each other. Aliphatic structures, because of their possible microbial or algal origin, are likely to exist as physically distinct components that are mixed with others derived predominantly from vascular plant remains, namely, the aromatic moieties.

These aromatic structures themselves appear to be quite different from those of lignin, their presumed source. The small amounts of phenolic and methoxyl-substituted aromatic groups indicate that if lignin is the source, then it has undergone substantial modification whereby most of these oxygen-bearing functional groups have been eliminated, probably by oxidative processes.

Carboxyl, carbons, as mentioned above, display a rather strong intensity in the CP/MAS ^{13}C-NMR spectra at approximately 175 ppm. Though carboxylic acid groups are most likely responsible for a major portion of the intensity in this region, ester and amide carbons also have chemical shifts in this range. Consequently, it is not entirely correct to assign the peak at 175 ppm to carboxyl groups alone. Nevertheless, Hatcher et al. [6] have integrated this region of the spectrum (160 to 190 ppm) in order to compare the carboxyl carbon content calculated by NMR with the carboxyl content and total acidities calculated by wet-chemical methods. The results of this comparison indicate that the carboxyl content of humic acids calculated by NMR agrees more with the corresponding value of total acidity than with the carboxyl content determined by chemical methods. This agreement implies either that the carboxyl content calculated by NMR is high owing to interfering groups such as amides and esters, or that the wet-chemical measurement of carboxyl content is in error. If we accept the latter explanation for these discrepancies, then we can partially explain the discrepancy mentioned earlier between phenolic contents and the apparent lack of phenolic carbon in the NMR spectra. The phenolic content of these humic acids is calculated as a difference between total acidity and carboxyl contents [5]. If the latter measurement were inaccurately low, then the phenolic contents would be overestimated by these wet-chemical methods. Such discrepancies are in need of further study.

Peat

Humic substances in peat compose a dominant fraction of the organic matter. Peat forms in a depositional environment where the water level is at or near the top of the peat profile. The contributing flora must be capable of sustaining continual inundation, and the peat itself forms in an aquatic milieu where the availability of oxygen for use by organisms capable of degrading the organic matter is limited. Anaerobic conditions generally prevail within peat deposits. Therefore, the level of degradative activity is lower than that of soils, and the organic matter tends to be preserved better.

In literature on soil, peat is considered as a separate soil group; however, humic substances formed in peat have been thought to be similar to those of soil [76,87]. Humic substances in peat have always been assumed to be highly aromatic because they are thought to be derived from lignin residues of vascular plants. In this study, several peat samples were examined by NMR and other methods to shed more light on their chemical structure and their relationship to soil and aquatic humic substances.

Fulvic Acids. CP/MAS ^{13}C-NMR spectra of fulvic acids isolates of peat have low signal-to-noise ratios, as evidenced by the spectrum of fulvic acids from a Minnesota sedge peat (Cedar Creek Lake) shown in Figure 8. However, one good spectrum was obtained for fulvic acids from a peat from the Okefenokee swamp of Georgia.

The CP/MAS spectrum of this fulvic acid is dominated by signals characteristic of polysaccharides and polyuronic acids. Peaks at 72, 105 and 175 ppm are resonance bands of carbohydrates and carboxyl or amide carbons. In fact, this spectrum is remarkably similar to the CP/MAS spectrum of an aquatic sedimentary fulvic acid (Figure 2) that was earlier suggested to be composed of polyuronic acids. Additional peaks at 130 and 150 ppm in the spectrum of fulvic acids from the Okefenokee peat delineate the presence of aromatic and oxygen-substituted aromatic carbons, respectively, that are probably derived from lignin residues. Thus, fulvic acids from the Okefenokee peat sample are predominantly polysaccharides, and contain smaller amounts of aromatic structures. Fulvic acids from the Minnesota peat (Cedar Creek Lake) also display a strong resonance for polysaccharides in their spectrum, even though the signal-to-noise ratio is poor. The IR spectra of these fulvic acids confirm the presence of polysaccharides with a strong band at 1050 cm^{-1}. The band at 1710 cm^{-1} also suggests that these fulvic acids contain carboxyl groups whose presence is also indicated in the NMR spectra (Figure 8). Such groups could be associated with the polysaccharides, thus suggesting that the fulvic acids contain polyuronic acids.

Humic Acids. In peat, humic acids are easily distinguished from fulvic acids, as observed in some representative CP/MAS ^{13}C-NMR spectra shown in Figure 8. Strong aromatic resonances (110 to 160 ppm) are observed in the spectra of all humic acids from peat; the major band is at 130 ppm and a lesser band is at 150 ppm. A strong peak at 150 ppm is characteristic of lignin and, in humic acids from peat, is most likely derived from lignin

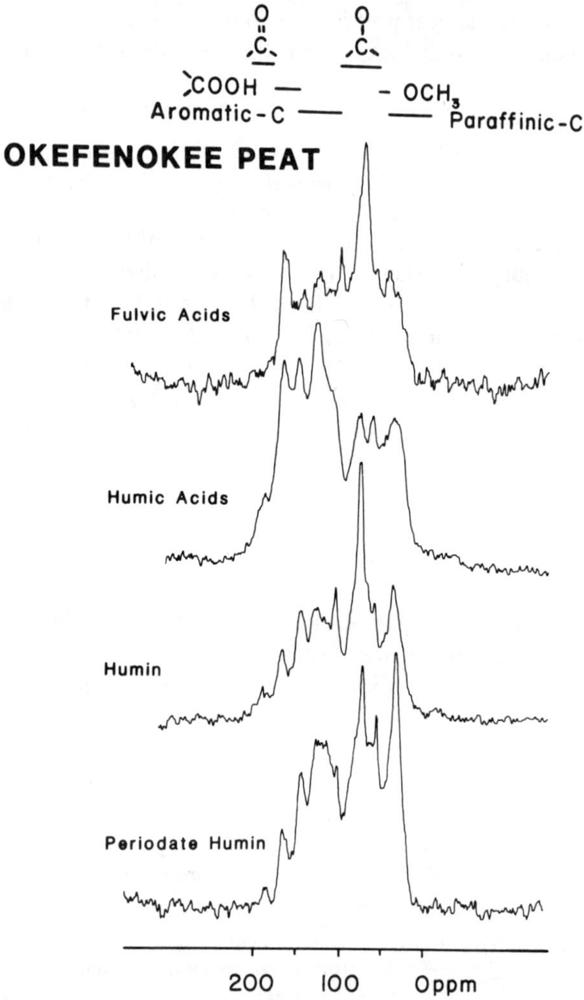

Figure 8. CP/MAS ^{13}C-NMR spectra of fulvic acids, humic acids and humin from Minnesota. A spectrum of the Okefenokee peat humin treated with paraperiodate is

residues. Unlike spectra of soil humic acids (Figure 7) spectra for humic acids in peat contain a distinct peak for oxygen-substituted aromatic carbons at 150 ppm. The aromatic spectral region does not resemble that of soil humic acids; this suggests that peat humic acids may be structurally different than those of soils. Other peaks, typically related to lignin-derived

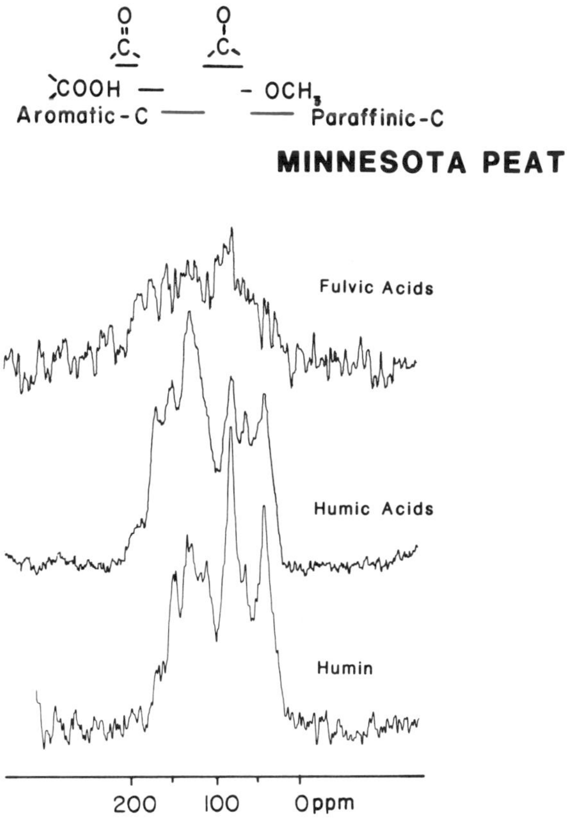

peat samples collected from the Okefenokee Swamp, Georgia, and Cedar Creek Lake in also shown.

structures, are present at 130 ppm, 120 ppm and 55 ppm [79]. Whereas the first two are aromatic peaks, the third is related to methoxyl carbon. This methoxyl peak is definitely more distinct in spectra of these humic acids than in spectra of humic acids of soils, an indication that aromatic rings in peat humic acids are more highly substituted by methoxyl groups.

Carbohydrates or oxygen-substituted aliphatic carbons do not appear to be major contributors to humic acids in peat; both the CP/MAS spectra (Figure 8) and infrared spectra confirm these findings. The intensity of peaks at 72 ppm (Region II) in CP/MAS spectra of all peat humic acids is variable (19 to 27%). These variations may simply reflect the presence or availability of carbohydrates in the peat at the particular horizon sampled. Peat collected from Mangrove Lake was buried for about 10,000 years and carbohydrates are not major components [43,44]; the humic acids reflect this in the low intensity of carbohydrate peaks in their CP/MAS spectra.

A peak at 30 ppm, denoting resonances of paraffinic carbons, is observed in nearly all the CP/MAS spectra of humic acids from peat (Region I). In spectra of humic acids of soils and aquatic sediments, this band was attributed to contributions from microbial or algal sources [4] and it is likely that it has a similar origin in this instance. Humic acids from the Rice Lake peat of Minnesota apparently contain the greatest concentration of paraffinic substances (24%). In humic acids from other peat samples, the amounts of paraffinic carbons appear to be rather constant. The lowest percentage of paraffinic structures (13%) is found in humic acids from a sawgrass peat collected in Mangrove Lake, Bermuda. These particular humic acids appear to be composed mostly of aromatic structures probably derived from lignin residues and aliphatic substances whose NMR spectral characteristics are poorly resolved.

The CP/MAS spectra of humic acids of the peat samples studied display a signal for carboxyl (Region IV) and carbonyl carbons (Region V). These functional groups are characteristic of humic acids from many peat deposits (see Swain [45]). These functional groups appear to compose approximately 2 to 14% of the total carbons. Thus, the carboxyl content of peat humic acids is generally less than that of humic acids in soils, indicating possibly that humic acids from peat may be less oxidized.

Elemental compositions of humic acids of the peat samples are similar to those as published by Manskaya and Drozdova [87]. The humic acids examined in this study do not show any significant variability in the elemental data, attesting to the insensitivity of such methods for structural determinations. The atomic H:C ratios are generally 1.1 or less and are indicative of a high degree of aromaticity. The yields of humic acids based on a moisture- and ash-free weight are also within the ranges reported for peat [87].

In summary, humic acids from peat are highly aromatic, with aromaticities ranging from 40 to 45%. As expected, CP/MAS ^{13}C-NMR spectra have shown that humic acids formed from terrestrial vascular plants in peat are highly aromatic. Such quantitative investigations were not possible until the advent of NMR methods. However, one observation should be mentioned;

there appears to be some variation in aromaticities of humic acids from peat, and this variation is related to the presence of paraffinic structures. These macromolecular structures, because they are presumed to be derived from a source other than vascular plants [4], probably exist as distinct and physically separate structural entities.

Humin. Because the content of mineral matter in peat is comparatively lower than that of marine sediments, humin is isolated as the residual fraction without treatment with HF and HC1. Polysaccharides, particularly cellulose, are expected to contribute significantly to humins of peats because of the presence of undegraded plant remains. The CP/MAS ^{13}C-NMR spectra of humins (Figure 8) confirm these expectations, as carbohydrate peaks at 72 and 105 ppm dominate most spectra. These observations are additionally confirmed by IR spectra of humins.

Sizable peaks are observed in the CP/MAS spectra for paraffinic (Region I, 0 to 50 ppm) and aromatic carbons (Region III, 110 to 160 ppm). The humin from the Okefenokee peat sample was treated with sodium para-periodate to remove some of the cellulose (method of Ritchie and Purves [88]), and the spectrum of the residue is shown in Figure 8. Methoxyl (55 ppm) and aromatic carbons (110 to 160 ppm) give rise to distinct and resolved peaks as evidence of the presence of lignin residues in humin. The relative proportions of such structures vary considerably from one humin to the next in inverse proportion to the amounts of paraffinic carbons, which are represented by a conspicuous peak centered at 30 ppm. The shape and chemical shift of this paraffinic carbon peak is similar to that observed for humins and humic acids from aquatic sediments. Consequently, this paraffinic structure should have a similar complex, highly branched, and cross-linked macromolecular structure. Although infrared spectra provide little evidence for the presence of such structures, the IR spectra of humins from peat indicate that they may be present. A small peak at 1540 cm^{-1} is thought to be that of the amide II band; its presence in infrared spectra of aquatic humins was earlier mentioned to be a characteristic feature of such humins. Also, the peaks assigned to aliphatic C-H stretching frequencies (2920 cm^{-1} and 2840 cm^{-1}) appear more intense than similar peaks commonly observed for soil humic acids. Elemental data give no clue as to the presence of such structures because of the presence of carbohydrates that also have aliphatic structures. This clearly demonstrates the value of CP/MAS ^{13}C-NMR in identifying paraffinic structural components of humins.

The NMR spectra also display signals in the carboxyl or amide region (Region IV) and carbonyl region (Region V), though intensities are clearly reduced as compared to the corresponding signals of NMR spectra of humic acids. If lignin is assumed to contribute to humin of peats, carboxyl carbons would not be expected (see Sarkanen and Ludwig [78] and

Figure 6) unless the lignin were oxidized. Aquatic humins, however, have a measurable concentration of carboxyl groups. It is entirely possible that the carboxyl groups of peat humins are associated with the paraffinic structures. Because humins of peat were not treated with acid as were humins of aquatic sediments, some proteinaceous materials may be present. However, proteins are thought to be degraded rapidly in peat [87]–certainly more rapidly than carbohydrates (including cellulose)–and are not expected.

The presence of large amounts of paraffinic structures in peat was not observed before the advent of NMR, mainly because of the lack of other adequate methods for detection of these substances. Consequently, CP/MAS ^{13}C-NMR techniques have made a significant contribution to the discovery of these materials. Considering the fact that they constitute a sizable fraction of the entire humin fraction of algal sapropels where lignin-derived substances are not important contributors, we conclude that their origin is likely to be from algae or possibly the microorganisms responsible for the degradation of algae and cellulose. In peat, such as that of The Everglades, contributions from vascular plants rich in lignin are present in addition to algal or microbial flora, and the humin fraction is more aromatic than it is in marine humins derived entirely from algal or microbial residues. No significant paraffinic structures are observed in lignin [4]. Consequently the relative proportions of lignin-derived and paraffinic components in humin of peat should provide a quantitative estimate of the contributions from vascular plants and algal or microbial residues, respectively.

In summary, humins of peats appear to be composed of at least three structural entities: (1) residual carbohydrates from undegraded plant matter, (2) aromatic structures most likely derived from lignin residues and (3) paraffinic structures probably derived from microbial or algal residues. The chemical structure of paraffinic residues is thought to be highly branched and cross-linked.

The presence of cellulose and other carbohydrates in peat is well known [87]. Lucas [89] examined several peats and determined that some of the polysaccharides were generally typical of the macrovegetation on site, but microbial polysaccharides and polyuronides contributed substantially to the peat. The presence of polyuronides as major components of fulvic acids is certainly consistent with these findings.

Vascular plants generally provide the cellulose or holocellulose that forms such an integral part of the peat in the upper layers [90], and also the lignin, one of the primary source materials for humic substances. Humification, as applied to soils, generally involves the conversion of lignin to humic acids and humin. Soil scientists often class soil organic matter according to its degree of humification: they use the amounts of humic and fulvic acids produced as guides. Thus, the greater the amounts of humic and fulvic acids, the greater is the degree of humification. This scheme is appropriate in soils because most of the organic matter is eventually converted oxidatively to these humic substances, and very little of the original lignin is preserved. In peats, however, the degree of oxidation is limited, and anaerobic conditions prevail only a few centimeters below the peat surface. Consequently, a large portion of original plant lignin may be preserved, and only some may be oxidatively altered. When samples of peat are fractionated into fulvic acids, humic acids, and humin, both altered and unaltered lignin residues may appear in the various fractions. It is difficult to determine by CP/MAS ^{13}C-NMR whether humic acids or humins contain altered or unaltered lignin. The presence of intense signals for oxygen-substituted aromatic carbons and methoxyl carbons for both humic acids and humin in peat is a strong indication that original, unaltered lignin residues are present in large amounts. Similar peaks were not observed in spectra of humic and fulvic acids of soils. It appears then that the anaerobic environment of peats has influenced the aromatic structures of humic substances to resemble their parent material, namely lignin.

CONCLUSIONS

Nuclear magnetic resonance has provided a new dimension in the study of the origin and chemical structure of humic substances. Detailed, macromolecular, direct and nondestructive analyses of the chemical structures of these substances have never previously been possible, to the extent presently available by NMR, primarily because other techniques for structural analysis have been insensitive, destructive or indirect. Examination of humic substances from a variety of depositional environments has expanded the knowledge of their chemical structural composition. Such knowledge is the key to our ability to determine the mechanisms by which they are formed from plant residues.

In sedimentary deposits such as soils, peats, sapropels and marine sediments, humic substances appear to be composed of several mechanically mixed structural components that are derived from at least two major sources of plant remains–aquatic plants and vascular plants. One component, presumed to be derived from aquatic plants, has a macromolecular chemical structure that is predominantly aliphatic. Both NMR and chemical anslyses indicate that its structure is highly branched, paraffinic and highly cross-linked. Carboxyl, carbohydrate, ether and nitrogenous functional groups are clearly associated with these molecules. Such structures appear to be ubiquitous in humic acids and humin.

Another structural component of humic substances, observed mostly in peats and soils, is the one derived from vascular plant remains. Lignin, the insoluble macromolecular and predominantly aromatic substance that is ubiquitous in vascular plants, is presumed to be the source of this aromatic component of humic substances. Such structures are chemically distinct from those produced by aquatic plants. Though the paraffinic structures that are so common in aquatic sedimentary humic substances are also observed in the terrestrial sediments, the major components of humic substances in terrestrial sediments are the aromatic structures. The NMR spectra indicate that, in soils, these structures are not as highly substituted by oxygen-bearing functional groups (phenols) as has previously been thought, and their carboxyl contents are generally greater than expected on the basis of wet-chemical determinations. The NMR spectra indicate that, in peats, aromatic structures are more highly substituted by oxygen-bearing functional groups, and these structures resemble lignin, their probable progenitor.

ACKNOWLEDGMENTS

We graciously thank those who kindly provided samples; Bernd R. T. Simoneit, J. W. Farrington, Pete Zubovic, Morris Schnitzer, Ron Malcolm, Greg Diachenko, Z. S. Altschuler, and C. Blaine Cecil. Special thanks go to the late Bob Rowan, Mark Mattingly, Dave VanderHart and Bill Earl, who made available their time and NMR spectrometers so that we could obtain some of the data required for this study. We also thank Mrs. Marta Krasnow for CHN analyses. Larry W. Dennis and Gary E. Maciel acknowledge financial support from the U.S. Geological Survey Grant No. 14-08-001-G638.

REFERENCES

1. Forsman, J. P. "Geochemistry of Kerogen," in *Organic Geochemistry*, I. A. Breger, Ed. (Oxford: Pergamon Press, 1963), p. 148.
2. Stuermer, D. H., I. R. Kaplan and K. E. Peters. "Source Indicators of Humic Substances and Proto-Kerogen. Stable Isotope Ratios, Elemental Compositions and Electron Spin Resonance Spectra," *Geochim. Cosmochim. Acta* 42:989-997 (1978).
3. Hatcher, P. G., R. Rowan and M. A. Mattingly. "^{1}H and ^{13}C NMR of Marine Humic Acids," *Org. Geochem.* 2:77-85 (1980).
4. Hatcher, P. G., G. E. Maciel and L. W. Dennis. "Aliphatic Structure of Humic Acids: A Clue to Their Origin," *Org. Geochem.* 3:43-48 (1981).
5. Schnitzer, M., and S. U. Khan. *Humic Substances in the Environment* (New York: Marcel Dekker, Inc., 1972), p. 327.

6. Hatcher, P. G., M. Schnitzer, L. W. Dennis and G. E. Maciel. "Aromaticity of Humic Substances in Soils," *J. Soil Sci. Soc. Am.* 45:1089-1094 (1981).
7. Breger, I. A. "Origin and Classification of Naturally Occurring Carbonaceous Substances," in *Organic Geochemistry*, I. A. Breger, Ed. (Elmsford, NY: Pergamon Press, Inc., 1963), p. 50.
8. Cane, R. F. "The Origin and Formation of Oil Shale," in *Oil Shale*, T. F. Yen and G. V. Chilingarian, Eds. (Amsterdam: Elsevier, 1976), p. 26.
9. Stuermer, D. H., and J. R. Payne. "Investigation of Seawater and Terrestrial Humic Substances with Carbon-13 and Proton Nuclear Magnetic Resonance," *Geochim. Cosmochim. Acta* 40:1109-1114 (1976).
10. Schnitzer, M., and S. U. Khan, Eds. *Soil Organic Matter* (New York: Elsevier, 1978) p. 319.
11. Hatcher, P. G., D. L. VanderHart and W. L. Earl. "Use of Solid-State ^{13}C NMR in Structural Studies of Humic Acids and Humin from Holocene Sediments," *Org. Geochem.* 2:87-92 (1980).
12. Vila, F. J. G., H. Lentz and H. D. Lüdemann. "FT-^{13}C Nuclear Magnetic Resonance Spectra of Natural Humic Substances," *Biochem. Biophys. Res. Comm.* 72:1063-1969 (1976).
13. Wilson, M. A., and K. H. Goh. "Proton-Decoupled Pulse Fourier-Transform ^{13}C Magnetic Resonance of Soil Organic Matter," *J. Soil Sci.* 28:645-652 (1977).
14. Wilson, M. A. ^{13}C-n.m.r. of Humic Substances," *Plant and Soil* 46:287-289 (1977).
15. Lentz, H., H. D. Lüdemann and W. Ziechmann. "Proton Resonance Spectra of Humic Acids from the Solum of a Podzol," *Geoderma* 18:325-328 (1977).
16. Ruggiero, P., F. S. Interesse and O. Sciacovelli. "[^{1}H] and [^{13}C] NMR Studies on the Importance of Aromatic Structures in Fulvic and Humic Acids," *Geochim. Cosmochim. Acta.* 43:1771-1775 (1979).
17. Wilson, M. A., A. J. Jones and B. Williamson. "Nuclear Magnetic Resonance Spectroscopy of Humic Materials," *Nature* 276:487-489 (1978).
18. Newman, R. H., K. R. Tate, P. F. Barron and M. A. Wilson. "Towards a Direct Non-Destructive Method of Characterizing Soil Humic Substances Using ^{13}C N.M.R.," *J. Soil Sci.* 31:623-631 (1980).
19. Hatcher, P. G., I. A. Breger and Mattingly, M. A. "Structural Characteristics of Fulvic Acids from Continental Shelf Sediments," *Nature* 285:560-562 (1980).
20. Wilson, M. A. "Applications of Nuclear Magnetic Resonance Spectroscopy to the Study of the Structure of Soil Organic Matter," *J. Soil Sci.* 32:167-186 (1981).
21. Dereppe, J. M., C. Moreaux and Y. Debyser. "Investigation of Marine and Terrestrial Humic Substances by ^{1}H and ^{13}C Nuclear Magnetic Resonance and Infrared Spectroscopy," *Org. Geochem.* 2:117-124 (1980).
22. Pines, A., M. G. Gibby and J. S. Waugh. "Proton Enhanced Nuclear Induction Spectroscopy. A Method for High Resolution NMR of Dilute Spins in Solids," *J. Chem. Phys.* 56:1776-1777 (1972).
23. Pines, A., M. G. Gibby and J. S. Waugh. "Proton Enhanced NMR of Dilute Spins in Solids," *J. Chem. Phys.* 59:569-590 (1972).

24. VanderHart, D. L., and H. L. Retcofsky. "Analysis of the Aromatic Content of Whole Coals by High Power Proton Decoupled C-13 NMR," *Proc. Coal Chem. Symp.* (Stanford, CA: Stanford Research Inst., 1976).
25. Schaefer, J., E. O. Stejskal and R. Buchdahl. "High Resolution Carbon-13 Nuclear Magnetic Resonance Study of Some Solid Glassy Polymers," *Macromolecules* 8:291-296 (1975).
26. Lippmaa, E., M. Alla and T. Tuhern. "High-Resolution ^{13}C NMR in Solids with Polarization Transfer, Decoupling, and Sample Spinning at the Magic Angle," Proc. XIX Congress Ampere, Heidelberg (1976).
27. Lippmaa, E. T., M. Alla, M. A. Pehk and G. Engelhardt. "Solid-State High Resolution NMR Spectroscopy of Spin 1/2 Nuclei (^{13}C, ^{29}Si, ^{119}Sn) in Organic Compounds," *J. Am. Chem. Soc.* 100:1929-1931 (1978).
28. Garroway, A. N., W. B. Moniz and H. A. Resing. "C-13 NMR in Cured Epoxies: Benefits of Magic Angle Rotation," Division of Organic Coatings and Plastics, Preprints, 172nd American Chemical Society Meeting 36:133 (1976).
29. Bartuska, V. J., G. E. Maciel, J. Schaefer and E. O. Stejskal. "Prospects for Carbon-13 Nuclear Magnetic Resonance Analysis of Solid Fossil Fuel Materials," *Fuel* 56:354-358 (1977).
30. Maciel, G. E., V. J. Bartuska and F. P. Miknis. "Characterization of Organic Material in Coal by Proton-Decoupled ^{13}C Nuclear Magnetic Resonance with Magic Angle Spinning," *Fuel* 58:391-394 (1979).
31. Miknis, F. P., M. Sullivan, V. J. Bartuska and G. E. Maciel. "Cross Polarization Magic-Angle Spinning ^{13}C NMR Spectra of Coals of Varying Rank," *Org. Geochem.* 3:19-28 (1981).
32. Zilm, K. W., R. J. Pugmire, D. M. Grant, R. E. Wood and W. H. Wiser. "A Comparison of the Carbon-13 NMR Spectra of Solid Coal and their Liquids Obtained by Catalytic Hydrogenation," *Fuel* 58:11-16 (1979).
33. Sullivan, M. J. and G. E. Maciel. "Spin Dynamics in the ^{13}C NMR Analysis of Coal by Cross-Polarization and Magic-Angle Spinning," *Anal. Chem.*, in press.
34. Maciel, G. E., M. J. Sullivan, L. Petrakis and W. D. Grandy. "^{13}C Nuclear Magnetic Resonance Characterization of Coal Macerals by Magic-Angle Spinning," *Fuel*, in press.
35. Resing, H. A., A. N. Garroway and R. N. Hazlett. "Determination of Aromatic Hydrocarbon Fraction in Oil Shale by ^{13}C NMR with Magic Angle Spinning," *Fuel* 57:450-454 (1978).
36. Vitorovic, D. D. Vucelic, M. J. Gasic, N. Juranic and S. Macura. "Analysis of the Organic Matter of Oil Shales by Nuclear Magnetic Resonance," *Org. Geochem.* 1:89-91 (1978).
37. Miknis, F. P., G. E. Maciel and V. J. Bartuska. "Cross Polarization Magic-Angle Spinning ^{13}C NMR Spectra of Oil Shales," *Org. Geochem.* 1:169-176 (1979).
38. Hatcher, P. G., and P. A. McGillivary. "Sewage Contamination in the New York Bight: Coprostanol as an Indicator," *Environ. Sci. Tech.* 13:1225-1229 (1979).
39. Simoneit, B. R. T., M. A. Mazurek, S. Brenner, P. T. Crisp and I. R. Kaplan. "Organic Geochemistry of Recent Sediments from Guaymas Basin, Gulf of California," *Deep Sea Res.* 26A:879-891 (1979).

40. Degens, E. T., and D. A. Ross. "Chronology of the Black Sea Over the Last 25,000 years," *Chem. Geol.* 10:1-16 (1972).
41. Simoneit, B. R. T. "The Black Sea, a Sink for Terrigenous Lipids," *Deep Sea Res.* 24:813-830 (1977).
42. Simoneit, B. R. T. "Sources of Organic Matter in Oceanic Sediments," PhD Dissertation, University of Bristol, England (1975).
43. Hatcher, P. G. "A Study of the Organic Geochemistry of Mangrove Lake, Bermuda," MS Thesis, University of Miami, Coral Gables, FL (1974).
44. Hatcher, P. G. "The Organic Geochemistry of Mangrove Lake, Bermuda," NOAA Prof. Paper No. 10 (1978), pp. 1-92.
45. Swain, F. M., Ed. *Non-Marine Organic Geochemistry* (Cambridge: University of Cambridge Press, 1970), p. 445.
46. Spackman, W., A. D. Cohen, P. H. Given and D. J. Casagrande. "The Comparative Study of the Okefenokee Swamp and the Everglades–Mangrove Swamp–Marsh Complex of Southern Florida," Field Guidebook for Geol. Soc. Am. Meet., Miami, FL (1974).
47. Cohen, A. D. "Petrology of Some Holocene Peat Sediments from the Okefenokee Swamp–Marsh Complex of Georgia," *Geol. Soc. Am. Bull.* 84:3867-3878 (1974).
48. Breger, I. A., M. Krasnow and J. C. Chandler. "Peat from the Everglades of Florida: A Study of the Origin of Coal and of Natural Gas," Abstracts of 10th Intl. Congress on Sedimentology, Jerusalem, July, 1978, p. 85.
49. Matsuda, K., and M. Schnitzer. "The Permanganate Oxidation of Humic Acids Extracted from Acid Soils," *Soil Sci.* 114:185-193 (1972).
50. Khan, S. V., and M. Schnitzer. "Further Investigations on the Chemistry of Fulvic Acid, a Soil Humic Fraction," *Can. J. Chem.* 49:2302-2309 (1971).
51. Malcolm, R. L. "Method and Importance of Obtaining Humic and Fulvic Acids of High Purity," *J. Res. U.S. Geol. Survey* 4:37-40 (1976).
52. Ortiz de Serra, M. I., and M. Schnitzer. "The Chemistry of Humic and Fulvic Acids Extracted from Argentine Soils - I. Analytical Characteristics," *Soil Biol. Biochem.* 5:281-286 (1973).
53. Chen, Y., and M. Schnitzer. "Chemical and Physical Characteristics of Humic and Fulvic Acids Extracted from Soils of the Mediterranean Region," *Geoderma* 20:87-107 (1978).
54. Schnitzer, M., and S. I. M. Skinner. "Gel Filtration of Fulvic Acid, a Soil Humic Compound," in *Isotopes and Radiation in Soil Organic Matter Studies* (Vienna: International Atomic Energy Agency, 1968), p. 41-55.
55. Kwak, J. C. T., R. W. P. Nelson and S. D. Gamble. "Ultrafiltration of Fulvic and Humic Acids, a Comparison of Stirred Cell and Hollow Fiber Techniques," *Geochim. Cosmochim. Acta* 41:993-996 (1977).
56. Hartman, S. R., and E. L. Hahn. "Nuclear Double Resonance in the Rotating Frame," *Phys. Rev.* 128:2042-2053 (1962).
57. Stejskal, E. O., and J. Schaefer. "Removal of Artifacts from Cross-Polarization NMR Experiments," *J. Magn. Reson.* 18:560-563 (1975).
58. Earl, W., and D. L. VanderHart. "Observations in Solid Polyethylenes by ^{13}C NMR with Magic Angle Sample Spinning," *Macromolecules* 12:762-767 (1979).
59. Maciel, G. E., M. J. Sullivan, N. Szeverenyi and F. P. Miknis. "^{13}C NMR

on Solid Samples and its Application to Coal Science," Chemistry and Physics of Coal Utilization–1980, American Institute of Physics Conference Proceedings No. 70, Eds. B. R. Cooper and L. Petrakis, pp. 68-81, New York, 1980.
60. Wind, R. A., J. Trommel and J. Smidt. "Proton Dynamic Polarization and Proton-Carbon Cross Polarization in Coal," *Fuel* (in press).
61. Senesi, N., and M. Schnitzer. "Free Radicals in Humic Substances," in *Environmental Biogeochemistry and Geomicro-Biology, Vol. 2: The Terrestrial Environment*, W. E. Krumbein, Ed. (Ann Arbor, MI: Ann Arbor Science Publishers, Inc., 1978), p. 467.
62. Hatcher, P. G. "The Origin, Composition, Chemical Structure, and Diagenesis of Humic Substances, Coals, and Kerogens as Studied by Nuclear Magnetic Resonance," PhD Dissertation, University of Maryland, (1980).
63. Ishiwatari, R. "Chemical Nature of Sedimentary Humic Acids," Proc. International Meeting on Humic Substances, Nieuwersluis, Pudoc, Wageningen (1972), p. 87.
64. Steelink, C. "What is Humic Acid?," *J. Chem. Ed.* 40:379-384 (1963).
65. Rashid, M. A., and L. H. King. "Major Oxygen-Containing Functional Groups Present in Humic and Fulvic Acid Fractions Isolated from Contrasting Marine Environments," *Geochim. Cosmochim. Acta* 34:193-201 (1970).
66. Ernst, W. "Nachweis, Identifizierung und quantitative Bestimung von Uronsauren im Polysaccharidfraktionen mariner Sedimente der Deutschen Bucht," *Veroff. Inst. Meeresforsch., Bremerh.* 10:81-92 (1966).
67. Wise, L. E., and E. C. Jahn. *Wood Chemistry* (New York: Rheinhold Publishers, 1952).
68. Nissenbaum, A., and I. R. Kaplan. "Chemical and Isotopic Evidence for the *in situ* Origin of Marine Humic Substances," *Limnol. Oceanogr.* 17:570-582 (1972).
69. Huc, A. Y., and B. Durand. "Occurrence and Significance of Humic Acids in Ancient Sediments," *Fuel* 56:73-80 (1977).
70. Debyser, Y., and F. Gadel. "Etude Geochimique des Composes Humiques et des Kerogenes," in *Geochimie Organique des Sediments Marins Profonds, Orgon I, Mer de Norvege*, (Paris: Editions du Centre National de la Recherche Scientifique, 1977), p. 247.
71. Pelet, R., and Y. Debyser. "Organic Geochemistry of Black Sea Cores," *Geochim. Cosmochim. Acta*, 41:1575-1586 (1977).
72. Stevenson, F. J., and K. M. Goh. "Infrared Spectra of Humic Acids and Related Substances," *Geochim. Cosmochim. Acta.* 35:471-483 (1971).
73. Voelter, W., and E. Breitmaier. "Der Einfluss des Anomer Verknuptten Adeninrestes auf die ^{13}C-chemischen Verschiebungen von Pyranosen," *Org. Mag. Res.* 5:311-319 (1973).
74. Degens, E. T., J. H. Reuter and K. N. F. Shaw. "Biochemical Compounds in Offshore California Sediments and Sea Waters," *Geochim. Cosmochim. Acta.* 28:45-63 (1964).
75. Bordovskiy, O. K. "Accumulation of Organic Matter in Bottom Sediments," *Mar. Geol.* 3:33-82 (1964).
76. Kononova, M. *Soil Organic Matter,* 2nd ed. (Oxford: Pergamon Press, (1966).
77. Sackett, W. M. "The Depositional History of the Isotopic Organic Carbon Composition of Marine Sediments," *Mar. Geol.* 2:173-185 (1964).

78. Sarkanen, K. V., and C. H. Ludwig, Eds. *Lignins, Occurrence, Formation, Structure, and Reactions* (New York: John Wiley & Sons, Inc., 1971).
79. Bartuska, V. J., G. E. Maciel, H. I. Bolker and B. Fleming. "Structural Studies of Lignin Isolation Procedures by ^{13}C NMR," *Holtzforschung* 34:214-217 (1980).
80. Maciel, G. E., V. J. Bartuska, J. J. H. Ackerman, D. J. O'Donnell, and B. L. Hawkins. "^{13}C NMR Study of Four Lignins in the Solid and Solution States," *Makromol Chem.* Vol. 182:2297-2304, 1981.
81. Ortiz de Serra, M. I., and M. Schnitzer. "The Chemistry of Humic and Fulvic Acids Extracted from Argentine Soils–II. Permanganate Oxidation of Methylated Humic Acids," *Soil Biol. Biochem.* 5:287-296 (1973).
82. Schnitzer, M. "Recent Findings on the Characterization of Humic Substances Extracted from Soils from Widely Differing Climatic Zones," in *Soil Organic Matter Studies, Vol. II* (Vienna: International Atomic Energy Agency, 1977), p. 117.
83. Swincer, G. D., J. M. Oades and D. J. Greenland. "Studies on Soil Polysaccharides. I. The Isolation of Polysaccharides from Soil," *Aust. J. Soil Res.* 5:211-224 (1968).
84. Swincer, G. D., J. M. Oades and D. J. Greenland. "Studies on Soil Polysaccharides. II. The Composition and Properties of Polysacharrides in Soils under Pasture and under a Fallow–Wheat Rotation," *Aust. J. Soil Res.* 6:225-235 (1968).
85. Lowe, L. E. "Carbohydrates in Soil," in *Soil Organic Matter*, M. Schnitzer, and S. U. Khan (New York: Elsevier, 1978), p. 65.
86. Mendez, J., and G. Ogner. "Permanganate Oxidation Products of Algal Humic-Like Substances," *Chem. Geol.* 27:265-270 (1979).
87. Manskaya, S. M., and T. V. Drozdova. *Geochemistry of Organic Substances* (Oxford: Pergamon Press, 1968).
88. Ritchie, P. F., and C. B. Purves. "Periodate Lignins: Their Preparation and Properties," *Pulp Paper Mag. Canada* 48:74-82 (1947).
89. Lucas, A. J. "Geochemistry of Carbohydrates in Some Organic Sediments of the Florida Everglades," PhD dissertation, The Pennsylvania State University (1970).
90. Waksman, S. A., Ed. *Humus* (Baltimore, MD: Williams and Wilkins Press, 1936).

CHAPTER 4

MAGNETIC RESONANCE STUDIES OF HUMATES AND RELATED MODEL COMPOUNDS

Cornelius Steelink, Michael A. Mikita* and Kevin A. Thorn

Department of Chemistry
University of Arizona
Tucson, Arizona

ABSTRACT

A number of humates and fulvates have been examined by ^{1}H-NMR, ^{13}C-NMR and ESR spectroscopy. Enhancement of the hydroxyl ^{13}C-NMR absorbances was achieved by methylating the humates with ^{13}C enriched methylating agents. Sharp bands in the 59 ppm region are characteristic of the 6-OCH_3 groups in carbohydrates; sharp peaks in the 51 ppm region represent aliphatic ester groups. Phenolic ether and aromatic ester groups can also be detected.

The ^{13}C-NMR spectra of alkaline solutions of humates, fulvates, tannins, lignins and flavonoids all show significant differences from the corresponding neutral spectra. The ESR spectra of these alkaline solutions all show a high free radical content. These paramagnetic radical species cause distortions of the aromatic absorptions in the ^{13}C NMR spectra. Correlations of structure and distortion are discussed.

INTRODUCTION

We have examined humic materials from a variety of well defined sources by proton nuclear magnetic resonance (^{1}H-NMR) and carbon-13 nuclear

*Current address: Exxon Research and Engineering Co., Coal Research Laboratory, Baytown, Texas.

magnetic resonance (^{13}C-NMR) spectroscopy. The purpose of this study was to identify structural features common to fulvic and humic acids. In particular, we wished to elucidate the structures of the hydroxyl functionality in humic materials; i.e., the types and abundances of carboxyl, phenolic, aliphatic and carbohydrate OH groups. So many of the unique physical and chemical properties of humates and fulvates are governed by hydroxyl functionality [1].

A second objective of this study was to determine the effect of paramagnetic species on the NMR spectra of polyphenolic compounds. Recently, we reported [2] dramatic changes in the ^{13}C-NMR spectra of naturally occurring aquatic compounds when they were placed in alkaline solutions. A similar phenomenon was reported for ^{1}H-NMR spectra of fulvates by Ruggiero [3]. Strong electron spin resonance (ESR) signals were exhibited by these alkaline solutions due to semiquinone radical anions. We decided to examine this behavior in a variety of compounds, including fulvates, humates, lignins, tannins and flavonoids.

Recently, we [4,5] have developed a method for enhancing OH absorptions in the ^{13}C-NMR spectra of fulvates and humates. Phenolic, carboxylic and aliphatic OH group absorbances are enhanced by this method. Further structural information can be gleaned by combining the spectral data from ^{1}H-NMR and ^{13}C-NMR data of the humic substrates.

EXPERIMENTAL

Sources of Humates and Model Compounds

The Suwannee humic and fulvic acids were isolated from the Suwannee River in Florida and provided to us by E. M. Thurman of the U.S. Geological Survey (USGS), Denver, Colorado. The Biscayne humic and fulvic acids were isolated from the Biscayne Florida Aquifer and provided to us also by Dr. Thurman. The Mollisol soil humic acid was extracted from a Mollisol, dark black chernozem, in central Alberta, Canada, and provided to us by M. Schnitzer, Soil Research Institute, Agriculture Canada, Ottawa, Ontario, Canada. The Black Lake humic and fulvic acids were isolated from the Black Lake near Elizabethtown, North Carolina, and provided to us by R. F. Christman, Department of Environmental Sciences and Engineering, University of North Carolina at Chapel Hill. The Aldrich humic acid was purchased as the sodium salt from Aldrich Chemical Company, Inc., Milwaukee, Wisconsin. The Fluka Humic Acid was purchased through Tridom Chemical Inc., Hauppauge, New York. The Kraft Lignin was purchased from Westvaco, Charleston Research Center, North Charleston, South Carolina. Tannic acid was obtained from MCB, Norwood, Ohio. Rutin was obtained from Nutritional Biochemicals Corporation, Cleveland, Ohio. Quercetin was purchased from Weyerhauser Timber Co., Longview, Washington. Quercitrin (Dihydrate) was obtained from Trans World Chemicals, Inc., Washington, D.C.

Reagents

N-methyl-^{13}C-N-nitroso-*p*-toluene sulfonamide and ^{13}C-iodomethane were purchased from Merck, Sharp, and Dohme, Canada, Ltd., Montreal, Quebec. Tetrabutylammonium hydroxide, 40 wt% in water, was obtained from Aldrich Chemical Company, Inc., Milwaukee, Wisconsin.

NMR Spectra

The ^{1}H-NMR spectra were recorded on a Varian Associates EM-360 spectrometer. Chemical shifts were assigned relative to internal tetramethylsilane (TMS).

The ^{13}C-NMR spectra were recorded either on a Varian FT80A Fourier transform (FT) NMR spectrometer operating at 20.00 MHz or a Bruker Instruments WM-250 FT NMR spectrometer operating at 62.90 MHz in the ^{1}H broad band decoupler mode. Methylated samples were recorded in $CDCl_3$ solution; underivatized samples were measured in DMSO-d_6. These deuterated solvents served as the internal ^{2}H lock. The spectra were obtained using 16K data points over a spectral width of 200 ppm. A pusle delay of from 10 to 20 sec was employed in generating the spectra of the methylated humic and fulvic acids. Chemical shifts were assigned relative to internal TMS.

ESR Spectra

Electron spin resonance measurements were performed on a Varian Associates E-3 spectrometer. The instrument was operated at 9.5 GHz with a frequency modulation of 100 KHz.

Methylation Procedure

The Black Lake and Suwannee humic and fulvic acids were permethylated as previously described by Mikita [4] and by Wershaw [5], using ^{13}C-enriched diazomethane followed by ^{13}C-enriched methyl iodide/NaH. The Aldrich humic acid was methylated using the procedure of Liotta et al. [6] using tetrabutylammonium hydroxide and unenriched iodomethane.

The Aldrich humic acid was prepared for methylation as follows. Five grams of the sodium salt was dissolved in distilled deionized water and then acidified to pH 1.0 with concentrated HCl. The precipitated humic acid was allowed to settle, filtered with the aid of a vacuum pump, and washed until free of chloride ion. The humic acid was then redissolved in 0.1 N NH_4OH, reprecipitated with HCl, filtered and then washed again. This procedure was repeated five times. The elementary analysis of the free acid form of the Aldrich humic acid is presented in Table I.

Where ^{13}C-enriched methylating reagents were used, anywhere from 50 to 100 mg of permethylated sample was obtained for ^{13}C-NMR spectral analysis. With the Aldrich humic acid, approximately 300 mg of methylated product was obtained from 500 mg of starting material. All methylated samples were redissolved in $CDCl_3$ for NMR analysis.

Methanol Extraction

One gram each of the Aldrich and Fluka humic acids were extracted with methanol for 24 h in a Soxhlet apparatus. In the methanol-soluble fractions 0.46 g of Aldrich and 0.27 g of Fluka were obtained. After removal of the solvent, the soluble and insoluble fractions were dried in an oven at 100 °C overnight, stored in a dessicator, and then dissolved in DMSO-d_6 for ^1H- and ^{13}C-NMR analysis.

Elementary Analyses

The elementary analyses of the humic and fulvic acids are presented in Table I.

RESULTS AND DISCUSSION

^{13}C-NMR Studies of Underivatized Humic Materials

The ^{13}C-NMR spectra of DMSO-d_6 solutions of humic and fulvic acids all show broad bands very similar to the solid-state, cross polarization magic

Table I. Elementary Analyses of Humic and Fulvic Acid Samples

Sample	%C	%H	%N	%S	%Ash
Suwannee fulvic acid	51.3	4.32	0.56		0.05
Suwannee humic acid	51.2	4.38	0.56	0.61	0.05
Biscayne fulvic acid	55.4	4.2	1.8	1.1	0.4
Biscayne humic acid	58.3	3.4	5.8	1.4	0.1
Black Lake fulvic acid	49.57	4.39	2.00		5.03
Black Lake humic acid	50.71	7.05	5.84		12.03
Contech fulvic acid	50.9	4.5	0.3		1.0
Aldrich humic acid	58.88	4.01	0.97		3.66
Fluka humic acid	50.2	4.7	0.0	7.8	7.8
Mollisol humic acid	56.4	5.5	4.1	1.1	1.0

angle spinning (CP/MAS) spectra (Figures 1, 2 and 3). These bands represent aliphatic (40 ppm), aliphatic OH (79 ppm), aromatic (129) and carboxyl (172 ppm). Methanol extracts of humates reveal additional structure (Figures 4 and 5), including methyl ester and methyl aromatic ethers in the 50 to 60 ppm region. It appears that some methylation of the acidic and phenolic groups has occurred in the extracting process. The ^{1}H-NMR spectrum indicates that methanol preferentially removes aromatic and aliphatic fractions of these substances at the expense of the carbohydrate fraction.

^{13}C-NMR Spectra of Permethylated Humates and Fulvates

When humic substances are permethylated with ^{13}C-enriched reagents, strong bands appear in the 50 to 60 ppm region of the spectrum. These bands represent $COOCH_3$, aromatic OCH_3 and aliphatic OCH_3 groups and therefore reveal the identity and approximate abundance of OH functionality in these materials. Figure 6 shows the spectra of a soil humic acid, an aquatic fulvic acid and some reference compounds. The spectra of other fulvates and humates are shown in Figures 7, 8 and 9. Approximate relative abundances can be calculated from these spectra (Table II). The carboxyl/phenol ratios

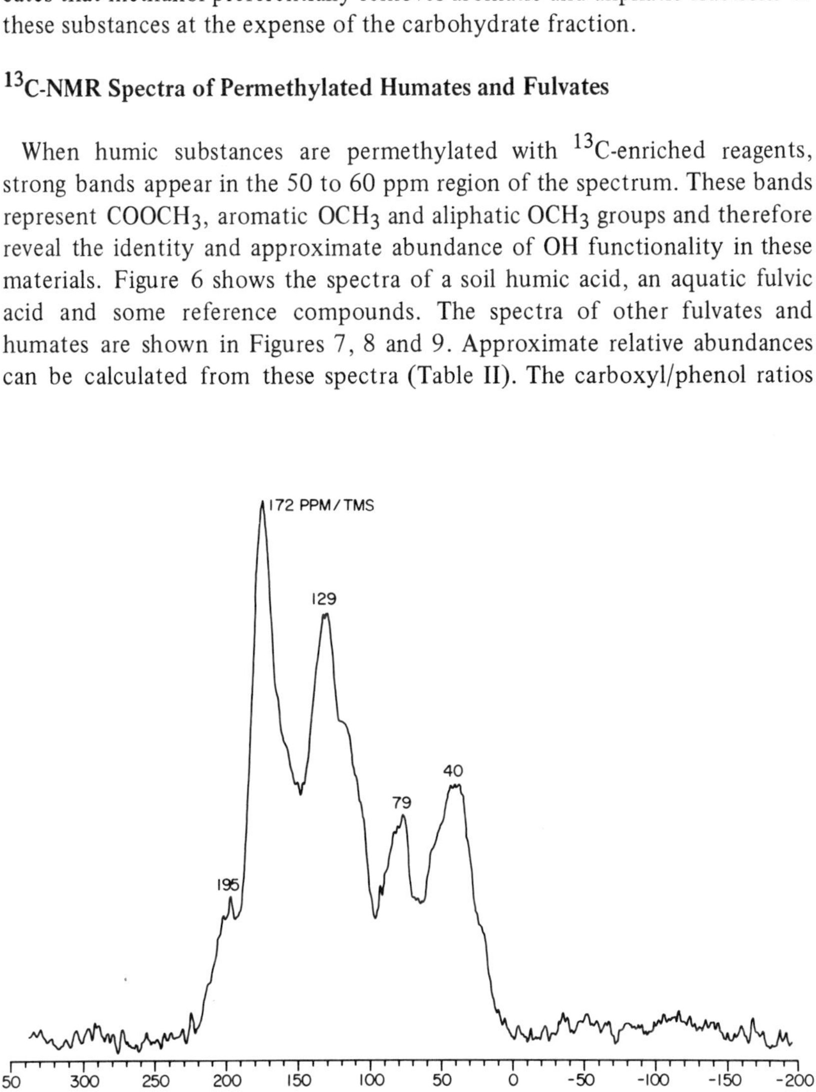

Figure 1. Solid-state ^{13}C-NMR spectrum of Contech (soil) fulvic acid.

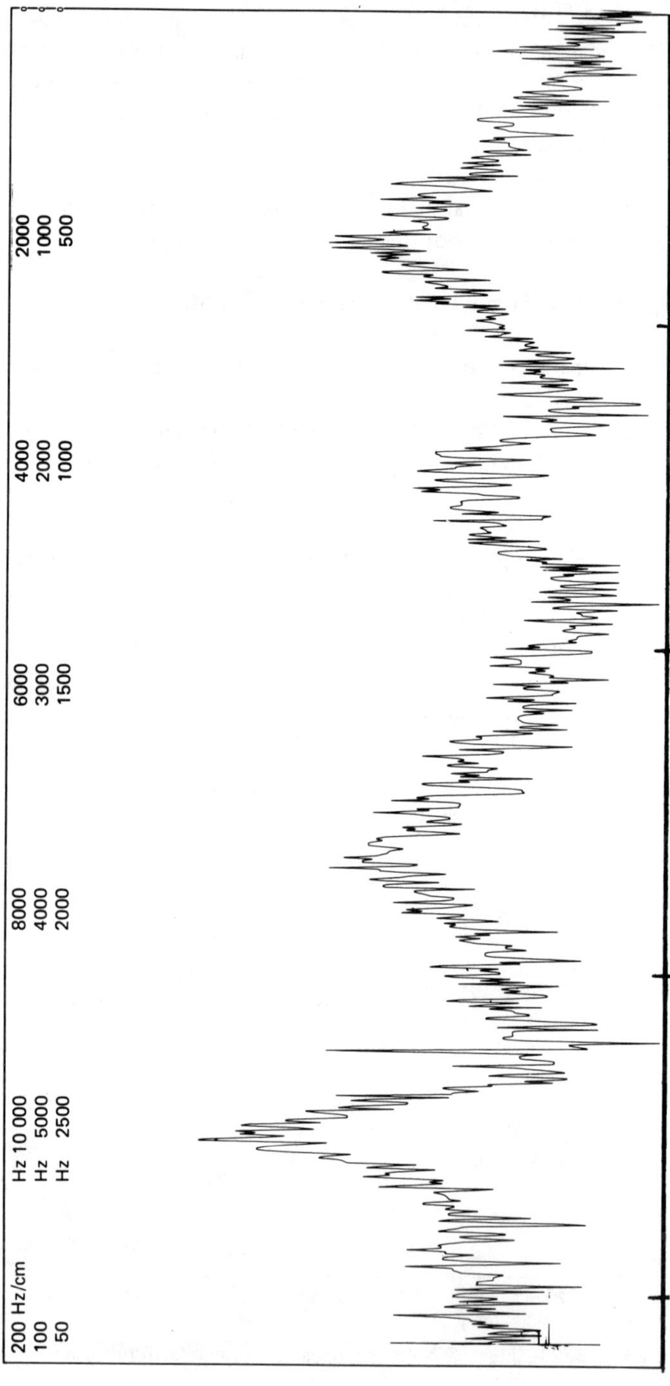

Figure 2. ^{13}C-NMR spectrum of Contech fulvic acid in D_2O. Scale in ppm relative to TMS.

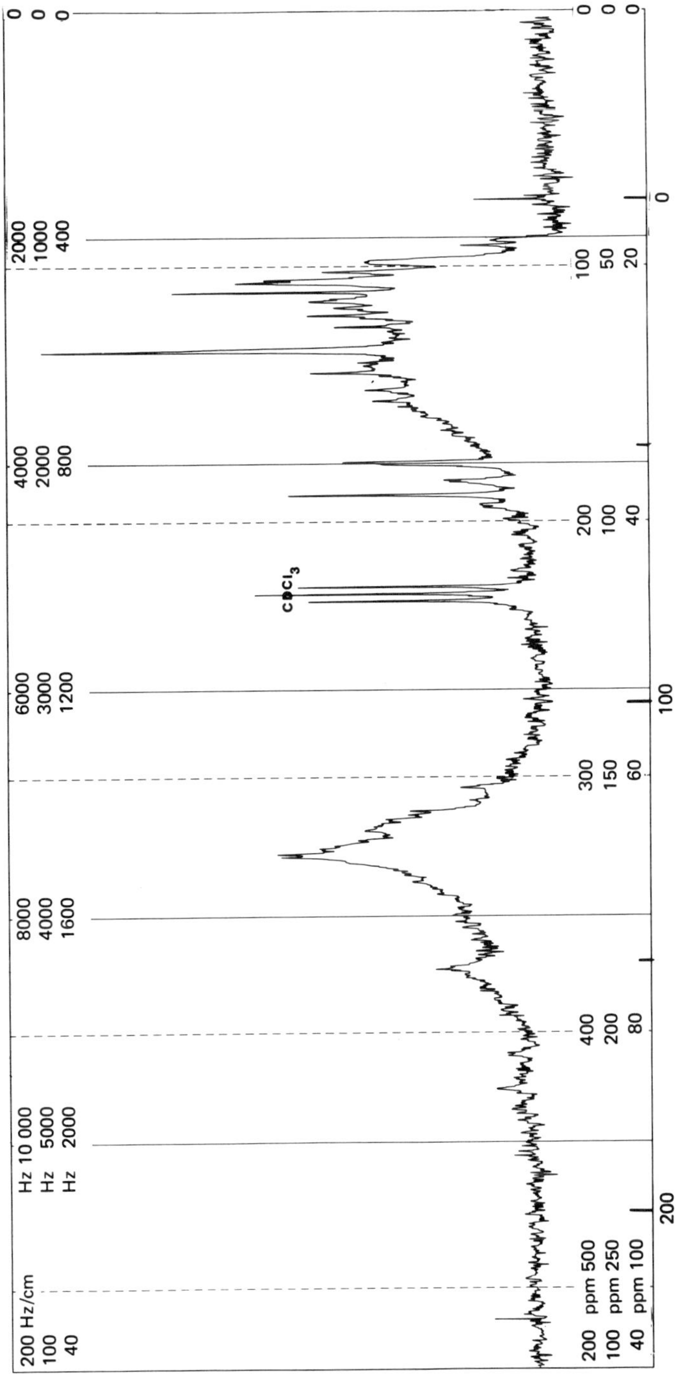

Figure 3. ^{13}C-NMR spectrum of hydrogenated Fluka humic acid in $CDCl_3$. Scale in ppm relative to TMS.

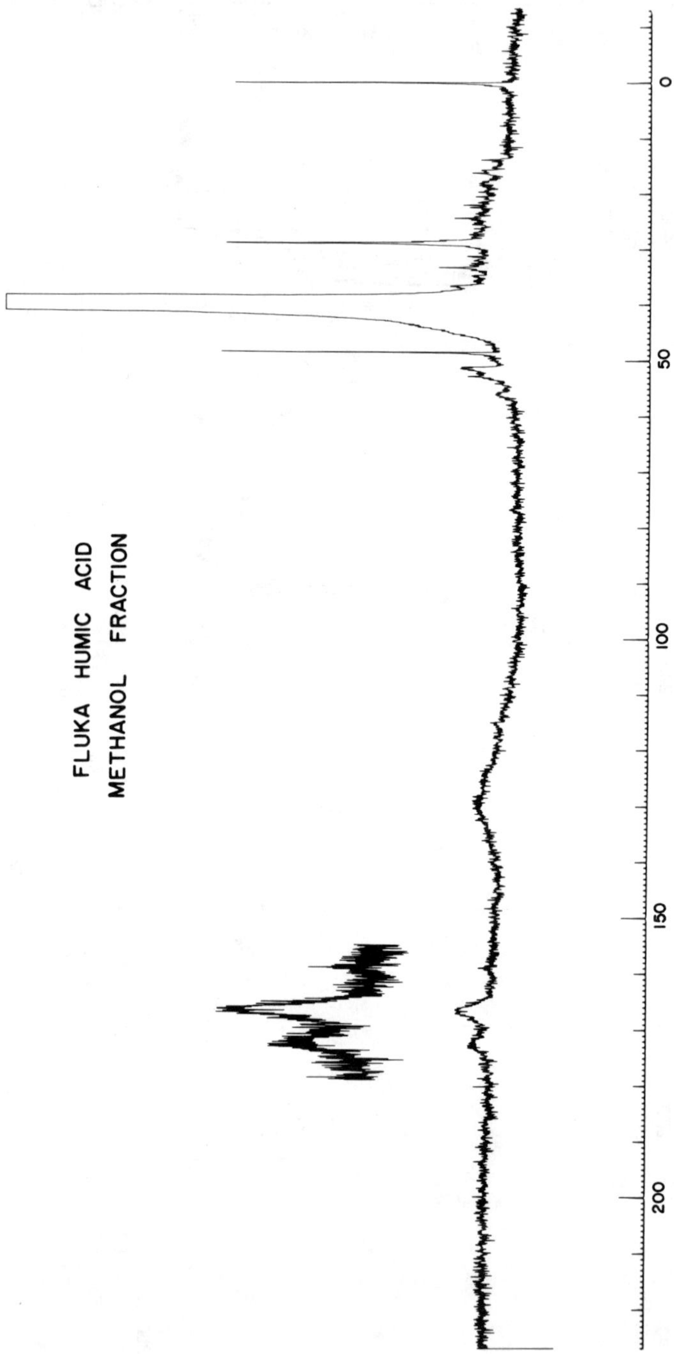

Figure 4. ^{13}C-NMR spectrum of methanol extract of Fluka humic acid in DMSO-d_6. Scale in ppm relative to TMS.

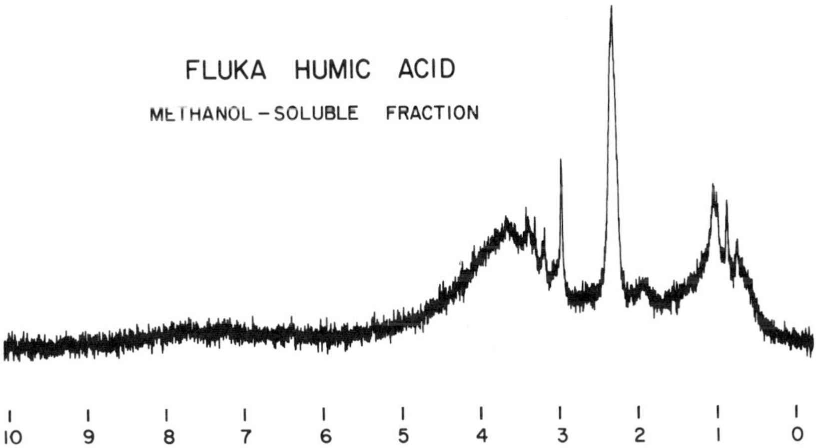

Figure 5. ^{1}H-NMR spectrum of methanol extract of Fluka humic acid in DMSO-d_6. Scale in ppm relative to TMS.

compare favorably with the titration data (Table III) reported for these samples.

Of particular interest are the two sharp bands that appear in many of these spectra (Table IV). One band at 50.8 ppm is characteristic of aliphatic $COOCH_3$; the other at 59 ppm is characteristic of 6-OCH_3 of carbohydrates. Broader bands in the 52 ppm region represent a cluster of aromatic $COOCH_3$ groups, while bands in the 55 to 56 ppm region represent aromatic OCH_3. Thus, it is possible to obtain a direct measure of carbohydrate content from these ^{13}C-enriched samples, as well as an estimate of aliphatic and aromatic

Table II. Relative Abundance of Hydroxyl Functionality in ^{13}C-NMR Spectra of Humic and Fulvic Acids

Sample	Carbohydrate	Phenolic	Carboxylic Acid
Suwannee fulvic acid	0.76	0.61	1.00
Suwannee humic acid	1.70	0.84	1.00
Black Lake fulvic acid	0.90		1.00
Black Lake humic acid	0.88	0.45	1.00
Biscayne fulvic acid	0.97	0.59	1.00
Biscayne humic acid	1.05	0.58	1.00
Mollisol humic acid	1.66	0.79	1.00
Contech fulvic acid	0.43	0.40	1.00

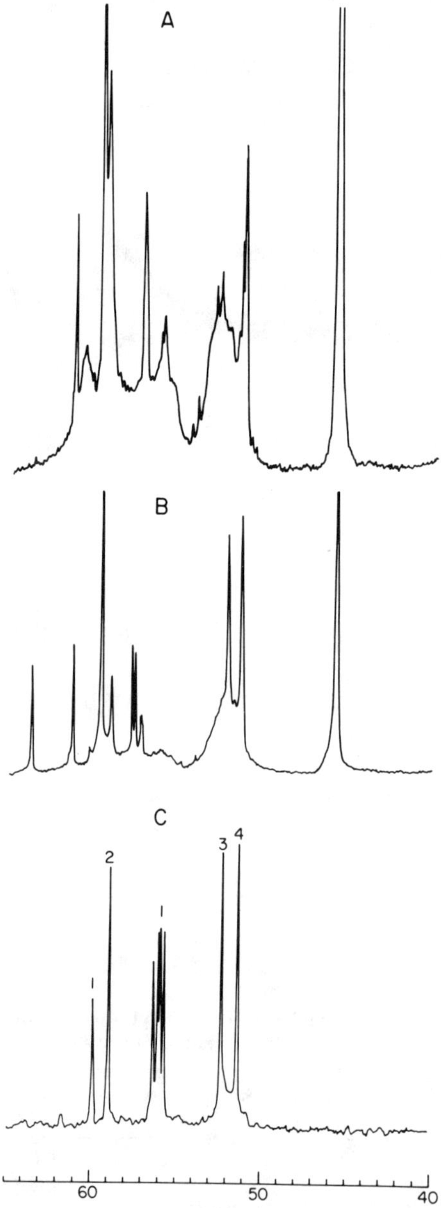

Figure 6. ^{13}C-NMR spectrum of ^{13}C-permethylated humic and fulvic acids and reference compounds in $CDCl_3$. Scale from 40 to 65 ppm relative to TMS. (A) Mollisol soil humic acid. (B) Biscayne aquifer fulvic acid. (C) Mixture of reference compounds: (1) pentamethylquercetin, (2) diglyme, (3) methylbenzoate, (4) methyl pentanoate.

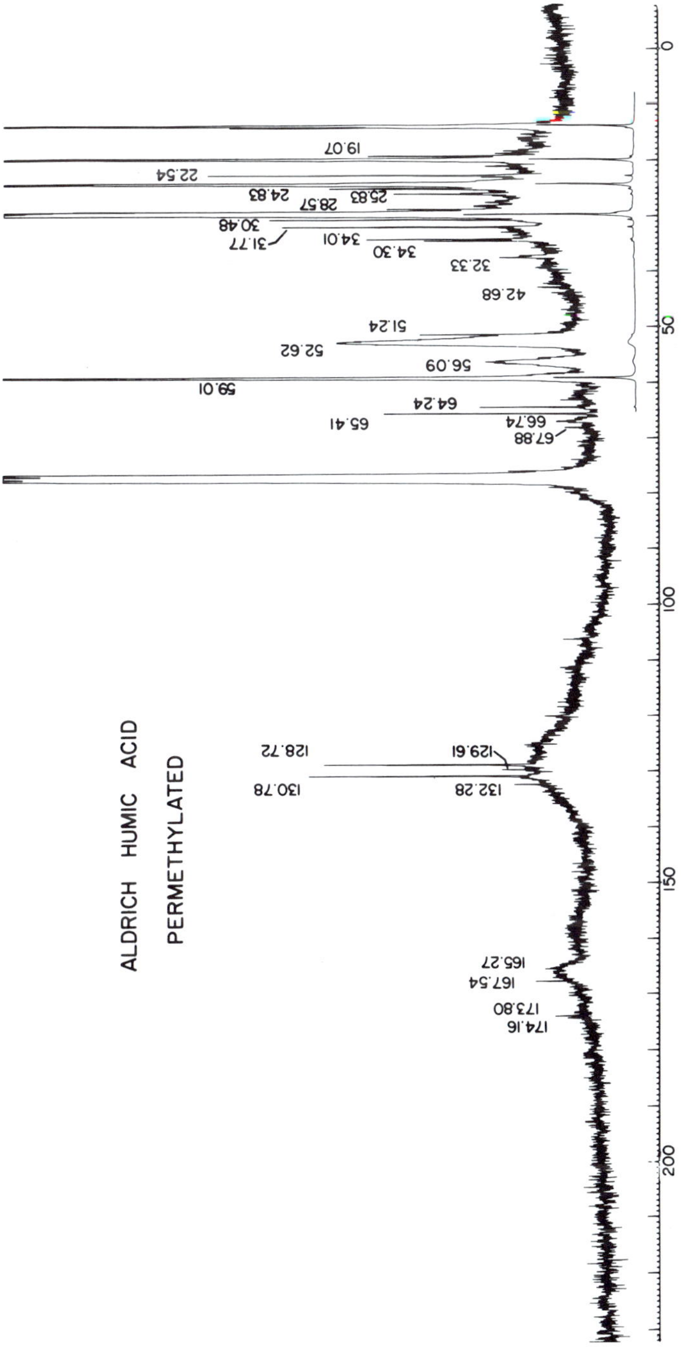

Figure 7. ^{13}C-NMR spectrum of permethylated Aldrich humic acid in $CDCl_3$. Scale in ppm relative to TMS. The resonances at 29.6, 24.0, 19.7 and 13.6 ppm are due to tetrabutylammonium salts.

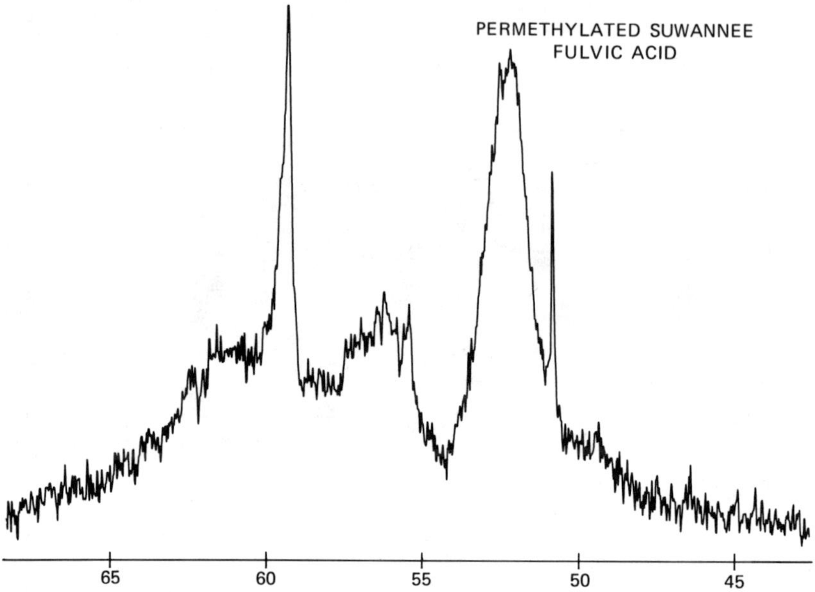

Figure 8. ^{13}C-NMR spectrum of permethylated Suwannee fulivc acid in $CDCl_3$. Scale in ppm relative to TMS.

carboxyl content. There are additional enhanced bands in the 61 to 64 ppm region and the 30 to 45 ppm region which may be diagnostic. We are currently studying them.

Recently Sigleo [7] reported that marine fulvic acids may be composed entirely of polysaccharide residues. This prompted us to look at the ^{13}C-NMR spectrum of permethylated glucuronic acid. The spectrum was surprisingly similar to a spectrum of an aquatic fulvic acid measured in this laboratory

Table III. Titration Data for Humic and Fulvic Acid Samples

Sample	Carboxylic Acid (meq/g)	Phenol (meq/g)	Carboxylic Acid + Phenol (meq/g)
Suwannee fulvic acid	6.0	2.1	8.1
Suwannee humic acid	4.7	2.3	7.0
Biscayne fulvic acid	6.3	1.8	8.1
Biscayne humic acid	5.0	1.6	6.6
Contech fulvic acid	7.7	3.3	11.0
Mollisol humic acid	4.5	2.1	6.6

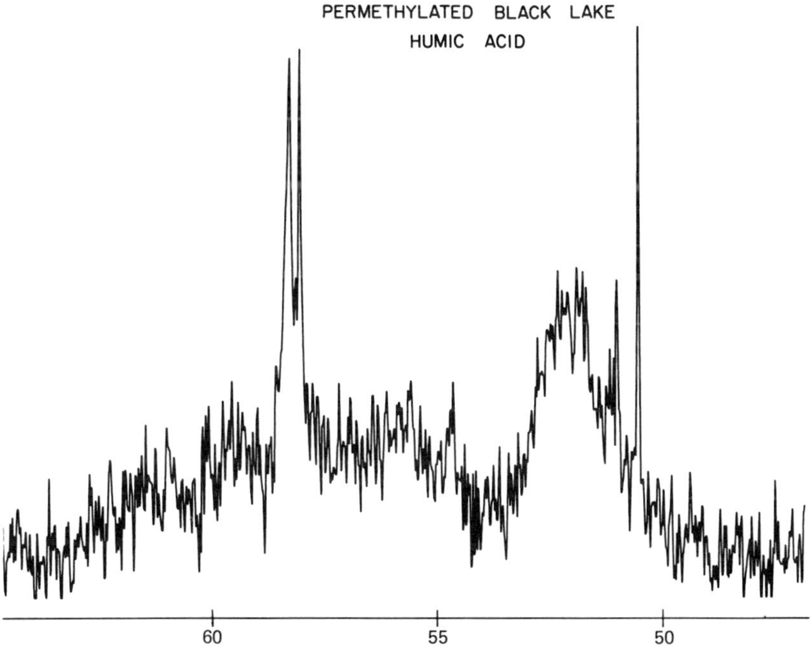

Figure 9. ^{13}C-NMR spectrum of permethylated Black Lake humic acid in $CDCl_3$. Scale in ppm relative to TMS.

(Figure 10). The anomeric methoxyl group absorbs in the same region as phenolic OCH_3. One could imagine a mixture of simple sugars that would yield a spectrum similar to a fulvic acid; however, elemental composition and chemical properties would be quite different from classic fulvates.

Table IV. Chemical Shifts of Sharp Bands in the ^{13}C-NMR Spectra of Permethylated Humic and Fulvic Acids[a]

Sample	Carbohydrate Methyl Ether	Methyl Ester
Suwannee fulvic acid	59.19	50.80
Suwannee humic acid	59.18, 59.12	50.82
Black Lake fulvic acid	59.12, 59.04	50.78
Black Lake humic acid	59.15, 58.90	50.84
Aldrich humic acid	59.01	51.24
Contech fulvic acid	59.25, 59.15	50.90

[a]Chemical shifts in ppm relative to TMS. Spectra determined in $CDCl_3$.

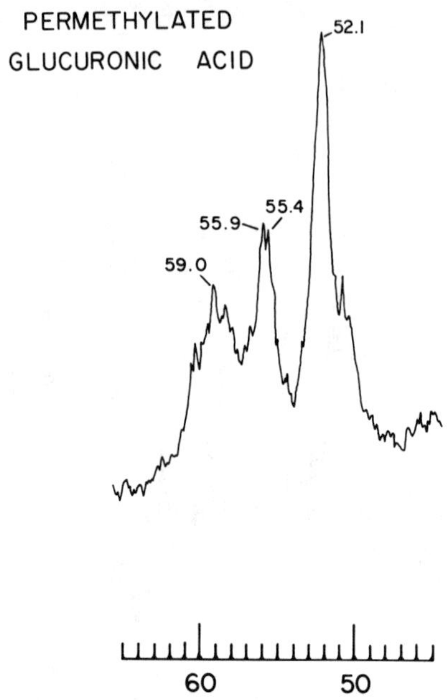

Figure 10. ^{13}C-NMR spectrum of permethylated glucuronic acid in $CDCl_3$. Scale in ppm relative to TMS.

Effect of Base on The ^{13}C-NMR Spectra of Polyphenols

A number of naturally occurring phenolic macromolecules exist as stable free radicals in the solid and solution forms. Humates, fulvates, tannins and lignins are all free radicals; furthermore, in the presence of base, these compounds form more radical (paramagnetic) species [8,9]. This phenomenon has been attributed to a quinone–hydroquinone complex [8]. Paramagnetic species alter ^{13}C-NMR spectra [10]. Since many published spectra of humic materials are taken in basic solutions, we decided to compare the neutral and alkaline spectra of a group of naturally occurring polyphenolic compounds in order to assess and interpret spectral distortions.

The ^{13}C-NMR spectrum of soil fulvic acid in NaOD was indeed different from the spectrum taken in neutral solution (Figure 11). This solution also exhibited a strong ESR signal (Figure 12). In order to examine this effect more precisely, we selected some model compounds which contained phenolic and carbohydrate groups and which had well defined spectra. Rutin, the 3-beta-rutinoside of quercetin, and quercitrin, the 3-alpha-rhamnoside of quercetin, were chosen as models.

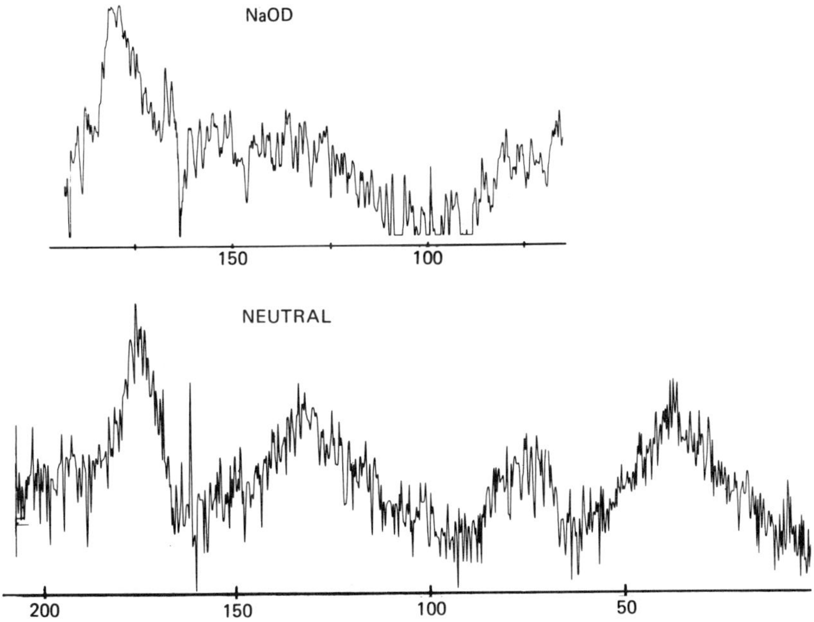

Figure 11. ^{13}C-NMR spectra of Contech fulvic acid in D_2O and D_2O/NaOD. Scale in ppm relative to TMS.

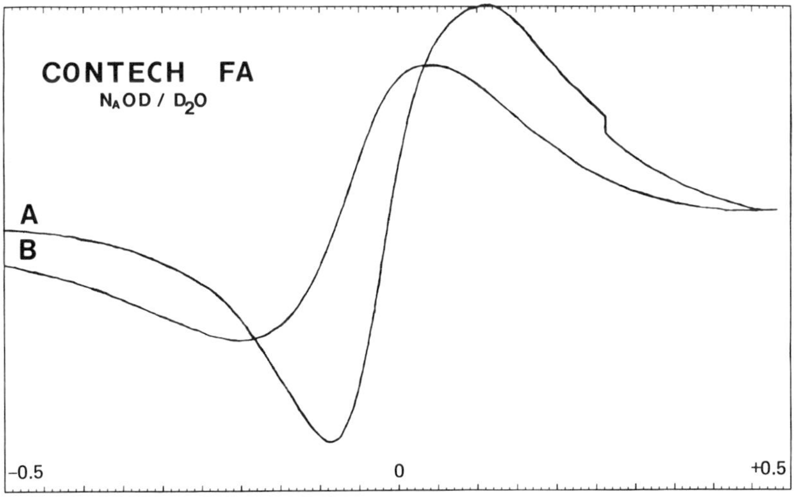

Figure 12. ESR spectrum of Contech fulvic acid in NaOD. (A) Immediately after mixing fulvic acid with NaOD, (B) 48 hr after mixing.

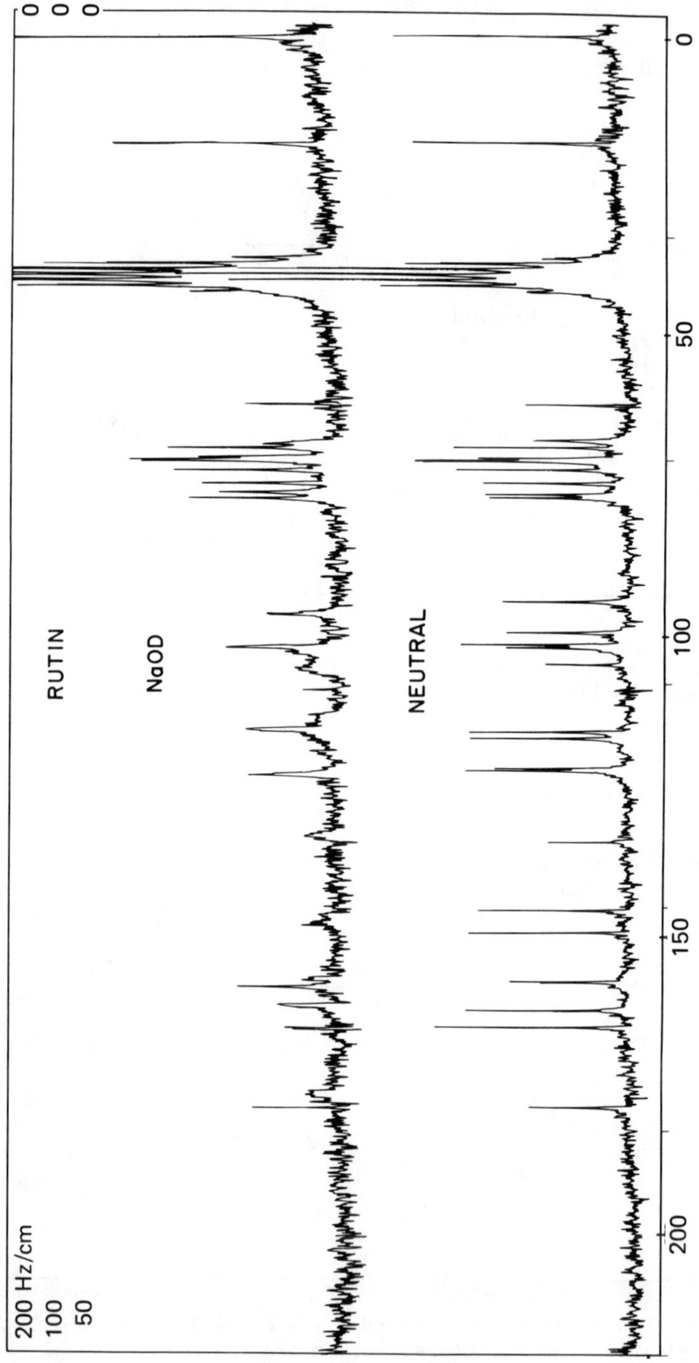

Figure 13. ^{13}C-NMR spectra of rutin in DMSO-d_6 and in DMSO-d_6/NaOD. Scale in ppm relative to TMS.

A well defined (one line per carbon atom) ^{13}C-NMR spectrum was obtained for rutin in DMSO-d_6. After addition of NaOD, the following changes were observed: broadening, loss of intensity and total collapse of some of the bands in the 100 to 175 ppm region (Figure 13). On the other hand, signals from the rutinoside portion (21 to 80 ppm) were enhanced and only slightly broadened.

An 8-line ESR signal was exhibited by this alkaline solution, identical to the ESR spectrum of 3,4-dihydroxybenzoic acid (Figure 14) taken under identical conditions. Thus, the radical anion was located in the B ring of the rutin. This was reflected in the ^{13}C-NMR spectrum, in which the bands for

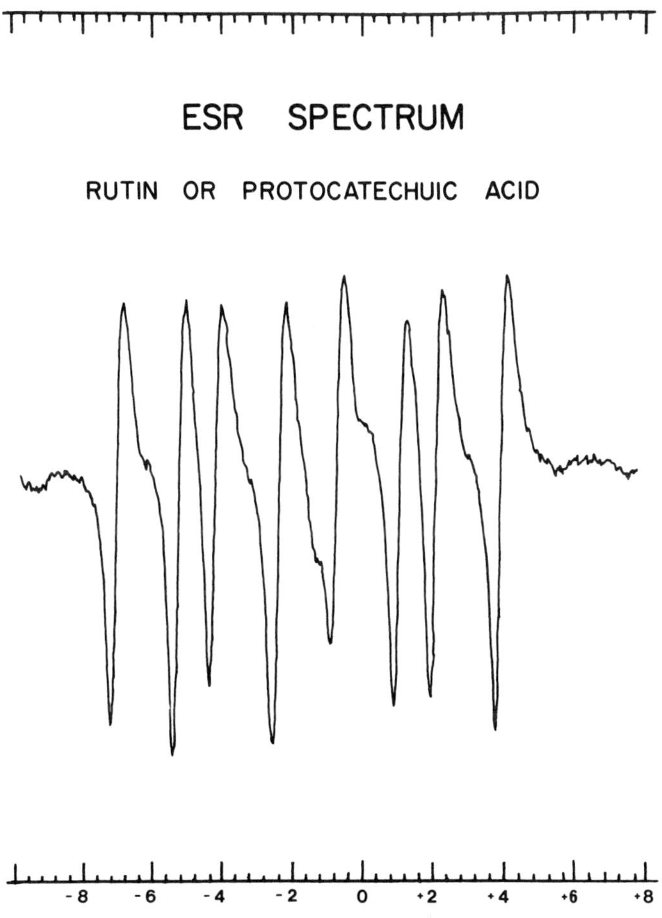

Figure 14. ESR spectrum of rutin in NaOD.

the C′-3 and C′-4 carbons had collapsed. Sugars in NaOD gave spectra identical to those in neutral solution. Mixtures of flavone aglycones and sugars in NaOD showed only spectral changes in the aglycone (aromatic) portion (Figure 15).

Thus, the carbohydrate moiety, whether alone or in glycosidic linkages, is unaffected by base or phenoxy radical. Quercitrin also showed the same broadening and diminution of NMR bands as did rutin; however, the changes were more pronounced in the A ring than in the B ring (Figure 16). This may be due to the steric effect of the carbohydrate moiety. In rutin, the carbohydrate group is relatively close to the A ring, whereas in quercitrin it reacts sterically with the B ring [11].

Tannins and lignins are constituents of many aquatic systems. They are less chemically defined than the simple flavonoids, but they are polyphenols with some carbohydrate character. Both Kraft lignin and tannic acid underwent changes in the aromatic portion of the ^{13}C-NMR spectra when placed in NaOD (Figures 17 and 18). Although not as pronounced as with flavones, the spectra revealed similar distortions. In the spectrum of tannic acid, the intensity of the carbohydrate band (~70 ppm) is still maintained.

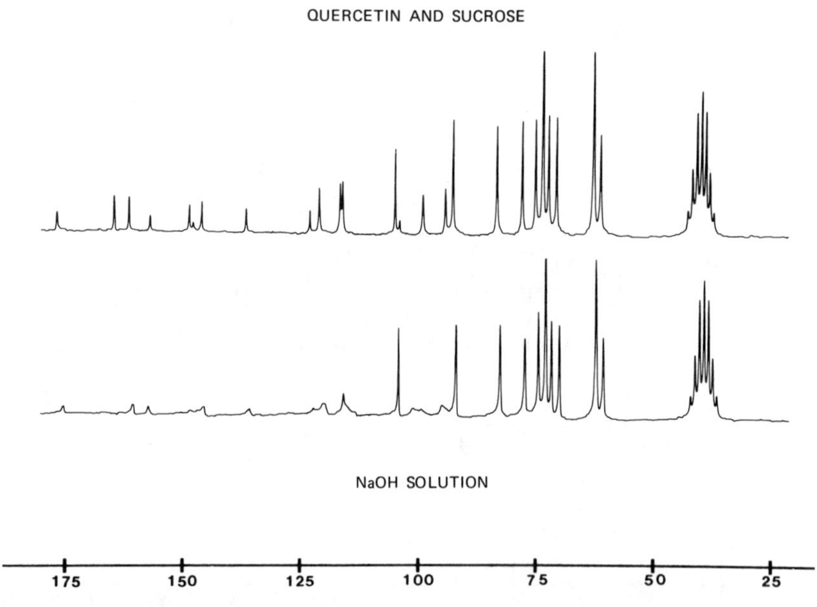

Figure 15. ^{13}C-NMR spectra of sucrose and quercetin in DMSO-d_6 and DMSO-d_6/NaOH. Scale in ppm relative to TMS.

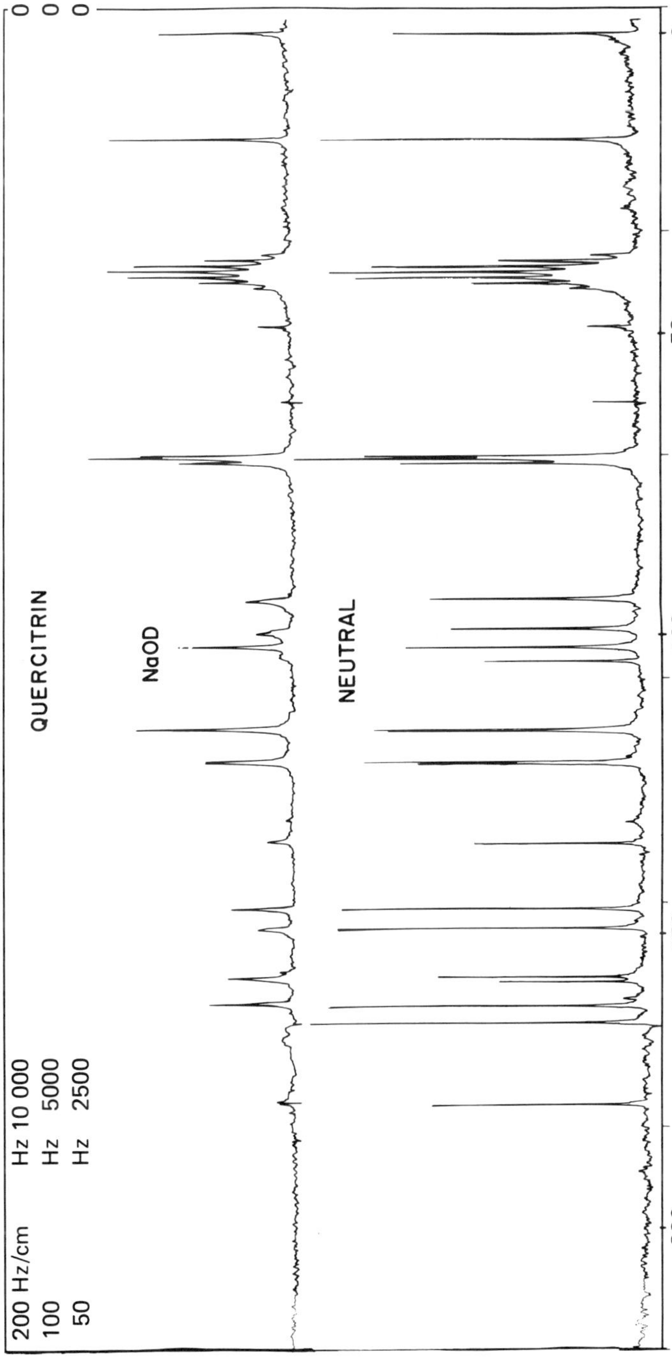

Figure 16. ^{13}C-NMR spectra of quercitrin in DMSO-d_6 and DMSO-d_6/NaOD. Scale in ppm relative to TMS.

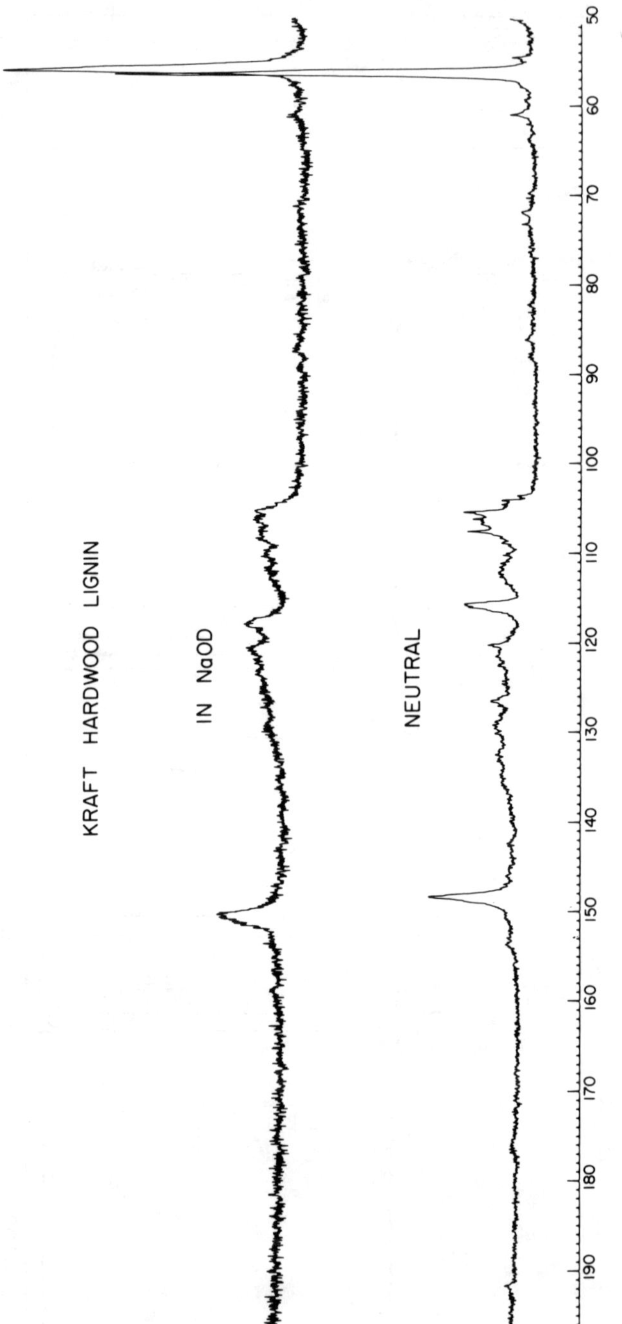

Figure 17. ^{13}C-NMR spectra of Kraft lignin in DMSO-d_6 and DMSO-d_6/NaOD. Scale in ppm relative to TMS.

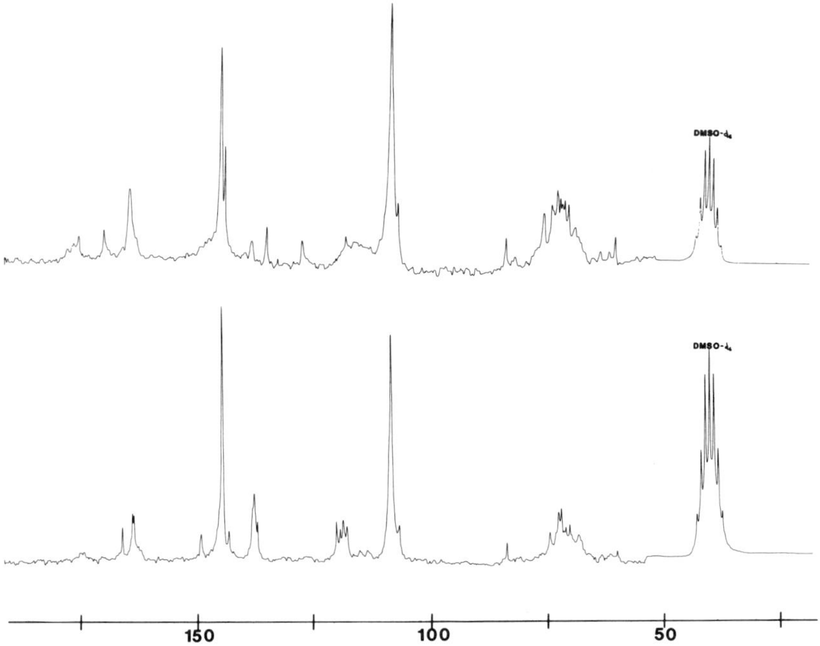

Figure 18. ^{13}C-NMR spectra of tannic acid in DMSO-d_6 and DMSO-d_6/NaOD. Scale in ppm relative to TMS.

The compounds with the broadest ^{13}C-NMR spectra would be expected to show less obvious changes in base than the model substances cited above. As expected, soil fulvic acid (Figure 11) was moderately distorted in the aromatic region in basic solutions.

If one relies on the ^{13}C- or ^{1}H-NMR spectra for estimating phenolic or aromatic character, then one has to avoid using the salt or basic form of humates, fulvates and other polyphenolic compounds.

CONCLUSIONS

The NMR data of permethylated humic materials reveal three major groups of hydroxyl functionality: carboxy, phenolic and carbohydrate moieties. Within these groups are sharp peaks for aliphatic carboxyl and 6-hydroxyl carbohydrates; these occur in humic and fulvic samples from a wide variety of sources. Humic acids appear to have the highest carbohydrate content. ^{1}H-NMR can supplement ^{13}C-NMR information. The extent of substitution on the aromatic ring may be estimated from the ^{1}H-NMR. Our data indicate that fulvic acids have more highly substituted aromatic rings than humic acids.

The addition of sodium hydroxide to a variety of naturally occurring aquatic substances causes the formation of semiquinone radical anions. These paramagnetic species distort the ^{13}C-NMR spectra and cause the collapse or diminution of aromatic resonances.

ACKNOWLEDGMENT

The authors would like to thank George R. Aiken, Robert L. Wershaw, Michael Thurman and Ronald L. Malcolm of the U. S. Geological Survey in Denver, Colorado, for their help and collaboration in this work. We are indebted to Dr. Kenner Christiansen of the University of Arizona Chemistry Department for several of the NMR spectra presented here.

This work was supported in part by the Office of Water Research and Technology, A-094, U. S. Department of the Interior, Washington, D.C., as authorized by the Water Research and Development Act of 1978, and in part by the U.S. Environmental Protection Agency, Office of Research and Development, EPA No. CR-807363.

REFERENCES

1. Steelink, C. "Humates and Other Natural Organic Substances in the Aquatic Environment," *J. Chem. Ed.* 54:599-603 (1977).
2. Steelink, C., M. A. Mikita and R. L. Wershaw. "Characterization of Functional Groups in Humic and Fulvic Acids," *Book of Abstracts of the 181st ACS National Meeting,* Atlanta, Georgia, 1981.
3. Ruggiero, P., F. S. Interesse, L. Cassidei and O. Sciacovelli. "^{1}H NMR Spectra of Humic and Fulvic Acids and Their Peracetic Oxidation Products," *Geochim. Cosmochim. Acta.* 44:603-609 (1980).
4. Mikita, M. A., C. Steelink and R. L. Wershaw. "Carbon-13 Enriched Nuclear Magnetic Resonance Method for the Determination of Hydroxyl Functionality in Humic Substances," *Anal. Chem.* 53:1715-1717 (1981).
5. Wershaw, R. L., M. A. Mikita and C. Steelink. "Direct ^{13}C NMR Evidence for Carbohydrate Moieties in Fulvic Acids," *Environ. Sci. Technol.* 15(12):1461-1463 (1981).
6. Liotta, R., K. Rose and E. Hippo. "O-Alkylation Chemistry of Coal and Its Implication for the Chemical and Physical Structure of Coal," *J. Org. Chem.* 46(2):277-283 (1980).
7. Sigleo, A. C. U.S. Geological Survey, Reston, Virginia. Personal Communication.
8. Steelink, C. "Free Radical Studies of Lignin, Lignin Degradation Products and Soil Humic Acids," *Geochim. Cosmochim. Acta.* 28:1615-1622 (1964).

9. Tollin, G., and C. Steelink. "Biological Polymers Related to Catechol: Electron Paramagnetic Resonance and Infrared Studies of Melanin, Tannin, Lignin, Humic Acid and Hydroxyquinones," *Biochim. Biophys. Acta.* 112:377-379 (1966).
10. LaMar, G. N., W. DeW. Horrocks and R. H. Holm, Eds. *NMR of Paramagnetic Molecules* (New York: Academic Press, Inc., 1973).
11. Markham, K. R., and B. Ternai. "^{13}C NMR of Flavonoids–II," *Tetrahedron* 32(21):2607-2612 (1976).

CHAPTER 5

OXIDATIVE DEGRADATION STUDIES ON AQUATIC HUMIC SUBSTANCES

J. H. Reuter

School of Geophysical Sciences
Georgia Institute of Technology
Atlanta, Georgia

M. Ghosal, E. S. K. Chian and M. Giabbai

School of Civil Engineering
Georgia Institute of Technology
Atlanta, Georgia

ABSTRACT

Aquatic humus from the Satilla River in south Georgia was studied by analyzing oxidative degradation products obtained under conditions of varying severity of reaction. Under the mildest degradation conditions employing potassium permanganate and 18-crown-6 in an aprotic solvent, oxalic acid was the major reaction product. From the high ratio of oxalic acid vs aryl compounds, it may be concluded that Satilla River aquatic humus is highly aliphatic. The predominance of oxalic acid among the aliphatic compounds indicates that the aliphatic component is either highly unsaturated or highly substituted with oxygen-containing functions, or some combination of both. Phenolic acid moieties appear to dominate benzene-carboxylic acid structural units in the aromatic component.

Increasingly severe oxidation conditions (aqueous permanganate, higher temperatures) lead to a progressive change of the product mixture makeup toward the predominance of structures stabilized by a $4n + 2$ overlapping π-electron system, mainly a series of benzene carboxylic acids. Structural interpretations based on results from these latter experiments would introduce an unjustified bias favoring a greater degree of aromaticity for Satilla

River aquatic humus. Neglecting the effect of the conditions of oxidative degradation when interpreting structures of humic substances from the resulting product mixtures can therefore lead to serious misinterpretations.

INTRODUCTION

The chemistry of humus in rivers and lakes has become the focus of interest for, inter alios, geochemists interested in the geochemical carbon cycle, and environmental chemists concerned with the preparation of safe drinking water, since it has been recognized that aquatic humus constitutes the bulk of the organic compounds in natural waters. On the basis of striking similarities between aquatic humus and soil humus, a soil origin for at least part of the aquatic humus has been suggested [1–3]. Data from some southeastern United States coastal plain streams show that organic carbon concentrations increase with sharp increases in river discharge [4–6]. This suggests that the bulk of the organic carbon, and specifically the humic substances, is derived from the watershed soils. Recently, Wilson et al. [7] came to the conclusion that much of the aquatic humus from a lake water is allochthonous, i.e., derived from the surrounding peat soils, rather than autochthonous, i.e., derived from the abundant aquatic plant material.

Studies of soil humic substances by oxidative degradation have led to structural models that strongly emphasize the aromatic character of the humic molecules [8,9]. However, recent advances in nondestructive methods for structural investigations of humic materials by ^{13}C-NMR have suggested that humic substances from a variety of sources (peat, soil, freshwater, marine sediments) exhibit a wide range of structural composition [7,10,11] with aromaticities (fa) as low as 0.10 and as high as 0.41.

In the present study we have investigated the effects of varying conditions of oxidative degradation of aquatic humus from the Satilla River in south Georgia on the composition of the degradation products. We have attempted to determine to what extent the experimental conditions introduce a bias into the relative abundances of aromatic vs aliphatic components. These investigations will hopefully contribute to our understanding of the molecular nature of aquatic humic substances.

EXPERIMENTAL

Samples

River water humic substances were isolated from the Satilla River using a modification of a previously described adsorption technique which utilizes

XAD-7 resin followed by elution with triethylamine [12]. Excess triethylamine was removed in a rotary evaporator and the residual solution was continuously extracted with chloroform to remove neutral molecules. The resultant aqueous solution was reacted with BioRad AG 50WX8 cation exchange resin (H^+-form) and then freeze-dried. About 70% of Satilla River humic substances was recovered. The elemental composition of the aquatic humus is as follows: 50.77% carbon, 3.81% hydrogen, 0.79% nitrogen and 2.28% ash.

Methylation

An ice-cold solution of ethereal diazomethane (prepared from Diazald, Aldrich Chemical Co.) and KOH was added to a cold suspension of humus (1 g) in methanol (20 ml). The solution was allowed to stand for 3 hr, and was then methylated again. The process was repeated until the IR spectrum confirmed that there were no more unreacted acidic functional groups. Usually, two methylations were required. The residue (1.18 g) after evaporation of the solvent was redissolved in chloroform and filtered. Elemental composition: 57.46% carbon, 5.48% hydrogen. 2.37% nitrogen and 0% ash.

Oxidation

Experiment 1. A sample of aquatic humus (0.5 g) was dissolved in a minimum amount of water. To this solution, aqueous potassium permanganate solution (4%, 125 ml) was added, and the mixture was heated slowly. The reaction was initially strongly effervescent from evolving CO_2. After gas evolution had subsided, the reaction mixture was refluxed for 8 hr. Excess permanganate was then destroyed with methanol and the mixture was filtered. The black residue (MnO_2) was washed with dilute NaOH, the filtrate acidified to pH 2 and continuously extracted for 4 hr each with ethylacetate (300 ml) four times. The organic solvent was separated, dried (Na_2SO_4) and evaporated to leave a residue which was methylated with ethereal diazomethane. Yield was 0.060 g.

Experiment 2. An aliquot of aquatic humus (0.5 g) was methylated with diazomethane and oxidized with aqueous permanganate (4%, 250 ml) following the procedure described for Experiment 1. Yield was 0.095 g.

Experiment 3. $KMnO_4$ (6 g) was added to a stirred solution of crown ether (18-crown-6, 1.0 g) in methylene chloride (200 ml) and stirred for 2 hr to make a homogeneous solution. An aliquot of aquatic humus (0.5 g) was methylated, the methylated product dissolved in methylene chloride and filtered. This solution was added to the crown ether/permanganate solution and refluxed for 3 hr. After filtering the reaction mixture, the residue was washed with Na_2SO_3 solution, chloroform, 10% NaOH solution and water, in that sequence. The filtrate (CH_2Cl_2 solution) was washed with Na_2SO_3 followed by NaOH (10%) and water. All the aqueous washings were combined and acidified. Nitrogen was bubbled through the solution to drive off SO_2. The combined aqueous solutions were continuously extracted for a period of 4 hr each with ethylacetate. The extract was dried over Na_2SO_4, evaporated and the residue methylated with diazomethane in ether. The ether was evaporated, the residue was taken up in chloroform and filtered. The product yield after solvent removal was 0.354 g.

Gas Chromatography/Mass Spectrometry

The GC/MS analysis was performed on a Hewlett-Packard 5830A gas chromatograph, equipped with capillary column injector port, interfaced to a FINNIGAN 4000 mass spectrometer, equipped with Incos Data System. The separation of the organic components was carried out on a 30-m X 0.25-mm-i.d. glass capillary column, deactivated according to Grob and Grob [13] and coated by the static method [14] with approximately 0.25 μm film thickness of SE-54 silicone gum phase (Applied Science; State College, PA).

Ultrahigh-purity helium (AIRCO) was employed as carrier-gas. Aliquots of 0.1 to 0.2 μl of the solution were injected in splitless mode and the chromatography was operated under the oven temperature program from 40° (3 min) to 300°C at 6°C/min. The glass capillary column was directly connected to the ionization source by means of 40-cm X 0.1-mm-i.d. fused silica tubing that operated as sample transfer line. The mass spectrometer was operated under the following conditions:

Ionization mode	EI
Emission current	0.5 mA
Electron energy	70 eV
Preamp sensitivity	10^{-7} A/V
Electron multiplier	1500 V
Mass range	41-500 amu
Scan rate	1 scan/sec

Perfluorotributylamine (FC43) was used initially to tune the ion source; 20 to 30 ng/μl of decafluorotriphenylphosphine (DFTPP) was subsequently injected through the GC in order to check the tune obtained, the column and the transfer line performances.

Chemical ionization mass spectra were obtained on a Varian MAT 112S mass spectrometer operated under the following conditions:

Ionization mode	CI
Reagent gas	Isobutane
Reagent gas pressure	1 Torr
Emission current	0.7 mA
Electron energy	95 eV
Mass range	80–800 amu
Scan rate	1 sec/mass decade

Identification of Oxidation Products

The INCOS data system was used to match the mass spectra of the eluting GC peaks with those of authentic samples stored in the library. Two indices, FIT and PURITY, are provided by this system as measures of agreement between sample and reference mass spectra. A FIT index of ⩾850 (out of 1000), and a PURITY index of ⩾750 (out of 1000) were accepted as a good match. In addition, a visual check, subtraction of the library spectrum from the sample spectrum and logical fragmentation patterns were the criteria examined for the tentative identification of the mass spectra. *Relative* quantitation was done automatically by programming for a quantitation of the total ion chromatogram. Relative amounts were expressed in terms of peak area of compound as percentage of the total chromatographic peak area from the total ion chromatogram, which was corrected for contributions from identified artifacts introduced during sample derivatization. The relative amounts provide a reasonable basis for comparing compound ratios (see below) obtained from the same sample under varying oxidation conditions, but under identical extraction conditions. No attempts were therefore made to provide absolute quantitative data by using either internal or external standards. Figures 1, 2 and 3 are examples of typical reconstructed ion chromatograms of each of the 3 experiments. Tables I, II and III identify chromatographic peaks according to scan numbers on the reconstructed ion chromatograms. Structures confirmed by matching our GC/MS parameters with authentic specimens have been marked by asterisks. All others are based on tentative identifications.

The same method has been followed whenever comparisons between runs have been made. Several mass spectra have been checked on the Cornell University Probability Based Matching (PBM) System and the Self-Training Interpretive and Retrieval System (STIRS). Several structures and substructural units have been reasonably suggested by the system.

RESULTS AND DISCUSSION

Since humic substances have been shown to contain a range of phenolic acid moieties [8], chemical protection of these moieties from oxidative degradation by methylation is necessary for their analytical recovery.

However, the methylated aquatic humus is insoluble in the aqueous medium used in Experiment 2, and therefore the reaction mixture is non-homogeneous: the substrate to be oxidized is a solid. This is definitely a disadvantage with regard to reaction kinetics, and may affect both product yield and spectrum. We thought it necessary to choose rather harsh oxidation conditions, i.e., excess permanganate and reflux temperature, to drive the reaction forward.

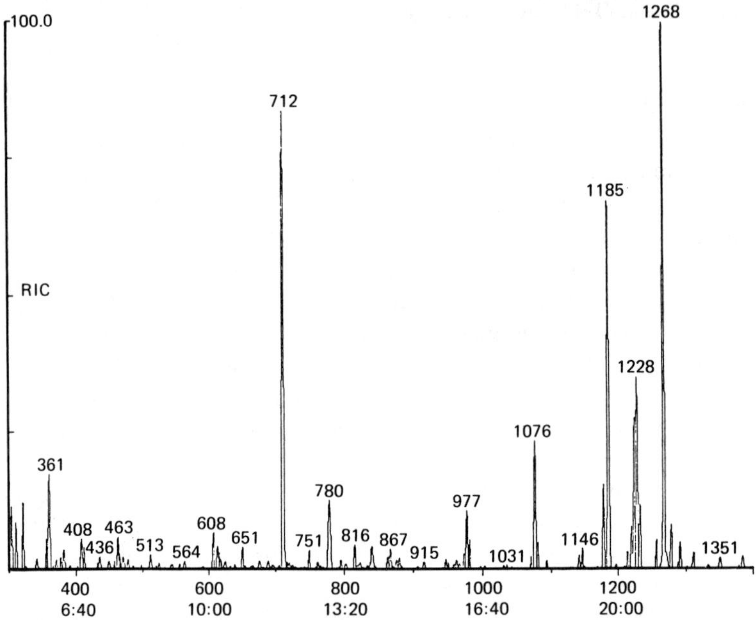

Figure 1.a. Reconstructed ion chromatogram of the permanganate oxidation product of aquatic humus in Experiment 1.

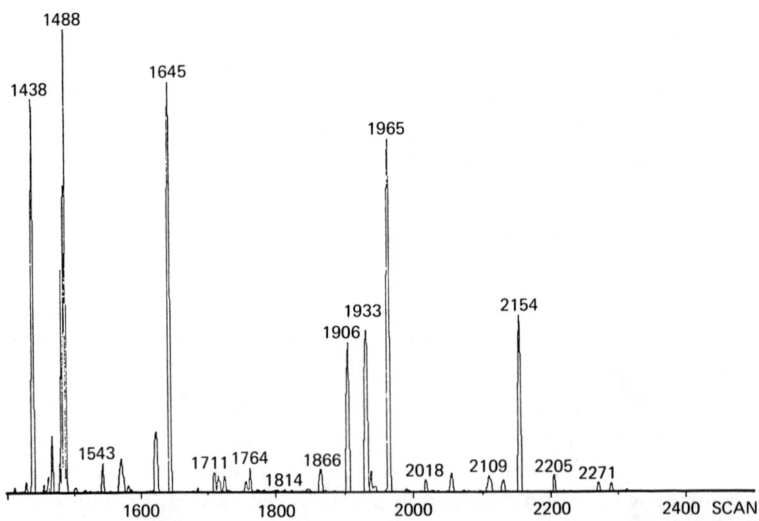

Figure 1.b. Reconstructed ion chromatogram of the permanganate oxidation product of aquatic humus in Experiment 1.

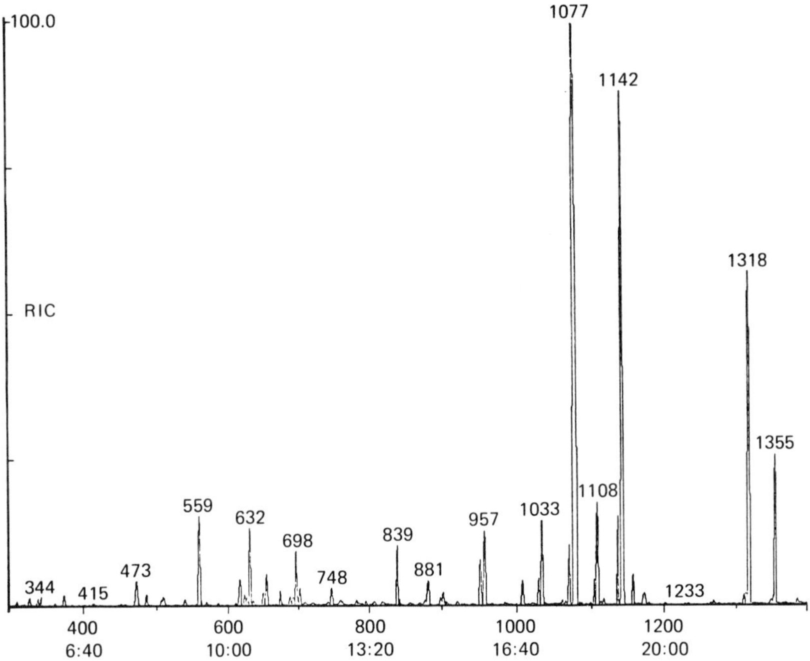

Figure 2.a. Reconstructed ion chromatogram of the permanganate oxidation product of methylated aquatic humus in Experiment 2.

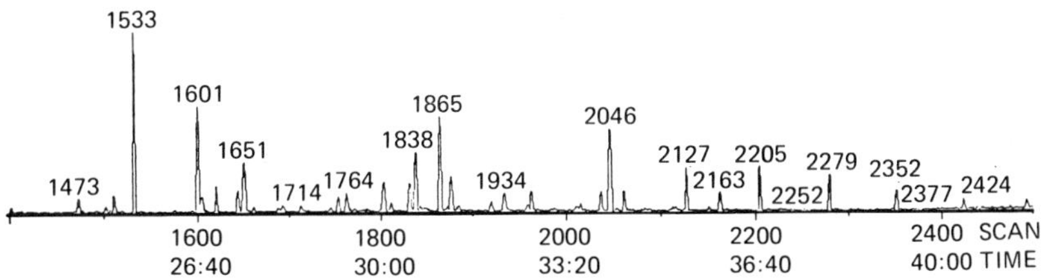

Figure 2.b. Reconstructed ion chromatogram of the permanganate oxidation product of methylated aquatic humus in Experiment 2.

The alternative to the above choice is to conduct the oxidation reaction in a homogeneous system, i.e., where both substrate and oxidant are in solution. To accomplish this, one may choose either an aqueous or an aprotic solvent system. For the aqueous system this means foregoing the protection of the aquatic humus—in particular, of the phenolic moieties in the aquatic humus—

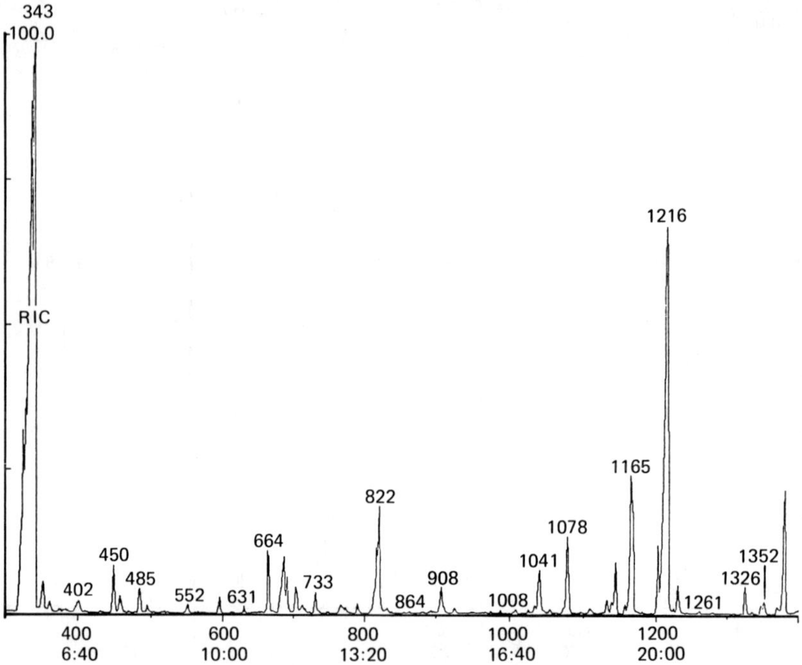

Figure 3.a. Reconstructed ion chromatogram of the oxidation product of methylated aquatic humus with Crown-ether permanganate in Experiment 3.

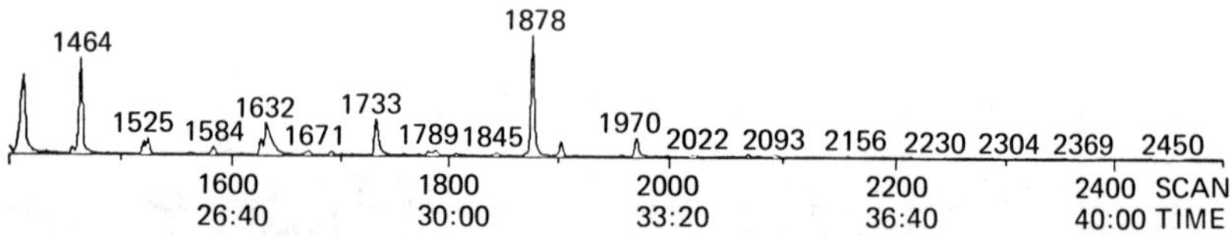

Figure 3.b. Reconstructed ion chromatogram of the oxidation product of methylated aquatic humus with Crown-ether permanganate in Experiment 3.

by methylation. Experiment 1 was therefore conducted with the unprotected aquatic humus.

The methylated aquatic humus is soluble in an aprotic solvent. However, an additional complication is the solubilization of the oxidant, for which the addition of a complexing agent, such as crown ether, is required. For Experiment 3, we chose the crown ether 18-crown-6 in methylene chloride. This

Table I. Permanganate Oxidation Products of Aquatic Humus in Experiment 1[a]

Scan Number	Structure	Scan Number	Structure
*383	$(COOMe)_2$	*1185	CH_2COOMe | CHCOOMe | CH_2COOMe
*712	$(CH_2COOMe)_2$		
751	CH_2COOMe | CHCOOMe Me	*1228 1278 *1292	
779	CH_2COOMe Me | CCOOMe Me	*1624 1645 *1711	
*780	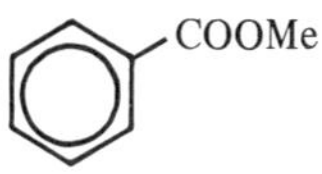		
842	$CH_2(CH_2COOMe)_2$	1906 *1932 1965	
1132		*2154	
1178	CH_2COOMe | CHCOOMe MeO	2314	

[a]Asterisks denote structures confirmed by matching with authentic samples.

solvent was preferred because of its low boiling point (40 °C). Besides working with a homogeneous system, where both substrate and oxidant are in solution, the choice of an aprotic solvent offers another advantage: oxidation products, such as diprotic acids, are insoluble and precipitate as their potassium salts. They are thus protected from further attack by permanganate [15]. Table IV summarizes the experimental parameters.

In Experiments 1 and 2, carbon dioxide evolved in a strongly effervescent reaction during the initial stages of the reaction, signalling the loss of carbon through either oxidation or decarboxylation or both. The product yields (Table IV) for Experiments 1 and 2 reflect this initial carbon loss, although further losses may have occurred through absorption of unreacted humic residues on the manganese dioxide solids formed during the reaction.

The oxidative product with the lowest molecular weight found in all three experiments was oxalic acid, the relative amounts of which are given in

Table II. Permanganate Oxidation Products of Methylated Aquatic Humus in Experiment 2[a]

Scan Number	Structure	Scan Number	Structure
*226	$(COOMe)_2$	*1070	CH_2COOMe – $CHCOOMe$ – CH_2COOMe
486	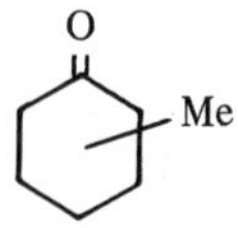	1077	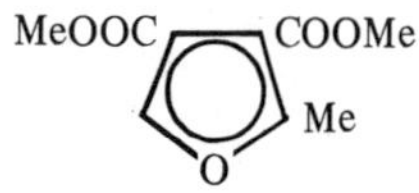
*559	$(CH_2COOMe)_2$		
617	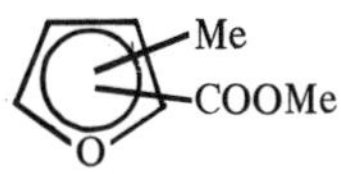	*1108 1157 *1172	
632	$CH_3(CH_2)_6COOMe$	1403 1473	
*651	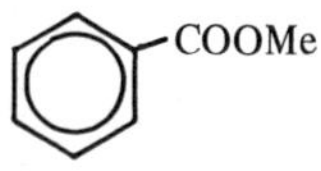	*1511 1533 *1690	
656	CH_2COOMe – $(Me)_2C$–$COOMe$	1644 1714 1755 1764	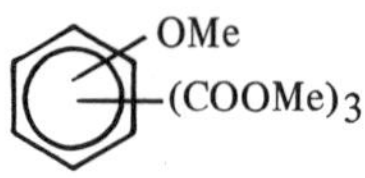
881	$(CH_2CH_2COOMe)_2$	1804 *1832 1865	
1008		1920	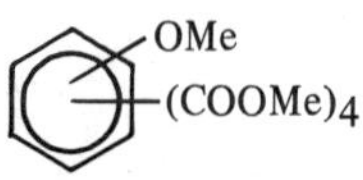
1033	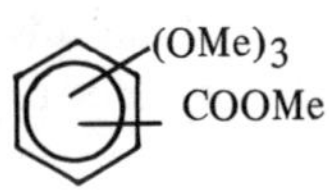	*2061	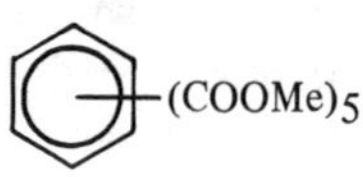

[a]Asterisks denote structures confirmed by matching with authentic samples.

Table V. Experiment 3 shows the highest yield of oxalic acid. If it can be assumed that the primary diprotic acid oxidation products are removed from the aprotic solvent by precipitation as potassium salts, and are thus protected from further degradation, one has to conclude that oxalic acid is the major *primary* oxidation product formed from Satilla River aquatic humus in

Table III. Crown Ether Permanganate Oxidation Products of Methylated Aquatic Humus in Experiment 3[a]

Scan Number	Structure	Scan Number	Structure
*343	$(COOMe)_2$	*1174 1224 *1237	
*485	$CH_2(COOMe)_2$	1326	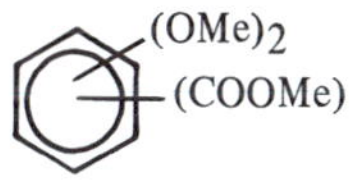
596	$CH_3COCH_2CH_2COOMe$	1457 1489 1525	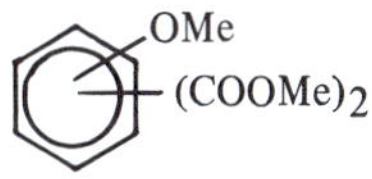
664	$(CH_2COOMe)_2$	*1563 1584 *1627	
704	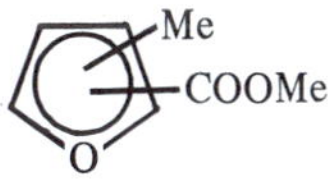	1692 1789 1808	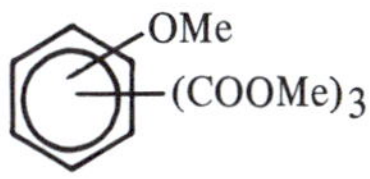
706	CH_2COOMe | Me–CHCOOMe	1845 *1873 1903	
714	$Me(CH_2)_6COOMe$	1882	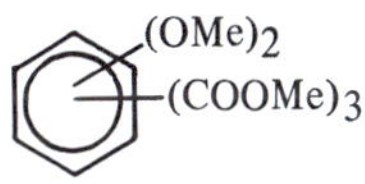
*731	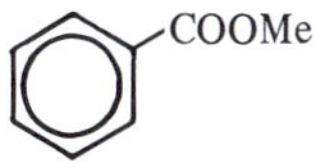	1957 2022	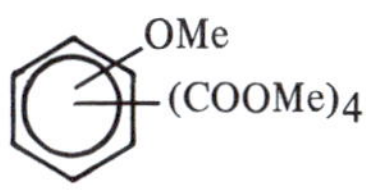
733	CH_2COOMe Me | >CCOOMe Me	2011	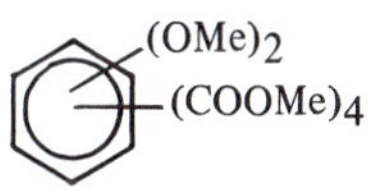
791	$CH_2(CH_2COOMe)_2$	*2093	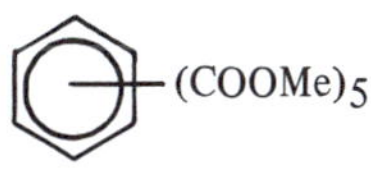
1078			
*1132	CH_2COOMe | CHCOOMe | CH_2COOMe		
1144	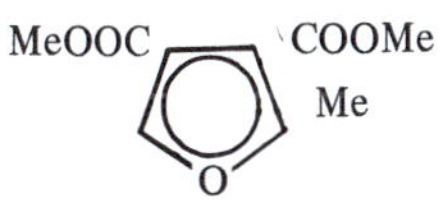		

[a]Asterisks denote structures confirmed by matching with authentic samples.

Table IV. Reaction Parameters and Product Yields

	Experiment		
	1	2	3
Sample weight (g)	0.5	0.5	0.5
Sample pretreatment	none	methylation	methylation
Oxidant	$KMnO_4$	$KMnO_4$	crown ether-$KMnO_4$
Solvent	H_2O	H_2O	CH_2Cl_2
Reaction temperature (°C)	>100	>100	~40
Reaction time (hr)	8	8	3
Product weight (g)	0.060	0.095	0.354

Table V. Aliphatic Oxidation Products

	Experiment		
Compound	1[a]	2[a]	3[a]
Oxalic acid, dimethyl ester	0.5	7.6	59.2
Malonic acid, dimethyl ester	0	0	0.74
Succinic acid, dimethyl ester	11.60	2.25	2.42
Methoxysuccinic acid, dimethyl ester	1.52	0	0.29
Methyl succinic acid, dimethyl ester	0.02	0	0.49
2,2-Dimethylsuccinic acid, dimethyl ester	0.89	0.73	0.15
Pentanoic acid, 4-oxo, methyl ester	0	0	0.37
Pentanedioic acid, dimethyl ester	0.41	0	0.29
Hexanedioic acid, dimethyl ester	0	0.68	0
Propane-1,2,3-tricarboxylic acid, trimethyl ester	7.64	1.23	0.43
1,2,3,4-Cyclobutane tetracarboxylic acid, tetramethyl ester	1.20	0	0
Octanoic acid, methyl ester	0	1.70	0.39
Undecanoic acid, methyl ester	0	0.47	0

[a]Relative amounts expressed in terms of peak area of compounds as percentage of corrected total chromatographic peak area of total ion chromatogram.

Experiment 3. It follows that, if the primary oxidation step leads to aliphatic fragments as short as oxalic acid, the aliphatic component of the aquatic humus is either highly unsaturated or highly substituted with oxygen-containing functions, or some combination of both.

Oxalic acid as such is oxidized quantitatively by aqueous permanganate at temperatures ≥60 °C and at low pH. However, since during the oxidation reaction the pH of the product mixture increases to pH ~10, one of the conditions of quantitative oxidation is eliminated.

The difference in relative amounts of dimethyloxalate between Experiment 1 and 2 may be explained by assuming that a larger proportion of derivatized oxalic acid is produced in the product mixture of Experiment 2, since acidic functional groups of the parent material had been methylated before oxidation. Thus, one may expect the formation of the oxalic acid monomethylester MeOOC–COOH(I), which is more resistant to permanganate oxidation, since the ester would have to be hydrolyzed as a first step. Methylation of the acidic functional groups of the starting material would also prevent decarboxylation of compounds with the following substructure (II):

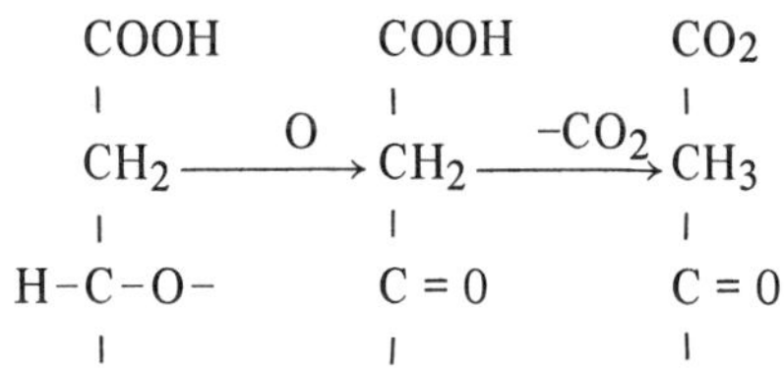

II

which, when protected by methylation, may produce (I) or malonic acid monomethylester.

The predominance of benzenecarboxylic acids (Table VI) in Experiment 1 is presumably caused by the high degree of oxidation, by which aromatic structures, stabilized by a 4n + 2 overlapping π-electron system, will be preferentially enriched in the product mixture. A relative degree of oxidation may thus be expressed in terms of the ratio (R_1) of the relative amounts of oxalic acid to benzenoid products (benzenecarboxylic and phenolic acids), if oxalic acid is taken as the primary and most abundant (see Experiment 3) oxidation product of the aliphatic component of the aquatic humus. This ratio of oxalic acid to benzenoid structures is lowest in Experiment 1 with $R_1 < 0.01$, and increases drastically to $R_1 = 0.50$ in Experiment 2 and further to $R_1 > 10$ in Experiment 3. The ratio R_1, it should be emphasized, is thus a characteristic of the experimental conditions, and not of the aquatic humus.

Among the benzene carboxylic acids (Table VI) there is a predominance of the di-, tri- and tetracarboxylic acids. Some researchers [16] have made similar observations, whereas others [17] find benzenepentacarboxylic acid to be the predominant structure. Benzenehexacarboxylic acid is only found in Experiment 1. Among the benzenetricarboxylic acids the 1-, 2-, 4-substitution pattern is the most prevalent.

The ratio of phenolic to benzenecarboxylic acids (R_2) may be taken as another measure of the degree of oxidation. In Experiment 1, $R_2 < 0.001$, increasing drastically to $R_2 = 0.34$ in Experiment 2 and to $R_2 = 3.48$ in Experiment 3. In other words, there is a dramatic increase in relative content of phenolic acids over benzenecarboxylic acids with increasingly milder reaction conditions (compare Table IV). The ratio R_2 then,one should note, is indicative only of the reaction conditions, and is not a sample characteristic.

Table VI. Benzenoid Oxidation Products

m(MeO)  (COOMe)$_n$		Experiment		
n	m	1[a]	2[a]	3[a]
1	0	0.98	0.29	0.15
2	0	6.15	3.51	0.07
3	0	16.21	4.22	0.47
4	0	24.11	2.79	0.43
5	0	5.39	0.44	0.08
6	0	0.08	0	0
1	1	0.05	0.59	2.9
1	2	0	0	0.29
1	3	0	1.62	0
2	1	0	0.34	0.52
2	3	0	0	0.04
3	1	0	1.27	0.23
3	2	0	0	0.12
4	1	0	0.06	0.06
4	2	0	0	<0.01

[a]Relative amounts expressed in terms of peak area of compounds as percentage of corrected total chromatographic peak area of total ion chromatogram.

Table VII shows the relative amounts of two furan derivatives. Of these two, 2-methyl-3,4-furandicarboxylic acid (dimethylester) occurs only in traces in Experiment 1, whereas it is a major constituent of the oxidation product mixture in Experiment 2 (~21% of the chromatographic peak area). It appears that this furan derivative is an artifact of the oxidation reaction, whose synthesis could be explained in the following fashion. The hypothetical humic substructure III yields upon methylation of the humic material substructure IV.

Table VII. Furanoid Oxidation Products

	Experiment		
	1[a]	2[a]	3[a]
2-Methylfuran-5-carboxylic acid, methylester	<0.01	0.80	0.62
2-Methylfuran-3,4-dicarboxylic acid, dimethylester	<0.01	20.81	1.96

[a]Relative amounts expressed in terms of peak area of compounds as percentage of corrected total chromatographic peak area of total ion chromatogram.

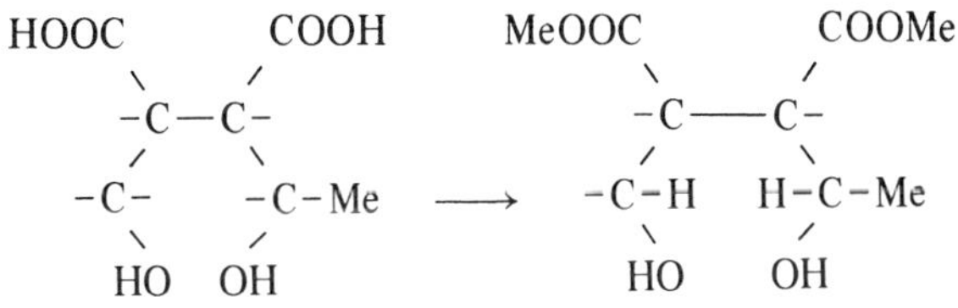

III IV

Oxidative attack by permanganate on substructure IV yields an intermediate (IVa) that will enolize and dehydrate when exposed to aqueous acid

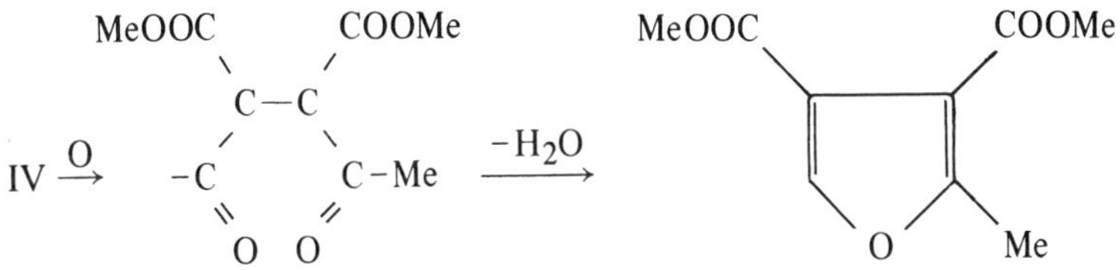

IVa

2-methyl-3,4-furandicarboxylic acid, dimethylester

to form 2-methyl-3,4-furandicarboxylic acid dimethylester. This proposed reaction pathway thus indicates the possible existence of substructures such as III in Satilla River aquatic humus. If the aquatic humus is either unmethylated or is not quantitatively methylated prior to oxidation, substructure III will lead to entirely different products through decarboxylation

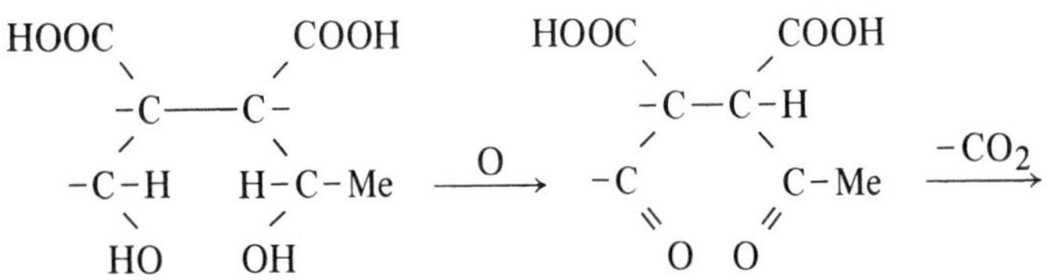

III IIIa

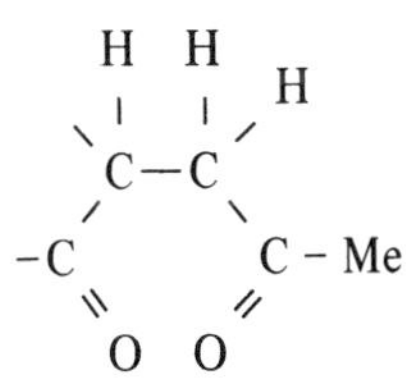

IIIb

and further degradation into smaller fragments. The occurrence of small amounts of 4-oxo-pentanoic acid (Table V) supports the hypothetical assumption of substructure III and its oxidative derivative IIIb, to which 4-oxo-pentanoic acid is related.

2-Methylfuran-5-carboxylic acid, on the other hand, occurs only in small quantities (Table VII). Its possible origin from carbohydrate reaction products during the synthesis of the humic substances can be reasonably assumed.

In general, the aliphatic products of oxidative degradation of Satilla River aquatic humus are dominated by short-chain diprotic acids (Table V). Higher homologs are present only in comparatively insignificant amounts (e.g., pentanedioic and hexanedioic acids), and succinic acid becomes dominant only under the more severe oxidative conditions of Experiment 1, where oxalic acid is eliminated (see discussion above). It appears that the increase in relative abundance of propane-1,2,3-carboxylic acid in Experiment 1 is controlled by the same factors.

The distribution spectrum of aliphatic degradation products clearly indicates the importance of aliphatic structures in Satilla River aquatic humus. However, since only short carbon chains are produced even under the mildest oxidation conditions (Experiment 3), one may reasonably assume that aliphatic carbon chains without unsaturated bonds or oxygen-containing functional groups do not extend beyond 5-carbon units, as exemplified by the production of propane-1,2,3-tricarboxylic acid and the shorter diprotic acids. This is a reasonable assumption on the basis of the atomic oxygen-to-carbon ratio for this aquatic humic material.

The tentative identification of 1,2,3,4-cylobutane tetracarboxylic acid is surprising because its structure has no commonly occurring precursor among the generally assumed source materials for humic substances.

Table VIII lists a number of compounds generated in Experiment 3 that have been tentatively identified as polyethylene glycols. Nine compounds have been found that can be divided into two series on the basis of their molecular weights (as obtained by GC/MS in the CI mode). The molecular weight increases in each series by 44, whereas the two series are separated from each other by a difference of 30. Thus, the two series can be described by the molecular formulae $(CH_2CH_2O)_n$ and $CH_2O(CH_2CH_2O)_n$, where n assumes values from 4 to 8.

It should be pointed out that these compounds occur only in the crown ether permanganate oxidations of aquatic humus. They are not detected in the corresponding blank runs. One of the compounds that can be described by $(CH_2CH_2O)_6$ has the same molecular weight as the crown ether 18-crown-6 (MW 264). However, a comparison of the mass spectrum of 18-crown-6 with that of compound number 5 (Table VIII) shows that the two compounds are not identical. It should also be emphasized that compounds 7, 8 and 9 (Table VIII) have considerably higher molecular weights than 18-crown-6,

Table VIII. Oxidation Products of Experiment 3, Tentatively Identified as Polyethylene Glycols

		$(CH_2CH_2O)_n$	$CH_2O(CH_2CH_2O)_n$
Compound Number	MW	n	n
1	176	4	
2	206		4
3	220	5	
4	250		5
5	264	6	
6	294		6
7	308	7	
8	338		7
9	352	8	
18-crown-6	264	6	

thus excluding the possibility that these compounds are derived by simple degradation of the crown ether.

In order to decide whether these compounds are by-products of the crown ether or whether they are authentic oxidation products of the humic substrate, the chemical structures of these unknowns need to be elucidated. To this date, polyethylene glycols have never been reported as degradation products of humic substances.

SUMMARY AND CONCLUSIONS

The study of aquatic humus by oxidative degradation with permanganate requires certain choices to be made regarding degradation conditions.

1. The oxidation of underivatized aquatic humus is carried out in an aqueous system, buffered by alkali if required to assure complete solution of the higher-molecular-weight humic acid fraction. The disadvantage of this approach is that certain sensitive reaction sites (e.g., phenolic functions) remain vulnerable to oxidative attack.
2. The aquatic humus is methylated to provide protection of certain sensitive moieties. However, methylation renders the aquatic humus insoluble in aqueous systems. Oxidation can then be carried out in two possible ways: (a) in an aqueous system, which has the disadvantage of being nonhomogeneous, since the substrate to be oxidized is a solid, and (b) in an aprotic solvent system, which requires the addition of a complexing agent to solubilize permanganate. The latter system offers the advantage that hydrophilic primary oxidation products (e.g., diprotic acids) precipitate out of solution and are thus protected from further oxidative degradation.

Our investigations show that oxidation of aquatic humus from the Satilla River under the above described conditions leads to qualitatively and quantitatively different spectra of degradation products depending on the choice of conditions. Under condition 2b oxalic acid is the major reaction product and the ratio of aryl compounds to oxalic acid is very low. Among the aryl compounds there is a predominance of phenolic acids over benzenecarboxylic acids.

Under condition 2a the relative oxalic acid yield is drastically decreased, whereas aryl compounds increase in relative proportion. Benzenecarboxylic acids now predominate over phenolic acids.

Under condition 1 oxalic acid is only a minor component of the reaction mixture, and succinic acid becomes the predominant aliphatic product. Aryl components are predominant, but consist almost exclusively of benzenecarboxylic acids.

Because of the high degree of preservation of the primary oxidation product oxalic acid, oxidation condition 2b can be considered the most protective. Judging from the ratio of aryl compounds to oxalic acid, Satilla River aquatic humus is highly aliphatic. The predominance of oxalic acid among the aliphatic compounds indicates that the aliphatic component of Satilla River aquatic humus is either highly unsaturated or highly substituted with oxygen-containing functions, or some combination of both. Phenolic acid moieties appear to dominate benzenecarboxylic acid structural units in the aromatic component.

Oxidation conditions 2a and 1 lead to a progressive shift of the composition of the product mixture toward predominance of structures stabilized by a $4n + 2$ overlapping π-electron system. Using the experimental results obtained under these conditions for a structural interpretation of Satilla River aquatic humus would introduce a bias toward a greater degree of aromaticity. Structural interpretations of humic substances based on a product spectrum obtained by oxidative degradation have in many instances neglected to take into account the conditions of degradation. These interpretations have thus led to structural models with a strong emphasis on aromatic character, contradicted by more recent structural findings by nondestructive (NMR) analyses on comparable samples of humic substances.

ACKNOWLEDGMENTS

This work was supported by National Science Foundation Grant CME 80-08683. We wish to thank Lawrence E. Abbey for providing high-resolution EI and CI GC/MS.

REFERENCES

1. Beck, K. C, .J. H. Reuter and E. M Perdue. "Organic and Inorganic Geochemistry of Some Coastal Plain Rivers of the Southeastern United States," *Geochim. Cosmochim. Acta* 38:341-364 (1974).
2. Weber, J. H. and S. A. Wilson. "The Isolation and Characterization of Fulvic Acid and Humic Acid from River Water," *Water Res.* 9:1079-1084 (1975).
3. Malcolm, R. L. and W. H. Durum. "Organic Carbon and Nitrogen Concentrations and Annual Organic Load of Six Selected Rivers of the United States," *U.S. Geol. Surv. Water Supply Paper 1817-F* (1976).
4. U.S. Geol. Survey, *Water Resources Data for Georgia* (1976).
5. U.S. Geol. Survey, *Water Resources Data for Georgia* (1977).
6. U.S. Geol. Survey, *Water Resources Data for Georgia* (1978).
7. Wilson, M. A., P. F. Barron and A. H. Gillam. "The Structure of Freshwater Humic Substances as Revealed by ^{13}C-NMR Spectroscopy," *Geochim. Cosmochim. Acta* 45:1743-1750 (1981).
8. Schnitzer, M. and S. U. Khan. *Humic Substances in the Environment* (New York: Marcel Dekker, Inc., 1972).
9. Schnitzer, M. and S. U. Khan. "Humic Substances: Chemistry and Reactions." In *Soil Organic Matter (Developments in Soil Science),* (New York: Elsevier, 1978) pp. 1-64.
10. Hatcher, P. G., R. Rowan and M. A. Mattingly. "^{1}H and ^{13}C-NMR of Marine Humic Acids," *Org. Geochem.* 2:77-85 (1980).
11. Newman, R. H., K. R. Tate, P. F. Barron and M. A. Wilson. "Towards a Direct Non-Destructive Method of Characterizing Soil Humic Substances Using ^{13}C-NMR," *J. Soil Sci.* 31:623 (1980).
12. Perdue, E. M. "Solution Thermochemistry of Humic Substances–II. Acid-Base Equilibria of River Water Humic Substances." In *Chemical Modeling in Aqueous Systems.* E. A. Jenne, Ed., 94-114. ACS Symp. Ser. No. 93. Am. Chem. Soc. (1979).
13. Grob, K. and G. Grob. "Deactivation of Glass Capillaries by Persilylation," *J. High Resol. Chromat. and Chromat. Comm.* 3:197-198 (1980).
14. Giabbai, M., M. Schoults and W. Bertsch. "Static Coating of Glass Capillary Columns. Some Practical Observations," *J. High Resol. Chromat. and Chromat. Comm.* 1:277-278 (1978).
15. Sam, D. J. and H. E. Simmons. "Crown Polyether Chemistry," *J. Am. Chem. Soc.* 94:4024-4025 (1972).
16. Ogner, G. and E. T. Gjessing. "Permanganate Oxidation of Methylated and Non-Methylated Aquatic Humus in Norway," *Geoderma* 14:139-145 (1975).
17. Liao, W., R. F. Christman, D. J. Johnson and D. S. Millington. "Structural Characterization of Aquatic Humic Material," *Environ. Sci. Technol.* in press (1982).

CHAPTER 6

CROSS-COUPLING OF HUMUS CONSTITUENTS AND XENOBIOTIC SUBSTANCES

Jean-Marc Bollag

Laboratory of Soil Microbiology, Department of Agronomy
The Pennsylvania State University
University Park, Pennsylvania

ABSTRACT

During humus formation reactive compounds are linked together either through enzymatic activity or through abiotic chemical reactions, resulting in complexes of polymerized molecules. If humic acids are formed by the polymerization of phenols and other chemicals, it can be expected that reactive products include not only naturally occurring compounds, but also xenobiotics or pollutants. We isolated an extracellular enzyme, a laccase, from the fungus *Rhizoctonia praticola* and extracted a catalyst from soil; both are capable of polymerizing various phenols. Substrates tested included phenolic lignin derivatives such as vanillic acid, vanillin, ferulic and syringic acid, and a number of xenobiotics such as chlorinated phenols, naphtholic compounds and halogenated anilines. The combined incubation of phenols occurring in soil and certain xenobiotic compounds with the fungal enzyme resulted in the formation of cross-coupling products. Oligomeric products were isolated and then characterized primarily by mass spectrometry; certain oligomers were identified by additional analytical techniques. The reaction can be considered a model for explaining a probable binding mechanism of pollutants or other xenobiotics to humic material.

INTRODUCTION

The fate of man-made chemicals or xenobiotics is currently of major concern, since the pollution of our environment constitutes a serious threat to the lives of many other organisms in addition to human beings. The inability to detect a certain xenobiotic or pollutant does not necessarily imply its disappearance, since conventional analytical procedures may not be adequately developed for determining its existence. Incorporation into humic material may prevent detection of a chemical once it is introduced into soil.

The reaction which involves the chemical binding of xenobiotic compounds to humus substances can be caused by oxidative coupling. This process, in which phenols are linked together or react with various other chemicals, is important in the synthesis of many naturally occurring substances such as lignins, melanins, alkaloids and humic material. It can be catalyzed or caused by various enzymes (e.g., phenoloxidases [1]), inorganic chemicals (e.g., ferric chloride, cupric hydroxide [2]), clay [3] and soil extracts [4], or it can take place by autoxidative processes [5].

In soil many compounds possessing phenolic characteristics and originating from lignin or other organic material are essential constitutents in the formation of humus [6]. Since many agricultural and industrial chemicals or their degradation products are also substituted phenols, it is not surprising that they may be incorporated into organic matter during the humification process.

Although adsorption or bonding of xenobiotics to organic matter can occur by several mechanisms [7], such as van der Waal's forces, hydrogen bonding, hydrophobic bonding, charge transfer, ion exchange or ligand exchange. Chemical alteration and subsequent covalent bonding usually produce more persistent complexes which may constitute a part of the so-called "bound residues" in soil.

Oxidative coupling appears to be a major mechanism causing polymerization during humification, and it is this process which we studied under laboratory conditions to demonstrate the basic reactions which can take place between humus constituents and man-made chemicals.

OLIGOMERIZATION REACTIONS OF XENOBIOTICS AND NATURALLY OCCURRING PHENOLS

Oxidative coupling reactions with phenols or aromatic amines result in the formation of oligomeric or polymeric products. In the oxidative coupling of a phenol, for example, aryloxy or phenolate radicals are formed by removal of an electron and a hydrogen ion from the hydroxyl group. The resulting phenolate radicals then couple to yield stable dimerized and polymerized products.

The occurrcncc of oxidativc coupling reactions mediated by chemical catalysts or enzymes in soil can be deduced from analyses of humus substances that constitute a complex mixture of polymerized molecules [6]. However, the chemical identification of these high-molecular-weight molecules in the natural environment is very difficult and sometimes impossible, and therefore it appears necessary to study the basic reactions with single reactants using isolated enzyme systems or other catalysts.

In our laboratory we isolated an extracellular enzyme from the fungus *Rhizoctonia praticola* which was characterized as a laccase [8], and we succeeded in preparing and purifying a soil extract [4], both of which catalyzed oxidative coupling. In order to measure the efficiency of these catalysts, 2,6-dimethoxyphenol was used as a substrate, and the formation of a dimeric quinone, namely 3,5,3′,5′-tetramethoxydiphenoquinone (Figure 1), was followed in a spectrometric assay [4,9].

The laccase from *R. praticola* was purified by ion-exchange column chromatography with diethylaminoethyl-cellulose followed by gel filtration with Sephadex G-200. The fungal enzyme was assayed with a large number of xenobiotic and naturally occurring phenols as substrates, and it was possible to determine the formation of oligomeric products by mass spectrometry. Table I shows oligomers of halogenated phenols and various naphtholic compounds which were formed after incubation with the isolated laccase from *R. praticola*.

Similar results were obtained when humus constituents were incubated with the fungal enzyme (Table II). Vanillic acid [11,12], vanillin [12], orcinol [12] and syringic acid [12,13] were selected for studying the formation of oxidative coupling products generated by the laccase of

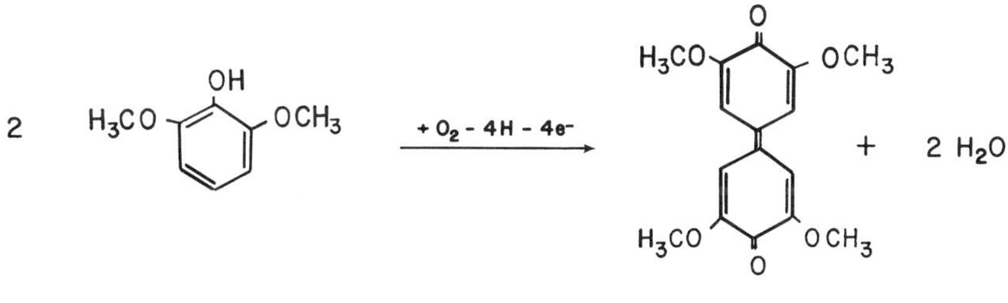

Figure 1. Oxidative coupling of 2,6-dimethoxyphenol to the extended quinone 3,5,3′,5′-tetramethoxydiphenoquinone.

Table I. Oligomeric Products of Various Xenobiotic Phenols Formed after Incubation With a Laccase from the Fungus *Rhizoctonia praticola*

	Oligomeric Products (m/z Value of Molecular Ion)				
	Dimer	Trimer	Tetramer	Pentamer	Reference
2-Chlorophenol	254	380	506		9
4-Chlorophenol	254	380	472 506		9
2,4-Dichlorophenol	322 302	482 448	642 608	802 768	11
4-Bromo-2-chlorophenol	410				9
4-Chloro-2-methylphenol	282	422	494 528	600 634	10
1-Naphthol	286	428	570	712	9
2-Naphthol	286	428	570		9
4-Chloro-1-naphthol	332	336	478		9
4,4′-Bi-1-naphthol	570				9

R. praticola. In general, we detected only oligomers with molecular ions of up to 1000, which was the limit of detection for the available mass spectrometric technique,* but it was very clear that higher oligomers and polymers were formed. The size of the majority of the oligomers could be regulated by varying incubation time of the substrate with the enzyme or with the amount of applied enzyme units used.

We also succeeded in extracting phenoloxidase-like catalysts from soil that were capable of polymerizing phenolic substances in a manner analogous to the isolated fungal laccase [4,14]. This catalyst was extracted from soil with a citrate buffer (0.05 *M*, pH 6.0) and was subsequently purified by ion exchange chromatography (DEAE-cellulose) and gel filtration with Sephadex G-50.

When guaiacol, a lignin derivative, was incubated with the partially purified soil-extract, various oligomers ranging from dimers to pentamers were determined by mass-spectrometric analysis (Table III). The products

*An AEI MS-902 mass spectrometer with sample introduction by direct insertion probe was used for electron impact mass spectrometric analysis (70 eV) at a temperature varying from 150 to 350 °C, or a Finnigan 3200 instrument with methane as the reagent gas was used for chemical ionization analysis.

Table II. Oligomeric Products of Phenolic Humus Constituents after Incubation With a Laccase from the Fungus *Rhizoctonia praticola*

	Oligomeric Products (m/z Value of Molecular Ion)					
	Dimer	Trimer	Tetramer	Pentamer	Hexamer	Reference
Vanillic acid	290	412	534	656		11
	304					
	334	456	578	700		
Vanillin	302					12
Orcinol	246	368				12
Syringic acid	306	458	610	762	914	13
	334	486	638	790	942	
	350	502	654	806	958	
	364	516	668	820	972	

Table III. Oligomeric Products Obtained after Incubation of Guaiacol with a Partially Purified Soil-Extract [15]

Acetylated Oligomeric Products (m/z Value of Molecular Ion)				
Monomer	Dimer	Trimer	Tetramer	Pentamer
166	288	410	532	654
	330	452	574	696
		494	616	738
			658	780

were reductively acetylated, which allowed an easier characterization [14]. No change of guaiacol was evident in the boiled soil extract control. Essentially the same mass spectral results were obtained by Gierer and Opara [15] in their study of the products formed during the catalytic oxidation of guaiacol by a laccase from *Neurospora crassa* and a commercial horseradish peroxidase.

CROSS-COUPLING OF AROMATIC COMPOUNDS

Since many xenobiotic compounds, particularly pesticides, yield phenol- or aniline-like chemicals analogous to naturally occurring compounds, it is not surprising, as discussed previously, that these substances can be incorporated by cross-coupling into humic material. In a study by Wolf and Martin [16] it was shown that ^{14}C-ring-labeled 2,4-D (2,4-dichlorophenoxyacetic acid) and chloropropham (isopropyl 3-chlorophenylcarbamate) or perhaps their transformation products, were incorporated into humic polymers produced by fungi. Similarly, Mathur and Morley [17] observed that the insecticide methoxychlor [1,1,1-trichloro-2,2-di-(4-methoxyphenyl)ethane] was incorporated into a humic acid analogue during the chemical synthesis. Both research groups ascertained that the pesticides were bound to the humus-like product, but the nature of the chemical bonding was not determined.

Our approach with the laccase from *R. praticola* provided clear chemical evidence for the formation of oligomeric cross-coupling of hybrid products, and some of our results will be described.

Cross-Coupling Between Phenols

To illustrate the formation of a hybrid product from two phenolic compounds after incubation with the laccase from *R. praticola*, 2,4-dichlorophenol and 4-bromo-2-chlorophenol were selected [18]. Both chemicals are intermediary degradation products of phenoxyalkanoate herbicides. Figure 2 shows the three dimeric products which could be isolated from an enzyme assay, with the two halogenated phenols as substrates. The molecular weights of these compounds were determined by mass spectrometry as 322, 410 and 336, which correspond with the respective dimers of each of the phenols, and with a hybrid formed from both. Gas chromatography/mass spectrometric analysis of these products and of their methylated derivatives lent support to the structural assignments.

Most of our work related to cross-coupling of phenols was done with phenolic humus constituents and 2,4-dichlorophenol [12]. When orcinol, syringic acid, vanillic acid or vanillin were incubated together with 2,4-dichlorophenol and the fungal laccase, the formation of cross-coupling products could be determined (Table IV). The characterization of the hybrid products was based primarily on mass spectrometric analysis, whereby the appearance of a cluster of peaks typical for the isotopic pattern of a certain number of chlorine atoms was an important indicator. Figure 3 presents a scheme of hybrid oligomers found in an enzyme assay with 2,4-dichlorophenol and syringic acid. It should be emphasized that the structures presented are only hypothetical. Mass spectrometric analysis does not allow the determination of the sequence of phenol units and the position (*ortho-ortho*, *ortho-para*, etc.) or type (carbon-carbon, carbon-oxygen) of coupling. However, the structures shown are probable model

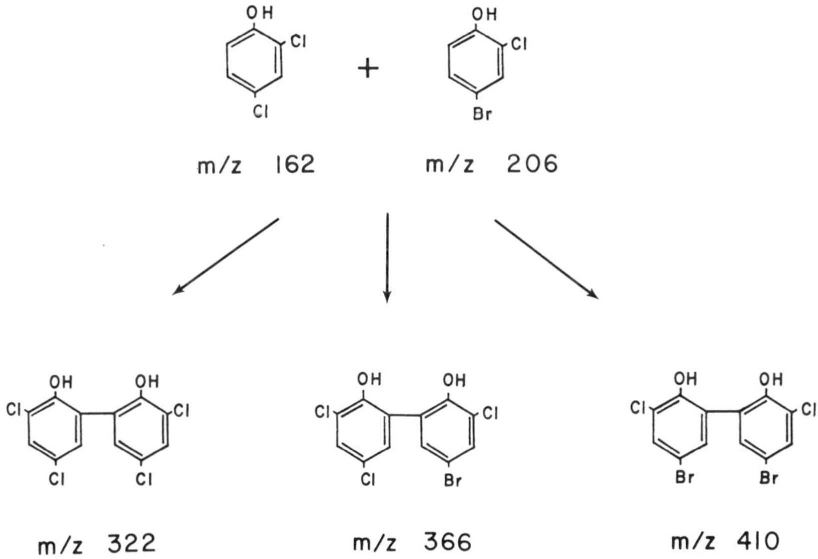

Figure 2. Scheme of oxidative coupling products resulting from the simultaneous incubation of 2,4-dichlorophenol and 4-bromo-2-chlorophenol with a laccase from *R. praticola* [19].

Table IV. Formation of Hybrid Products after Incubation of Phenolic Humus Constituents and 2,4-Dichlorophenol in the Presence of a Laccase from *R. praticola* [12]

	Hybrid Products (m/z Value of Molecular Ion)			
Substrates	Dimer	Trimer	Tetramer	Pentamer
2,4-Dichlorophenol + Orcinol	284			
2,4-Dichlorophenol + Syringic acid	314	466 510	618 626	770
2,4-Dichlorophenol + Vanillic acid	284	444 450		
2,4-Dichlorophenol + Vanillin	312			

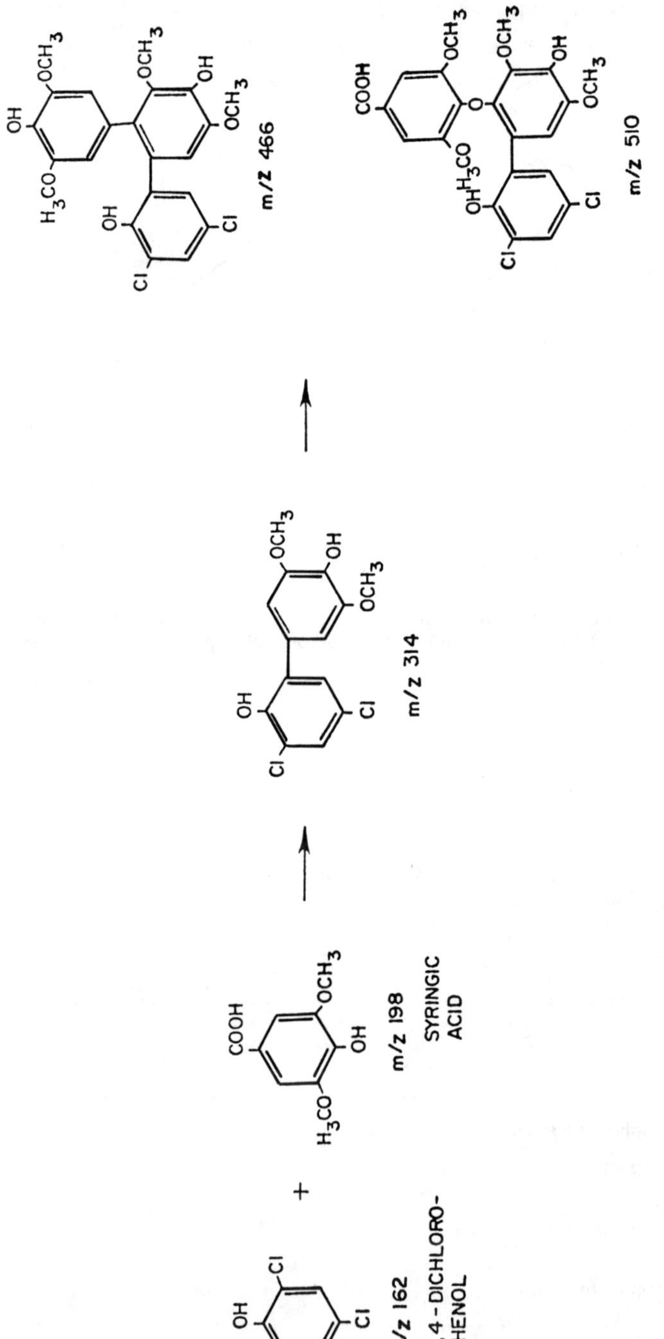

Figure 3. Suggested hybrid products resulting from the combined incubation of syringic acid and 2,4-dichlorophenol with a laccase from *R. praticola* [12].

representations based on the known molecular weights of the products and on the chlorine isotope patterns.

Cross-Coupling Between Anilines

The formation of mixed azobenzenes from differently substituted aniline molecules has been reported in experiments with soil. Bartha [19] described the formation of 3,3′,4-trichloro-4′-methylazobenzene from the anilines derived from the herbicides propanil [N-(3,4-dichlorophenyl)-propionamide] and solan [N-(3-chloro-4-methylphenyl)-2-methylpentanamide]; 3,4-dichloroaniline and 3-chloro-4-methylaniline, released from propan and solan, respectively, combined to form the asymmetric azobenzene. Kearney et al. [20] reported the condensation in soil of 3-chloroaniline and 3,4-dichloroaniline to form the hybrid product 3,3,4′-trichloroazobenzene (Figure 4). The catalysts responsible for these two condensation reactions in soil have not been elucidated.

The acaricide chlordimeform [N-(4-chloro-*o*-tolyl)N′,N′-dimethylformamidine] is degraded to 4-chloro-*o*-toluidine in soil. Iwan et al. [21] were able to isolate and identify several oxidative coupling products of this aniline (dimers and trimers), and they attributed the coupling reactions to the activity of soil peroxidases.

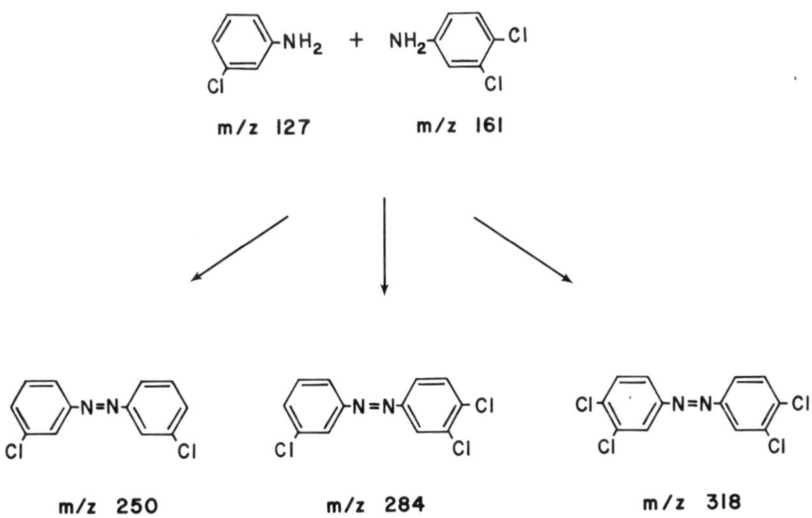

Figure 4. Proposed condensation products of 3-chloroaniline and 3,4-dichloroaniline found in soil [20].

Cross-Coupling Between Phenols and Anilines

The oxidation and coupling of aromatic compounds is known to occur in enzymatic or nonenzymatic systems. The chemical reaction of the condensation of aniline with quinone in a weak acid solution to dianilinoquinone was first described by Willstätter and Majima [22]. Suida and Suida [23] studied the reaction of *p*-benzoquinone with different anilines under various conditions. They determined that *p*-benzoquinone with anilines in a mild acid solution formed only "mono-anilidoquinones," whereas in an ethanolic solution the formation of "dianilidoquinones" occurred.

Our investigation with phenols and various anilines showed that the cross-coupling catalyzed by the fungal enzyme had both enzymatic and nonenzymatic causes [24]. When the laccase from *R. praticola* was incubated with 2,4-dichlorophenol and various halogenated anilines, hybrid oligomers were formed. Mass spectrometric analysis indicated the formation of two types of trimers (Figure 5): one trimer consisted of a phenoxyquinone dimer coupled with an aniline molecule; the other was a trimer composed of a quinone coupled with two aniline molecules. Since the incubation of anilines alone with the fungal enzyme did not cause the formation

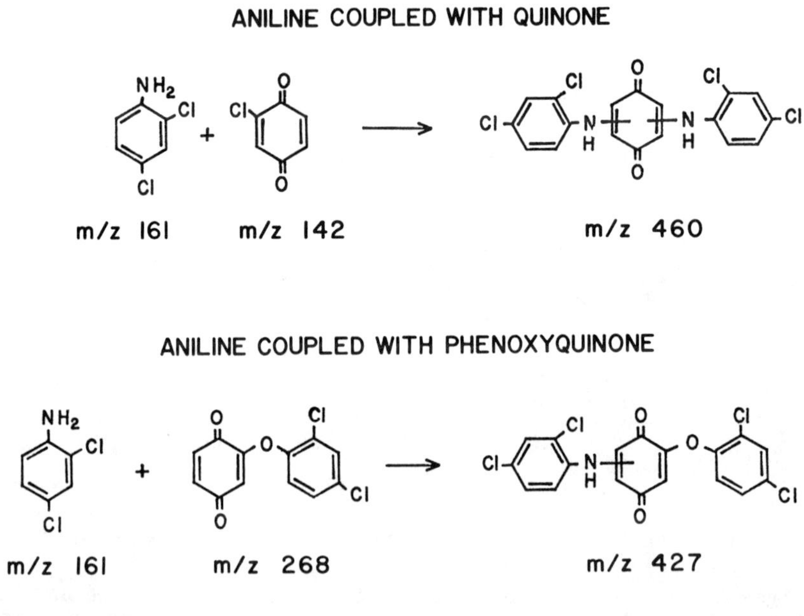

Figure 5. Hybrid product of 2,4-dichloroaniline and two chlorinated quinones [24].

of oligomeric products, it could be assumed that the coupling of aniline to enzymatic products of 2,4-dichlorophenol has to be of chemical origin. This assumption was proven when the quinone products formed in an enzyme assay with 2,4-dichlorophenol as substrate were found to react with the anilines in the absence of the enzyme [24].

The coupling of various anilines with phenolic humus constituents has also been demonstrated (Bollag, Minard and Liu, unpublished data). For instance, 2,4-dichloroaniline was incubated with the laccase and vanillic acid, and a number of hybrid products were determined (Table V). Similar results were obtained with other phenolic humus constituents, such as ferulic acid, syringic acid or protocatechuic acid, when they were combined in an enzyme assay with 4-chloroaniline, 3,4-dichloroaniline or 2,6-diethylaniline (Bollag, Minard and Liu, unpublished data).

Cross-Coupling Between Phenols and Other Chemicals

The formation of humic-like substances by condensation of aromatic polyphenols with amino acids and peptides has been shown by several researchers. Haider et al. [25] studied the linkage of amino acids and peptides with phenols during oxidation with phenol oxidases. The binding to nitrogen of amino acids was found mostly in the oxidation products of dihydroxyphenols like catechol, protocatechuic acid, and hydroquinone. Ladd and Butler [26] synthesized phenolic polymers with silver oxide as a catalyst. They polymerized *p*-benzoquinone or catechol and found that amino acids, peptides or proteins could be incorporated in the coupling reaction.

Table V. Hybrid Products of 3,4-Dichloroaniline and Vanillic Acid Formed after Incubation with a Laccase from *R. praticola*

Substrate	Molecular Ion of Hybrid Product (m/z)	Oligomer	Suggested Composition[a]
Vanillic acid + 3,4-Dichloroaniline	403	Trimer	2 V[b] (334) + 1 DCA[b] – 2 CO_2 – 2 H
	447	Trimer	2 V (290) + 1 DCA – 2 H
	449	Trimer	2 V (290) + 1 DCA
	463	Trimer	2 V (304) + 1 DCA
	560	Tetramer	2 V (334) + 2 DCA – 2 CO_2 – 4 H
	608	Tetramer	2 V (290) + 2 DCA

[a]m/z of the vanillic acid-dimers 290, 304 or 334.
[b]V = vanillic acid
DCA = 3,4-dichloroaniline.

The formation of dark-colored nitrogenous humic-like substances in microbial cultures was often observed and has been summarized in the reviews by Flaig et al. [6] and Martin et al. [27]. Küster [28] first reported that the formation of these substances in microbial cultures results from the activity of phenol oxidases. Under nonenzymatic oxidizing conditions Flaig and Riemer [29] showed a reaction between 2-hydroxy-6-methyl and 2-hydroxy-5-methylbenzoquinone-1,4 with glycine using polarographic techniques.

Bondietti et al. [30] determined the binding of monomeric and polymeric amino sugars, D-glucosamine and chitosan [(a polymer of 1→4)-linked glucosamine units] to phenols or phenol mixtures in the presence of a commercial mushroom phenol oxidase or trihydroxyphenols, which readily autoxidize. These conclusions were obtained from respirometric studies and from using ^{14}C-labeled glucosamine and chitosan which were incorporated into the dark phenol polymers. Glucosamine reacted with *o*-diphenols and to a lesser extent with *p*-diphenols, but phenols containing one methylated hydroxyl group (e.g., vanillic or syringic acid) did not undergo a cross-coupling reaction.

SIGNIFICANCE OF CROSS-COUPLING BETWEEN XENOBIOTICS AND NATURALLY OCCURRING AROMATIC CHEMICALS

Humic substances are formed to a large extent by random polymerization of phenolic compounds (oxidized to radicals) which react mainly with nitrogenous substances derived from proteins [6]. The general conception is that phenols originating primarily from higher plants are the major building blocks for humic substances. It is not surprising that aromatic intermediates which resemble naturally occurring chemicals and are formed during the transformation of xenobiotics can also be incorporated into soil humus through enzymatic or abiotic chemical oxidative coupling reactions. However, our knowledge of the environmental consequences of xenobiotic incorporation into soil humus is limited. The synthetic chemicals may not lose their identity and properties through covalent binding to humic material. It is difficult to assess if and when these chemicals will be released from humus, and therefore if they can cause a delayed pollution problem.

In order to clarify the fate of bound material, most experiments were performed with radiolabelled pesticides, and their determination was done by combustion or through strong hydrolytic techniques [16,17,31]. However, this approach does not allow conclusions concerning the actual identity of the bound chemicals.

Our experiments with a fungal enzyme and a catalyst extracted from soil provide additional evidence that oligomerization and polymerization reactions of xenobiotic and naturally occurring phenols take place. Further, the chemical determination of the products clearly indicates the multitude of possible molecular configurations. Of particular interest appear to be the isolation and identification of cross-coupling or hybrid products, since they may not only suggest a mechanism for possible incorporation into humic material, but they may also be more resistant to further transformations and therefore constitute a new group of environmentally alien products. The experiments described have to be considered a model approach for explaining the binding of xenobiotics to humic substances, but they can provide a basis for investigating the highly complex problem of chemical identification of 'bound residues' in the natural environment.

ACKNOWLEDGMENT

The preparation of this article was supported in part by the Pennsylvania Agricultural Experiment Station and by a research grant from the Environmental Protection Agency (R 808165).

REFERENCES

1. Sjoblad, R. D., and J.-M. Bollag. "Oxidative Coupling of Aromatic Compounds by Enzymes from Soil Microorganisms," In: *Soil Biochemistry, Vol. 5*, E. A. Paul and J. N. Ladd, Eds. (New York: Marcel Dekker, Inc., 1981), pp. 113-152.
2. Taylor, W. I., and A. R. Battersby, Eds. *Oxidative Coupling of Phenols.* (New York: Marcel Dekker, Inc., 1967).
3. Wang, T. S. C., S. W. Li and Y. L. Ferng. "Catalytic Polymerization of Phenolic Compounds by Clay Minerals," *Soil Science* 126:15-21 (1978).
4. Suflita, J. M., and J.-M. Bollag. "Oxidative Coupling Activity in Soil Extracts," *Soil Biol. Biochem.* 12:177-183 (1980).
5. Haider, K., J. P. Martin, and Z. Filip. "Humus Biochemistry," In: *Soil Biochemistry, Vol. 4*, E. A. Paul and A. D. McLaren, Eds. (New York: Marcel Dekker, Inc., 1975), pp. 195-244.
6. Flaig, W., H. Beutelspacher and E. Rietz. "Chemical Composition and Physical Properties of Humic Substances," In: *Soil Components, Vol. 1*, J. E. Gieseking, Ed. (New York: Springer-Verlag, Inc., 1975), pp. 1-211.
7. Khan, S. U. "The Interaction of Organic Matter with Pesticides," In: *Soil Organic Matter*, M. Schnitzer and S. U. Khan, Eds. (Elsevier Scientific Publishing Co., 1978).

8. Bollag, J.-M., R. D. Sjoblad and S.-Y. Liu. "Characterization of an Enzyme from *Rhizoctonia praticola* which Polymerizes Phenolic Compounds," *Can. J. Microbiol.* 25:229-233 (1979).
9. Sjoblad, R. D., and J.-M. Bollag. "Oxidative Coupling of Aromatic Pesticide Intermediates by a Fungal Phenol Oxidase," *Appl. Environ. Microbiol.* 33:906-910 (1977).
10. Bollag, J.-M., R. D. Sjoblad and R. D. Minard. "Polymerization of Phenolic Intermediates of Pesticides by a Fungal Enzyme," *Experientia* 33:1564-1566 (1977).
11. Bollag, J.-M., S.-Y. Liu and R. D. Minard. "Enzymatic Oligomerization of Vanillic Acid," *Soil Biol. Biochem.* 14:157-163 (1982).
12. Bollag, J.-M., S.-Y. Liu and R. D. Minard. "Cross-Coupling of Phenolic Humus Constituents and 2,4-Dichlorophenol," *Soil Sci. Soc. Am. J.* 44:52-56 (1980).
13. Liu, S.-Y., R. D. Minard and J.-M. Bollag. "Oligomerization of Syringic Acid, a Lignin Derivative, by a Phenoloxidase," *Soil Sci. Soc. Am. J.* 45:1100-1105 (1981).
14. Suflita, J. M., and J.-M. Bollag. "Polymerization of Phenolic Compounds by a Soil-Enzyme Complex," *Soil Sci. Soc. Am. J.* 45:297-302 (1981).
15. Gierer, J., and A. E. Opara. "Studies on the Enzymatic Degradation of Lignin. The Action of Peroxidase and Laccase on Monomeric and Dimeric Model Compounds," *Acta Chem. Scand.* 27:2909-2922 (1973).
16. Wolf, D. C., and J. P. Martin. "Decomposition of Fungal Mycelia and Humic-Type Polymers containing Carbon-14 from Ring and Side-Chain Labeled 2,4-D and Chlorpropham," *Soil Sci. Soc. Am. Proc.* 40:700-704 (1976).
17. Mathur, S. P., and H. V. Morley. "Incorporation of Methoxychlor-^{14}C in Model Humic Acids Prepared from Hydroquinone," *Bull. Environ. Contam. Toxicol.* 20:268-274 (1978).
18. Bollag, J.-M., S.-Y. Liu and R. D. Minard. "Asymmetric Diphenol Formation by a Fungal Laccase," *Appl. Environ. Microbiol.* 38:90-92 (1979).
19. Bartha, R. "Pesticide Interaction Creates Hybrid Residue," *Science* 166:1299-1300 (1969).
20. Kearney, P. C., J. R. Plimmer and F. B. Guardia. "Mixed Chloroazobenzene Formation in Soil," *J. Agr. Food Chem.* 17:1418-1419 (1969).
21. Iwan, J., G.-A. Hoyer, D. Rosenberg and D. Goller. "Transformations of 4-chloro-*o*-toluidine in Soil: Generation of Coupling Products by One-Electron Oxidation," *Pestic. Sci.* 7:621-631 (1976).
22. Willstätter, R., and R. Majima. "Zur Kenntnis der Oxydation von Anilin," *Ber. Dtsch. Chem. Ges.* 1910:2588-2593 (1910).
23. Suida, H., and W. Suida. "Über Anilidochinone," *Liebigs Ann. Chem.* 416:113-163 (1918).
24. Liu, S.-Y., R. D. Minard and J.-M. Bollag. "Coupling Reactions of 2,4-Dichlorophenol with Various Anilines," *J. Agric. Food Chem.* 29:253-257 (1981).
25. Haider, K., L. R. Frederick and W. Flaig. "Reactions Between Amino Acid Compounds and Phenols During Oxidation," *Plant and Soil* 22:49-64 (1965).

26. Ladd, J. N., and J. H. A. Butler. "Comparison of Some Properties of Soil Humic Acids and Synthetic Phenolic Polymers Incorporating Amino Derivatives," *Aust. J. Soil Res.* 4:41-54 (1966).
27. Martin, J. P., K. Haider and E. Bondietti. "Properties of Model Humic Acids Synthesized by Phenoloxidase and Autoxidation of Phenols and Other Compounds Formed by Soil Fungi," In: *Humic Substances, their Structure and Function in the Biosphere,* D. Povoledo and H. L. Golterman, Eds. (Wageningen, The Netherlands: Pudoc, 1975), pp. 171-186.
28. Küster, E. "Humusbildung und Phenoloxydasen bei Streptomyceten," *Z. Pflanzenernähr. Düng. Bodenk.* 69:137-142 (1955).
29. Flaig, W., and H. Riemer. "Polarographische Untersuchungen zum Verhalten von Trihydroxytoluolen bei der Reaktion mit Glycin unter oxydierenden Bedingungen," *Liebigs Ann. Chem.* 747:81-84 (1971).
30. Bondietti, E., J. P. Martin and K. Haider. "Stabilization of Amino Sugar Units in Humic-Type Polymers," *Soil Sci. Soc. Amer. Proc.* 36:597-602 (1972).
31. Katan, J., and E. P. Lichtenstein. "Mechanisms of Production of Soil-Bound Residues of [^{14}C] Parathion by Microorganisms," *J. Agric. Food Chem.* 25:1404-1408 (1977).

CHAPTER 7

BULK CHEMICAL AND SPECTROSCOPIC PROPERTIES OF MARINE AND TERRESTRIAL HUMIC ACIDS, MELANOIDINS AND CATECHOL-BASED SYNTHETIC POLYMERS

John R. Ertel and John I. Hedges

School of Oceanography
University of Washington
Seattle, Washington

ABSTRACT

Marine and terrestrial humic acids differ in elemental, isotopic and spectroscopic properties. To ascertain if these characteristics result from known differences in the biochemical constituents of the indigenous plant material, sedimentary marine and terrestrial humic acids were compared with model humic acids that represent polymerization of major types of biological source material of their respective environments. Melanoidins, made from glucose reacted with either alanine or ammonia, and phenolic humic acids, made from catechol and the aforementioned nitrogen sources, were analyzed for their elemental compositions and their infrared (IR), ultraviolet-visible (UV-visible), fluorescence and electron spin resonance (ESR) spectroscopic properties, together with representative marine and terrestrial humic acids. As compared to their terrestrial counterparts, marine humic acids contain: (1) higher hydrogen and nitrogen concentrations, (2) less highly conjugated or aromatic structural units, and (3) greater levels of slightly altered carbohydrate, protein and pigment residues. The synthetic humic acids have elemental and spectroscopic properties generally similar to the natural geopolymers. The melanoidins more closely resemble the marine humic acids, whereas the synthetic phenolic polymers resemble the terrestrial humic acids. However, from the spectroscopic techniques employed in this study, the phenolic and melanoidin synthetic humic acids display no unique characteristics that would unambiguously indicate their presence in any of our natural humic acids.

INTRODUCTION

Humic acids are a ubiquitous component of soils and lacustrine, nearshore and deep-sea sediments. Although marine and terrestrial humic acids fit the same operational definition, these geopolymers possess some fundamental differences in elemental, isotopic and spectroscopic properties [1,2]. These dissimilarities are thought to arise from differences in the biochemical components found in the indigenous plants that are available for the formation of humic substances in these environments [3]. Lignin, considered a major source of the phenolic building blocks of terrestrial humic acids [4], is found only in vascular plants, which predominate on land but are absent in the open ocean.

Because marine humic acids are predominantly autochthonous [1,2], other biochemical source material and reactive pathways have been proposed to explain their presence in marine sediments. Melanoidins, the dehydrative condensation products of sugars and amino acids [5], have been suggested as a likely model for marine humic acids [3] because of: (1) the similarity of the "Maillard" reaction products to humic acids [6]; (2) the general abundance of the protein and carbohydrate precursors in the marine plants; and (3) the spontaneity of the "browning" reaction under a wide range of natural conditions.

We are presently testing the hypothesis that marine (and possibly terrestrial) humic acids contain a quantitatively significant melanoidin component. Our experimental approach is to compare the chemical and spectroscopic characteristics of representative marine and terrestrial humic acids with each other and with parallel suites of melanoidin and phenolic humic acids synthesized in the laboratory under controlled conditions. The synthetic polymers are produced by reacting either glucose (hypothesized marine pathway) or catechol (hypothesized terrestrial pathway) with either ammonia or the amino acid alanine, which together represent major forms of nitrogenous substances found in natural environments that are apparently involved in humic acid formation.

Although obviously an oversimplification, synthetic humic acids provide a means of recognizing compositional characteristics that are precursor-related and that might be otherwise overlooked. In addition, the effects of reactant ratios, reaction conditions and the extent of reaction on these characteristics can be systematically determined and compared. Extensive and detailed characterizations of synthetic humic acids made with phenol and/or quinone precursors have been published [7,8]. However, comparatively little compositional information has been reported for melanoidins [1,9,10]. Few direct comparisons have been made of melanoidin and phenolic polymers made with the same nitrogenous reactants under identical reaction conditions. As a result it is presently not known by what chemical properties naturally occurring melanoidins might be recognized, if in fact they do exist, and to what extent these properties might depend on their environment of formation or specific biochemical precursors.

This chapter compares the elemental compositions and the IR, UV-visible, fluorescence and ESR spectroscopic properties of seven humic acids from soils or marine sediments with ten synthetic humic acids made from simple biochemicals representing the proposed melanoidin and phenolic pathways.

SAMPLE LOCATIONS

Washington Slope Sediment (WS)

Recent, fine grain sediment was collected with a 20- by 30-cm box core from the Washington continental slope (46° 42.0′ N; 125° 10.3′ W) approximately 78 km offshore. At this water depth (1004 m) an oxygen minimum zone impinges on the slope, resulting in the preservation of high levels [3.5% organic carbon (OC)] of predominantly marine-derived organic matter as shown by lignin and carbon isotopic analyses [11]. The top 20 cm of sediment was homogenized for humic acid extraction.

Gulf of Mexico Sediment (GM)

The surface sediment (top 10 cm) was collected with a Smith-MacIntyre grab sampler from the Texas continental shelf (27° 16.5′ N; 96° 23.12′ W; approximately 87 km south of Baffin Bay, TX) in 128 m of water. Organic matter in sediments in this general area (1% OC) is essentially marine-derived [12].

Saanich Inlet Sediment (SI)

Saanich Inlet is an anoxic fjord of Vancouver Island, B.C., whose organic geochemistry has been extensively studied [13]. The sediments are highly reducing and rich in ammonia and organic material (3.3% OC). The top 20 cm of sediment was collected from the central basin (230 m water depth) with a gravity core and was homogenized before use.

Pacific Pelagic Sediment (PP)

Humic acids were extracted from the top 4 cm of a 1-m^2 box core collected at a water depth of 4400 m from the abyssal plain of the Equatorial Pacific (15° 7.0′ N; 125° 56.0′ W) as part of the NOAA DOMES project (site C). The surface sediment has less than 0.4% organic carbon.

Florida Humate-Cemented Sands (FH)

Dark, organic-rich (4.5% OC) sand was collected at a depth of 20 to 40 cm from the banks of the Intracoastal Waterway near Destin, FL near site "6" described by Swanson and Palacas [14]. The immediate area was devoid of vegetation, although wood fragments were found in the sample. Previous analyses of humic acids from this general area have been done by both Hoering [15] and Hedges [16].

Beaverhills Soil (BH)

Humic substances extracted from the Ah horizon of Beaverhills soil in central Alberta, Canada have been extensively studied by Schnitzer and his colleagues [17]. Our Ah soil sample (4.5% OC) was taken from the 2- to 27-cm depth interval at a site described by Dudas and Pawluk [18]. This sample was homogenized and sieved (1.65 mm) before extraction.

Washington Peat Bog Sediment (WP)

Black, organic-rich sediment (26.3% OC) was collected from a reducing freshwater bog at a water depth of 10 cm. This bog is part of the Steele Lake peat area, near Federal Way, WA. Relatively root-free sediment from the top 4 cm was used.

EXPERIMENTAL

Humic Acid Extraction and Purification

Wet soil or sediment samples were extracted for 24 hr on a platform shaker with 0.1 *N* NaOH and 0.1 *M* $Na_4P_2O_7$ (200 g wet weight/1) under a nitrogen atmosphere. Samples were adjusted to 1 *M* KCl to flocculate clays and centrifuged at 350 *g* for 10 min (or 13,000 *g* for 30 minutes for clay-rich samples) to remove the particulates. The supernatants were combined, filtered through glass fiber filters (Gelman AE) and acidified to pH 1 to precipitate humic acids.

After 12 hr the humic acids were separated from the fulvic acids by centrifugation (13,000 *g* for 30 min). The humic acids were redissolved in 0.1 *N* NaOH by ultrasonication, centrifuged at 21,000 *g* for 1 hr, filtered through glass fiber filters and reprecipitated with 6 *N* HCl. The humic acids were separated by centrifugation (21,000 *g* for 1 hr) and the redissolution/reprecipitation repeated twice, the latter time using 1 *N* NaOH. Finally, the particulate humic acids were twice suspended in distilled water, sonicated and centrifuged (21,000 *g* for 30 min) to remove salts. Humic acids were recovered by freeze-drying.

Melanoidin and Phenolic Humic Acid Synthesis

Reagent grade chemicals were refluxed for 72 hr in 0.1 *M* KH_2PO_4/NaOH buffer (pH 8), according to the molar ratios in Table I. The reaction mixture was aerated as much as was possible without excess foaming. The pH was measured and adjusted to 8 with 10 *N* NaOH every hour for the first 24 hr, after which little pH change occurred. Aliquots were removed at 12, 24 and 48 hr and precipitated with concentrated HCl to check for humic acid formation. After 72 hr all solutions were cooled, filtered through glass fiber filters and acidified to pH 1. All the synthetic humic acids were isolated and purified in exactly the same manner as their natural counterparts.

Instrumental Analysis

The carbon, hydrogen and nitrogen contents of the humic acids were determined in triplicate with a Carlo Erba Model 1106 elemental analyzer. Average reproducibilities (standard deviation as a percentage of the mean) for all humic acids were: 1.0% for carbon, 1.8% for hydrogen and 4.0% for nitrogen. Ash was taken to be the total residue left upon heating at 650°C for 12 hr.

Carbon isotopic compositions were assayed by Coastal Science Laboratories, Port Aransas, TX with a MicroMass Model 600 isotope ratio mass spectrometer according to the procedure of Soffer [19]. The $\delta^{13}C$ values are reported relative to the PDB standard with a precision of 0.2‰. Nitrogen isotopes were analyzed by S. Macko (University of Texas) and are reported relative to atmospheric nitrogen with a precision of 0.3‰.

Infrared spectroscopy was performed with a Perkin Elmer Model 137 spectrophotometer using KBr disks (1.25 wt% humic acid). UV-visible spectra (900 to 200 nm) were recorded on a Beckman Acta II spectrophotometer in 1-cm quartz cells at concentrations of 250 and 25 μg/ml in 0.05 *N* $NaHCO_3$.

Table I. Yields of Synthetic Humic Acids

Synthetic Humic Acids[a]	Moles Reactant[a]				Humic Acid Yield			
						% Reactants		
	G	C	A	N	Total Weight (g)	Total Carbon	Glucose/ Catechol Carbon	Total Nitrogen
G	1.0				5.27	4.50	4.50	
$GA_{1:1}$	0.5		0.5		2.68	2.53	3.80	1.80
$GA_{1:9}$	0.1		0.9		0.31	0.46	2.50	0.18
$GN_{1:1}$	0.5			0.5	2.38	3.27	3.27	2.00
$GN_{1:9}$	0.1			0.9	0.39	3.06	3.06	0.24
C		1.0			0.13	0.10	0.10	
$CA_{1:1}$		0.5	0.5		0.17	0.17	0.26	0.046
$CA_{1:9}$		0.1	0.9		0.12	0.16	0.88	0.056
$CN_{1:1}$		0.5		0.5	0.30	0.47	0.47	0.156
$CN_{1:9}$		0.1		0.9	0.43	3.03	3.03	0.090

[a]Abbreviations: G, glucose; C, catechol; A, alanine; N, ammonia.

Fluorescence measurements were made on a Farrand Mark I spectrofluorometer with a corrected excitation module using 5-nm slit widths. Excitation and emission scans (50 nm/min) were run at humic acid concentrations of 25 μg/ml in 0.05 N $NaHCO_3$ adjusted to a pH of 9.5 with 0.1 N NaOH. Fluorescence intensity was calibrated against a 15 μg/l solution of quinine sulfate in 0.1 N H_2SO_4 that gave an intensity of 0.63 (excitation wavelength = 350; emission wavelength = 450).

The ESR spectra of most of the humic acids were determined on a Varian E-3 spectrometer, with 100 KHz modulation and a nominal operating frequency of 9.5 GHz. Humic acids were run as finely ground powders in quartz tubes; and the g values and spins/g were measured relative to standards of DPPH diluted with KCl (0.02 to 0.002%).

RESULTS AND DISCUSSION

Humic Acid Synthesis

Yields of all the synthetic humic acids were less than 5% of the carbon and 2% of the nitrogen in the reactants (Table I). The melanoidins showed relatively constant percentage conversion of glucose to humic acid carbon despite differences in molar ratios or nitrogen source. Ammonia appears to be as efficient in the formation of melanoidins as amino acids and possibly more efficient for the phenolic polymers. For both melanoidin and phenolic polymers ammonia produced a greater amount of base-insoluble residue.

Catechol yielded significantly less humic acid product than glucose under the reaction conditions employed (see EXPERIMENTAL). This is to be expected since catechol requires oxidation to quinone prior to polymerization [4] whereas melanoidin formation is a nonredox controlled condensation [5]. Much greater yields for the catechol polymers could be realized by the addition of oxidizing agents [7] or clays [20], or by increasing the pH [4].

Humic acids were produced in all the melanoidin reactions within 24 hr and in some cases as early as 12 hr (Figure 1). Elemental analysis of some of these intermediate humic acids indicates that with increased heating time melanoidin humic acids lose oxygen and nitrogen relative to carbon (Figure 1) while H/C ratios remain constant (not shown). Similar changes in elemental composition occur during thermal alteration of marine [21] and terrestrial [22] humic acids and during humic-kerogen diagenesis [2]. Prolonged thermal treatment of melanoidins results in significant conversion to base-insoluble residue [23].

At the termination of the reaction only the GA1:1 polymer appears

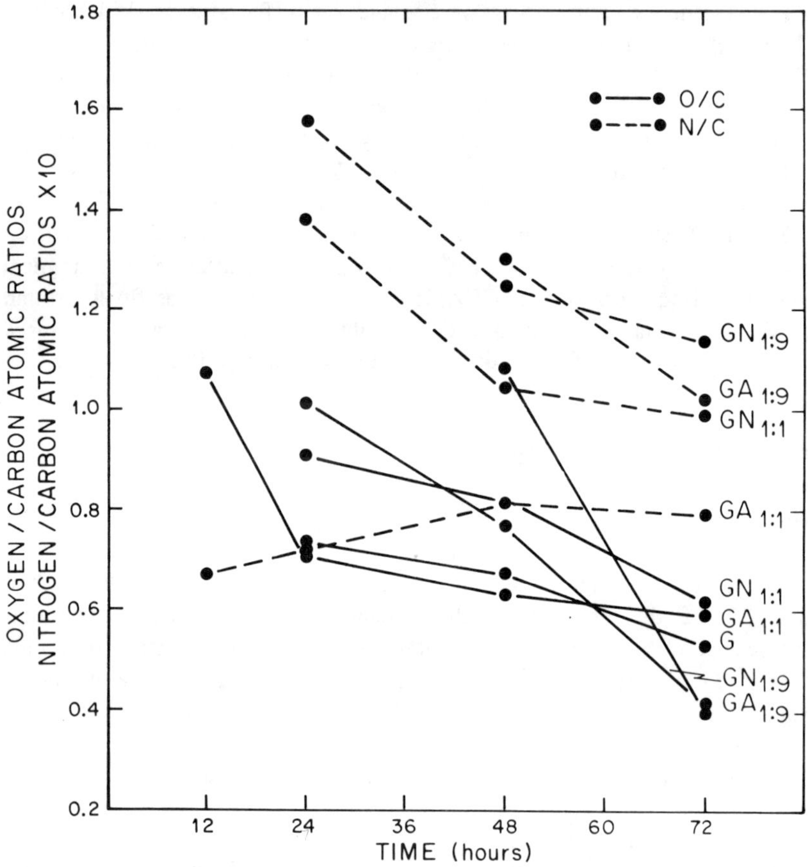

Figure 1. Elemental ratios of glucose-based synthetic humic acids plotted vs length of reaction time. For abbreviations, see Table I.

to have reached a constant composition. The phenolic polymers show the same loss of oxygen and nitrogen, although low overall yields prevent detailed analysis of the intermediate humic acids. Thus melanoidin humic acids encompassing a wide range of elemental compositions can be synthesized from the same biochemicals simply by changing their molar ratios or the duration of reaction.

Isotopic and Elemental Analyses

Humic acids extracted from marine sediments have $\delta^{13}C$ values (Table II) typical of marine plankton [24], suggesting that the bulk carbon is

Table II. Elemental Compositions of Natural Humic Acids

	% Composition				Atomic Ratios		Isotope Data	
Samples[a]	C[b]	N[b]	H[b]	Ash	C/N	H/C	$\delta^{13}C$[c]	$\delta^{15}N$[d]
				Marine Humic Acids				
WS	47.22	4.13	4.77	2.47	13.3	1.21	−21.6	+4.0
GM	54.94	4.53	5.59	1.51	14.1	1.22	−20.6	+4.0
SI	56.15	5.34	6.18	0.08	12.3	1.32	−21.2	+3.7
PP	53.73	5.63	5.93	0.33	11.1	1.32	−20.8	nd
Mean	53.01	4.91	5.62		12.7	1.27	−21.1	+3.9
± s.d.	3.99	0.70	0.61		1.3	0.06	0.4	0.2
				Terrestrial Humic Acids				
FH	57.72	0.36	3.10	2.65	106.4	0.65	−23.0	+4.1
BH	48.46	2.90	3.14	1.34	19.5	0.78	−25.8	+4.7
WP	49.77	3.13	4.63	1.41	18.5	1.12	−28.7	+5.7
Mean	51.98	2.13	3.62		48.1	0.85	−25.8	+4.8
± s.d.	5.01	1.54	0.87		50.4	0.24	2.9	0.8

[a]Abbreviations: WS, Washington slope; GM, Gulf of Mexico; SI, Saanich Inlet; PP, Pacific pelagic; FH, Florida humate; BH, Beaverhills soil; WP, Washington peat.
[b]Corrected for ash.
[c]Relative to PDB.
[d]Relative to atmospheric nitrogen.

autochthonous. The terrestrial humic acids show the characteristic greater depletion in ^{13}C, though covering a wider range of values [24]. The nitrogen isotopic compositions of our marine and terrestrial humic acids are not appreciably different.

Our marine and terrestrial humic acids have C, H and N compositions (Table II) typical of their respective environments with the marine humic acids being enriched in hydrogen and nitrogen [2]. The marine humic acids show no significant C, H and N compositional differences between reducing (SI and WS) and oxidizing (PP and GM) sediments. The only terrestrial humic acid from a reducing environment (WP) has a higher H/C ratio than the soil samples, although it has a typical N/C ratio.

For the melanoidins (Table III), an increase in the mole fraction of nitrogen in the reaction mixture results in a greater incorporation of nitrogen in the humic acid polymer [10]. Also, the higher the proportions of nitrogen in the polymer, the more carbon-rich the humic acid becomes at the expense of oxygen. This oxygen loss could result from nitrogen-catalyzed dehydration of the melanoidin polymer, as has been reported previously by Hodge [5]. As compared to the starting materials, the

Table III. Elemental Compositions of Synthetic Humic Acids

	% Composition				Atomic Ratios		
Samples[a]	C	N	H	O[b]	C/N	H/C	O/C
G	56.26	0.00	4.46	39.28		0.95	0.53
$GA_{1:1}$	51.25	4.69	4.25	39.81	12.7	0.99	0.58
$GA_{1:9}$	57.96	6.90	4.80	30.34	9.8	0.99	0.39
$GN_{1:1}$	49.70	5.79	4.22	40.29	10.0	1.00	0.61
$GN_{1:9}$	57.22	7.59	4.60	30.59	8.8	0.96	0.40
C	55.82	0.00	2.09	42.09		0.45	0.57
$CA_{1:1}$	55.33	1.91	2.32	40.44	33.9	0.50	0.55
$CA_{1:9}$	53.38	5.79	3.12	37.71	10.8	0.70	0.53
$CN_{1:1}$	56.49	3.60	3.17	36.74	18.3	0.67	0.49
$CN_{1:9}$	50.65	2.57	2.24	44.54	22.9	0.53	0.66

[a]Abbreviations: see Table I.
[b]Oxygen by difference.

H/C and O/C ratios of the humic acids are reduced by about one-half, which corresponds to a 2 to 1 loss of hydrogen relative to oxygen. This result is consistent with the condensation and dehydration mechanism known for melanoidin formation [5].

The catechol-based polymers are not as consistent in their elemental compositions as the melanoidins, possibly due to variability in the required oxidative initiation. The H/C ratios are about half that of melanoidins and also of the starting materials. Thus it appears that these synthetic humic acids reflect the greater aromaticity of their starting material. The O/C ratios increase in the phenolic polymers relative to catechol by 1.5 to 2 times, suggesting that atmospheric oxygen is being incorporated into the polymers. This result is also consistent with the observation that phenolic polymer formation is typically an oxidative reaction.

The natural humic acids and the model polymers are within the same range of elemental compositions (Figure 2) with the exception of the higher H/C ratios of the marine and WP humic acids. The high H/C ratios of the natural humic acids could result from incorporation of lipids, as has been shown to occur readily with synthetic melanoidins [23]. However, an extraction of the SI humic acid with benzene/methanol did not appreciably lower its high H/C ratio, indicating that, if present, the lipid material is strongly associated with the humic acid.

Infrared Spectroscopy

The natural and synthetic humic acids yield overall similar IR spectra (Figure 3), with major bands at 3450 cm^{-1} (O–H stretch of phenols

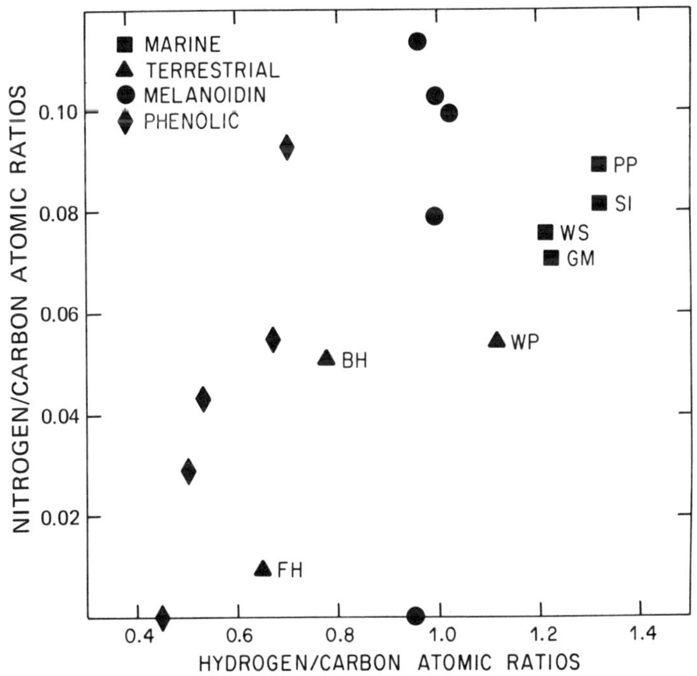

Figure 2. N/C vs H/C atomic ratios of the natural and synthetic humic acids. For abbreviations, see Table II.

or alcohols), 1710 cm^{-1} (C=O stretch of acids, aldehydes and ketones), 1620 cm^{-1} (C=C stretch of unsaturated and aromatic groups) and 1210 cm^{-1} (C–O stretch of ethers, esters and phenols). Similar spectra have previously been reported for melanoidins [9,10,15], melanins [25] and synthetic phenolic humic acids [9]. However, characteristic differences exist among our samples in minor bands and also in the position and shape of the major bands.

Marine humic acids show strong aliphatic C–H stretching bands (2960, 2980 cm^{-1}), indicative of methyl and methylene groups, which could account for their high H/C ratios. Similar absorption bands are present in the WP humic acid, which likewise has a high H/C value. Marine humic acids are further distinguished by bands at 1640 and 1520 cm^{-1}, which are characteristic of the amide linkages of proteins, and also by absorption at 1040 cm^{-1}, which is typical of the C–O stretch of carbohydrates. These results support previous reports that sedimentary humic acids may contain lipid, protein and carbohydrate residues [26].

Terrestrial and phenolic humic acids have broader O–H absorption

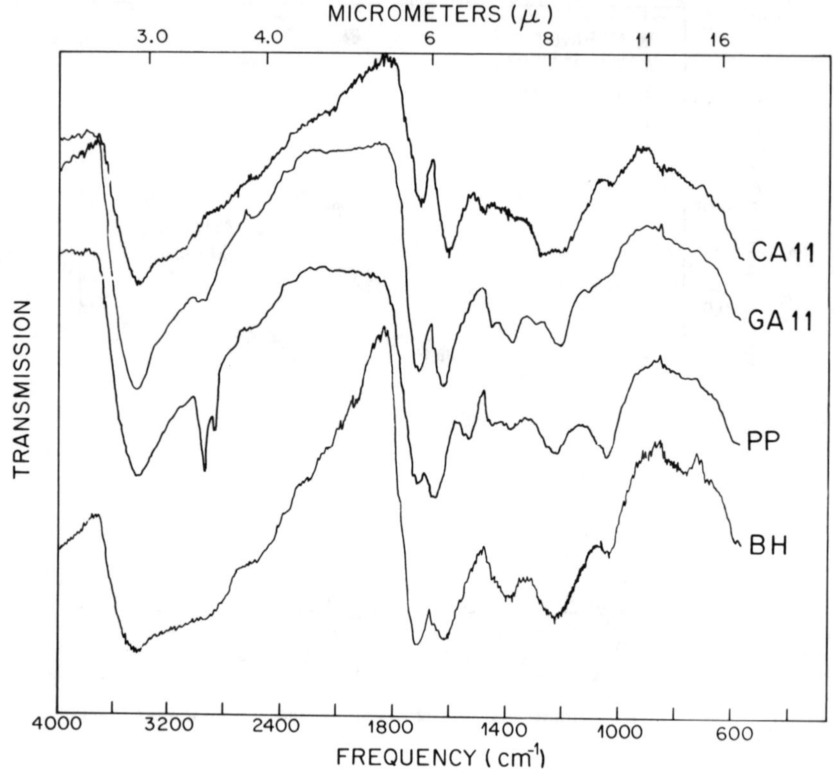

Figure 3. Representative IR spectra of phenolic (CAl:1), melanoidin (GAl:1), marine (PP) and terrestrial (BH) humic acids. For abbreviations see Tables I and II.

bands than the marine or melanoidin humic acids and no distinct aliphatic C–H bands (except WP). Of the synthetic polymers only G and GA humic acids show minor aliphatic absorption bands. For the G humic acid methyl side groups could result from methyl furfural formation and polymerization. The GA polymers could also reflect this source or incorporation of amino acid methyl groups.

UV-Visible Spectroscopy

With the exception of the marine samples, the UV-visible spectra for our humic acids are featureless and monotonically increasing with decreasing wavelength. The marine humic acids show distinct absorption bands at 665 and 405 nm, which are more intense in the humic acids from reducing sediments (SI and WS). Povoledo et al. [26] reported identical absorption bands in lake humic acids that were isolated similarly to our humic acids, and they attributed the peaks to pheophytin pigments. Ishiwatari [27] was able to remove the shoulder at 410 nm of a lake humic acid fraction by extraction with methanol. To determine if these chromophores are bound to the marine humic acids, powdered SI humic acid was exhaustively extracted with a benzene/methanol mixture. Figure 4 indicates that though the extract showed strong absorption at 405 and 665 nm, only a small portion of the total chromophores was removed by extraction. These results suggest that pigments are to a large extent strongly associated with the marine humic acids.

The slope of the absorption curve as measured by the ratio of absorbance at 465 and 665 nm (E4/E6) is greater for the melanoidins than the phenolic humic acids. A plot of E4/E6 vs H/C (Figure 5) yields a strong correlation (r^2 = 0.85) for the synthetic and terrestrial humic acids. The magnitude of the E4/E6 ratio has been determined to be inversely

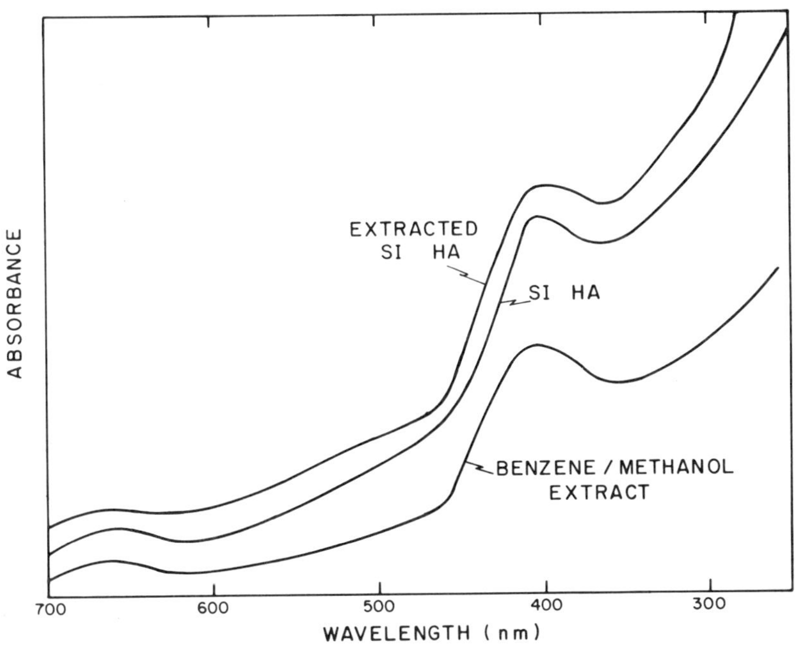

Figure 4. UV-visible spectra of the SI marine humic acid, the benzene/methanol-extracted SI humic acid and the benzene/methanol extract. Concentrations are equivalent for the SI HA and the extracted SI humic acid and are more concentrated by a factor of 10 for the benzene/methanol extract.

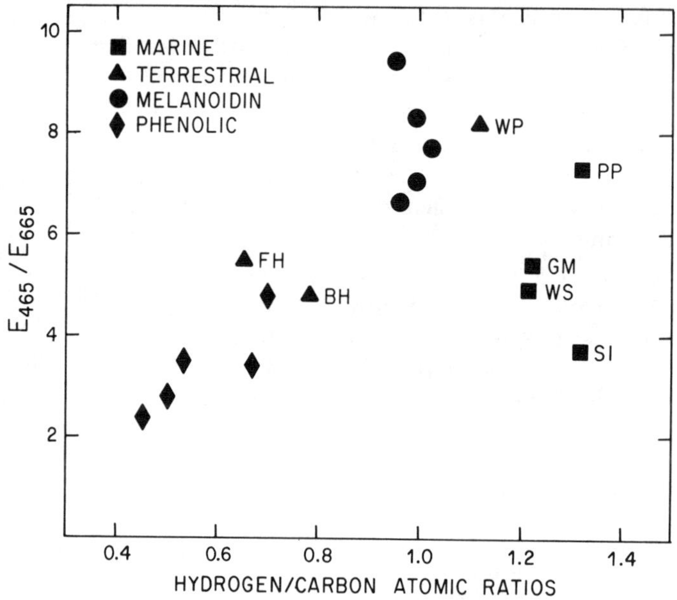

Figure 5. E4/E6 ratios vs H/C atomic ratios for the natural and synthetic humic acids.

related to either the degree of condensation of aromatic groups [28] or the molecular weight [29]. With our synthetic humic acids the differences in the aromaticity based on the starting material and the correlation with H/C both suggest that the major factor affecting the E4/E6 ratio is the degree of chromophore conjugation.

The marine humic acids have low E4/E6 ratios compared to their H/C ratios. Similar results have been obtained by Nissenbaum and Kaplan [1]. The use of E4/E6 to measure the slope implies that no specific absorption bands are present, which is not valid for the marine humic acids. Marine humic acids could have low E4/E6 ratios due to pigment absorption at 665 nm, although differences in molecular weight cannot be ruled out.

Fluorescence Spectroscopy

Fluorescence intensities of the humic acids (Table IV) are low compared to naturally occurring dissolved humic substances [30] and also to initial melanoidin reaction products [31]. Stewart and Wetzel [32] have shown that high-molecular-weight humic substances absorb strongly in the UV but fluoresce weakly. Although the fluorescence intensity of the humic acids examined here differ by a factor of 20, no relationship is apparent with regard to source or other polymer characteristics.

The wavelengths of excitation and emission maxima have been used to characterize aquatic humic substances [30]. Excitation maxima fall in the range of 325 to 427 nm and emission maxima are generally in the violet (450 nm) to green (520 nm) portion of the visible spectrum [30]. A plot of the excitation wavelength maxima vs the emission wavelength maxima is presented in Figure 6. Regardless of the source, most natural and synthetic humic acids have a blue fluorescence maxima (490 nm). Exceptions to this pattern are the humic acids from reducing sediments (SI, WS and WP) and the glucose polymer whose emission maxima are shifted to lower wavelengths (460 to 470 nm). The excitation maxima of the marine humic acids from oxidizing sediments (PP and GM) occur at the lowest wavelength of the samples measured and at the excitation maxima reported for Baltic Sea humic substances [30]. Since corrected excitation fluorescence spectra are equivalent to absorbance spectra [33], the low excitation wavelength maxima in the marine humic

Table IV. Spectroscopic Data for Natural and Synthetic Humic Acids

	UV-VIS Data		Fluorescence Data				
Samples[a]	E_4/E_6[b]	Abs. peaks	Fl. int.[c]	Exc. λ_{max}	Emi. λ_{max}	Exc. peaks	Emi. peaks
WS	4.91	405;665	0.54	365	465	395;405	405
GM	5.40	405;665	1.25	325	490	395;405	365
SI	3.68	405;665	0.47	350	465	405	395
PP	7.39	405	3.27	325	485	405	365
FH	5.47		2.26	350;405	490		510
BH	4.79		1.53	350;400	490		
WP	8.22		0.75	350	460	400	405
G	9.35		0.71	365	470	395	430
$GA_{1:1}$	8.27		0.98	400	490		
$GA_{1:9}$	7.06		1.24	400	490		
$GN_{1:1}$	7.66		0.54	400	490		
$GN_{1:9}$	6.63		nd	nd	nd		
C	2.43		0.19	385	485		
$CA_{1:1}$	2.76		0.23	390	490		
$CA_{1:9}$	4.77		0.62	385	490		
$CN_{1:1}$	3.39		0.23	395	490		
$CN_{1:9}$	3.48		0.39	385	490		

[a]Abbreviations: see Tables I and II.
[b]Absorbance at 465 nm divided by absorbance at 665 nm.
[c]Fluorescence intensity of a 25 μg/ml solution of humic acid relative to a 15 μg/1 quinine sulfate standard which yields an intensity of 0.63.

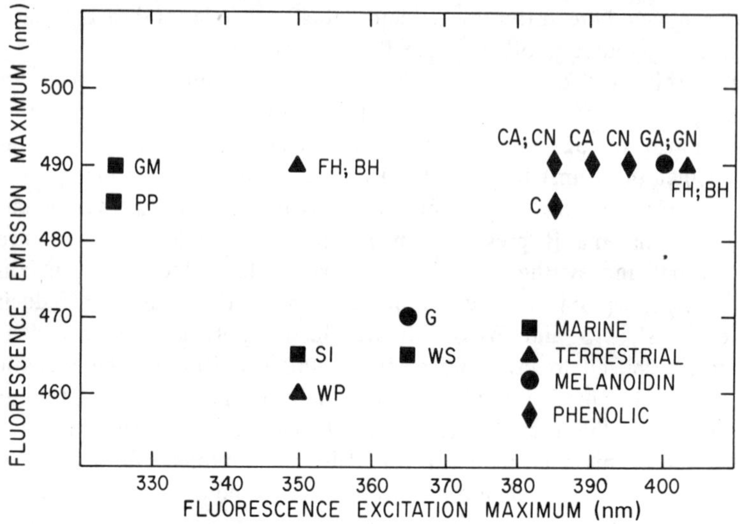

Figure 6. Fluorescence emission wavelength maximum vs excitation wavelength maximum for the natural and synthetic humic acids. FH and BH are presented twice due to dual maxima. For abbreviations, see Tables I and II.

acids is another indication of the lower levels of conjugated chromophores in these polymers.

The soil humic acids produce broad excitation maxima, with distinct peaks at 350 and 400 nm. Previous reports [34] of fluorescence of humic and fulvic acids indicated an excitation maxima around 465 nm when the emission wavelength was fixed at 520 nm. Examination of the BH humic acid at different excitation wavelengths (Table V) reveals a shift to higher emission wavelength maxima with increasing excitation wavelength such that the above excitation/emission wavelength relationship would be attained, though not at the maximum fluorescence intensity.

With the exception of the glucose (G) humic acid, melanoidin and phenolic humic acids possess similar featureless spectra, with excitation maxima in the 385 to 400 nm range. All natural humic acids also show excitation peaks within this range, although not always at the maximum excitation wavelength. To determine if the excitation/emission absorption bands are due to relatively nonpolar components associated with the humic acids, the benzene/methanol-extracted SI humic acid was also analyzed by fluorescence spectroscopy. No differences were observed in the fluorescence intensity, wavelength maxima or absorption peaks between the extracted and nonextracted humic acid even though this

Table V. Emission Maxima and Fluorescence Intensity at Different Excitation Wavelengths for BH Humic Acid

Excitation Wavelength	Emission λ Maxima	Fluorescence Intensity
325	495	1.23
350	490	1.29
375	490	1.17
400	490	1.29
425	500	1.20
450	515	0.90
475	525	0.90

sample is particularly rich in pigments. These results, together with the UV-visible analysis, indicate that the spectroscopic characteristics of the humic acids are not caused by weakly associated, nonpolar constituents.

ESR Spectroscopy

Electron spin resonance has been used extensively to determine the concentration and character of organic free radicals in marine [2] and terrestrial [35] humic substances, in biological polymers [36], and in synthetic humic acid analogs [8]. Results of the ESR measurements for most of our humic acids are presented in Table VI.

Spin concentrations and *g* values of all the humic acids are within the range of reported values [2,35] and show no correlation with elemental composition or any spectroscopic data. Line widths are larger for the melanoidin and marine humic acids than for the terrestrial or catechol-based polymers. The ESR spectra for some terrestrial and synthetic humic acids (Table VI) have shoulders indicative of more than one type of free radical structure [36]. In general the melanoidin and phenolic humic acids possess similar ESR characteristics which also closely resemble those of the natural humic acids.

CONCLUSION

Among the samples examined here, marine humic acids could generally be distinguished from terrestrial humic substances by elemental and spectroscopic properties. Compared to the terrestrial humic acids, the marine humic acids have: (1) higher N/C and H/C atomic ratios, (2) higher $^{13}C/^{12}C$ ratios, (3) lower excitation wavelength maxima (from

Table VI. ERS Data for Natural and Synthetic Humic Acids

Samples[a]	Line Width (gauss)	g Values	Spins/g $\times 10^{-17}$	Peak[b] Shape
WS	5.30	2.0028	2.30	
GM	4.85	2.0038	2.59	
SI	5.30	2.0032	1.94	
FH	3.80	2.0034	1.67	sh
BH	4.05	2.0034	4.71	
WP	3.85	2.0038	0.38	sh
G	4.85	2.0034	0.30	sh
$GA_{1:1}$	6.25	2.0038	1.77	
$GA_{1:9}$	6.20	2.0038	4.10	
$GN_{1:1}$	6.15	2.0038	2.12	
$GN_{1:9}$	6.10	2.0035	5.41	
$CN_{1:1}$	4.20	2.0034	12.83	sh
$CN_{1:9}$	3.85	2.0039	4.44	sh

[a]Abbreviations: see Tables I and II.
[b]A shoulder on derivative peak.

oxidizing sediments), (4) greater ESR line widths and (5) a higher abundance of recognizable biochemical constituents, as evidenced by specific absorption bands in IR and UV-visible spectra. Other measured characteristics such as nitrogen isotopic composition, fluorescence intensity, spin concentration and g values showed no significant source-dependent differences.

The biochemical components of marine humic acids were most prevalent in samples from reducing sediments where microbial activity is known to be relatively slow. Protein and carbohydrate residues are not likely to remain in the humic acid fraction after repeated dissolution/precipitation cycles unless they are partially altered or strongly associated with the geopolymers. Also, the lipid and pigment fragments found in the marine humic acids are not readily extracted with relatively nonpolar solvents and thus, likewise appear to be an integral part of the humic acid isolate. In general, by spectroscopic analysis marine humic acids from both oxidizing and reducing sediments appear to be more structurally similar to biological material than do the terrestrial humic acids. Included in the marine humic acid suite is a sample from the deep-sea (PP) which is potentially the oldest humic acid studied. Thus, preservation of the partially altered biochemical component of humic acids could be related to the extent of microbial utilization, which is less in reducing than oxidizing sediments and possibly in marine than terrestrial environments.

Melanoidin and phenolic humic acids could be synthesized readily from simple biochemicals, and have elemental compositions overlapping those of the natural polymers. Changes in atomic ratios in the course of their synthesis indicate that melanoidin formation occurs via dehydration, whereas phenolic polymerization requires oxidation. The consistently lower H/C ratios of the phenolic polymers as compared to the melanoidins appears to reflect the aromaticity of the phenolic precursor, although H/C ratios decrease by one-half during both melanoidin and phenolic formation. The greater aromaticity of the phenolic humic acids is also indicated by their lower E4/E6 ratios as compared to the melanoidins. Ammonia and amino acid are equivalent as reactants with regard to both obtained yields of humic acids and measured properties of the final polymers. This result is particularly significant for reducing marine environments where ammonia in pore waters typically has higher concentrations and lower turnover rates than amino acids [37].

The synthetic humic acids have generally the same spectroscopic properties as their natural counterparts. Similar to the marine humic acids, melanoidins display narrow O–H stretch bands but they necessarily lack the specific absorption bands or spectroscopic features indicative of intact biochemical units. With the spectroscopic techniques employed here, neither the melanoidin nor phenolic humic acids possess unique spectroscopic characteristics that alone could demonstrate their presence as components of either our marine or terrestrial humic acids. Further characterization of these natural and synthetic humic acids by their pyrolytic and oxidative degradation products is currently in progress; this should more critically evaluate the melanoidin hypothesis.

ACKNOWLEDGMENTS

Support for this research was provided by National Science Foundation grants OCE 78-18260 and OCE 80-23970. We thank R. Clark, National Marine Fisheries and R. Carpenter, University of Washington, for use of analytical equipment, and J. Weikel and R. Citterman for assistance in the humic acid extractions. This is contribution 1278 from the School of Oceanography, University of Washington.

REFERENCES

1. Nissenbaum, A., and I. R. Kaplan. "Chemical and Isotopic Evidence for the In Situ Origin of Marine Humic Substances," *Limnol. Oceanogr.* 17:570-582 (1970).

2. Stuermer, D. H., K. E. Peters and I. R. Kaplan. "Source Indicators of Humic Substances and Proto-Kerogen: Stable Isotope Ratios, Elemental Compositions and Electron Spin Resonance," *Geochim. Cosmochim. Acta* 42:989-997 (1978).
3. Nissenbaum, A. "The Organic Geochemistry of Marine and Terrestrial Humic Substances," in *Advances in Organic Geochemistry 1973*, F. Biernner and B. Tissot, Eds. (Paris: Editions Technip, 1974), pp. 39-52.
4. Flaig, W., H. Beutelspacher and E. Rietz. "Chemical Composition and Physical Properties of Humic Substances," in *Soil Components Volume 1*, J. E. Gieseking, Ed. (New York: Springer-Verlag, 1975), p. 1.
5. Hodge, J. E. "Chemistry of Browning Reactions in Model Systems," *Agric. Food Chem.* 1:928-943 (1953).
6. Enders, C. and K. Theis. "Die Melanoidine and ihre Beziehung zu den Huminsauren," *Brennst.-Chem.* 19:402-407 (1938).
7. Ladd, J. N. and J. H. A. Butler. "Comparison of Some Properties of Soil Humic Acids and Synthetic Phenolic Polymers Incorporating Amino Derivatives," *Aust. J. Soil Res.* 4:41-54 (1966).
8. Mathur, S. P. and M. Schnitzer. "A Chemical and Spectroscopic Characterization of Some Synthetic Analogues of Humic Acids," *Soil Sci. Soc. Am. J.* 42:591-596 (1978).
9. Goh, K. M. and F. J. Stevenson. "Comparison of Infrared Spectra of Synthetic and Natural Humic and Fulvic Acids," *Soil Sci.* 112:392-400 (1971).
10. Hedges, J. I. "The Formation and Clay Mineral Reactions of Melanoidins," *Geochim. Cosmochim. Acta* 42:69-76 (1978).
11. Hedges, J. I. and D. C. Mann. "The Lignin Geochemistry of Marine Sediments from the Southern Washington Coast," *Geochim. Cosmochim. Acta* 43:1809-1818 (1979).
12. Hedges, J. I. and P. L. Parker. "Land-Derived Organic Matter in Surface Sediments from the Gulf of Mexico," *Geochim. Cosmochim. Acta* 40:1019-1029 (1976).
13. Brown, F. S., M. J. Baedecker, A. Nissenbaum and I. R. Kaplan. "Early Diagensis in a Reducing Fjord, Saanich Inlet, B.C.–III Changes in Organic Constituents of Sediments," *Geochim. Cosmochim. Acta* 36:1185-1204 (1972).
14. Swanson, V. E. and J. G. Palacas. "Humate in Coastal Sands of Northwest Florida," *Geological Survey Bulletin* 1214-B (Washington, DC: U.S. Govt. Printing Office, 1965).
15. Hoering, T. C. "A Comparison of Melanoidin and Humic Acid," *Carnegie Inst. Washington Yearb.* 72:682-690 (1973).
16. Hedges, J. I. "Lignin Compounds as Indicators of Terrestrial Organic Matter in Marine Sediments," PhD Thesis, University of Texas, Austin (1975).
17. Schnitzer, M. and S. U. Khan. *Humic Substances in the Environment* (New York: Marcel Dekker, Inc., 1972).
18. Dudas, M. and S. Pawluk. "Chernozem Soils of the Alberta Parklands," *Geoderma* 3:19-36 (1970).
19. Soffer, Z. "Preparation of Carbon Dioxide for Stable Carbon Isotope Analysis of Petroleum Fractions," *Anal. Chem.* 52:1385-1388 (1980).
20. Wang, T. S., S. W. Li and Y. L. Ferng. "Catalytic Polymerization of Phenolic Compounds by Clay Minerals," *Soil Science* 126:15-21 (1978).

21. Ishiwatari, R., M. Ishiwatari, B. G. Rohrback and I. R. Kaplan. "Thermal Alteration Experiments on Organic Matter from Recent Marine Sediments in Relation to Petroleum Genesis," *Geochim. Cosmochim. Acta* 41:815-828 (1977).
22. Schnitzer, M. and S. I. M. Skinner. "Free Radicals in Soil Humic Compounds," *Soil Sci.* 108:383-390 (1969).
23. Larter, S. R. and A. G. Douglas. "Melanoidins–Kerogen Precursors and Geochemical Lipid Sinks: A Study Using Pyrolysis Gas Chromatography (PGG)," *Geochim. Cosmochim. Acta* 44:2087-2095 (1980).
24. Degens, E. T. "Biogeochemistry of Stable Carbon Isotopes," in *Organic Geochemistry*, G. Eglinton and M. T. J. Murphy, Eds. (Berlin: Springer, 1969), p. 304.
25. Filip, Z., K. Haider, H. Beutelspacher and J. P. Martin. "Comparison of IR-Spectra from Melanins of Microscopic Soil Fungi, Humic Acids and Model Phenol Polymers," *Geoderma* 11:37-52 (1974).
26. Povoledo, D., D. Murray and M. Pitze. "Pigments and Lipids in the Humic Acids of Some Canadian Lake Sediments," in *Humic Substances–Their Structure and Function in the Biosphere*, D. Povoledo and H. L. Golterman, Eds. (Wageningen, The Netherlands: Pudoc, 1972), p. 233.
27. Ishiwatari, R. "Chemical Characterization of Fractionated Humic Acids from Lake and Marine Sediments," *Chem. Geol.* 12:113-126 (1973).
28. Kononova, M. M. *Soil Organic Matter* (London: Pergamon Press, Inc., 1966), p. 400.
29. Chen, Y., N. Senesi and M. Schnitzer. "Information Provided on Humic Substances by E4/E6 Ratios," *Soil Sci. Soc. Am. J.* 41:352-358 (1977).
30. Larson, R. A. and A. L. Rockwell. "Fluorescence Spectra of Water-Soluble Humic Materials and Some Potential Precursors," *Arch. Hydrobiol.* 86:416-425 (1980).
31. Adhikari, H. R. and A. L. Tappel. "Fluorescent Products in a Glucose-Glycine Browning Reaction," *J. Food Sci.* 38:486-488 (1973).
32. Stewart, A. J. and R. G. Wetzel. "Fluorescence: Absorbance Ratios–A Molecular-Weight Tracer of Dissolved Organic Matter," *Limnol. Oceanogr.* 25:559-564 (1980).
33. Porro, T. J., R. E. Anacreon, P. S. Flandreau and I. S. Fagerson. "Corrected Fluorescence Spectra of Polynuclear Materials and Their Principal Applications," *J. AOAC* 56:607-620 (1973).
34. Ghosh, K. and M. Schnitzer. "Fluorescence Excitation Spectra of Humic Substances," *Can. J. Soil Sci.* 60:373-379 (1980).
35. Riffaldi, R. and M. Schnitzer. "Electron Spin Resonance Spectroscopy of Humic Substances," *Soil Sci. Amer. Proc.* 36:301-305 (1972).
36. Steelink, C. and G. Tollin. "Free Radicals in Soil," in *Soil Biochemistry*, A. D. McLaren and G. H. Peterson, Eds. (New York: Marcel Dekker, Inc., 1967), p. 287.
37. Hendrichs, S. "Biogeochemistry of Dissolved Free Amino Acids in Marine Sediments," PhD Thesis, Woods Hole Oceanographic Institution, (1980).

CHAPTER 8

USE OF ULTRAVIOLET SPECTROSCOPY, GEL FILTRATION, PYROLYSIS/MASS SPECTROMETRY AND NUMBERS OF BENZOATE-METABOLIZING BACTERIA IN THE STUDY OF HUMIFICATION AND DEGRADATION OF AQUATIC ORGANIC MATTER

H. De Haan

Limnological Institute
Tjeukemeer Laboratory
Oosterzee, The Netherlands

ABSTRACT

This chapter describes the results of UV spectroscopy, Sephadex gel filtration and Curie-point pyrolysis/mass spectrometry of fulvic acids from some Dutch and Finnish lakes. It appears that differences in the color and the trophic level of the aquatic systems studied are reflected in the ratio of the UV absorbances at 250 and 365 nm (E_2/E_3), the molecular weight distribution and the composition of their fulvic acids. Thus fulvic acids from strongly humified and oligotrophic waters are characterized by relatively low E_2/E_3 ratios (~4.0), a relatively high molecular weight (>5000) and a high content of phenolic polymers.

Benzoate-metabolizing bacteria were found to partly degrade fulvic acids. This process, performed with fulvic acids and bacteria isolated from either Dutch or Finnish water, involves an increase in E_2/E_3 ratio and a decrease in the mean molecular weight of the fulvic acids. These parameters enabled us to establish the relative stage of humification or mineralization of the organic matter in the lakes by simple and rapid UV spectroscopy.

The proportion of benzoate-metabolizing bacteria within the total heterotrophic flora seems to depend on the concentration of fulvic acids and the trophic level of the aquatic system studied. Therefore, the proportion of these bacteria may be indicative of the relative rate of degradation of fulvic acids in different waters.

INTRODUCTION

It has long been realized that humic substances determine to a large extent the limnological features of colored aquatic ecosystems [1]. However, despite many efforts, our knowledge of the structure, properties and role of humic substances in natural waters is still only fragmentary. We do know that humic substances are generally acidic [2], attenuate light [3-5], chelate metals and interact with simple organic molecules [6]. Since the latter two properties involve hazardous heavy metals and herbicides, they have environmental implications and have attracted many scientists. Thus our knowledge of the function of humic substances in lakes has increased in the last few decades, but the underlying chemical mechanisms are still almost unknown. This paucity of knowledge is mainly due to their complex chemical nature, which seems resistant to most analytical techniques.

It is therefore not surprising that so far the chemical characterization of humus formation and degradation in lakes could only be done by rather nonspecific and rough estimates. In fact we still lack methods to establish whether humus formation or degradation dominates in brown-colored lakes. Probably, modern techniques such as Curie-point pyrolysis/mass spectrometry [7,8] and ^{13}C-NMR spectroscopy can help elucidate the role of humic substances in the C cycle and give more insight into their structure.

In 1968 we started a study on the role of humic substances, i.e., fulvic acids, in the C cycle of the humic and eutrophic Lake Tjeukemeer, The Netherlands. During this study Sephadex gel filtration, UV spectroscopy and Curie-point pyrolysis/mass spectrometry were used to characterize fulvic acids. Furthermore, we investigated the resistance of fulvic acids to microbes enriched on simple organic substrates which, as far as known, structurally resembled fulvic acids. This study was extended to oligotrophic humic Finnish lakes. In this chapter the main results are summarized, and some generalizations on the application of the methods used in this study of humification of lakes are attempted.

SEPHADEX GEL FILTRATION AND UV SPECTROSCOPY

So far, Sephadex gel filtration has been the most successful and applied fractionation technique of humic substances from freshwater [9-14]. It appeared from precipitation and solubilization experiments at acidic and basic pH that the brown organic matter in Tjeukemeer consists almost entirely of fulvic acids. Sephadex G-25 gel filtration at pH 7.0 in a phosphate buffer (I = 0.02) of a freeze-dried sample of filtered (<0.2 μm) lake water results in three main fulvic acid fractions (I, II and III, Figure 1). All three fractions absorb UV light, fluoresce and contain phenolic groups, amino acids and carbohydrates [15,16]. However, the ratios of absorbances at 250 and 365 nm, phenolic groups and fluorescence of the fulvic acid fractions I, II and III appear to differ (Table I). As the elution volume of the fractions increases, i.e., their molecular weight decreases, they fluoresce relatively more and contain more phenolic groups. The ratio of the UV absorbances at 250 and 365 nm of fulvic acid fractions I and II increases with decreasing molecular weight, whereas that of fraction III, having the lowest molecular weight, decreases again.

In principle the three ratios listed in Table I all can be used to determine relative concentrations of fulvic acid fractions I, II and III, i.e., the molecular weight distribution in a fulvic acid mixture. Since the ratio of the UV absorbances at 250 and 365 nm (E_2/E_3) is the easiest to measure,

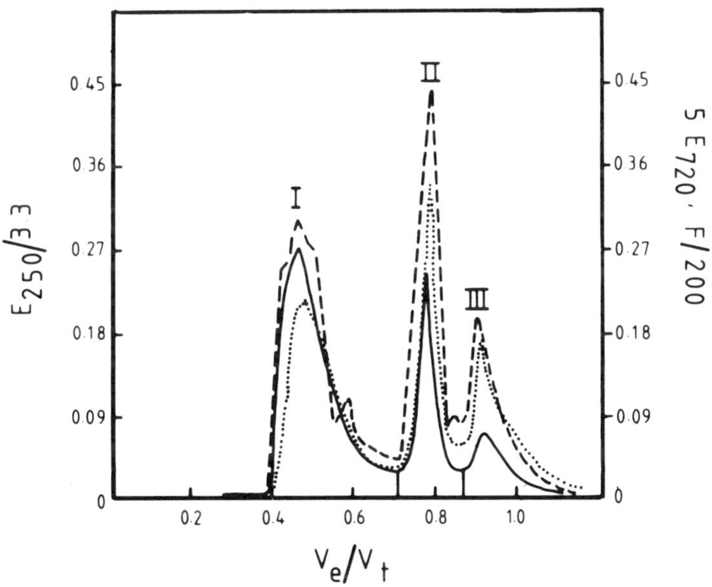

Figure 1. Sephadex G-25 gel filtration in phosphate buffer (pH = 7.0; I = 0.02) of fulvic acids from Tjeukemeer. Fulvic acids in the effluent were measured by absorbance at 250 (E_{250}) and fluorescence (λ_{ex} = 365 nm; λ_{em} = 470 nm). Phenolic groups (E_{720}) were determined with the Folin-Ciocalteu reagent. E_{250} (———); E_{720} (-----); F (.....); V_e, elution volume; V_t, total gel volume.

Table I. Ratios of Absorbances, Phenolic Groups and Fluorescence of Fulvic Acid Fractions I, II and III from Tjeukemeer[a]

Fraction	E_2/E_3	E_7/E_2	F/E_2
I	5.69	0.086	0.016
II	6.36	0.136	0.045
III	4.96	0.181	0.066

[a]E_2/E_3, absorbances at 250 and 365 nm; E_7/E_2, phenolic groups and absorbance at 250 nm; F/E_2, fluorescence and absorbance at 250 nm.

we choose this parameter for further study. Measurements during several years revealed that the E_2/E_3 ratio in Tjeukemeer varies seasonally between 5.5 and 8.5 (Figure 2). The lowest E_2/E_3 ratio is always found during autumn and winter when the water surplus in the peaty polders surrounding the lake leads to an influx of fulvic acid-rich drainage water. This influx of drainage water is reflected in high fulvic acid concentrations as measured as absorbance at 365 nm (E_{365}) in the lake during this period. When evaporation in Tjeukemeer and its drainage area exceeds rainfall, usually in summer, transparent water from IJsselmeer, which is fed by a branch of the river Rhine, is transported to Tjeukemeer. Consequently, the fulvic acid concentration decreases and the E_2/E_3 ratio increases (Figure 2). As expected, the results of monthly performed Sephadex

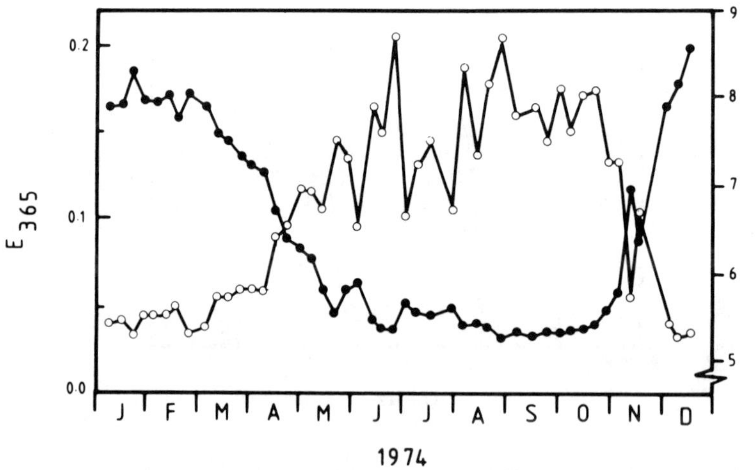

Figure 2. Fulvic acid concentration (E_{365}) and ratio of absorbances at 250 and 365 nm (E_2/E_3) of Tjeukemeer water during 1974. E_{365} (•); E_2/E_3 (0).

G-25 filtrations showed that in winter Tjeukemeer water contains relatively more (~60%) high-molecular-weight fulvic acid fraction I than in summer (~45%) (Table II). Thus relatively high colored water is characterized by low E_2/E_3 ratios and a high proportion of high-molecular-weight fulvic acid. Therefore the E_2/E_3 ratio reflects the humification of Tjeukemeer water. On the contrary, assuming that fulvic acid degradation coincides with a decrease in molecular weight, an increasing E_2/E_3 ratio indicates that degradation occurs. Thus the E_2/E_3 ratio can be used as a predictor of humification in Tjeukemeer.

To see whether the E_2/E_3 ratio and the molecular weight distribution of fulvic acids can be generally used to identify the stage of humification of freshwaters, some Finnish lakes were investigated too. For these investigations we applied sequential Sephadex G-50 and G-25 gel filtration in order to achieve a better resolution of especially high-molecular-weight materials. The Sephadex elution patterns of the concentrated samples from all water types, as listed in Table III, showed the same four fractions I, II, III and IV as demonstrated in Figure 3 for two of the waters studied. Besides fraction I, having lignin-like properties,

Table II. Relative Contents of Fulvic Acid Fractions I, II and III and Ratios of Absorbances at 250 and 365 nm of Tjeukemeer Water During 1974

Date	% I	% II	% III	E_2/E_3
29-1	68	19	13	5.52
25-2	63	24	13	5.33
25-3	64	24	12	5.75
23-4	59	28	13	6.53
21-5	48	28	24	7.49
18-6	50	27	24	7.62
16-7	51	24	24	7.49
13-8	44	24	32	7.35
24-9	41	27	32	7.51

Table III. Relative Contents of Brown Organic Matter Fractions I, II + III and IV and Ratios of Absorbances at 250 and 365 nm of Some Finnish Waters

Water	% I	% II + III	% IV	E_2/E_3
Unpolluted Lake Saimaa	9	65	26	5.98
Polluted Lake Saimaa	16	67	17	5.09
Paper industry effluent	18	65	17	5.26
Lake Saanjärvi	8	47	45	7.50
Pond Kaakonlampi	5	76	19	4.34

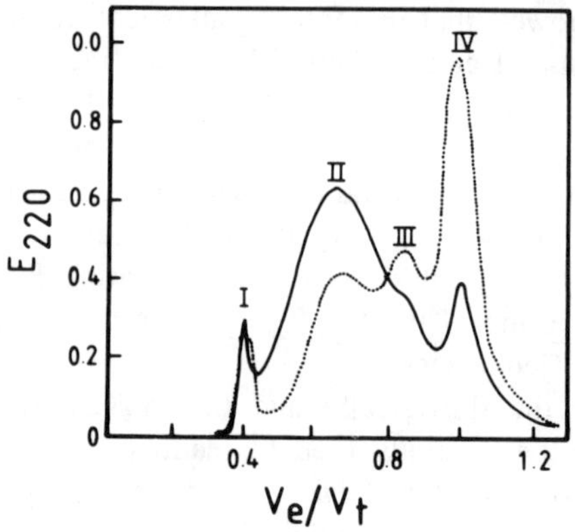

Figure 3. Sequential Sephadex G-50 and G-25 gel filtration in phosphate buffer (pH = 7.0; I = 0.02) of brown organic matter from Saanjärvi (..........) and Kaakonlampi (———). Effluent was monitored at 220 nm. V_e, elution volume; V_t, total gel volume.

the other fractions are spectroscopically very similar to the fulvic acid fractions from Tjeukemeer. The molecular weight distribution of the four fractions in the different lakes appear to vary considerably (Table III). Thus Kaakonlampi, a very brown-colored pond (~200 mg Pt/l), contains almost twice as much high-molecular-weight fulvic acids as oligotrophic and clear (~10 mg Pt/l) Lake Saanjärvi. The high mean molecular weight of the fulvic acids in Kaakonlampi, as in Tjeukemeer in winter, is reflected in a low E_2/E_3 ratio (Table III). On the other hand, the E_2/E_3 ratio of the hardly humified water from Lake Saanjärvi is rather high. Obviously, the molecular weight distribution and the E_2/E_3 ratio reflect the humification of Finnish lakes also, suggesting a general applicability of this parameter as a measure of humification of freshwaters.

The Sephadex elution pattern of fulvic acids from slightly colored (~35 mg Pt/l) Lake Saimaa water consists of the same four fractions as those of pond Kaakonlampi and Lake Saanjärvi. The relatively high proportion of high-molecular-weight fulvic acids and the corresponding low E_2/E_3 ratio are in between those of the bog pond Kaakonlampi and clear Lake Saanjärvi (Table III). Therefore the humification increases in the series Saanjärvi, Saimaa and Kaakonlampi.

Remarkably, Sephadex gel filtration revealed that the organic matter in paper and pulp mill effluents, as discharged into Lake Saimaa, consists of the same four fractions as found in the Finnish lakes studied (Figure 3 and Table III). Apparently, the wood processing industry yields by-products resembling the products of natural humification, indicating that the birch and fir woods are a major source of the fulvic acids in Finnish lakes. The high proportion of fraction I in the paper and pulp mill effluents is characteristic (Table III). Since this fraction has lignin-like properties and lignin is a major constituent of wood, this is not surprising. The relatively high proportion of high-molecular-weight fulvic acids in the waste water is reflected in the low E_2/E_3 ratio. As expected, discharge of this wastewater into Lake Saimaa lowers the E_2/E_3 ratio (Table III).

In soil science the ratio of absorbances at 465 and 665 nm (E_4/E_6) is used for the characterization of humic substances. In general, progressive humification is indicated by decreasing E_4/E_6 ratios [17]. However, owing to the relatively low fulvic acid concentrations, filtered humic waters hardly absorb at these wavelengths. Therefore the E_4/E_6 ratio is not applicable in limnology. But our studies show that the E_2/E_3 ratio, instead of the E_4/E_6 ratio is an effective index of humification of lakes.

CURIE-POINT PYROLYSIS/MASS SPECTROMETRY

Though, as discussed earlier, Sephadex gel filtration and UV spectroscopy can be used to indicate the stage of humification, these techniques do not give much insight into the chemistry of the humification process. To get a better chemical understanding of the humus dynamics in natural waters, degradation techniques giving specific products are required. Preferably, such a technique should be quantitative, making intersample comparison possible. Curie-point pyrolysis/mass spectrometry, as recently developed by Meuzelaar et al. [18-21] to characterize complex synthetic and natural polymers, might fulfill these conditions. The essentials of this technique are a very rapid pyrolysis (1 sec) in vacuum at 510 °C and on-line analysis of the pyrolyzate by low-voltage (14 eV) mass spectrometry through a mass range of m/z 15 to 160. The low ionizing voltage should minimize further fragmentation, thus avoiding the formation of less characteristic fragments in the lower mass range. All successive mass scans of a particular pyrolyzate are summed, giving the ultimate pyrolysis mass spectrum (pyrogram) of that sample. Pyrograms of fulvic acid fractions I, II and III from Tjeukemeer appear to be highly reproducible and characteristic [7,8] (Figure 4) and indicate the presence of phenolic polymers, polysaccharides and proteinaceous material. Compared with soil fulvic acids, they seem to be richer in aliphatic components.

Further humification or degradation of fulvic acids may coincide with small structural changes. Therefore, close comparison of pyrograms of related fulvic acid samples, which generally differ only subtly in the relative intensities of the various mass peaks, is required. For this purpose

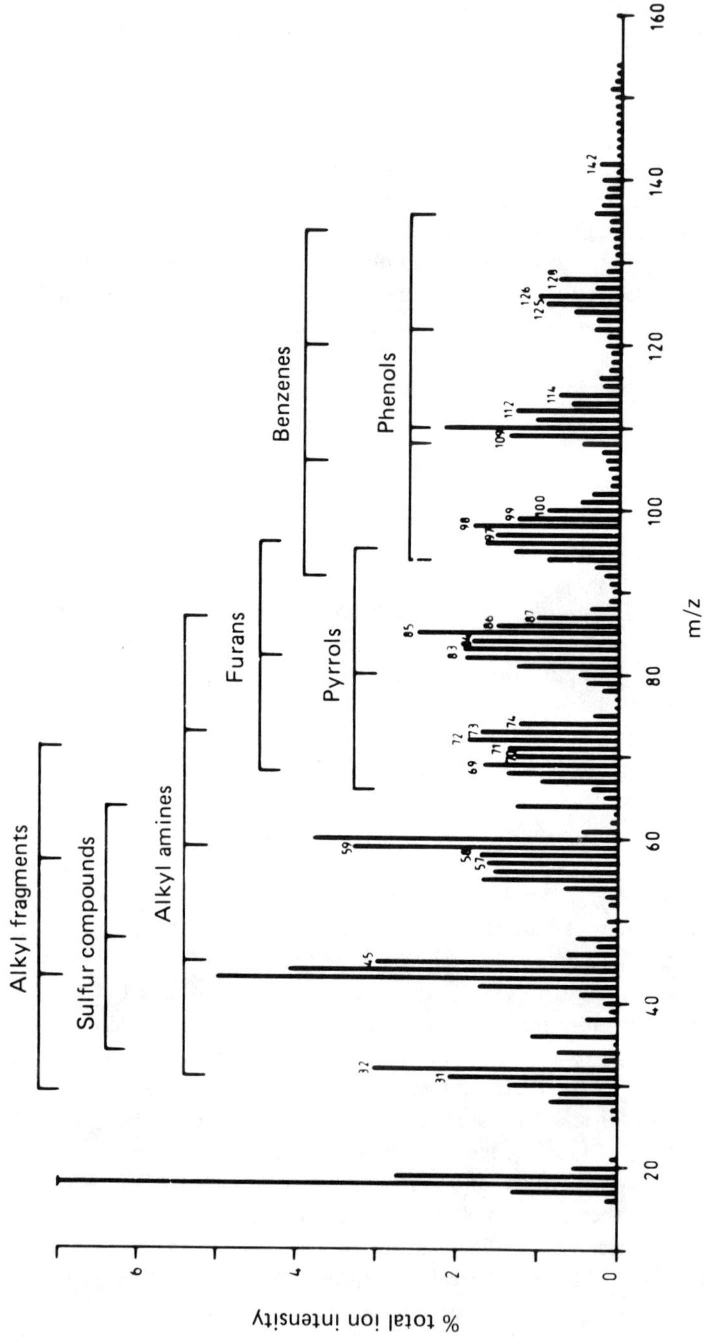

Figure 4. Typical pyrogram of fulvic acid fraction I from Tjeukemeer. Characteristic ion series for different classes of compounds are indicated.

the multivariate data analysis technique, developed by Eshuis et al. [22], was used. Ultimately, this technique results in a distance matrix, graphically represented in a nonlinear map, in which each pyrogram is given by a point. The relative distance between points reflects the dissimilarity of the pyrograms. Figure 5 shows the nonlinear map of the pyrograms (in quadruplicate) of fulvic acid fraction I from Tjeukemeer isolated monthly during 1978. The quadruplicate pyrograms, represented by four interconnected dots, are relatively close together, indicating a good reproducibility. Furthermore, despite the variance between the pyrograms of four samples taken at the same station in September and at three different stations in January, the pyrograms are clearly divided into two subgroups. One group contains the pyrograms of fulvic acid fraction I isolated between January and May 1978; the other contains those from the rest of the sampling period. This result indicates that in 1978 the chemical composition of fulvic acids in Tjeukemeer in winter differs markedly from that in summer. Comparing this result with the fulvic acid concentration (E_{365}) in the lake during 1978 (Table IV), it is clear that in winter when the lake contains humus-rich drainage water, the chemical composition of fulvic acids is quite different from that in

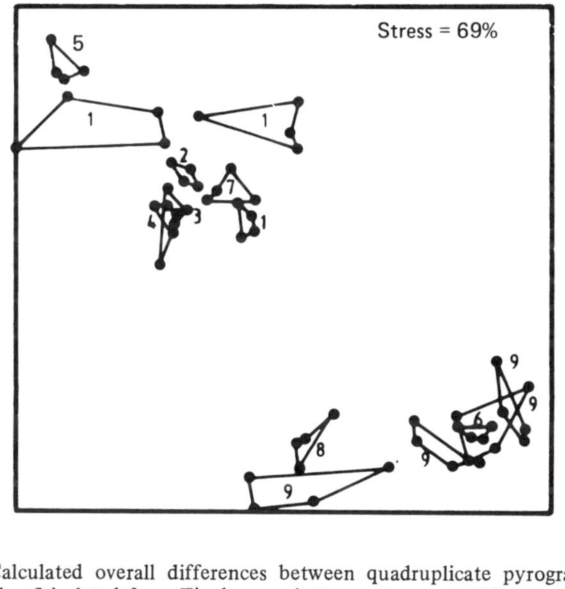

Figure 5. Calculated overall differences between quadruplicate pyrograms of fulvic acid fraction I isolated from Tjeukemeer between January and September 1978. The differences are presented as a distance matrix, in which each point represents one pyrogram. The relative distance between the points reflects the dissimilarity of the pyrograms. Numbers refer to the months of the year.

Table IV. Fulvic Acid Concentration (E_{365}), Ratio of Absorbances at 250 and 365 nm (E_2/E_3) and Relative Contents of Fulvic Acid Fractions I, II and III of Tjeukemeer Water Between January and September 1978

Date	E_{365}	E_2/E_3	% I	% II	% III
4-1	0.117	5.90	63	24	13
7-2	0.141	5.57	63	25	12
14-3	0.106	6.29	49	41	10
11-4	0.108	6.45	58	31	11
9-5	0.079	6.87	64	25	11
13-6	0.051	6.63	49	37	15
11-7	0.046	6.87	60	27	13
29-8	0.034	7.35	33	40	27
19-9	0.036	7.23	51	34	15

summer when it contains humus-poor IJsselmeer water. Apparently, chemical changes are involved in the seasonally varying humification, as indicated by the E_2/E_3 ratio and molecular weight distribution of the fulvic acids in Tjeukemeer (Table IV).

Detailed study of the pyrograms enables us to determine which masses, i.e., structural fragments, contribute to the observed compositional differentiation. Carbohydrate substances, proteins and aromatic fragments (phenols) indicative of fulvic acids and lignins appear to contribute the most and play important roles in the seasonal changes in the fulvic acid composition. The role of phenols is clearly demonstrated in Figure 6, in which the sum of three peaks typical for aromatic fragments in the monthly determined pyrograms is plotted against the fulvic acid concentration in the lake. In winter, when the Tjeukemeer contains humified drainage water, fulvic acids are relatively rich in aromates (phenols). Similar scatter diagrams for masses indicative of carbohydrates show that in summer the composition of fulvic acids is partly determined by production processes, which confirmed earlier studies on amino acids, carbohydrates and fulvic acids [16,23].

Nonlinear mapping of monthly isolated fulvic acid fractions I, II and III from Tjeukemeer demonstrated that relative to the high-molecular-weight fraction I, the composition of the low-molecular-weight fractions II and III changes more. Especially in summer the latter fractions show large compositional variation, suggesting that these are caused by temperature-dependent processes, e.g., microbial activity. This suggestion agrees with the general idea that high-molecular-weight fulvic acids are more resistant to microbial activity than low-molecular-weight fractions [17,24].

Combining the results of Curie-point pyrolysis/mass spectrometry, Sephadex gel filtration and UV spectroscopy, humification of lakes leads

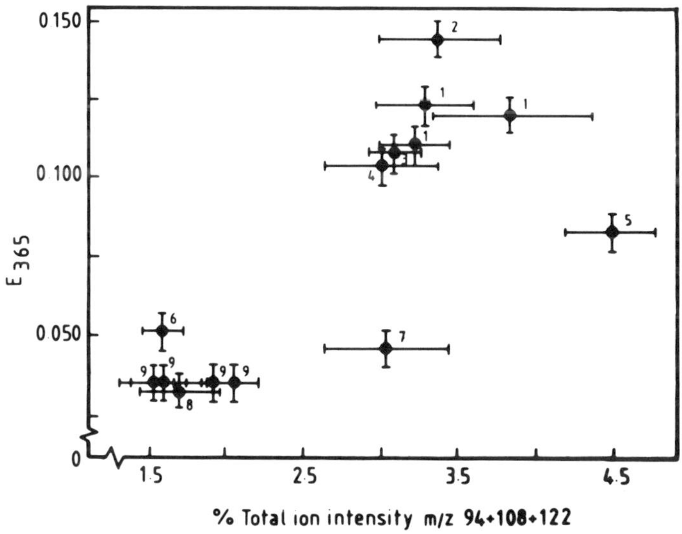

Figure 6. Diagram of the sum of the ion intensities at m/z 94, 108 and 122, i.e., phenols, in the pyrograms of fulvic acid fraction I and the fulvic acid concentration (E_{365}) in Tjeukemeer between January and September 1978. The ion intensities are averaged over four determinations. Numbers refer to the months of the year.

to increase of molecular weight and content of aromatic compounds (phenols), which is reflected in a decrease of the E_2/E_3 ratio.

Curie-point pyrolysis/mass spectrometry and multivariate and factor analysis of pyrograms can contribute considerably to the chemical understanding of the humification of freshwater. However, the technique has its restrictions and does not give a complete quantitative picture of the chemical composition of fulvic acids, mainly because of the difficulty in determining to what extent the pyrolysis products represent the original sample. Therefore it gives only fingerprints which of course can be compared quantitatively.

BENZOATE-METABOLIZING BACTERIA

In general, fulvic acids are believed to be refractory. However, many resistant substances such as herbicides, chlorinated aromates and alkyl benzenes are susceptible to biodegradation in the presence of readily degradable substrates. This process is called cometabolism [25]. Besides cometabolism, stimulation of already slowly proceeding fulvic acid degradation can occur by actively metabolizing bacteria. This effect is called the priming effect and is only shown by organisms able to grow with fulvic acids as the only C source [26]. The chance that a refractory substance can be cometabolized is greater with increasing structural resemblance to degradable substrates. Since Curie-point pyrolysis/mass spectrometry showed that aromates (phenols) are a major constituent of fulvic acids, benzoate metabolizing bacteria were studied for their ability to (co)metabolize fulvic acids.

When adding benzoate to Tjeukemeer water, bacterial growth develops and color changes occur which can be attributed to a minor extent to absorption of fulvic acids onto the surface of bacterial cells [27]. Because these results strongly suggest that the fulvic acids are partly (co)metabolized, bacteria enriched on benzoate were grown in synthetic media containing benzoate and different fulvic acid fractions as the only C source. Sephadex G-25 gel filtration of medium containing fulvic acid fraction I from Tjeukemeer and benzoate before and after growth of an *Arthrobacter* sp. demonstrated the disappearance of benzoate but also the formation of a low-molecular-weight fraction (II) (Figure 7A). Fraction II is spectroscopically very similar to fulvic acid fraction I, but has a higher E_2/E_3 ratio and shows more fluorescence (Table V), indicating that the *Arthrobacter* sp. metabolized fulvic acid fraction I, leading to the formation of lower-molecular-weight fulvic acid. Because the *Arthrobacter* is able to grow slowly on fulvic acid fraction I [27], this is an example of the priming effect of benzoate on fulvic acid degradation. Similar results were obtained with a medium containing benzoate and concentrated Tjeukemeer water (Figure 7B). This experiment in particular showed the formation of relatively low-molecular-weight fulvic

Table V. Ratios of Absorbances, Phenolic Groups and Fluorescence of Fulvic Acid Fractions I and II from Tjeukemeer and of Fulvic Acids (FA) from Lake Saimaa, Before and After Growth of Bacteria Enriched on Benzoate in Media with Benzoate and Different Fulvic Acids as Organic C Source[a]

	Before Growth		After Growth		
Ratio	I	FA	I	II	FA
E_2/E_3	3.4	4.3	3.3	3.7	4.5
E_7/E_2	1.1		1.6	7.0	
F/E_2	2.9		3.2	4.0	

[a] E_2/E_3, absorbances at 250 and 365 nm; E_7/E_2, phenolic groups and absorbance at 250 nm; F/E_2, fluorescence and absorbance at 250 nm.

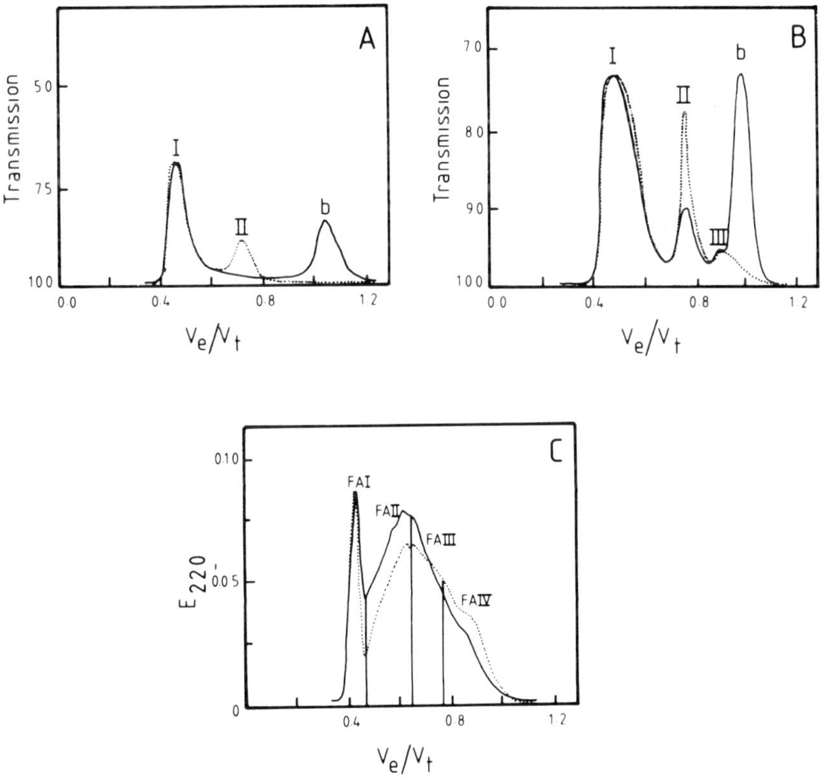

Figure 7. A: Sephadex G-25 gel filtration in phosphate buffer (pH = 7.0; I = 0.02) of medium containing fulvic acid fraction I from Tjeukemeer and benzoate as only organic C sources before (————) and after (·········) growth of an *Arthrobacter* sp. enriched on benzoate. B: Same, of a medium containing benzoate and concentrated Tjeukemeer water. C: Same, of a medium containing benzoate and concentrated Lake Saimaa water. After sterilization, this medium was inoculated with Lake Saimaa water. Transmission and absorbance were measured at 250 and 220 nm respectively. b, benzoate; V_e, elution volume; V_t, total gel volume.

acid fraction II from high-molecular-weight fulvic acid fraction I. This molecular weight change is reflected in an increase in the E_2/E_3 ratio (Table V), confirming the use of this ratio as a humification measure. Therefore, increasing E_2/E_3 ratios of Tjeukemeer water, indicating decreasing humification, are likely to be caused by benzoate-metabolizing bacteria. Incubations of benzoate-metabolizing bacteria in media containing benzoate and concentrated humic Finnish lake water (Figure 7C), showed that this is not specific for Tjeukemeer water, emphasizing that benzoate-metabolizing bacteria in general participate in the fulvic acid dynamics in lakes. These results suggest that aquatic fulvic acids are not as refractory as usually assumed.

It is clear from the Sephadex elution patterns before and after growth of benzoate-metabolizing bacteria (Figure 7) that the fulvic acids are only partly degraded. The degradation is performed by actively degrading specialized bacteria. However, it is reasonable to assume that other specialized bacterial species can attack the degradation products of the benzoate bacteria and parts of the fulvic acids not affected. Therefore the complete degradation of fulvic acids is the work of various groups of specialized bacteria. The rate of fulvic acid degradation is probably high and enhanced by cometabolism and priming effects in waters with a relatively high turnover of various classes of easily degradable substrates, i.e., productive lakes. In this respect the cometabolism of fulvic acids by a lactate-metabolizing *Pseudomonas* sp. from Tjeukemeer may be mentioned [28].

Since bacteria enriched on benzoate appear able to partly metabolize fulvic acids, their numbers in humic waters may be indicative of the rate of fulvic acid degradation. To test this hypothesis, numbers of benzoate-metabolizing bacteria were counted during prolonged periods in Dutch and Finnish aquatic systems. The results, as listed in Table VI, reveal some interesting aspects. Firstly, the trophic level of humic lakes seems to affect the proportion of benzoate-metabolizing bacteria tremendously, with respect to the total heterotrophic flora (BB/HB). Apparently, benzoate-like metabolism is a major bacterial activity in oligotrophic humic Lake Saimaa. On the other hand, in the eutrophic, humic Tjeukemeer this activity is of relatively minor importance. Secondly, in oligotrophic

Table VI. Trophic Level, Organic C Concentration, Color and Numbers of Benzoate-Metabolizing Bacteria (BB) with Respect to Organic C Concentration and Heterotrophic Flora (HB) in Finnish and Dutch Humic Lake Water

Lake and Sampling Period	Trophic Level	N[a]	mg C/1	mg Pt/1	BB/mg C	BB/HB
Tjeukemeer, summer	eutrophic	8	10.0	18	6×10^3	4×10^{-5}
Tjeukemeer, winter	hypertrophic	11	22.8	65	5×10^4	538×10^{-5}
Polder water, July–May	hypertrophic	14	43.9	128	1×10^6	777×10^{-5}
Lake Saimaa, January–July	oligotrophic	10	6.9	18	9×10^3	0.5
Polluted Lake Saimaa, January–May	enriched	10	9.6	23	3×10^6	1.0

[a]Mean number of fortnightly performed bacterial counts and chemical determinations.

Lake Saimaa the discharge of sewage enlarged the proportion of benzoate-metabolizing bacteria, suggesting an enhancement (priming effect) of the fulvic acid degradation. Therefore, as discussed earlier, the rate of fulvic acid degradation may be related to the trophic level of the aquatic system. However, the numbers of benzoate-metabolizing bacteria per unit of organic C in eutrophic Tjeukemeer and oligotrophic Lake Saimaa do not differ much and do not fit within this framework. Increase of the fulvic acid concentration in eutrophic Tjeukemeer does seem to enhance the rate of fulvic acid degradation, since the highest number of benzoate bacteria are observed in the hypertrophic and strongly humified drainage water from the polders (Table VI). Though we are aware of the speculative nature of these data, they may give an estimate of the relative rate of fulvic acid degradation. More study is required to test the value of this fulvic acid degradation measure. Moreover, the bacterial counts should be performed in combination with substrate uptake kinetics to see whether bacterial counts and benzoate uptake rates are really related.

CONCLUSIONS

The E_2/E_3 ratio is a simple and easy way to determine measure of the relative humification of colored lake water. Progressive humification is predicted by relatively low E_2/E_3 ratios and coincides with increasing molecular weight and aromaticity of fulvic acids, while increasing E_2/E_3 ratios should indicate fulvic acid degradation. These results were confirmed by incubation experiments with bacteria enriched on benzoate. These bacteria appear to partially metabolize fulvic acids, affecting the formation of relatively low-molecular-weight fulvic acids having a high E_2/E_3 ratio. Thus degradation of fulvic acids by benzoate-metabolizing bacteria may explain the increasing E_2/E_3 ratios and decreasing mean molecular weight of fulvic acids in Tjeukemeer usually observed in summer. Curie-point pyrolysis/mass spectrometry revealed that carbohydrates, proteins, alkyl and phenolic fragments are involved in the humification of aquatic fulvic acids. However, with this technique the structure of fulvic acids cannot be completely elucidated. The technique gives fingerprints which can be used to determine relative compositional variance of different fulvic acids.

Most likely, direct degradation of aquatic fulvic acids proceeds slowly. However, easily degradable substrates such as lactate and benzoate appear to stimulate the degradation rate of fulvic acids, either by a priming effect of an easily degradable substrate or by cometabolism by bacteria actively metabolizing fulvic acid-like substances, e.g., benzoate. These results suggest that the rate of fulvic acid degradation is connected with the productivity of colored lakes.

The proportion of benzoate-metabolizing bacteria with respect to the total heterotrophic flora seems to reflect the rate of fulvic acid degradation in humic lakes. Counts of these bacterial groups indicate that the fulvic acid concentration and the trophic level control the degradation rate of aquatic fulvic acids. Further investigation of these bacteria combined with substrate uptake studies will give more insight into the humus dynamics in freshwater ecosystems. These studies, together with modern analytical techniques such as Curie-point pyrolysis/mass spectrometry, will also increase the chemical knowledge of the processes involved. This knowledge is urgently needed for a better understanding of the role of fulvic acids in, for example, the speciation and bioavailability of metals and extracellular enzyme activity.

ACKNOWLEDGMENTS

The study of the Finnish lakes was performed during a period as guest-worker at the laboratory of the National Board of Waters of Finland at Kouvola. The Curie-point pyrolysis/mass spectrometrical study was a joint investigation of Drs. G. Halma (Agricultural University, Wageningen, The Netherlands), J. Haverkamp (FOM-Institute for Atomic and Molecular Physics, Amsterdam, The Netherlands) and H. De Haan. The technical assistance of Mr. T. De Boer and Mrs. A. G. Frank-Landman during gel filtration, UV spectroscopy and bacterial counting is greatly acknowledged.

REFERENCES

1. Ruttner, F. *Grundriss der Limnologie, 2 Auflage* (Berlin: W. De Gruyter & Co., 1952), pp. 189-195.
2. Scheffer, F., and B. Ulrich. *Lehrbuch der Agrikulturchemie und Bodenkunde. III Teil. Humus und Humusdüngung. Band I* (Stuttgart: I. F. Enke Verlag, 1960), Chapter 4.
3. Hutchinson, G. H. *A Treatise on Limnology. Volume I* (New York: Wiley & Sons, Inc., 1957), Chapter 17.
4. Golterman, H. L. *Physiological Limnology* (Amsterdam: Elsevier, 1975), Chapter 12.
5. Wetzel, R. G. *Limnology* (Philadelphia: W. B. Saunders Company, 1975), Chapter 17.
6. Schnitzer, M., and S. U. Khan. *Humic Substances in the Environment* (New York: Marcel Dekker, Inc., 1972), Chapters 6 and 8.

7. De Haan, H., T. De Boer and G. Halma. "Curie-Point Pyrolysis-Mass Spectrometry of Fulvic Acids from Tjeukemeer, The Netherlands," *Freshwat. Biol.* 9:315-317 (1979).
8. De Haan, H., G. Halma, T. De Boer and J. Haverkamp. "Seasonal Variations in the Composition of Fulvic Acids in Tjeukemeer, The Netherlands, as Studied by Curie-Point Pyrolysis-Mass Spectrometry," *Hydrobiologia* 78:87-95 (1981).
9. Gjessing, E. G. "Humic Substances in Natural Waters: Methods for Separation and Characterization," in *Chemical Environment in the Aquatic Habitat*, H. L. Golterman and R. S. Clymo, Eds. (Amsterdam: Noord-Hollandsche Uitgevers Maatschappij, 1967), pp. 191-201.
10. Shapiro, J. "Yellow Organic Acids of Lake Water: Differences in Their Composition and Behavior," in *Chemical Environment in the Aquatic Habitat*, H. L. Golterman and R. S. Clymo, Eds. (Amsterdam: Noord-Hollandsche Uitgevers Maatschappij, 1967), pp. 202-216.
11. Povoledo, D., and M. Gerletti. "Recent Studies on the Sedimentary Organic Matter from Lake Maggiore (Northern Italy)," *Mitt. int. Verein theor. angew. Limnol.* 14:145-154 (1968).
12. Pennanen, V. "Humus Fractions and Their Distribution in Some Lakes in Finland," in *Humic Substances. Their Structure and Function in the Biosphere*, D. Povoledo and H. L. Golterman, Eds. (Wageningen, The Netherlands: Pudoc, 1972), pp. 207-215.
13. De Haan, H. "Molecule-Size Distribution of Soluble Humic Compounds from Different Natural Waters," *Freshwat. Biol.* 2:235-241 (1972).
14. Hall, K. J., and G. F. Lee. "Molecular Size and Spectral Characterization of Organic Matter in a Meromictic Lake," *Water Res.* 8:239-251 (1974).
15. De Haan, H. "Some Structural and Ecological Studies on Soluble Humic Compounds from Tjeukemeer," *Verh. int. Verein theor. angew. Limnol.* 18:685-695 (1972).
16. De Haan, H., and T. De Boer. "Seasonal Variations of Fulvic Acids, Amino Acids and Sugars in Tjeukemeer, The Netherlands," *Arch. Hydrobiol.* 85:30-40 (1979).
17. Kononova, M. M. *Soil Organic Matter, Second English Edition* (New York: Pergamon Press, Inc., 1966).
18. Meuzelaar, H. L. C., and R. A. In 't Veld. "A Technique for Curie Point Pyrolysis Gas Chromatography of Complex Biological Samples," *J. Chromatogr. Sci.* 10:213-216 (1972).
19. Meuzelaar, H. L. C., and P. G. Kistemaker. "A Technique for Fast and Reproducible Fingerprints of Bacteria by Pyrolysis Mass Spectrometry," *Anal. Chem.* 45:587-590 (1973).
20. Meuzelaar, H. L. C., P. G. Kistemaker and M. A. Posthumus. "Recent Advances in Pyrolysis Mass Spectrometry of Complex Biological Materials," *Biom. Mass Spectrom.* 1:312-319 (1974).
21. Meuzelaar, H. L. C., P. G. Kistemaker, W. Eshuis and A. J. H. Boerboom. "Automated Pyrolysis Mass Spectrometry, Application to the Differentiation of Microorganisms," *Adv. Mass Spectrom.* 7:1452-1456 (1978).
22. Eshuis, W., P. G. Kistemaker and H. L. C. Meuzelaar. "Some Numerical Aspects of Reproducibility and Specificity," in *Anal. Pyrolysis*, C. E. R. Jones and C. G. A. Cramers, Eds. (Amsterdam: Elsevier, 1977), pp. 151-166.

23. De Haan, H., and T. De Boer. "A Study of the Possible Interactions Between Fulvic Acids, Amino Acids and Carbohydrates from Tjeukemeer, Based on Gel Filtration at pH 7.0," *Water Res.* 12:1035-1040 (1978).
24. Stabel, H. H., K. Moaledj and J. Overbeck. "On the Degradation of Dissolved Organic Molecules from Pluss-See by Oligocarbophilic Bacteria," *Arch. Hydrobiol. Beih.* 12:95-104 (1979).
25. Horvath, R. S. "Microbial Co-Metabolism and the Degradation of Organic Compounds in Nature," *Bacterial Rev.* 36:146-155 (1972).
26. Jager, G. "The Effect of Living Roots and the Rhizosphere Microflora on the Decomposition of Soil Organic Matter," *Colloq. Pediol.* 4:57-67 (1971).
27. De Haan, H. "Effect of Benzoate on Microbial Decomposition of Fulvic Acids in Tjeukemeer (The Netherlands)," *Limnol. Oceanog.* 22:38-44 (1977).
28. De Haan, H. "Effect of a Fulvic Acid Fraction on the Growth of a *Pseudomonas* from Tjeukemeer (The Netherlands)," *Freshwat. Biol.* 4:301-310 (1974).

CHAPTER 9

FLUORESCENCE PHENOMENA OF HUMIC MATTER OF AQUATIC ORIGIN AND MICROBIAL CULTURES

Simon A. Visser
Department of Soil Science
Laval University
Cité Universitaire
Québec, Canada

ABSTRACT

Highly purified molecular weight fractions were taken of fulvic and humic acids sampled during two seasons at various locations of a watershed within the Laurentide Provincial Park, Quebec. The excitation and emission spectra of these fractions were then investigated. To study the effect of humification on the spectra, humic material produced by a streptomycete culture and sampled at regular intervals during a 6-months incubation period was also investigated.

Of particular interest were the effects on the spectra of the molecular weight of the material, its functional groups, pH, temperature and redox potential.

Microbially formed humus fluoresced more strongly than aquatic humic compounds and its spectrum usually was also more differentiated than that of aquatic humus. The excitation maxima of these humic materials occurred at lower wavelengths than those reported in the literature for soil humic matter.

Lower-molecular-weight fractions normally had higher excitation and emission wavelengths and intensities than fractions of higher molecular weight. Indications were obtained that the intensity of fluorescence was more determined by the basic aromatic structure of the molecules than by their functional groups. There was, however, a positive correlation between the aromatic hydroxyl group content of the humic matter and the wavelength of the main excitation peak. The latter peak was normally more pronounced at higher pH values, whereas fluorescence intensity under these conditions was lower. Intensities of major excitation and emission peaks diminished with increasing temperature, whereas their wavelengths generally increased. A rising redox potential of the environment usually resulted in an increase in excitation and emission intensity.

INTRODUCTION

Although some authors have expressed doubt as to whether the fluorescence observed on solutions of humic substances is a property of the humic material itself or whether it is due to contaminating substances [1], a large number of data sustains that fulvic acid and humic acid solutions show fluorescence due to the properties of the humic matter itself [2–5]. The intensity, however, is on the whole weak. Wavelengths of maximum excitation and emission observed on humic matter of different type and origin are listed in Table I. Optimal concentration for fluorescence of humic compounds is normally found to be situated between 10 and 40 ppm. In this range a linear relationship exists between the intensity of fluorescence and the concentration of the humic compound [3,4,8].

Table I. Wavelengths of Maximum Excitation and Emission in the Fluorescence Spectra of Different Types of Humic Compounds

Type of Humic Compound	Wavelength Maximum Excitation Band (nm)	Wavelength Emission Maximum (nm)	Reference
Soil fulvic acids			
Assam soil	468–470	520	3,6
Podzol	360		5
Soil humic acids			
Various origins	470	500–540	2
Black cotton soil		500–540	7
Chernozem	465		5
Podzol	335	440	8
Swamp	~360	460	4
Swamp		~460	1
Aquatic humic matter			
	315	420–430	10
	361	460	11
	365,405,436	490	12
Aquatic fulvic acids	360–370	450–460	9
Aquatic humic acids	270	460	13

In theory the excitation spectrum of a substance should resemble the mirror image of its absorption spectrum; a lack of such a relation is usually the result of an appreciable change in the configuration of the molecule upon excitation [14]. Because for humic substances the absorption spectrum shows an almost continuous decline of the absorption intensity with increasing wavelength, it is possible that the fluorophores are present in such a small concentration that although they can produce a small but measurable amount of fluorescence, their contribution to absorption is negligible [2]. It has been suggested that the phenomenon of fluorescence in humic substances is either due to an aromatic nucleus substituted by at least one electron-donating group or to a conjugated unsaturated system capable of a high degree of resonance [7]. Fluorescing compounds can be classified according to their fluorophoric component on the basis of the following general structures: phenols, aromatic amines (electron-donating), polycyclic compounds and conjugated polyenenes. Although fluorescence, therefore, is usually exhibited by aromatic, cyclic or closed-ring structures, it is not entirely limited to these systems since some aliphatic compounds can also show fluorescence. Because they increase the transition probability between the singlet and ground state, certain electron-donating groups such as hydroxyl, methoxyl and some amines can enhance fluorescence of aromatic compounds [14]. Some of these groups and their effects are listed in Table II.

In this chapter the fluorescent properties of highly purified humic molecular weight fractions of aquatic and microbial origin are investigated. We also show how they are influenced by a number of conditions such as temperature, acidity and redox potential.

EXPERIMENTAL

Material

Humic matter of natural aquatic origin was investigated as well as humic material of microbial origin in various stages of humification.

Table II. Effect of Certain Substituents on the Fluorescence of Aromatic Compounds

Substituent	Effect on Emission Frequency	Effect on Emission Intensity
OH, OCH_3, OC_2H_5	decrease	increase
NH_2, NHR, NR_2	decrease	increase
COOH	decrease	large decrease
NO, NO_2		quenching

The aquatic humic material was obtained from water samples collected from a watershed comprised of a lake (Lake Pikauba), a river (River Chicoutimi) and a swamp situated in the Laurentide Park in the Precambrian Shield region of the province of Quebec. In order to study seasonal changes in the humic matter also, the sites were sampled at the end of the spring thaw (early May) and at the end of summer (mid September).

The humic material of microbial origin was obtained from a simple glucose-yeast extract medium, inoculated with an actinomycete isolated from a Quebec forest soil, and incubated under constant conditions of aeration, temperature (25°C) and pH (7.0). Samples were collected for recovery of fulvic acids (FA) and humic acids (HA) after 4, 7, 14, 28, 60 and 180 days of incubation.

The FA and HA were separated, purified and fractionated into eight molecular weight fractions by ultrafiltration as described by Oliver and Visser [15].

Measurements

Excitation and fluorescence spectra of humic compounds were obtained on a Perkin Elmer spectrofluorimeter model 204. This instrument, operating with a Xenon arc source (150 W), has two monochromators, each using the same type of grating (600 lines per mm). Its slit width is fixed and gives an effective spectral band width $\Delta\lambda$ of 10 nm. The K salt of each molecular weight fraction of the humic products was dissolved in water at a previously determined concentration of maximum fluorescence (Table III) and, except when stated otherwise, was investigated at pH 7.0 and at a temperature of 25 °C. For the excitation spectrum the emission wavelength was fixed at 500 nm, whereas for the emission spectrum the wavelengths of the peaks of the excitation spectrum were used as excitation wavelengths.

To investigate the effect of external factors on the excitation and fluorescence spectra, the following conditions were applied:

effect of acidity: pH 3, 4, 7 and 8 at a temperature of 25 °C
effect of temperature: 18, 25, 37 and 48 °C at a pH of 7.0
effect of redox potential: −1.17, 0.00, +0.50 and +2.50 V at 25 °C and pH 7.0

Treatment of Data

Intensities of both excitation and emission spectra have been expressed in relative quantum units, therefore making comparison possible between our data and those obtained by other authors, if likewise treated. Correction was achieved by using reference compounds with known quantum efficiencies: for the emission spectrum both 2-amino pyridine and quinine sulfate were used, whereas in the case of the excitation spectrum use was made of the

Al-chelate of 2,2'hydroxyl-1,l'azonaphthalene-4-sulphonic acid [16]. The corrected intensity values were divided by the concentration of the humic substance; this allows comparison between the (relative) intensity values of the different humic compounds.

RESULTS AND DISCUSSION

During the course of our investigation, it was found that in the excitation spectrum of the various humic substances investigated absorption peaks could occur in one or more of the following regions: 320–348, 340–352, 360–370, 372–390, 392–395, 400–425, 424–445 and 451–461 nm. Normally, one peak could be considered as the major peak with the second most important peak having an intensity of between 50 to 75% of the former. Absorption by other peaks was generally less than 30% of that by the major peak. In the following discussion, therefore, only the properties of the major and secondary excitation peaks (if present) will be investigated. In this context it would be interesting to speculate that the multiple excitation peaks of similar intensities obtained by Lévesque [3] on soil FA could be attributed either to the simultaneous presence of different humic matter mw fractions and, therefore, to their incomplete separation, or to a property inherent to soil humic matter as distinct from aquatic humus.

Table III indicates, for each group of humic material considered, the wavelengths and the intensities of the two major excitation peaks (if present) within the region of 320 to 461 nm, as well as of the principal emission peak. It can be deduced from the table that the two major excitation peaks had average wavelengths of 334±7 nm and 415±20 nm. Taking into consideration the full range over which individual wavelengths of these two peaks could shift, without both peaks ever being within the same wavelength interval, the following ranges were determined: range I, 320 to 370 nm, and range II, 372 to 461 nm. The emission spectrum normally consisted of one peak only, situated within the range of 370 to 550 nm.

It should be mentioned that because in Table III the intensities of the excitation and emission spectra are based on different scales, they are not expressed in the same relative quantum units, and are therefore not directly comparable. As very little difference in fluorescent properties was found between the molecular weight of samples within the molecular weight ranges of 750 to 75,000 (<75,000) and 75,000 to 300,000 (≥75,000), only the results obtained for these two sets of molecular weight groups are presented. Furthermore, because for microbial humic matter incubated up to 180 days no consistent effect of the degree of humification could be detected, whereas for humic material of natural aquatic origin its source and the season at which it was collected did not seem to have any explicit effect on the excitation and

Table III. Average Wavelengths and Relative Intensities of the Two Major Excitation Peaks and the Emission Peaks Obtained from Fulvic and Humic Acid Molecular Weight Fractions from Natural Aquatic Environments and Microbial Cultures

Humic Matter	Molecular Weight Range	Concentration Maximum Fluorescence (mg/l)	Wavelengths: Excitation Bands Range I (320–370 nm)	Wavelengths: Excitation Bands Range II (372–461 nm)	Wavelengths: Emission Band (370–550 nm)	Intensities: Excitation Bands Range I (320–370 nm)	Intensities: Excitation Bands Range II (372–461 nm)	Intensities: Emission Band (370–550 nm)
FA (aquatic)	<75,000	60		418 ± 7	512 ± 32		13.7 ± 5.8	17.3 ± 15.8
	⩾75,000	25	330 ± 2	414 ± 8	425 ± 27	4.3 ± 1.1	2.2 ± 0.3	3.2 ± 0.3
HA (aquatic)	<75,000	70		440 ± 14	533 ± 24		4.1 ± 1.0	14.5 ± 5.2
	⩾75,000	32	340 ± 14	410 ± 5	418 ± 25	2.3 ± 0.3	1.7 ± 0.3	1.6 ± 0.2
FA (microbial)	<75,000	60	336 ± 1	417 ± 9	482 ± 49	37.1 ± 16.2	17.8 ± 10.8	36.3 ± 10.6
	⩾75,000	35	334 ± 4	415 ± 8	456 ± 68	2.5 ± 0.6	1.6 ± 0.2	2.3 ± 0.6
HA (microbial)	<75,000	40	335 ± 3	426 ± 25	438 ± 32	41.0 ± 20.1	19.9 ± 11.9	39.3 ± 10.9
	⩾75,000	16	331 ± 3	380 ± 4	413 ± 36	2.3 ± 0.3	1.8 ± 0.4	1.9 ± 0.2

fluorescence spectra, the respective data of each of these groups of FA and HA molecular weight fractions were pooled for the investigation of the effects of the pH, temperature and redox potential on the fluorescence spectrum.

The concentration of aquatic and microbial humic matter giving maximum fluorescence varied, as indicated in Table III, according to the type, origin and molecular weight of the material investigated, between 16 and 70 ppm. For soil humic acids maximum fluorescence was observed at a similar concentration range of between 20 and 85 ppm [3,8].

Table III indicates that fractions with molecular weight <75,000 normally have excitation and emission wavelengths and intensities exceeding those of higher-molecular weight humic matter. With respect to the wavelengths, the difference between the major excitation peaks of material of the two molecular weight ranges is normally small (between 2 and 4 nm). In the case of the emission spectrum the average differences are much larger: around 25 nm for the microbial humic matter and of the order of 100 nm in the case of the natural samples. These shifts in excitation wavelength with molecular weight should be attributed to rearrangements within the humic molecules with changing molecular weight and modifications in the degree of association between their groups.

Ghosh and Schnitzer [5] reported that soil FA showed a typical major excitation band at 360 nm. We find for FA and HA of microbial origin a band in the corrected spectrum at a slightly lower wavelength (~335 nm), which is particularly pronounced in the lower-molecular-weight fractions (<75,000). It is absent in the lower-molecular-weight fractions of humic matter of aquatic origin. The band reported by Ghosh and Schnitzer [5] for both soil FA and soil HA at 465 nm is in the corrected spectra of our humic products consistently situated at a lower wavelength (~420 nm). Furthermore, according to Jackson [17] autochthonous aquatic humus has a more intense fluorescence than its terrigenous counterpart; it seems therefore that humic matter of aquatic origin normally fluoresces more intensely but at lower wavelengths than humic compounds originating from soils. It should be verified whether these are intrinsic properties of the two types of humic material or whether the difference is due to the presence of larger amounts of impurities, such as heavy metals, in soil humus. It is known that these products frequently have a quenching effect on fluorescence.

Concerning the intensities, the average differences between the <75,000- and the ≥75,000-molecular-weight fractions for the major excitation and for the emission peaks are very similar: for samples of microbial origin the excitation and emission intensities encountered in the <75,000-molecular-weight fractions are between 10 to 20X larger than those of the ≥75,000-molecular-weight fractions; in the case of samples of natural origin the peaks of the lower-molecular-weight fractions are normally between 2 to 10X more intensive (Figure 1). Lévesque [3], as well as Hall and Lee [18], also reported

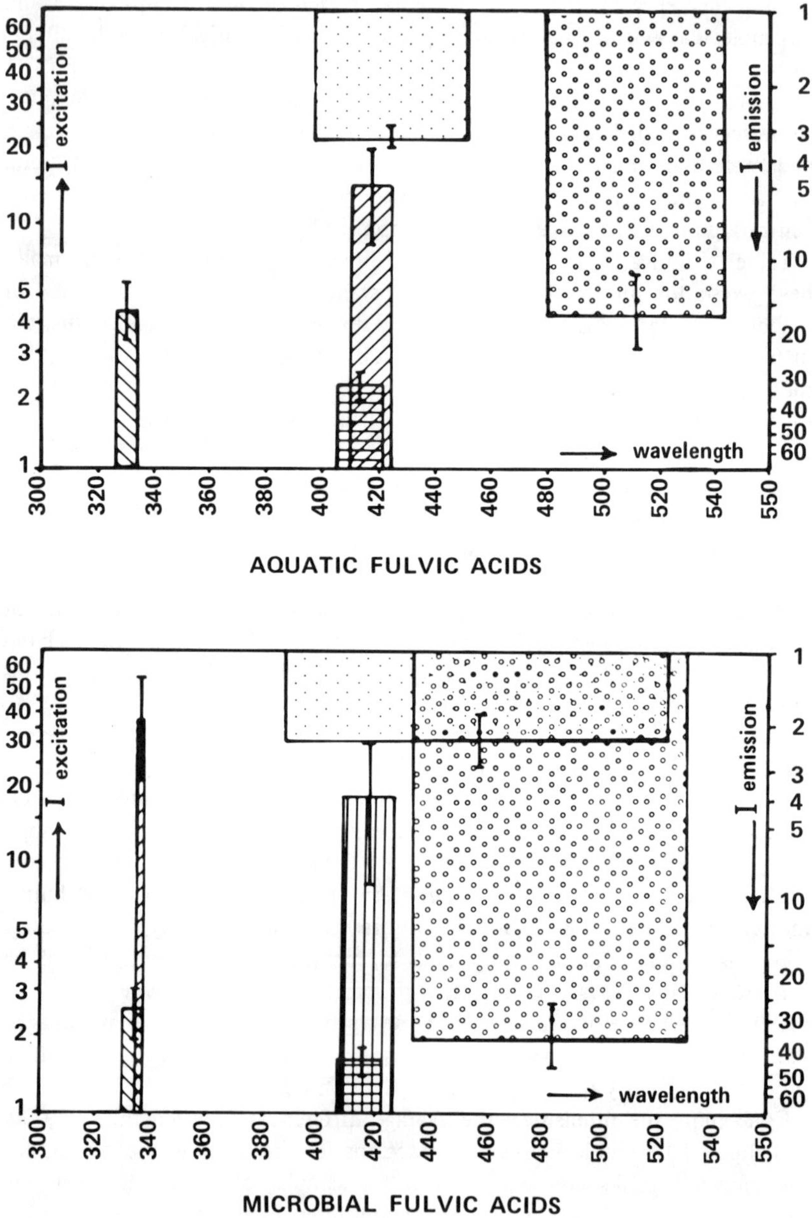

Figure 1. Intervals of wavelengths and relative intensities of the two major excitation tions (<75,000 and ⩾75,000) isolated from natural aquatic environments and from

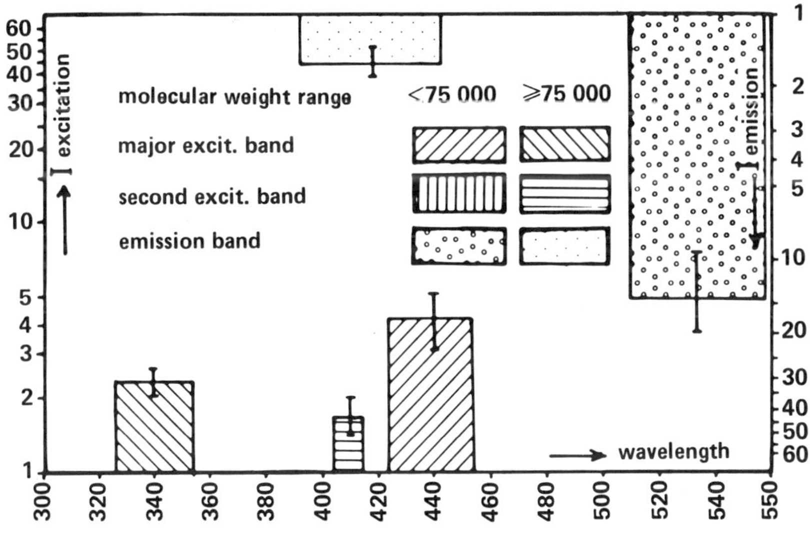

AQUATIC HUMIC ACIDS

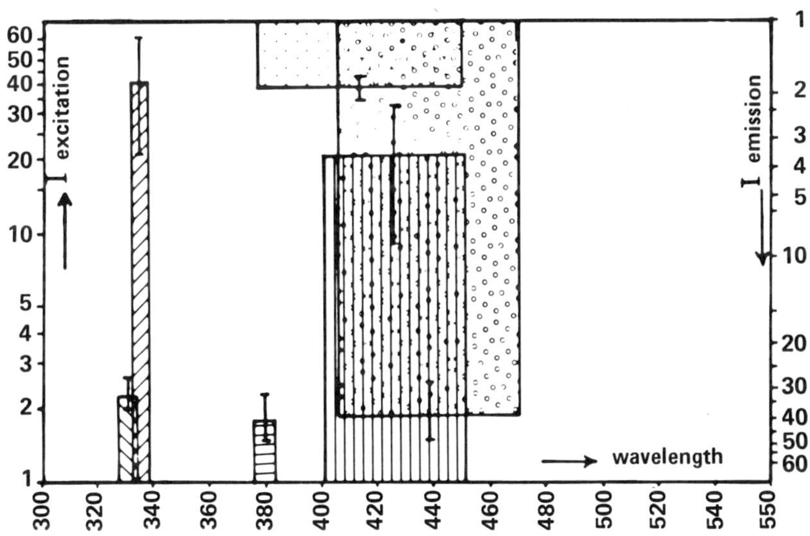

MICROBIAL HUMIC ACIDS

peaks and the emission peak obtained from fulvic and humic acid molecular weight frac- microbial cultures.

for soil humic compounds that a more intensive fluorescence was observed in the case of the lower-molecular-weight humic fractions.

With respect to differences in fluorescence intensity between FA and HA, Ghosh and Schnitzer [5] reported for soil humic matter that FA fluoresced more intensively than HA. They ascribed this to the presence in the HA of more absorbing centers and to the possible involvement of fluorophores in the cross-linking of the HA structure. We find a similar phenomenon in the case of humic matter of aquatic origin, whereas microbially produced FA and HA do not normally show it. Another difference which was evident between the two types of humic matter was that within the $<$75,000-molecular-weight range, material of natural origin had normally excitation and emission bands with higher wavelengths and lower intensities than microbially produced material. In the case of molecular weight fractions $\geqslant$75,000, excitation and emission intensities for material of both origins were rather similar; it follows that the bulk of the microbially formed humic matter fluoresced more strongly than the humic compounds of aquatic origin.

Finally, perusal of the data of Table III shows that although the wavelengths of the excitation and emission peaks vary with the type of humic material investigated, acceptable average wavelengths of maximum excitation and emission would be respectively 360 nm and 500 nm. It is suggested that these wavelengths be used for the quantitative determination by fluorescence of humic matter in aquatic environments.

For the investigation of the effect of the pH on the excitation and emission spectra of the humic compounds, solutions containing 100 mg/l of the various FA and HA molecular weight fractions were adjusted with dilute HCl or KOH to pH values of 3, 4, 7 and 8 respectively. The final concentrations were then adjusted with distilled water to the same values as indicated in Table III. At these concentrations the FA and HA of both aquatic and microbial origin remained in solution over the entire pH range investigated. The correlation coefficients for the pH and the intensities and wavelengths of the excitation and emission peaks (Table IV) suggest that, for the types of humic matter investigated, the main excitation peak often becomes more intense with increasing pH whereas, in particular for microbial humic material, the intensity of the emission band weakens. This is more clearly illustrated in Figure 2, in which the linear regression curves are presented for the significant correlations indicated in Table IV. Ghosh and Schnitzer [5] also noticed an increase in excitation intensity with pH; according to them this was the result of the ionization of functional groups and molecular dissociation.

With respect to the emission spectrum, Müller-Wegener [4] found that the pH-dependence of the emission intensity varied with the type of humic substances being investigated. This effect of the pH on the emission intensity would be associated in particular with the ionization of phenolic hydroxyl

Table IV. Correlation Coefficients (N = 48) Expressing the Relation Between pH and the Intensity and Wavelength of Excitation and Emission Peaks Obtained on Fulvic and Humic Acids from Natural Aquatic Environments and from Microbial Cultures[a]

	Wavelengths			Intensities		
	Excitation Bands		Emission Band	Excitation Bands		Emission Band
	Range I (330–340 nm)	Range II (400–461 nm)	(370–550 nm)	Range I (330–430 nm)	Range II (400–461 nm)	(370–550 nm)
FA (aquatic)	A	+0.21	+0.15	A	+0.45***	–0.09
HA (aquatic)	A	+0.21	A	A	+0.09	A
FA (microbial)	+0.21	+0.09	+0.17	–0.27	+0.04	–0.44**
HA (microbial)	–0.40**	+0.06	+0.05	+0.05	+0.46***	–0.43**

[a]* = $p < 0.05$; ** = $p < 0.01$; *** = $p < 0.001$; A = absence of major peaks.

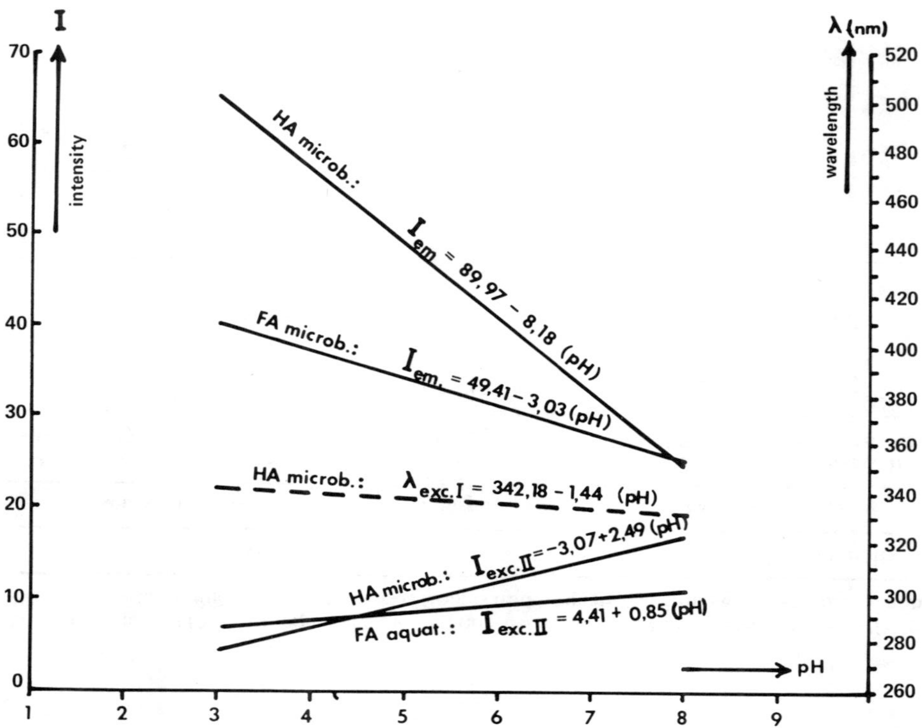

Figure 2. Linear regression curves calculated for significant relationships between pH conditions within the range pH 3 to 8 and the wavelengths and intensities of the major excitation peaks and the emission peak obtained on fulvic and humic acids from natural aquatic environments and from microbial cultures.

groups: the higher the content of phenolic hydroxyl, the larger the dependance of the emission intensity on the pH. An increase in the content of phenolic OH groups in humic substances would, according to some researchers, generally result in an increase in the intensity of the excitation band and its wavelength [19]. The carboxyl group, on the other hand, would have a weakening effect on the emission intensity [20], which would be at a minimum when the COOH group is oriented para to the OH group [6]. The generally observed weak fluorescence of humic matter could therefore be partly the result of the relative position of these two functional groups in the humic compound. Our observations do not entirely support the importance of the phenolic hydroxyl group in excitation and emission intensity phenomena in humic matter of aquatic and microbial origin. Within the pH range of 3 to 8, excitation and emission intensities of the HA of microbial origin were found to be more pH-sensitive than the intensities of the FA. Considering the phenolic OH content of the FA and HA (Table V), this throws some serious doubt on the contribution of this group to the fluorescence intensities of these humic compounds. On the other hand, the phenolic OH content of the humic matter was found to be positively correlated to the wavelength of the main excitation peak (Table VI). The wavelength and the phenolic OH were related as follows:

$$\lambda_{nm} = 245.85 + 134.31\,(\phi\mathrm{OH})_{meq/g}$$

This observation on the effect in humic matter of microbial and aquatic origin of phenolic OH on the main excitation wavelength is in agreement with similar findings by Müller-Wegener [8] on soil humic matter.

The data from the fluorescence spectra obtained on the FA and HA of microbial origin at temperatures of 18, 25, 37 and 48 °C (pH 7.0) indicate that with increasing temperature the intensities of the major excitation and emission bands diminish, whereas their wavelengths normally increase

Table V. Average ϕOH Content of the Fulvic and Humic Acids of Microbial and Aquatic Origin [22]

	Average ϕOH Content (meq/g)	
Origin of Humic Matter	**Fulvic Acids**	**Humic Acids**
Aquatic	1.18 ± 0.27	0.88 ± 0.18
Microbial	0.76 ± 0.24	0.60 ± 0.14

Table VI. Wavelengths of the Main Excitation Band and ϕOH Contents of Fulvic and Humic Acid Molecular Weight Fractions of Aquatic and Microbial Origin[a]

Type of Humic Matter	Molecular Weight Range	λ_{excit}[b] (nm)	ϕOH[c] (meq/g)
FA (aquatic)	<75,000	418 ± 7	1.32 ± 0.24
	⩾75,000	330 ± 2	0.95 ± 0.12
HA (aquatic)	<75,000	440 ± 14	0.97 ± 0.16
	⩾75,000	340 ± 14	0.73 ± 0.09
FA (microbial)	<75,000	336 ± 1	0.77 ± 0.24
	⩾75,000	334 ± 4	0.75 ± 0.23
HA (microbial)	<75,000	335 ± 3	0.61 ± 0.15
	⩾75,000	331 ± 3	0.58 ± 0.14

[a] $r = 0.73$; $p = 99.3\%$.
[b] From Table III.
[c] From Visser [22].

(Table VII and Figure 3). The lower fluorescence intensity at higher temperatures would be the result of the increased possibility of the excited molecule to become deactivited by means other than fluorescence. As Figure 3 indicates, the temperature effect was normally very similar for the FA and HA investigated.

To investigate the effect of the redox potential on the fluorescence spectrum of humic compounds of aquatic and microbial origin, solutions of the various FA and HA molecular weight fractions were treated with low concentrations of SO_2, H_2O_2 and O_3 until the redox potential had reached values of respectively −1.17 V, +0.5 V and +2.5 V. With the possible exception of the ozone treatment, these mild reductions and oxidations of the humic matter are normally reversible processes [21]. The fluorescence spectra on the humic matter at different redox potentials were obtained at 25 °C and pH 7.0. The positive correlation coefficients of Table VIII and the regression lines of Figure 4 indicate that the intensities of excitation and emission peaks of humic matter have a tendency to increase with rising redox potential of the environment.

CONCLUSIONS

- Humic matter of aquatic origin fluoresces more strongly, but at lower wavelengths, than soil humic compounds. Humus from the microbial cultures produced the strongest fluorescence.

Table VII. Correlation Coefficients (N = 48) Expressing the Relation Between Temperature and the Intensity and Wavelength of Excitation and Emission Peaks Obtained on Fulvic and Humic Acids from Microbial Cultures[a]

	Wavelengths			Intensities		
	Excitation Bands		Emission Band	Excitation Bands		Emission Band
	Range I (320–348 nm)	Range II (392–461 nm)	(450–550 nm)	Range I (320–348 nm)	Range II (392–461 nm)	(450–550 nm)
FA (microbial)	−0.09	+0.30*	+0.29*	−0.40**	−0.37**	−0.39**
HA (microbial)	A	+0.35*	+0.28*	A	−0.32*	−0.31*

[a] * = $p < 0.05$; ** = $p < 0.01$; A = absence of major peaks.

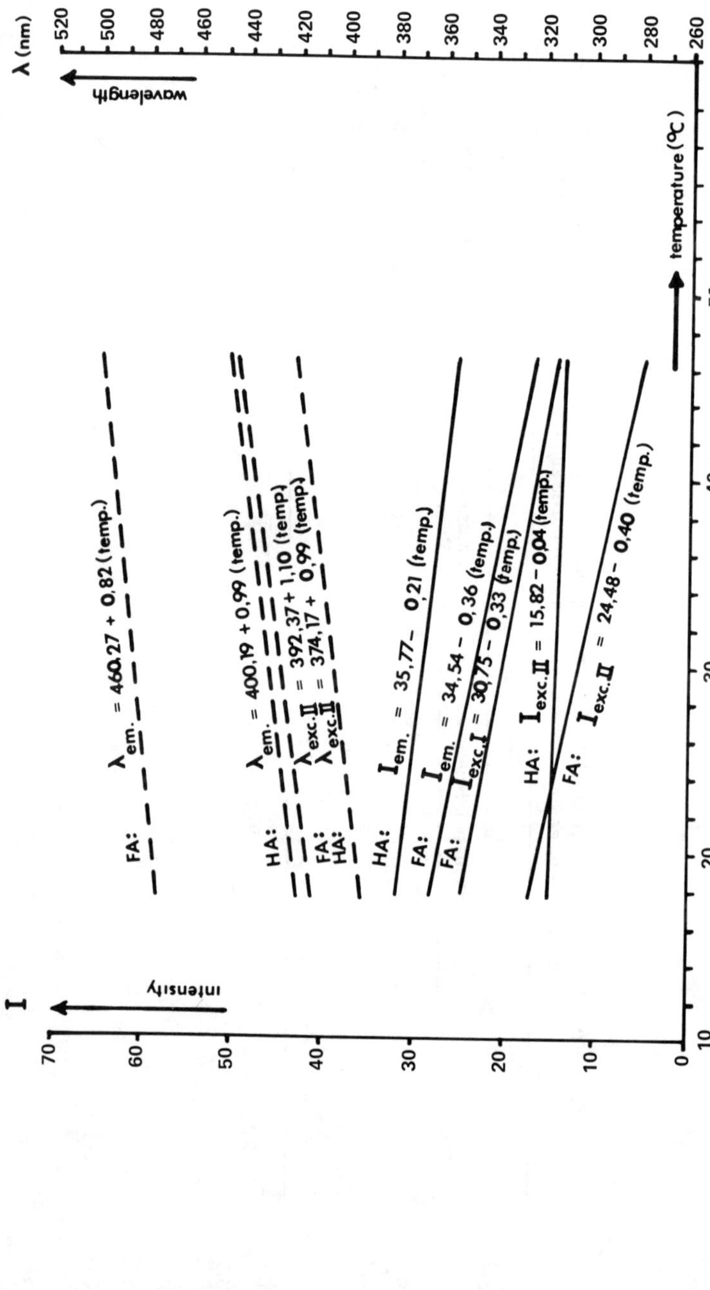

Figure 3. Linear regression curves calculated for significant relationships between the temperature within the range 18 to 48 °C and the wavelengths and intensities of the major excitation peaks and the emission peak obtained on fulvic and humic acids from microbial cultures.

Table VIII. Correlation Coefficients (N = 36) Expressing the Relation Between Redox Potential and the Intensity and Wavelength of Excitation and Emission Peaks Obtained on Fulvic and Humic Acids from Natural Aquatic Environments and from Microbial Cultures[a]

	Wavelengths			Intensities		
	Excitation Bands		Emission Band	Excitation Bands		Emission Band
	Range I (320–348 nm)	Range II (392–461 nm)	(450–550 nm)	Range I (320–348 nm)	Range II (392–461 nm)	(450–550 nm)
FA (aquatic)	A	+0.17	−0.34*	A	+0.27	+0.25
HA (aquatic)	A	+0.10	A	A	+0.33*	A
FA (microbial)	−0.24	A	−0.13	+0.32*	A	+0.22
HA (microbial)	−0.44**	A	+0.29	+0.15	A	+0.11

[a] * = $p < 0.05$; ** = $p < 0.01$; A = absence of major peaks.

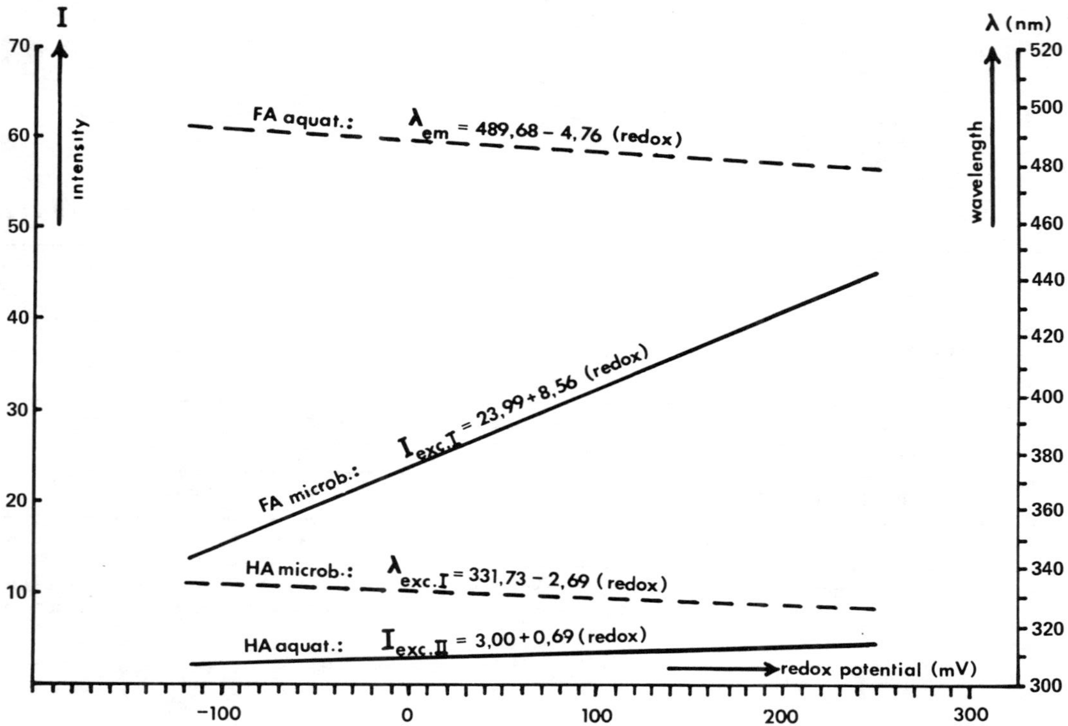

Figure 4. Linear regression curves calculated for significant relationships between the redox potential within the range −1.17 to +2.5 V and the wavelengths and intensities of the major excitation peaks and the emission peak obtained on fulvic and humic acids from natural aquatic environments and from microbial cultures.

- Maximum fluorescence of humic matter occurs at concentrations of between 16 and 70 mg/1, the actual value depending on the type, origin and molecular weight of the material investigated.
- The degree of humification (at least during the early stages of the process) and the season during which the humic material was collected did not appear to have any consistent effect on the excitation and emission spectra of FA and HA.
- Humic matter normally shows two major excitation bands: one at 334 ± 7 nm and the other at 415 ± 20 nm.
- FA and HA fractions with molecular weights <75,000 normally have excitation and emission wavelengths and intensities exceeding those of higher-molecular-weight material.
- The phenolic content of humic matter was found to be positively correlated to the wavelength of the main excitation peak.
- Rises in pH and redox potential and a lowering of the temperature of the environment all result in an increase in excitation intensity.
- The emission of humic matter intensifies with diminishing pH and temperature and with increasing redox potential. A cooler environment will also result in lower excitation and emission wavelengths.

REFERENCES

1. Kleist, H., and D. Mücke. "Optische Untersuchungen an Huminsäuren," *Albrecht-Thaer-Archiv* 10:471-484 (1966).
2. Datta, C., K. Ghosh and S. K. Mukherjee. "Fluorescence Excitation Spectra of Different Fractions of Humus," *J. Indian Chem. Soc.* 48:279-287 (1971).
3. Lévesque, M. "Fluorescence and Gel Filtration of Humic Compounds," *Soil Science* 113:346-353 (1972).
4. Müller-Wegener, U. "Fluoreszenzspektroskopische Untersuchungen an Huminsäuren," *Z. Pflanzenernaehr. Bodenkd.* 140:563-570 (1977).
5. Ghosh, K., and M. Schnitzer. "Fluorescence Excitation Spectra of Humic Substances," *Can. J. Soil Sci.* 60:373-379 (1980).
6. Banerjee, S. K., and S. K. Mukherjee. "The Spectral Analysis and Chromatographic Studies of Fulvic Acids Obtained from the Humin Fraction of Soil Humus," *J. Indian Soc. Soil Sci.* 21:389-393 (1973).
7. Seal, B. K., K. B. Roy and S. K. Mukherjee. "Fluorescence Emission Spectra and Structure of Humic and Fulvic Acids," *J. Indian Chem. Soc.* 41:212-214 (1964).
8. Müller-Wegener, U. "Fluorescencespectroscopic Investigations of Natural Humic Acids," *Chem. Erde* 36:352-354 (1977).
9. Ghassemi, M., and R. F. Christman. "Properties of the Yellow Organic Acids of Natural Waters," *Limnol. Oceanog.* 13:583-597 (1968).
10. Almgren, T., B. Josefsson and G. Nyquist. "A Fluorescence Method for Studies of Spent Sulfite Liquor and Humic Substances in Sea Water," *Analytica Chimica Acta* 78:411-422 (1975).

11. Christman, R. F., and M. Ghassemi. "Chemical Nature of Organic Color in Water," *J. Amer. Water Works Assoc.* 58:723-741 (1966).
12. Black, A. P., and R. F. Christman. "Characteristics of Colored Surface Waters," *J. Amer. Water Works Assoc.* 55:753-760 (1963).
13. Brun, G. L., and D. L. D. Milburn. "Automated Fluorometric Determination of Humic Substances in Natural Water," *Analytical Letters* 10:1209-1219 (1977).
14. Hercules, D. M. *Fluorescence and Phosphorescence Analysis* (New York: Interscience, 1967), p. 258.
15. Oliver, B. G., and S. A. Visser. "Chloroform Production from the Chlorination of Aquatic Humic Material: The Effect of Molecular Weight, Environment and Season," *Water Research* 14, 1137-1141 (1980).
16. White, C. E., and R. J. Argauer. *Fluorescence Analysis* (New York: Marcel Dekker, Inc., 1970), p. 389.
17. Jackson, T. A. "Humic Matter in Natural Waters and Sediments," *Soil Science* 119:56-64 (1975).
18. Hall, K. J., and G. F. Lee. "Molecular Size and Spectral Characterization of Organic Matter in a Meromictic Lake," *Water Res.* 8:239-251 (1974).
19. Schnitzer, M., and S. U. Khan. *Humic Substances in the Environment* (New York: Marcel Dekker, Inc., 1972), p. 327.
20. Rao, C. N. R. *Ultra-Violet and Visible Spectroscopy.* (London: Butterworths, 1967), p. 210.
21. Senesi, N., Y. Chen and M. Schnitzer. "The Role of Free Radicals in the Oxidation and Reduction of Fulvic Acid," *Soil Biol. Biochem.* 9:397-403 (1977).
22. Visser, S. A. "Acidic Functional Group Content of Aquatic Humic Matter: Its Dependence upon Origin, Molecular Weight and Degree of Humification of the Material," *J. Env. Sci. Health* (in press).

CHAPTER 10

CHARACTERIZATION OF AMBIENT LEVELS OF ULTRAVIOLET-ABSORBING DISSOLVED HUMIC MATERIALS IN NATURAL WATERS BY AQUEOUS LIQUID CHROMATOGRAPHY

Wayne S. Gardner and Peter F. Landrum

U.S. Department of Commerce
National Oceanic and Atmospheric Administration
Great Lakes Environmental Research Laboratory
Ann Arbor, Michigan

ABSTRACT

Ambient levels of ultraviolet-absorbing dissolved organic materials of river, lake and sediment pore water filtrates were fractionated into chemically distinct groups by high-performance, distilled water, size exclusion chromatography. When the mobile phase was distilled water, fractionation apparently depended on compound polarity, as well as on molecular weight. Water samples (0.5 ml) were analyzed directly on a TSK 3000 sw size exclusion column with ultraviolet (UV) detection. The chemical character of the three (or four) resolved peaks was examined by pretreatments, including ultracentrifugation; passage through cation exchange, anion exchange, and reverse phase columns; metal addition; and dialysis. The first two peaks could be removed by anion exchange (but not by reverse phase or cation exchange) pretreatment. Components of the third peak were not quantitatively removed by anion exchange, cation exchange, or reverse phase pretreatments. Peak 4, found only in sediment pore water, was removed by reverse phase pretreatment but not by anion or cation exchange. The addition of Na, Ca or Cu to filtrates resulted in removal of peaks 1 and 2, with a concomitant increase in peak 3. Peak 1 was removed progressively less efficiently by Cu, Ca and Na, but in contrast, peak 2 was removed by similar levels of each respective metal. Thus, when distilled water was used as the mobile phase, components of each UV-absorbing peak exhibited unique chemical characteristics.

INTRODUCTION

Knowledge of the chemical and physical properties of dissolved organic matter (DOM), as it exists in natural waters, is needed to understand how DOM affects the cycling of other chemicals. The complex nature of natural DOM [1] has prevented complete understanding of organic interactions with metals, nutrients and xenobiotics in aquatic systems. The bulk of DOM in natural waters, commonly referred to collectively as "aquatic humus" [2], is comprised of refractory, hydrophilic compounds resistant to biological, chemical and physical removal processes. The same properties that keep these stable organic materials in solution under natural conditions complicate the isolation of DOM for physical and chemical characterization. Analytical approaches to studying the composition and function of dissolved humic substances often involve techniques to concentrate and isolate DOM from water [3–5]. Unfortunately, these techniques may change the functional character of the component molecules, and thus bias interpretation of how the molecules interact under natural conditions. For example, polar hydrophilic compounds must be depolarized by decreasing the pH or by increasing the ionic strength of the solution before they can be removed from solution by sorption processes [3,5,6]. Evaporative concentration affects the chemistry of sample solutions by increasing the solute/water ratios, and may change the chemically active character of component organic compounds [7]. To understand the chemical and physical character of organic materials as they occur in the water, one should preferably examine DOM at ambient levels and minimize sample modification.

The purpose of this study was to characterize chemical and physical properties of UV-absorbing dissolved organic compounds at *ambient* concentrations in natural waters. High-performance aqueous liquid chromatography was chosen as the tool to examine natural levels of DOM in water for the following reasons: the compounds could be maintained in their natural solvent; UV detection could be used to measure ambient levels of responsive organic compounds; and pretreatment columns having different chemical and physical separation properties could be used to examine characteristics of component DOM compounds. Ultraviolet absorption does not respond equally to all organic compounds [8], but does measure a large fraction of DOM occurring in natural waters [8,9], and is a useful indicator of compound fractionation behavior on liquid chromatographic columns.

METHODS

Distilled water, size exclusion chromatography (DWSEC), coupled with various pretreatment techniques, was used to fractionate and characterize UV-absorbing DOM (UVDOM) in natural waters. An isocratic liquid chromatographic system was assembled from a high-performance pump (Beckman Model 110A), a Rheodyne 7125 injector, a TSK 3000 sw (50 cm × 7.5 mm o.d.) or TSK 2000 sw (50 cm × 7.5 mm o.d.) size exclusion column, and a Hitachi 100-10 variable wavelength detector (wavelength set of 254 nm) equipped with an Altex Spectrophotometer flow cell. The mobile phase, deionized distilled water, was pumped through the system at a rate of 1.0 ml/min. Sample filtrate (0.5 ml) was injected directly into the liquid chromatograph for sample analysis.

DWSEC recovery of river water UVDOM was examined by comparing total UV peak areas for identical samples passing through and bypassing the column respectively. The average UVDOM recovery (±SD) from a TSK 2000 sw column, previously equilibrated with river water injections, was 95 ± 8% (N = 5) of that obtained when the column was bypassed. In agreement with previous studies on sediment humic materials [10], this suggests that UVDOM is not irreversibly sorbed by TSK columns.

Pretreatments

Changes in DWSEC chromatograms resulting from different chemical or physical pretreatments provided insight about the physical-chemical nature (and about the chemically active sites) of the DWSEC-separated UVDOM peaks. Pretreatments included high-speed centrifugation; passage of the filtered water samples through columns having different chemical characteristics; metal addition; and dialysis.

Centrifugation

Samples were centrifuged for 2 or 4 hr at high speed (~130,000 *g* with a Beckman Airfuge) to establish whether filterable UVDOM in our samples occurred in dissolved or colloidal form. After centrifugation, the respective solutions in the upper or lower portions of the centrifuge tubes were combined and analyzed by DWSEC.

Chemically Specific Column Pretreatments

Filtrates were passed through reverse phase (Sep Pac C-18, Waters Associates), anion exchange (Whatman Partisil-10 SAX), or cation exchange [Benson

Co. (Reno, Nevada) D-X8.25] columns to determine dominant chemical characteristics of compounds in component peaks. The ion exchange columns were assembled in 1-ml syringes modified by replacing the plunger handles with flared Teflon®* inlet tubes. The resins (~0.2 ml) were held between the inlet and outlet openings with glass fiber plugs. The Sep Pac columns were regenerated with 4 ml of methanol, followed by 10 ml of distilled water (20 ml for later experiments). Sample and regenerating solvents were passed through the columns with a syringe. The first milliliter of sample eluate was discarded and the next two were collected for analysis. The anion and cation exchange columns were regenerated with 2 ml of 1 *N* NaCl solution, followed by 5 ml of distilled water.

Metal Addition

Solutions containing different concentrations (ranging from 0 to 0.1 meq/l) of NaCl, $CaCl_2$ or $CuCl_2$ in distilled water were added to an equal volume of sample filtrate to determine the effect of the respective metals on the chromatographic elution patterns of UVDOM. For each metal, progressively higher concentrations were passed through the column in a sequence to prevent analytical problems caused by column-metal interactions from previous metal additions.

Dialysis

Reversibility of metal binding was examined by subjecting metal-treated or untreated filtrates to dialysis with distilled water. Approximately 5 ml of solution was placed in dialysis tubing (Union Carbide, molecular weight cutoff 12,000) and dialyzed in 500 ml of stirred distilled water. The distilled water was changed every 2 hr, and the process repeated 3 times.

Sample Collection

Lake Michigan water samples were collected from a depth of 4 m with Niskin bottles. Water was collected from the Grand River (Michigan) about 1 km upstream from the river mouth. Lake Michigan sediments were collected about 16 km offshore from Grand Haven, Michigan, at a water depth of 80 m. The top layers of oxygenated sediments were scraped from the surfaces of Shipek grab samples. Pore water was removed from the sediments by centrifugation (10,000 *g* for 30 min at 5 °C). All samples were collected between June and October 1980.

---

*Registered trademark of E.I. duPont de Nemours and Company, Inc., Wilmington, DE.

RESULTS AND DISCUSSION

DWSEC Chromatographic Patterns for Untreated Waters

Direct DWSEC analysis of Grand River or Lake Michigan filtrates typically yielded three discrete UV-absorbing peaks (Figure 1). Although retention times were different, similar fractionations were obtained on TSK 3000 sw columns and TSK 2000 sw columns. The first broad peak, containing the largest fraction of UV-absorbing substances, was eluted shortly after the column void volume, and peaked at an elution volume of about 11 ml on the TSK 3000 sw column. Peak 2 was more discrete and eluted near the end of peak 1 at an elution volume of about 12 ml. Peak 3 was also discrete and

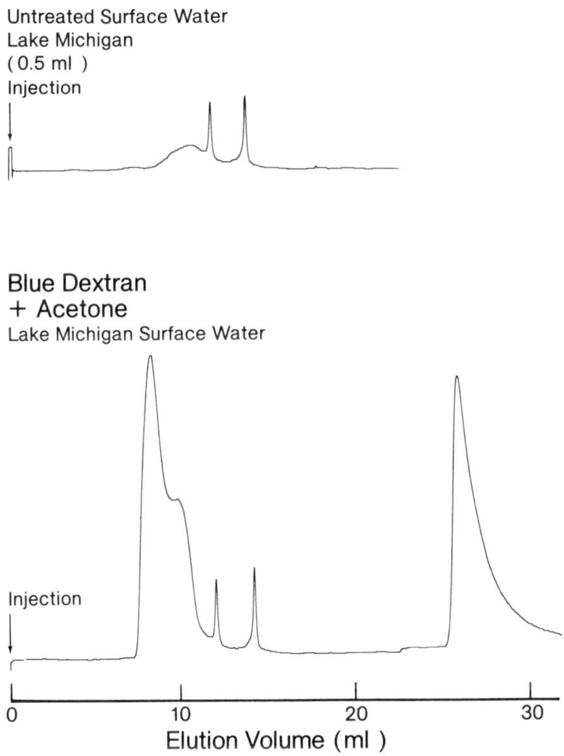

Figure 1. Chromatograms of 0.5-ml Lake Michigan filtrates injected directly onto a TSK 3000 sw size exclusion column and eluted with purified distilled water. Blue dextran (Pharmacia) and acetone were included in the second chromatogram to illustrate the void volume and total exclusion volume of the column.

eluted 14 ml after injection at about one-third of the distance between the column void volume and the total elution volume. In addition to these three peaks, sediment pore water often contained a fourth, smaller molecular-size component (peak 4) that eluted at 23 ml after sample injection. High-speed centrifugation (130,000 *g* for 2 and 4 hr) resulted in similar chromatograms for the upper and lower fractions from the centrifuge tubes and suggested that the UVDOM in the river and lake water was dissolved rather than colloidal.

A series of 10 samples, collected in June 1980 from a transect starting at the mouth of the Grand River and extending offshore into Lake Michigan, provided information about DOM composition and changes in UVDOM resulting from interactions between the river and lake water. The relative proportion of each sample derived from river water was estimated from the conductivity of the samples relative to river water and offshore lake water. The magnitude of total dissolved organic carbon (by the method of Menzel and Vaccaro [11]) and of component UVDOM peaks decreased linearly ($r > 0.90$) with decreasing conductivities. The composition of UVDOM appeared to be similar in all of these surface water filtrates. In each sample. the UVDOM (as eluted from the TSK 3000 sw column with distilled water) was distributed between three peaks (as previously reported for Lake Michigan and Grand River waters [12]). Total levels of dissolved organic carbon and UVDOM were much more abundant in the river water than in offshore lake water, but the relative distribution of components appeared similar in all of the filtrates.

Chemical Characterization of UVDOM by Pretreatment Columns

Chemical characteristics of the UVDOM fractions in the water samples were examined by passing portions of the filtrates through reverse phase, anion exchange, or cation exchange pretreatment columns before fractionation on the size exclusion column. By observing which peaks were reduced or removed by the respective pretreatments, we could make assumptions about probable chemical characteristics of component UVDOM compounds fractionated by DWSEC. All pretreatments were performed at ambient pH so that modification of original chemical properties would be minimized. DWSEC fractionation before and after pretreatment was used to assess the effect of pretreatment on each component peak. Representative results of these column pretreatments for sediment pore water are presented in Figure 2. Similar responses for peaks 1, 2, and 3 were observed for both river and lake water after column pretreatments.

Pretreatment with C-18 reverse phase column material removed a relatively small fraction (~20 to 30%) of the UVDOM appearing in peaks 1, 2, and 3. However, peak 4 was almost completely removed from the pore water

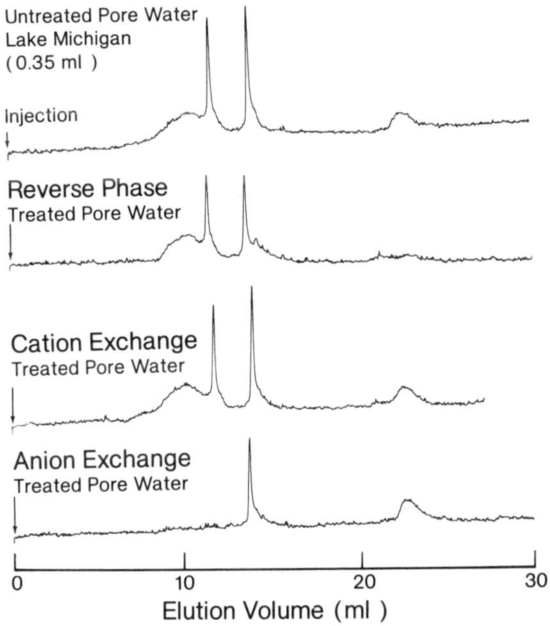

Figure 2. DWSEC chromatograms of sediment pore water before and after reverse phase, cation exchange and anion exchange treatment. The mobile phase was distilled water. (Flow rate = 1 ml/min.) See text for experimental detail.

(Figure 2). Cation exchange treatment removed little (<7%) of the UVDOM components appearing in chromatograms of any of the samples, but usually increased the size of peak 3 slightly. Passage of the sample filtrate through the anion exchange column removed all of the UVDOM from peaks 1 and 2. This quantitative removal by anion exchange and the low removal by the other treatments suggested a strong anionic character for the UVDOM contained in peaks 1 and 2 at ambient pH. This is consistent with the polyphenolic and polycarboxylic acid characteristics associated with organic humus [13,14]. The relatively low removal of these components by cation exchange or reverse phase treatments suggests that cationic sites and organic solubility are less important chemical features of these organic compounds in natural waters than are the active anionic sites. In these studies we made the general assumption that the pretreatments did not greatly affect the chemical nature of the components remaining in the eluant. The similarity in retention times and in the shapes of recovered peaks (e.g., Figure 2) suggested that this was probably a valid assumption for most peak components. An exception may have been peak 3. Peak 3 was not quantitatively removed by any of the pretreatment columns, but was slightly reduced in size by the reverse phase column and was usually enlarged by the cation exchange treatment. Peak 4 was not affected by anion or cation exchange resins, but was quantitatively removed by the reverse phase column. Peak 4, therefore, was likely comprised mainly of nonpolar, hydrophobic, organic compounds.

Separation Mechanisms for Size Exclusion Columns Eluted with Distilled Water

Pure distilled water was used as the mobile phase because it provided more effective resolution of UVDOM components on TSK columns than did the 0.1 *N* NaCl solution (Figure 3). Peaks eluted from river or lake water by 0.1 *N* NaCl appeared to be of smaller molecular size and were not effectively resolved from each other. Salts and buffers are commonly added to size exclusion mobile phases because they tend to reduce solute-column interactions due to charge effects that cause deviations from true molecular size fractionations [10]. On the other hand, a pure distilled water mobile phase has potential advantages for separating compounds having similar molecular sizes, but different chemical characteristics; it exploits potential sample-column interactions and thus can enhance resolution of chemically different components [12,15,16]. Solvent ionic strength, pH and compound composition are factors that can affect retention times of eluted molecules from size exclusion columns with distilled water [2,10,15], but the mechanisms causing deviations from true molecular weight fractionation are not completely

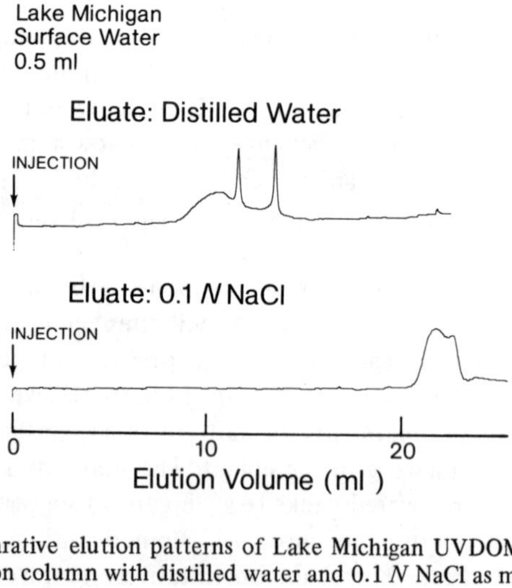

Figure 3. Comparative elution patterns of Lake Michigan UVDOM from a TSK 3000 sw size exclusion column with distilled water and 0.1 *N* NaCl as mobile phases.

understood. Column adsorption [15] and coulombic repulsion [10] have been proposed as possible mechanisms. Another potential mechanism may be a "solvent cage" (hydration) effect. A layer of water, due to hydrogen bonding, may surround charged compounds and increase the apparent size of the molecules [15]. Retention times of similarly sized compounds having different active groups are inversely related to ionization capability when the compounds are eluted from Sephadex with water [16]. Treating the UVDOM cation exchange sites with metals may reduce the apparent molecular size (or other ionic characteristics) of the organic molecules by eliminating their charge through formation of an ion pair. This could eliminate hydrogen-bonded solvent cages, reduce any charge repulsions and allow large molecules to fold, as occurs with proteins [17], to reduce the exposure of hydrophobic sites to the water. Which of the above factors would be important for dissolved organic materials would depend on their chemical composition. We conducted further studies to examine possible mechanisms causing these UVDOM retention time changes in the presence and absence of salts.

Possible reverse phase interactions with the column were investigated by analyzing sample filtrates on the TSK column with 30% methanol and 100% methanol, respectively, as the mobile phases. Elution with 30% methanol yielded peaks similar to those of distilled water elution, and indicated that column effects due to adsorption or reverse phase phenomena were likely not important. In contrast, 100% methanol eluted only one UVDOM peak similar to that observed with 0.1 *N* NaCl. This supported the hypothesis that hydrogen bonding, in a distilled water solvent matrix, could be an important factor regulating elution times.

Mechanisms of Metal Interactions

In agreement with previous work [10] the results with 0.1 *N* NaCl as the mobile phase demonstrated that longer retention times (smaller apparent molecular sizes) resulted when UVDOM was eluted with a salt solution rather than with pure distilled water. Unfortunately the mechanism of the "salt effect" was not elucidated by these experiments. For example, it was unclear whether the change in retention time was caused by an "ionic strength" effect on the column, or by interaction of the salt with the UVDOM constituents. In an attempt to differentiate between these processes (and to establish whether various UVDOM components responded differently to metals), we added metals (chloride forms) to river water filtrates and then subjected the samples to size exclusion chromatography with distilled water as the mobile phase. For water samples mixed 1:1 with 0.1 *N* NaCl and analyzed by TSK DWSEC, the UVDOM eluted as one discrete peak with a slightly longer retention value than peak 3 from the unreacted UVDOM. River water that was mixed 1:1 with 0.1 *N* solutions of divalent $CaCl_2$ and 0.1 *N* $CuCl_2$, respectively, yielded similar results. Thus, addition of any of the three salts effectively removed the resolution between UVDOM components and tended to increase the UVDOM retention volume, even when distilled water was used as the mobile phase.

The above experiments still did not definitively distinguish between metal-UVDOM and metal-column "ionic strength" interactions as possible mechanisms causing shifts in peak elution volumes. Although the mobile phase had zero ionic strength, the presence of added salts in the samples may have caused an "ionic strength" effect. Reducing the added metal concentrations, down to a range where "partial effects" could be observed, offered a means to differentiate between these two possible mechanisms and yielded more information about the chemical character of UVDOM peak components. If the retention time changes were due to ionic strength effects on the column, one would expect resolution to be decreased by the gradual merging of the three peaks into a single peak with increased levels of salt. On the other hand, if the elution volume changes were due to complexation of the metal with UVDOM components, then gradual increases of metal levels could change the relative distribution of the affected components, but would not cause the retention volumes to gradually merge.

To examine these questions, varying concentrations of NaCl, $CaCl_2$ and $CuCl_2$ (ranging from 0 to 0.1 N) were mixed with equal volumes of river water filtrate to establish the pattern of DWSEC peak conversion caused by metals. Two effects on component peaks were observed with increasing metal concentrations. First, a gradual increase in retention time of peak 3 was found with increasing concentrations of metal salts (e.g., Figure 4). This increase in retention time, possibly caused by ionic strength column effects, resulted in slightly greater resolution and did not cause the peaks to merge. The second and more dramatic effect was the gradual disappearance of peaks 1 and 2 with a corresponding increase in the size of peak 3 (e.g., Figure 4). When sufficient metals were added to sample filtrates, the magnitude of peaks 1 and 2 decreased exponentially with metal concentration (Figure 5).

We interpret these results to indicate the occurrence of interactions between the metals and UVDOM peaks 1 and 2. If, as we believe, compounds are separated (at least in part) on the basis of polarity, the complexes formed apparently were less polar than peaks 1 and 2 and appeared similar in ionic character to the original components of peak 3. This interpretation is consistent with our anion exchange pretreatment results; the removal of peaks 1 and 2 from solution by anion exchange suggested that components of these peaks have active cation exchange sites (e.g., Figure 2).

We examined the patterns of peak removal with additions of varying levels of sodium, calcium and copper salts to determine if both peaks would respond similarly to the three metals. The results (Figure 5) indicated that,

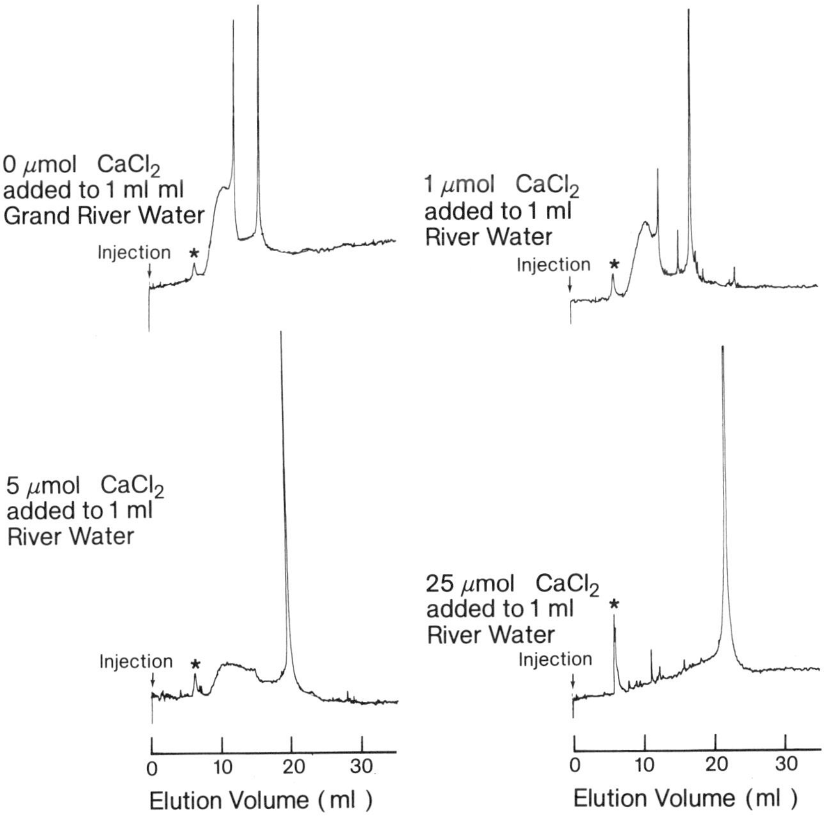

Figure 4. DWSEC elution patterns of UVDOM after reaction of Grand River filtrates with varying levels of $CaCl_2$. The mobile phase was distilled water. (Flow rate = 1 ml/min.) The first small peak in each chromatogram was an injection artifact and not part of the UVDOM from river water.

when the concentrations of metals added were sufficiently high to impact peaks 1 and 2, the peak removal effect was approximately linear with the logarithm of the increased metal concentration (Figure 5). This suggested a possible equilibrium reaction. Low levels (<1 meq metal per liter river water) generally had minimal effects on the peak sizes of the UVDOM components (an exception appeared to be the interaction of peak 2 with sodium). The level of each metal required to completely remove each respective peak was estimated by extrapolating the curve of peak height vs the logarithm of the metal concentration through the axis of zero peak height (Figure 5). This intercept provided an arbitrary measure of the "relative binding strength"

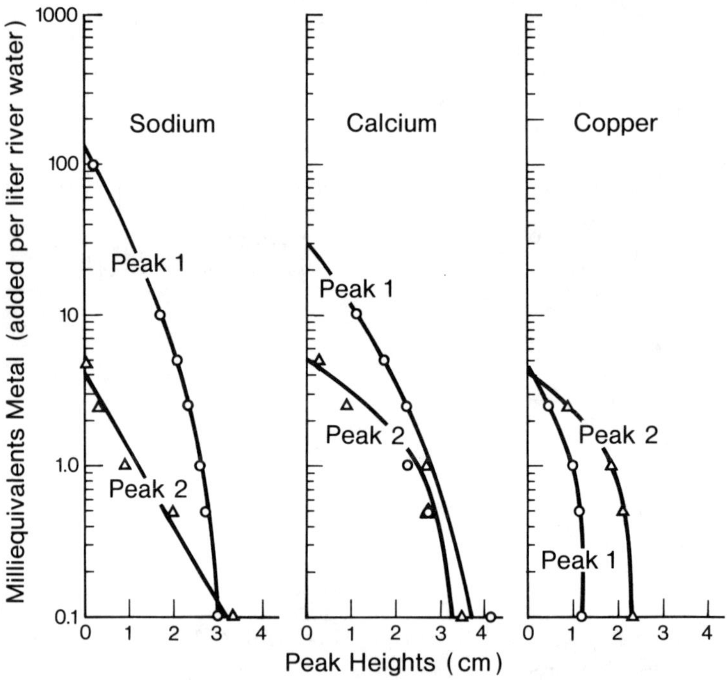

Figure 5. Relationships between added metal concentrations and peak heights for two ultraviolet-absorbing dissolved organic components isolated from river water with a TSK 3000 sw column. The mobile phase was distilled water. (Flow rate = 1 ml/min.) All chromatograms from treatment with a given metal were run on the same day and are therefore internally consistant. Peak heights for the different metal treatments are not directly comparable because chromatograms were run on different days under slightly different conditions.

(not to be confused with "metal binding capacity") of each peak for the respective metals [i.e., the concentration of metal (in solution) required to remove the peak was taken to be inversely proportional to the attraction or "binding strength" between the metal and the peak components]. Our preliminary comparisons of the "relative binding strengths" of the two peaks for the three respective metals (Figure 5) suggested interesting chemical differences between peaks 1 and 2. Whereas peak 2 was removed by similar concentrations (4 to 5 meq/l) of sodium, calcium or copper, peak 1 responded dramatically differently to the three metals. The peak was quantitatively removed by 5 meq/l copper, but required a concentration of 30 meq/l calcium and about 130 meq/l of sodium for complete removal (Figure 5). The differential binding of peaks 1 and 2 is consistent with fulvic acids in natural waters having two types of metal binding sites with different stability constants [18]. Interestingly, another study [12] suggested that metals (calcium and magnesium) may be associated with UVDOM peaks 2 and 3 for untreated river water, but not with peak 1. This could infer that a portion of the cationic exchange sites of peak 2 components were already filled.

The reversibility of the binding between the UVDOM and metals was examined to determine whether peaks 1 and 2 could be regenerated from the newly formed portion of peak 3. Separate aliquots of river water filtrate were reacted with each of the three metals and dialyzed. Unreacted river water was also dialyzed to determine the effect of dialysis on the original peak 3. Peak 1 was partially regenerated from all three metal-treated solutions. Peak 2 was barely detected in the dialyzed sodium- and calcium-treated solutions and was absent from the copper-treated solution. Dialysis caused partial removal of peak 3 from both untreated river water and the metal-treated waters, probably due to loss of organic components through the dialysis membrane and/or loss of metal from peak 3.

Analytical and Environmental Implications

The data presented above demonstrate that DWSEC can fractionate DOM in natural waters into chemically distinct components that likely function differently in aquatic systems. In contrast to size exclusion approaches attempting to minimize factors causing deviations from true molecular weight fractionations, use of distilled water as the mobile phase magnified these deviations and allowed separation of components that were not resolved on the basis of molecular weight alone. Examination of resolved UVDOM peaks, by various pretreatment methods, indicated that the compounds tend to elute from TSK columns in a sequence inversely related to their polarity when distilled water is used as the mobile phase. Similar results were observed when known compounds of similar molecular weight, but with different functional groups, were eluted from Sephadex with distilled water [16]. Our experiments indicate that, while DWSEC provides minimal information about the actual molecular weight spectrum of UVDOM, it provides an efficient and effective approach to fractionate UVDOM into chemically distinct groups that may function differently in water. Pretreatment of sample filtrates provided evidence that each of the DWSEC-resolved peaks contained compounds having different chemical properties than those in other peaks. Peaks 1 and 2 both contained strongly anionic compounds, but responded differently to treatment with various metals. Indirect evidence suggested that peak 3 may (at least in part) consist of organometallic complexes. Addition of metals caused a reduction in size of peaks 1 and 2 with a concomitant increase in that of peak 3. Dialysis resulted in removal of peak 3 and the partial recovery of peaks 1 and 2 from metal-treated filtrates. Peak 4, found in pore water, appeared to be the most hydrophobic of the examined components. It eluted later than the other peaks and was quantitatively removed by the reverse phase pretreatment. Its low abundance in surface waters may indicate that component compounds are formed more abundantly in the sediments than in surface waters. Alternatively, these compounds may be lost from the water column due to biological lability or susceptibility to physical-chemical removal processes.

The combination of DWSEC and sample pretreatment described above allows fractionation and characterization of ambient levels of UVDOM in natural waters. This approach does not determine exact structures of UVDOM components, but apparently isolates components having similar "active groups." This fractionation scheme, in conjunction with element-specific detectors, provides a potential mechanism to study associations of DOM with other chemicals (e.g., metals) in aquatic environments.

ACKNOWLEDGMENTS

We thank J. M. Malczyk for sample filtrate preparation; D. V. Morse, J. E. Grimes and W. R. Burns for help in sample collection; D. Passino (U.S. Fish and Wildlife Service) for the use of the refrigerated centrifuge; A. Corbin for doing many of the chromatographic analyses; and J. Alberts, B. J. Eadie, P. Meyer and S. Tarapchak for reviewing the manuscript.

REFERENCES

1. Jackson, T. A. "Humic Matter in Natural Water and Sediments," *Soil Sci.* 119:56-64 (1975).
2. Gjessing, E. T. *Physical and Chemical Characteristics of Aquatic Humus* (Ann Arbor, MI: Ann Arbor Science Publishers, Inc., 1976), 120 pp.
3. Mantoura, R. F. C., and J. P. Riley. "The Analytical Concentration of Humic Substances from Natural Waters," *Analytica Chimica Acta.* 76:96-106 (1975a).
4. Ishiwatari, R. H. Hamana and T. Machihara. "Isolation and Characterization of Polymeric Organic Materials in a Polluted River Water," *Water Res.* 14:1257-1262 (1980).
5. Thurman, E. M., and R. L. Malcolm. "Preparative Isolation of Aquatic Humic Substances," *Environ. Sci. Technol.* 15:463-466 (1981).
6. Kerr, R. A., and J. G. Quinn. "Chemical Studies on the Dissolved Organic Matter in Sea Water. Isolation and Fractionation," *Deep-Sea Res.* 22:107-116 (1975).
7. Means, J. L., D. A. Crerar and J. L. Amster. "Application of Gel Filtration Chromatography to Evaluation of Organo-Metallic Interactions in Natural Waters," *Limnol. Oceanog.* 22:957-965 (1977).

8. Steward, A. J., and R. G. Wetzel, "Asymmetrical Relationships between Absorbance, Fluorescence and Dissolved Organic Carbon," *Limnol. Oceanog.* 26:590-597 (1981).
9. Dobbs, R. A., R. H. Wise and R. B. Dean. "The Use of Ultraviolet Absorbance for Monitoring the Total Organic Carbon Content of Water and Waste Water," *Water Res.* 6:1173-1180 (1972).
10. Saito, Y., and H. Hayano. "Application of High-Performance Aqueous Gel Permeation Chromatography to Humic Substances from Marine Sediment," *J. Chromatog.* 177:390-392 (1979).
11. Menzel, D. W., and R. F. Vaccaro. "The Measurement of Dissolved Organic and Particulate Carbon in Sea Water," *Limnol. Oceanog.* 9:138-142 (1964).
12. Gardner, W. S., D. A. Yates and P. F. Landrum. "Fractionation of Metal Forms in Natural Waters by Size Exclusion Chromatography with Inductively Coupled Argon Plasma Detection," *Anal. Chem.* 54:1196-1198 (1982).
13. Christman, R. F., and M. Ghassemi. "Chemical Nature of Organic Color in Water," *J. Am. Water Works Assoc.* 58:723-741 (1966).
14. Christman, R. F. In: *Organic Matter in Natural Waters,* (Institute of Marine Science Occasional Publication No. 1., O. W. Hood, Ed., Univ. of Alaska, 1974), 625 pp.
15. Gelotte, B. J. J. "Studies on the Gel Filtration Sorption Properties on the Bed Material 'Sephadex'," *J. Chromatog.* 3:330-339 (1960).
16. Urano, K., K. Katagiri and K. Kawamoto. "Characteristics of Gel Chromatography Using Sephadex Gel for Fractionation of Soluble Organic Pollutants," *Water Res.* 14:741-745 (1980).
17. Lehninger, A. L. *Biochemistry.* (New York: Worth Publishers, Inc., 1970), 833 pp.
18. Mantoura, R. F. C., and J. P. Riley. "The Use of Gel Filtration in the Study of Metal Binding by Humic Acids and Related Compounds," *Analytica Chimica Acta.* 78:193-200 (1975b).

CHAPTER 11

INTERACTION OF METAL IONS WITH A WELL CHARACTERIZED FULVIC ACID

Cooper H. Langford

Chemistry Department
Concordia University
Montreal, Canada

Donald S. Gamble

CBRI Agriculture Canada
Ottawa, Canada

Alan W. Underdown and Stephen Lee

Chemistry Department
Carleton University
Ottawa, Canada

ABSTRACT

The organic chemistry of a well characterized fulvic acid is reviewed to show that it contains a mixture of phenol-carboxylate polyelectrolytes. Light scattering data emphasizing the polydisperse character of the mixture and the importance of aggregational processes under some circumstances are reported. From this base the equilibrium behavior of metal ion binding by a mixture is examined theoretically and experimentally. Several nonintuitive features are seen, including different results from different probes and highly variable apparent stoichiometry. It is shown that this can be explained by the subtle competition effects which arise between cations occupying sites on a mixture polyelectrolyte. Aggregation induced by metal ions is briefly explored, and the important redox properties of the FA are introduced.

INTRODUCTION

Our program is an attempt to make a systematic investigation of a single well characterized sample of fulvic acid (FA) with emphasis given to physical-chemical characteristics. The sample we are using was introduced by Morris Schnitzer, who gave us not only an extensive organic chemical characterization but also a firm foundation in the physical chemistry. The sample is from the Bh horizon of a Prince Edward Island podzol. The state of information on this sample at the outset of the work we are now reporting is well summarized by Gamble and Schnitzer [1].

Gamble and Schnitzer's work has established that mild degradative procedures yield a wide variety of aromatic phenol-carboxylates with benzene and napthalene aromatic nuclei, which unfortunately account for only a fraction of the total mass of the sample. This suggests a high degree of aromaticity and proves that the sample, which is characterized by a number average molecular weight less than 10^3, cannot be structurally homogeneous. There is no way to put all of these fragments back into a single representative molecule of molecular weight near 10^3. Similar conclusions were reached by Christman and Minear [2]. We begin, therefore, from the eminently reasonable proposition that a fulvic acid is a mixture. It is, however, a mixture notoriously resistant to unique fractionation, and is also an entity of significance to the binding and mobilization of metal ions, nutrient ions and pesticides in soils and water *as a mixture.* At this point, a methodological comment must be made. One can fully fractionate a natural mixture and then wonder if the clean chemistry of the components recovered suitably mimics natural behavior or has been modified by fractionation. At the other extreme, one can use raw natural materials in experiments and wonder whether the complexity of the sample renders all interpretation too ambiguous. Our use of "metal-free" fulvic acid clearly reflects a choice to follow a middle path that shares some of the advantages and the difficulties of the extremes.

The philosophy of our approach to any fulvic acid is that we seek to characterize the purified naturally occurring mixture in terms of its function—so far this has encompassed interaction with protons and metal ions—rather than structure. We will maintain a discreet silence concerning details of organic structure. We do so with the comforting observation that the definition of any fulvic acid depends on its acid character, and that the usual procedures for sample purification are ion exchange based and will tend to isolate samples from different origins that are functionally similar with respect to acid-base behavior and metal ion binding. We offer the hypothesis that the understanding gained of titrimetric behavior of a particular sample will be the information most readily generalizable to other samples and also most likely to facilitate predictions of behavior in metal and nutrient transport functions, i.e., extension to field problems.

What we do undertake is a systematic exploration of the consequences of a self-consistent mixture model. This will require formulation of binding equilibria in more general terms than has been common, and will also require attention to aggregational and colloidal aspects of the problem.

Our studies have encompassed four experimental approaches. First (and foremost) has been titration of the FA with OH^- or metal ions using ion-selective electrodes (ISEs), absorptivity and fluorescence as probes [3,4]. Second, we have carried out light scattering studies which were informed by the titrations, to monitor the role of aggregational processes [5]. Third, we have used the kinetics of the formation of easily detected complexes of metal ions (e.g., Fe^{3+}) with dye ligands to distinguish metal ions bound in species of significantly different lability [6]. Fourth, we have used water-nuclear magnetic resonance (NMR) relaxation time measurements to distinguish inner sphere from outer sphere complexation [7]. Currently, we are complementing some of the observations from the kinetic studies with electrochemical measurements.

THE MIXTURE CHARACTER–I

Schnitzer and Gamble [1] reported a functional group analysis for the present sample on a per gram basis as 3.3 meq phenolics, 7.7 meq carboxylate, 3.6 meq aliphatic alcohol, 2.5 meq ketone and 0.6 meq quinoid. This composition is accounted for quite plausibly by the wide variety of organic degradation products recovered [1]. If there is an error in earlier work it underrates the complexity of the sample. Schnitzer and Gamble [1] also report a number average aggregate weight of 951. (Later studies would, if anything, refine assumptions in the calculation and lower this value [8]). It is easily seen that no single polymer structure of molecular weight between 600 and 1000 could include all the fragments isolated. Thus we face the problem of describing a mixture. One component of such a mixture may be represented in Figure 1, although Schnitzer and Gamble [1] suggested an alternative way to build FAs as mainly H-bonded aggregates of their degradation products. This second possibility appears in Figure 2. One of our aims will be to evaluate the relative role of these two pictures in an overall description of the sample. (We do not mean the organic details, but rather the nature of the polymerization process.)

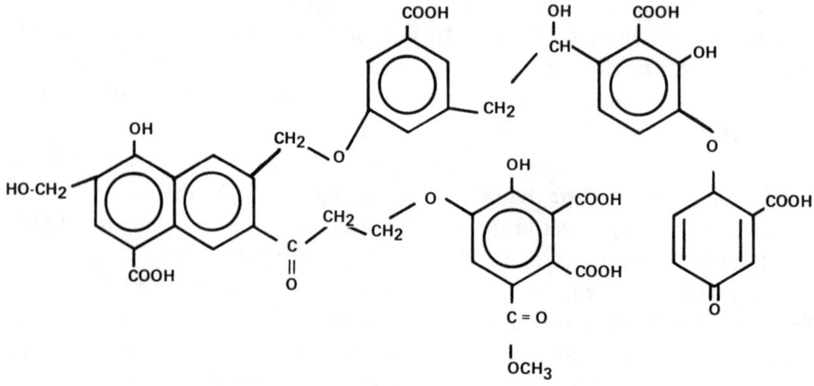

Figure 1. A speculative sketch of a plausible component of the fulvic acid mixture. The molecular weight is near 1000.

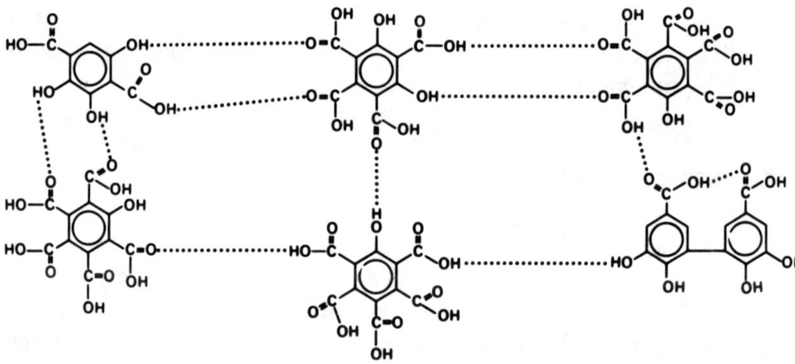

Figure 2. The proposal of Schnitzer and Gamble [1] for the liaisons assembling fulvic acid aggregates.

THE MIXTURE CHARACTER–II

Light scattering studies have been an essential tool of our recent work. Figure 3 shows that up to a concentration of 2 mg/ml of FA light scattering is a linear function of the FA sample concentrations under any particular set of conditions imposed by the medium, and does not reveal shifting aggregational equilibria that are concentration-dependent in this range. In contrast, Figure 4 shows that changing conditions of pH and ionic strength can significantly influence aggregation.

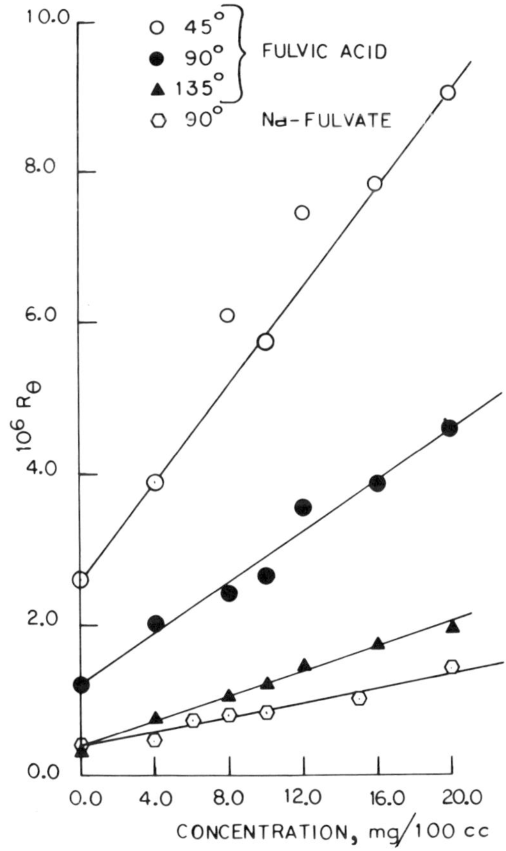

Figure 3. A plot of dependence of light scattering at various angles by the fulvic acid as a function of fulvic acid concentration.

Let's first look at pH-dependence. One might anticipate a breakup of the aggregates of the type shown in Figure 2 as anionic charge accumulates on the polymer, i.e., with accumulation of negative repulsion. The upper curves in Figure 4 show that this can happen only in the case of a sample which has not been subjected to filtration through a 0.2-μ polycarbonate filter. This means inclusion of larger particles. More than 75% of the optical absorbance of a solution passes filtration, but light scattering is significantly reduced and, as the lower curve in the figure shows, these smaller particles show an increase in light scattering as pH rises. This is what might be expected if anionic forms are more polarizable and scatter light better. (But, there is some time-dependence.) A "salting out" effect is seen for the neutralized polymer. At pH values associated with full protonation of carboxylate groups,

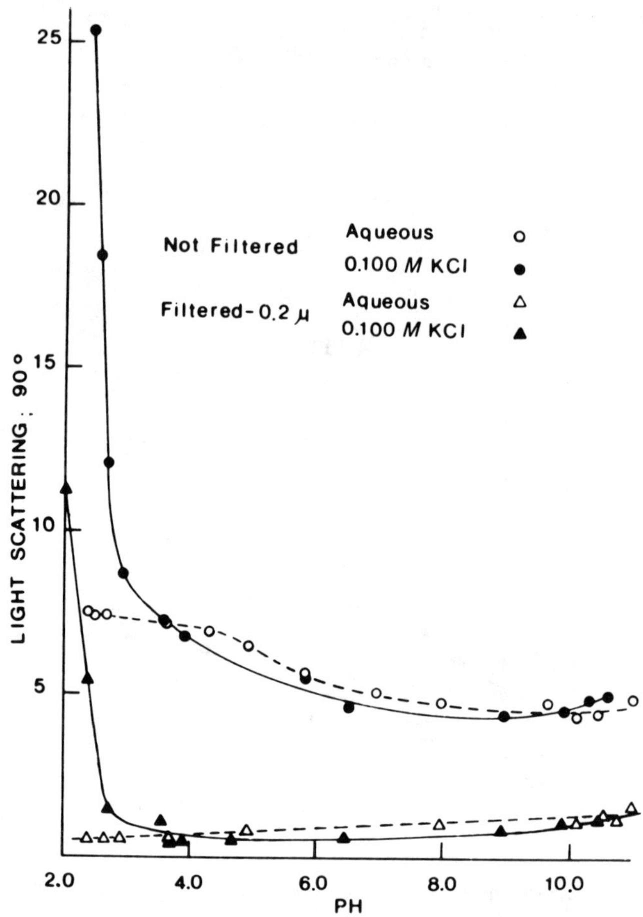

Figure 4. Right angle light scattering as a function of pH before and after filtration with a 0.2-μ polycarbonate filter. Seventy-five percent of the optical absorbance due to fulvic acid is still present after filtration.

0.1 *M* KCl induces extensive aggregation. We would only underline that this "salting out" does not produce a settleable precipitate.

On the basis of these experiments we propose the following reconciliation of Figures 1 and 2. Common small units of our FA sample are represented by Figure 1. The units shown in Figure 1 can aggregate in a manner suggested in Figure 2. That is, the Schnitzer-Gamble model expresses the essence of the structure of components of our mixture as long as we do not think of the building blocks as being too small. Units that are hydrodynamically independent in the mixture may range from very small to fairly large. Confirmation of this model comes from two experiments that concern aggregation. First, we find that number average aggregate weights are much less than weight average weights [5]. Also, Figure 5 shows that 90° light scattering induced by bound Cu^{2+} increases as sites for Cu^{2+} are progressively occupied. However, the measure of the particle size distribution that is most sensitive to the larger particles in the mixture–namely the shape-sensitive 45°/135° ($D_{45/135}$) scattering ratio–does not increase comparably. We see that Cu^{2+} ions are aggregating smaller components of the mixture, but little is happening to either size or shape of the larger components. The larger components are already aggregated after the manner of Figure 1 by H-bonds and/or hydrophobic interactions.

This section needs only one more comment in conclusion. Aggregational processes in the FA mixture are pictured in two dimensions in Figure 2. This is a defect of drawing on a page. The units are no doubt three-dimensional. Consequently, at some point aggregates approach the characteristics of water-swollen gels [9].

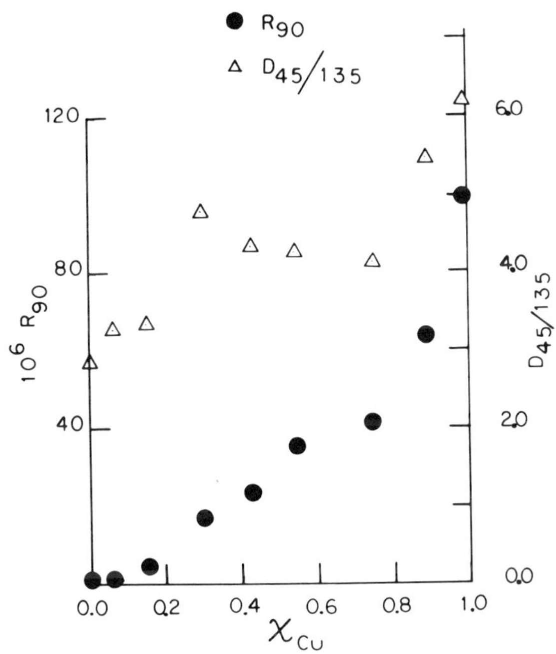

Figure 5. Comparison of right angle light scattering (90°) and the dispersity ratio $D_{45°/135°}$ as the sites on a fulvic acid bind Cu^{2+}. Fractions bound are according to results from ISE titrations.

METAL ION BINDING BY A MIXTURE LIGAND

It has often been suggested that humic substances are the most important organic ligands for metal ions in natural waters [10]. Consequently, models of natural water complexing must be based on treatments suitable to mixtures of the sort we are discussing. It is to be remembered, above all, that polyelectrolyte features will be important.

To introduce the subject, we must recognize that the various ways of monitoring binding of a metal ion do not necessarily "read" a mixture in the same way. Figure 6, which anticipates some results to be given below, shows a plot of three variables against the ratio of Cu^{2+} to FA. The first is a plot of the mole fraction, χ_c, of divalent ion sites occupied by Cu^{2+} as derived from an ion-selective electrode titration to be discussed below. Second, points are shown for the relative quenching, Q, of FA fluorescence by Cu^{2+}. Third, the effect of Cu^{2+} on optical absorbance at 465 nm is shown. No two of the three curves are isomorphous, and the absorbance seems especially to monitor more complex processes, including aggregation. The fluorescence quenching seems to monitor the smallest number of sites. It saturates at the lowest Cu^{2+} concentration. χ_c for ion exchange site occupancy saturates second. The behavior of absorbance at 465 relates to processes continuing after ion exchange sites are occupied, probably aggregation. However, it is noteworthy that the increases of A are not "apparent" absorbance attributable entirely to light scattering.

A suitable formalism for analysis of binding of an ion by a mixture was given by Gamble and Schnitzer [1] following Gamble's development of the theory of acid-base titration [11]. It is important to develop the theory in modest detail to highlight essentials. The case chosen is binding Cu^{2+} to a chelating site [3].

The mixture under analysis will be assumed to provide bidentate sites for binding of Cu^{2+} in four ways. These are: (1) monodeprotonated salicylate sites, (2) orthodicarboxylate sites, (3) two carboxylates on the same polymer molecule but not on the same ring, and (4) carboxylate groups from two hydrodynamically distinct ligands. (The fourth case can be a special sort of "chelate" site, which will receive more attention below.) Considering these four contributing types, we write for the mixture:

$$SH^{-} + Cu^{2+} = SCu + H^{+} \tag{1}$$

$$\overline{K}_4 = \frac{M_C M_H}{M_{SH} M_M} \tag{2}$$

where M denotes a molal concentration
subscript C refers to the complex

S is the ligand
subscript M designates free Cu^{2+} ion

The equilibrium function $\bar{K}_4$ is defined by a mass action expression applied to the mixture of chelating sites. Note that this is not an equilibrium constant.

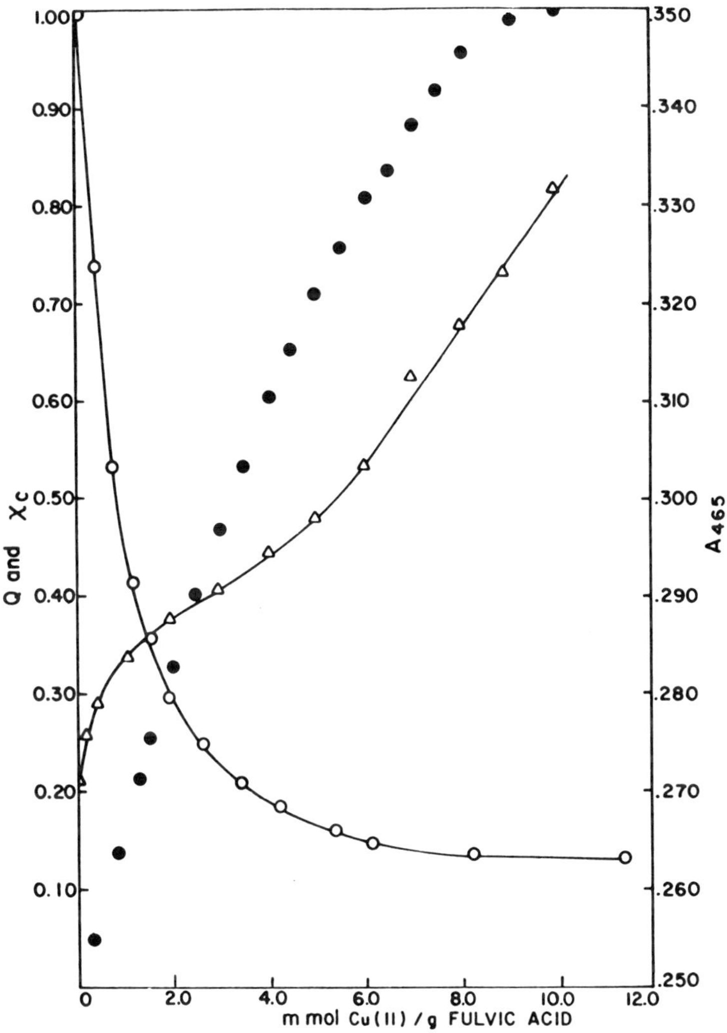

Figure 6. Comparison of three experimental probes of metal binding to the fulvic acid. (a) χ_C, the fraction of sites occupied by Cu^{2+} as calculated from ISE titration; Q, the relative quenching of FA fluorescence; A_{465}, the absorbance by fulvic acid at 465 nm as a function of Cu^{2+} added. See text for guide to curves.

If we now recognize that the bar over K_4 in $\overline{K}_4$ reflects averaging over components of the mixture, we could write an expression for any (ith) component of the mixture in the form:

$$K_{4i} = \frac{M_{Ci}M_H}{M_{SiH}M_M} \tag{3}$$

This expression is related to thermodynamic and/or conditional equilibrium constants because it deals with individual components of the mixture.

Material balances may be expressed by Equations 4 and 5

$$M_C = \sum_i M_{Ci} \tag{4}$$

$$M_{SH} = \sum_i M_{SiH} \tag{5}$$

Combining the above equations into Equation 2 (which is the experimentally accessible expression) Equation 6 can be derived.

$$\overline{K}_4 = \frac{\sum_i K_{4i}M_{SiH}}{\sum_i M_{SiH}} \tag{6}$$

This equation shows the meaning of $\overline{K}_4$. We see that the experimental function is a weighted average over constants for the components of the mixture.

Now, because fulvic acid is a polyelectrolyte, it is useful to factor the K_{4i} values for components into *intrinsic* (K_i^0) and *electrostatic* (ΔG_i^{el}) terms as in Equation 7.

$$K_{4i} = K_{4i}^0 e^{-(\Delta G_i^{el}/RT)} \tag{7}$$

If the variables in Equation 6 are transformed into mole fraction units, it may be shown that $\overline{K}_4$ is given by Equation 8 where χ_{SH} is the mole fraction of ligand sites in the monodeprotonated form.

$$\overline{K}_4 = \frac{1}{\chi_{SH}} \sum_i K_{4i}^0 e^{-\Delta G_i^{el}/RT} (\Delta\chi_{SH})_i \tag{8}$$

An examination of Equation 8 leads to the conclusion that there are three reasons why $\overline{K}_4$ is expected to be not only an average but also to be very well approximated by a continuous function of χ_{SH}. These three are: (1) the coefficient $1/\chi_{SH}$; (2) the relatively large number of individual K^0_{4i} values for which separate titration equivalence points cannot be identified; and (3) the increase in electrostatic effects as charge accumulates on the polymer molecules during titration. We postulate, therefore, that $\overline{K}_4 = \overline{K}_4(\chi_{SH})$. The final theoretical equation is thus Equation 9.

$$\overline{K}_4 = \frac{1}{\chi_{SH}} \int_0^{\chi_{SH}} K_4 d\chi_{SH} \tag{9}$$

Note that $\overline{K}_4$ is directly accessible from experiment but is *not* a constant in any sense. K_4 is the quantity related to thermodynamic or conditional equilibrium constants and Gibbs free energies and is a *differential* quantity. These matters are developed more fully in Gamble et al. [3].

STOICHIOMETRIC RESULTS OF ISE TITRATION

Variability of Binding Capacity

Turning to our largest collection of experimental results, we apply the point of view just explained to titrations of the fulvic acid sample with metal ions using ion-selective electrodes as monitors. The experiments with Cu^{2+} and Mg^{2+} may be approached at two levels. The first is stoichiometric; the second is thermodynamic. At the stoichiometric level we can plot the concentration of bound (total minus free) metal vs the total metal added. As this function approaches a limit, it defines the *apparent* number of binding sites or the "binding capacity." The term "apparent" deserves emphasis here because values are fairly sensitive functions of pH, ionic strength and concentration of fulvic acid. The point is illustrated in Figure 7, which shows the effect of ionic strength on Cu^{2+} binding. Other results for Cu^{2+} are summarized in Table I. We see that the apparent binding capacity reaches a maximum value approximating the theoretical limit for chelating sites *only* for high FA concentration at high pH (no protonation of sites) and low ionic strength. Decreasing pH, increasing ionic strength and decreasing total FA concentration all decrease the apparent binding capacity. It is also true that 10^{-3} *M* Mg^{2+} has a quite significant effect on the binding capacity for Cu^{2+}.

How are we to understand this "fickle" and changeable apparent stoichiometry? (As Table I suggests, Figure 7 has parallels in plots as a function of pH or fulvic acid concentration.) The answer will have to come from proper

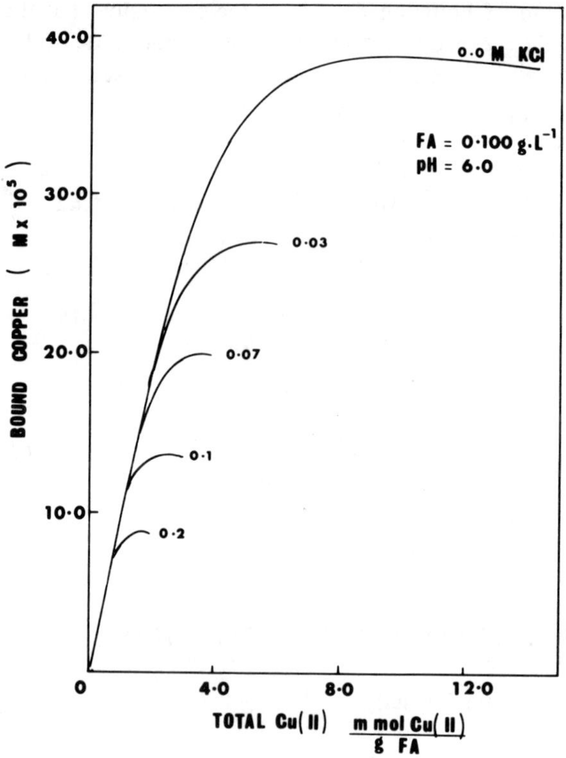

Figure 7. Plots of bound Cu^{2+} vs total Cu^{2+} at various ionic strengths as derived from ISE titration of the fulvic acid with Cu^{2+}.

Table I. Stoichiometry of Apparent Cu^{2+} Binding Capacity of Fulvic Acid

Concentration FA (mg/l)	pH	Ionic Strength	Binding Capacity [mmol/g (FA)]
100	6.00	~0	4.7
10	6.00	~0	3.3
100	3.50	~0	2.6
100	6.00	0.07	2.1
100	6.00	~0	3.6[a]

[a]In presence of 10^{-3} M Mg^{2+}.

understanding of the thermodynamics of binding of a metal ion by a mixture ligand. The values of K_4 must vary considerably as the sites become progressively occupied. Calculations show that the cation sites on the FA are generally occupied by one of the cations Cu^{2+}, Mg^{2+}, K^+ or H^+. The variable stoichiometry reflects the fact that sites with smaller K_4 do not show much selectivity for stronger binding ions and allow these ions to be displaced by the more weakly bound ions (e.g., K^+, Mg^{2+}). This point will be clarified when we have considered equilibria.

A short digression is important here. Many studies reported in the literature have sought to determine a parameter of natural water samples called complexing capacity. It must be realized from the present results that such a parameter is a very sensitive function of many factors (e.g., pH, which is usually fixed in the experimental determinations). We recommend that this parameter as determined by "blind" titration be abandoned.

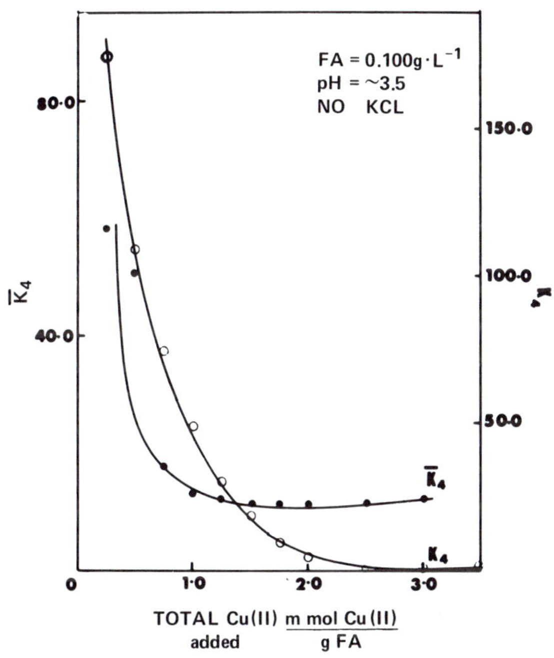

Figure 8. Variation of $\bar{K}_4$ and K_4 calculated according to Equations 2 and 9 for Cu^{2+} from data obtained by ISE titration. In this case pH is approximately 3.5, determined by the 0.100 g/l of FA.

EQUILIBRIUM FUNCTIONS FROM ISE TITRATIONS

Figure 8 shows that, at least for Cu^{2+}, the prediction required by the stoichiometric results is satisfied. In the course of a titration, the values of K_4 that are interrogated by the ISE cover a range of more than 10^2 as the weighted average measured, $\bar{K}_4$, varies by an order of magnitude. The point is reinforced by results at a different pH (6.0) shown in Figure 9, where free energies of binding are plotted against the fraction of apparent Cu^{2+} sites occupied.

Mg^{2+} is less strongly bound by the fulvic acid than Cu^{2+} in general. But, as Figure 10 reveals, the first sites of Mg^{2+} binding in the mixture ligand correspond to larger values of K_4 than those associated with weaker sites for

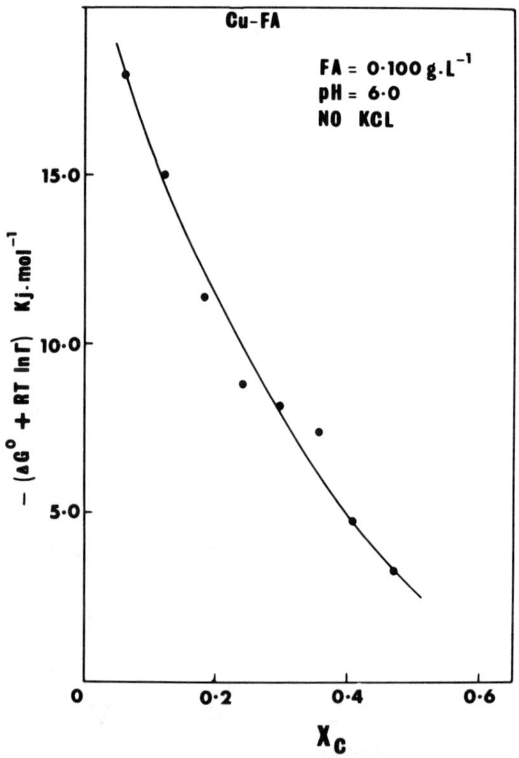

Figure 9. Results of titrations with Cu^{2+} at pH = 6.00 shown as the free energy of binding derived from the K_4 values of Equation 9. ΔG^0 is the intrinsic term. Γ represents the electrostatic term [3]. Convenient methods for these calculations appear in Christman and Minear [2].

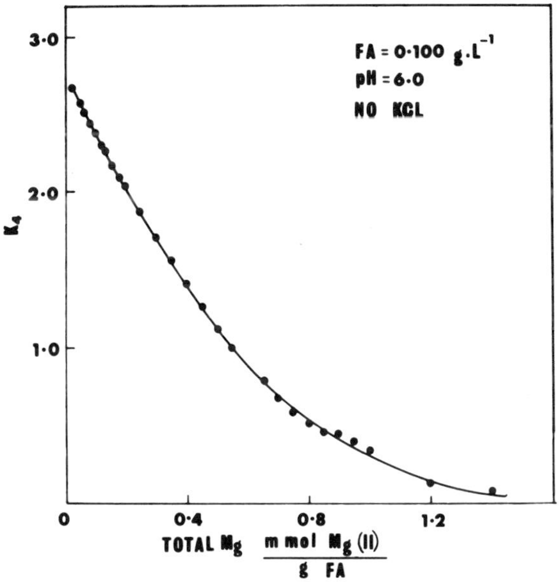

Figure 10. A plot of the differential binding constant, K_4, vs concentration of Mg^{2+}

Cu^{2+}. This means that there will be no simple calculation of the competition between Cu^{2+} and Mg^{2+} using single values of constants, K.

The question of variation of stoichiometry with changing conditions finds its explanation in the variability of K_4. However, this remained a bit opaque to us until we undertook calculations of the fraction of total carboxylate sites occupied by various cations (i.e., Cu^{2+}, K^+, H^+ and Mg^{2+}). It emerged that whenever enough cations were available, the total was constant. All sites bore cations. This makes clear the essential point: strongly bound ions (e.g., Cu^{2+}) compete well for initial selective sites, but at high coverage K_4 is smaller and selectivity is greatly reduced. There are no ions sufficiently strongly bound by these later sites (neither Cu^{2+} nor Fe^{3+}) to exclude the less strongly bound ions (such as Mg^+ or K^+). Therefore, the apparent stoichiometry is variable.

We must add that this explanation does not exhaust the unconventional features of FA as a ligand. Beyond the explanation of the variable stoichiometry of binding in terms of the variable K_4 characteristic of a mixture, there is the different behavior of absorbance (Figure 6) of FA as metal ions such as Cu^{2+} or Fe^{3+} (but notably not Mg^{2+}) bind. What is the relationship of aggregational phenomena to stoichiometry or equilibria of metal binding? The approach to an answer comes from examination of the light scattering data. Figures 11 and 12 show how right angle light scattering (R_{90}) varies as a function of bound Cu(II) or Fe(III). It is seen that for low coverage, the metal ions produce little aggregation (increase of R_{90}). However, as the apparent binding capacity is approached, the degree of aggregation increases sharply.

One interpretation would see this as little more than the "salting out" effect proposed as an interpretation of the left side of Figure 4. Such a view is incomplete because the effects are found at lower ionic strength *and* are selective (e.g., not produced by Mg^{2+}). At a recent meeting of L'Association Canadienne-Française Pour L'Avancement des Sciences, we proposed that the phenomenon observed by light scattering and from absorbance be called "pseudochelation." It is "pseudo" in the sense that the donor atoms come from two molecules–which confounds the definition of chelation. However, it is a chelate-like behavior in the sense that it is clear that interaction among

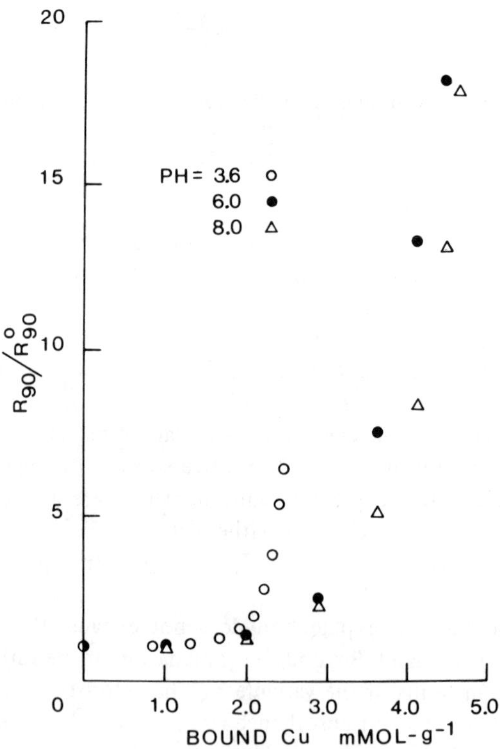

Figure 11. The behavior of relative right angle light scattering R_{90}/R^0_{90} at various pH values as bound Cu^{2+} increases. Bound Cu^{2+} is calculated from ISE titration.

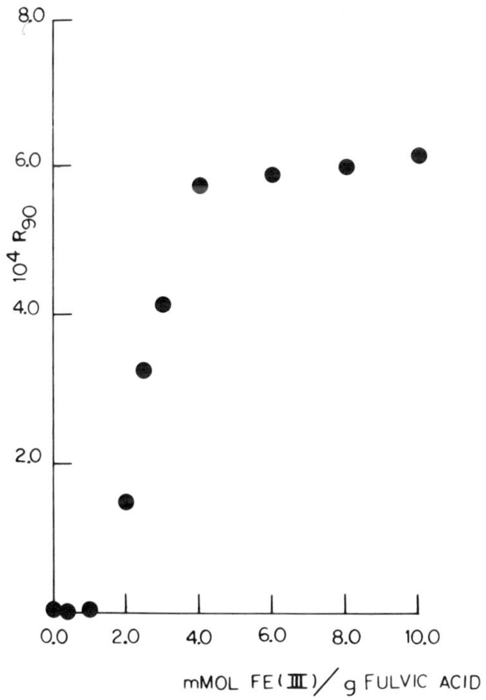

Figure 12. The increase of right angle light scattering by the fulvic acid as concentration of Fe(III) increases. pH $\simeq$ 3.5.

parts of the ligand molecules is important to metal-induced aggregation. It is clear that Figures 11 and 12 do not show a morphology similar to that for formation of simple ML_2 complexes. We suspect that it is H-bonding and/or hydrophobic interactions among the "ligand" parts that render the ligand in the aggregate chelate-like.

REDOX CHEMISTRY

Finally, the implications of the mixture model for the redox chemistry of FA should be addressed. The role of the present sample in reducing Fe has been discussed [6,12]. A field study of the correlation between metal ion concentrations and organic matter suggested that redox chemistry may play a larger role than complexation in metal ion mobilization [13]. Skogerboe and Wilson [14] recently discussed the probable role of kinetics in redox chemistry. We introduce the subject briefly here.

In Figure 13 the kinetics of the induced reaction of the FA sample with Fe(III), which follows from addition of an Fe(II) specific colorimetric reagent, is shown. The conversion to the Fe(II) complex is quantitative but not fast. The plot shows a large intercept at t = 0. This reflects the presence of a faster component. At a minimum, there are two kinetically distinguishable components in the reduction of Fe(III) by FA. In a cyclic voltammogram for FeFA, there is an irreversible peak. These preliminary experiments point to a complex situation. It seems unlikely that a treatment parallel to the mixture model for metal ion binding will be adequate for the metal fulvate redox chemistry. The kinetic factors may well dominate and are certainly important.

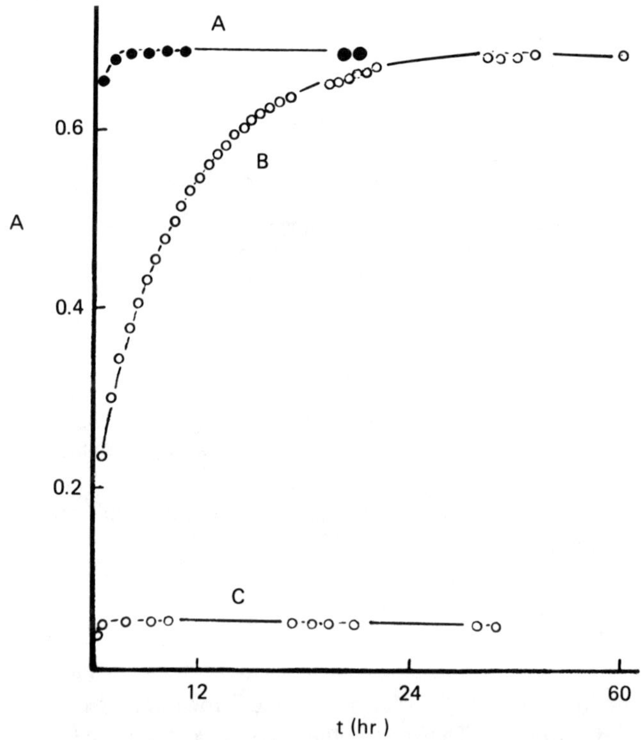

Figure 13. Kinetics of reaction of Fe(III) bound to FA with colorimetric reagent systems. The reagent systems are: A, Ferrozine plus hydroxylamine hydrochloride–an Fe(II) reagent plus a reducing agent; B, Ferrozine–an Fe(II) reagent *inducing* reduction of Fe by the FA; C, sulfosalicylic acid–an Fe(III) reagent. The Fe–FA systems were aged for 27 hr at pH = 6 and then mixed with colorimetric reagents. T = 25°C.

REFERENCES

1. Gamble, D. S., and M. Schnitzer in "Trace Metal and Metal-Organic Interactions in Natural Water," P. S. Singer, Ed. (Ann Arbor, MI: Ann Arbor Science Publishers, Inc., 1973).
2. Christman, R. F., and R. A. Minear in "Organic Compounds in Aquatic Environments," S. J. Faust and J. V. Hunter, Eds. (New York: Marcel Dekker, Inc., 1971).
3. Gamble, D. S., A. W. Underdown and C. H. Langford. *Anal. Chem.* 52:1901 (1980).
4. Underdown, A. W., C. H. Langford, and D. S. Gamble. *Can. J. Soil Sci.* 61:469 (1981).
5. Underdown, A. W., C. H. Langford and D. S. Gamble. *Anal. Chem.* 53 (1981), in press.
6. Wong, S. M., C. H. Langford and A. W. Underdown. *Can. J. Chem.* 59:181 (1981).
7. Gamble, D. S., C. H. Langford and J. P. K. Tong. *Can. J. Chem.* 54:1239 (1976).
8. Weber, J. H., and S. A. Wilson. Report #14, Water Resources Centre, Univ. of New Hampshire, Durum, NH, 1977.
9. Travers, C., and J. A. Marinsky. *J. Poly. Sci.* Part C, Symposium #47, 285 (1974).
10. Schnitzer, M., and S. U. Khan. *Humic Substances in the Environment,* (New York: Marcel Dekker, Inc., 1972).
11. Gamble, D. S., *Can. J. Chem.* 48:2662 (1970).
12. Langford, C. H., R. Kay, G. W. Quance and T. R. Khan. *Anal. Letters* 10:1249 (1977).
13. Khan, T. R. PhD Thesis, Carleton University, Ottawa, 1980.
14. Skogerboe, R. K., and S. M. Wilson. *Anal. Chem.* 53:228 (1981).

CHAPTER 12

PROTON AND CALCIUM COMPLEXATION OF FOUR FULVIC ACID FRACTIONS

Brian A. Dempsey* and Charles R. O'Melia

Department of Geography and Environmental Engineering
The Johns Hopkins University
Baltimore, Maryland

ABSTRACT

Fulvic acid (FA) from Lake Drummond, Virginia, was separated into four fractions based on desorption from a nonionic resin (XAD-7) with increasing pH (4, 7 and 10). The deprotonation of each fraction was determined as a function of pH (3 to 10) and I (0.01, 0.10 and 1.0) using the charge balance equation. The extent of calcium complexation of each fraction was determined as a function of pCa (3 to 4) and pH (5 to 9) at I = 0.1. Additionally, proton release during complexation of FA and Ca was measured.

The extent of proton complexation by the four FA fractions was similar, when expressed as meq of anionic charge generated per gram of total organic carbon (TOC). The calcium complexation reactions of the four FA fractions were quite different, however; the pH-dependence of the reaction was considerable for the FA fractions extracted at pH 4 (reaction stronger at pH 9 than at pH 5) and negligible for the FA fraction extracted at pH 10. Both proton and calcium complexation reactions were independent of FA concentration. For some experimental conditions, greater than 50% of the net negative charge on FA was neutralized by reaction with Ca, and the effect of this charge neutralization is demonstrated by the dramatic increase in the adsorption of FA on $MnO_x(s)$ in the presence of 5×10^{-3} *M* Ca. The measurement of Ca–FA complexation is discussed. Mechanisms for the reactions between Ca and FA are proposed.

---

*Address all correspondence to this author. Current address: Department of Civil Engineering, University of Missouri, Rolla, Missouri 65401.

OBJECTIVES

The goal of the research described in this chapter is to determine the processes by which humic substances are *removed* from natural waters, i.e., to characterize natural solid-liquid fractionation processes involving humic substances. The chapter is written with two principal objectives: (1) to present the preparative scale fractionation method used to isolate aquatic fulvic acids and (2) to present and interpret results of laboratory studies of the interactions among FA, H^+ and Ca^{2+}. In addition, an example of the effects of Ca on the adsorption of FA is given. Finally, results of model development for the FA–H–Ca system are presented and discussed.

INTRODUCTION

Fulvic acids are a group of acidic, polydisperse, randomly polymerized macromolecules. They are the predominant form of organic carbon in unpolluted waters. FA can affect the toxicity of organic and inorganic species by increasing the solubility or decreasing the toxicological activity of such species; they interfere with water treatment processes designed to remove other impurities from water; they are targets for removal *per se* because of the toxicity of chlorinated products or because of esthetic considerations.

Fulvic acids are primarily defined by the technique which is used for their extraction. The classical methods for the extraction of FA from other aquatic or soil components are based on the solubility of FA under specific solution conditions. However, separation of the polydisperse FA into more homogeneous fractions has typically been accomplished by size fractionation [gel permeation chromatography (GPC) or ultrafiltration (UF)]. Both GPC and UF techniques have been criticized because molecular weight as determined by these techniques is often much higher than molecular weight determined by other methods. However, although the basis for separation is uncertain, separations by GPC or UF have been successful in producing fractions of natural organic material that react differently in terms of metal complexing capacity [1–3], trihalomethane formation potential [4,5] and adsorption [5,6]. Mantoura and Riley [7] suggested that humic materials could be fractionated by successive desorption from XAD resin using a series of eluants having progressively higher pH values. Such a fractionation technique is based on solubility. MacCarthy et al. [8] have used pH gradient elution to fractionate analytical concentrations of aquatic humic materials. In this research, aquatic fulvic acid has been fractionated based on desorption from XAD-7 resin as a function of pH.

Much of the reactivity of humic materials is due to acidic functional groups that are predominantly carboxylic and secondarily phenolic. It is through these functional groups that FA complexes trace metals. Furthermore, the negative charge which is generated on these functional groups tends to stabilize FA in solution, since the solubility in water is increased and since adsorption onto environmentally available solids (which are predominantly negatively charged) is opposed by an electrostatic force. Schnitzer and Khan [9] have reported that model soil FA contains 8.2 meq of carboxylic groups per gram of humic material. Similar numbers have been reported for aquatic FA by Weber and Wilson [10]. Thus 12 mg/l of FA (about 6 mg/l as TOC) represents 10^{-4} *N* carboxylic acid, which can be compared to average river and lake water concentrations of 7.2×10^{-7} *M* of iron (Fe) and 3.8×10^{-4} *M* of calcium [11]. Carboxylic and phenolic concentrations in FA can be measured by several techniques. Perdue et al. [12] have critiqued and investigated techniques. In this research, potentiometric titrations have been used to determine the ionization of FA as a function of pH and ionic strength. Several models have been tested for description of the data.

Humic acid (HA) can be precipitated by the addition of unhydrolyzed di- or polyvalent cations [13–15]. HA and FA can be removed by reaction with hydrolyzed metal species [16–18]. There is a stoichiometry of reaction between humic material and cation, and restabilization sometimes occurs upon overdosing; these are indications that the removal of humic materials is sometimes due to charge neutralization. In engineered systems, either Fe or Al salts are typically added for the removal of humic materials. In lakes and rivers, it is unusual to find Fe or Al at normalities greater than the normality of humic materials. Additionally, hydrolysis reactions of Fe and Al are complex, both kinetically and thermodynamically, and the study of reactions between Fe or Al and humic materials must have some ambiguities. Ca, however, may be more concentrated than humic materials in aquatic systems ranging from estuary to lake to water treatment plant, and Ca does not hydrolyze significantly until pH exceeds 10. In addition, Ca appears to be more reactive than other alkaline earth or alkali metals towards HA [19]. The reactions between Ca and fractionated FA have been studied in this research, therefore, since Ca is a reactive, ubiquitous, concentrated cation whose reactions are more easily understood than are the reactions of Fe or Al.

REAGENTS AND INSTRUMENTATION

All chemicals used in this study were Analytical Reagent Grade or better. Water was distilled deionized. Ionic strength was adjusted to 0.01, 0.1 or 1.0 with KCl. Commercially prepared 0.1 *N* NaOH and HCl were calibrated with potassium hydrogen phthalate (KHP) and Na_2CO_3 for use in titration experiments. KHP was used as a TOC standard after drying at 103 °C for 2 h. Nitrogen (99.995% pure) was bubbled through NaOH solution and distilled deionized water to clean and humidify the gas. Experimental vessels were made of Pyrex glass. For the H-titration experiments, glassware was cleaned with a 30-min soak in 0.1 *N* NaOH to remove humic substances, a several-hour soak in 0.1 *N* HNO_3, and an overnight soak in distilled deionized water, followed by oven drying for nonvolumetric containers. For other experiments the soak in HNO_3 was extended overnight.

Determination of pH was made with a glass electrode and either a double junction sleeve-type or a single junction frit-type reference electrode. Ca was determined using a perm-selective specific ion electrode developed by Photovolt and a double junction electrode. Both Orion 701A and Fisher Accumet Model 620 pH/mV meters were used. The relative accuracy for the Orion 701A is ±0.001 pH, ±0.1 mV, or 1% of a mV reading, whichever is greatest. The Accumet 620 is slightly less accurate, and was used only to monitor pH while Ca was being monitored by the Orion 701A.

TOC was analyzed using either a Beckman 915 Carbon Analyzer in the TC mode after sparging of the acidified sample or a Dohrmann Envirotech DC54 Carbon Analyzer (in the seawater mode if I was greater than 0.01). A Perkin Elmer 560 flame atomic absorption spectrophotometer was used to measure metal concentrations while other UV-visible absorbance measurements were made using a Varian Techtron Model 635 spectrophotometer.

Amberlite XAD-7 resin was obtained from Rohm and Haas and was prepared for use by washing and decanting five times with 0.1 *N* NaOH. The resin was then Soxhlet-extracted sequentially with methanol, acetonitrile and diethyl ether for 24 hr each with a siphon every 20 min and then stored in methanol. Five hundred and fifty cm^3 of resin was packed from a methanol slurry into a column 55 cm in height and 5 cm in diameter. Pores were completely wetted using a vacuum and vibration during upflow washing with methanol. The resin was left in methanol 24 hr, and then washed with 100 bed volumes (V_B) of freshly distilled deionized water. During this last phase of cleaning the flow was adjusted so that resin fines were continuously being washed out. During subsequent downflow washing with water (30 V_B) the TOC in effluent was consistently 0.8 mg/l. The resin was washed with 5 V_B each of 1.25×10^{-3} *N* NaOH, 10^{-2} *N* HCl, 2.5×10^{-3} *N* NaOH and 10^{-2} *N* HCl. The effluent TOC after each of these elutions was 0.8, 0.2, 2.0 and 0.5 mg/l, respectively. During elution of FA the pH was never allowed to exceed 10, thus hydrolysis of the XAD-7 resin was not significant. Capacity, kinetics and peculiarities of the XAD series of resins have been discussed by Aiken et al. [20]. The superior capacity, kinetics and reversibility of XAD-7 over the other Amberlite resins studied by Aiken was rationale for the use of XAD-7 in these experiments.

Sephadex G-25 and G-50 resins were cleaned and prepared for use according to the manufacturer's instructions.

SOURCE AND PROCESSING OF FA

The source of the FA materials used in this study was Lake Drummond, Virginia. The water chemistry and flora, the morphology of the lake and drainage basin and its soil characteristics have been reviewed elsewhere [21]. Briefly, Lake Drummond is a circular lake with 13 km^2 of surface area. The drainage area is about 360 km^2 and is predominantly mucky peat of 1 to 5 m in depth. Dry and intermittently dry areas are mostly cane and wooded vegetation, while diatoms, chlorophytes and dinoflagellates are the predominant algal population within the lake. Lake water is derived from precipitation, several small tributaries from the Suffolk Scarp draining eastwards, and numerous drainage ditches. Lake Drummond was selected as the source of FA for these experiments because (1) the lake has a high concentration of TOC (typically 30 to 50 mg/l), (2) the humic material in Lake Drummond is predominantly FA, and (3) humic materials from Lake Drummond have been used in other studies of the nature of humic materials [22].

The Fractionation Strategy

It was decided that FA would be separated into fractions based on solubility in water as a function of pH. The general scheme of the fractionation procedure was (1) apply natural organic material at pH 2 to a column of XAD-7 resin and clean the adsorbed material with 0.01 *N* HCl followed by distilled deionized water using elution volumes such that material with a distribution coefficient (K_d) of 125 at pH 2 would be at least 50% retained [23], and (2) sequentially extract the adsorbed FA at pH 4, 7 and 10 using 0.1 or 1 *N* NaOH to adjust the pH and using the weak acid character of FA to achieve pH-buffering.

Concentration, Cleaning, Fractionation and Storage of FA

Sixty gallons of water was taken from the site described above on December 21, 1979, and transported to the laboratory in 5-gal plastic containers which had been washed with 0.1 *N* NaOH, 0.1 *N* HCl and distilled deionized water. Raw water characteristics are shown in Table I. The samples were purged with N_2, acidified to pH 2.0, allowed to sit 24 hr, filtered through a glass fiber filter to remove suspended solids and applied to 550 cm^3 of XAD-7 resin at 10 V_B/hr. A maximum of 100 V_B of acidified Lake Drummond water was applied per cycle. The column was subsequently eluted with 30 V_B of 10^{-2} *N* HCl and 20 V_B of distilled deionized water to remove inorganic material. During application of lake water, effluent samples were analyzed for TOC, absorbance (Abs), Fe_T and pH. Abs measurements were done at λ = 465 nm and pH = 7. Approximately 33% of TOC, 15% of Abs and 80% of Fe_T were eluted to waste during application and cleanup of

Table I. Chemical Characteristics of Lake Drummond

Parameter	Value
pH (average)	4.3
TOC (mg/l)	45
DO (mg/l)	4.7 to 12.4
Ca (mol/l)	3.5×10^{-4}
Mg (mol/l)	1.5×10^{-4}
Fe (mol/l)	3.2×10^{-5}
Mn (mol/l)	3.1×10^{-6}

Lake Drummond water. pH was steady at 2 during application and acid wash, and rose to 2.7 by the end of the water rinse.

To begin extraction of FA, the fluid level in the column was lowered until about 700 ml of liquid remained in the column, and 300 ml of distilled deionized water was placed in a beaker. The pH of the solution in this beaker was maintained at pH 4 by addition of 0.1 or 1.0 *N* NaOH by automatic titrator while the contents were recirculated through the column and beaker at 10 V_B/hr. During this time the air space in the beaker was kept under a positive pressure of nitrogen gas. The FA, which had adsorbed on the column at pH 2, desorbed until TOC and Abs plateaus were reached. This was assumed to be the equilibrium state for the new solution conditions. Once equilibrium conditions were achieved, the contents of the beaker were transferred to a storage container and the liquid level in the column was lowered so that it was just above the resin by eluting to the storage container. One V_B of distilled deionized water was then pumped onto the column while the eluant was added to the storage container. At this point distilled deionized water was added to the column to raise the liquid level, 300 ml of water was placed in the beaker and recirculation was renewed, with pH maintained at 4. The Abs was monitored and if, at equilibrium, the Abs was approximately equal to the plateau value from the first pH 4 extraction, the extract was removed in exactly the way described above. If, on the other hand, the Abs at plateau was less than 50% of the Abs of the initial pH 4 extraction, the pH was raised to 7 and allowed to equilibrate before removing the extract. The same criterion was applied to decide when to raise the pH to 10. The decision of when to raise the pH of the eluant is arbitrary, but experimentally the choice was not difficult because the drop in plateau Abs was normally quite dramatic. Because more FA was extracted at pH 4 than at the other pH values, the material that was extracted at pH 4 was split into two FA fractions. The material extracted initially at pH 4 was called FA #1. The material extracted at later pH elutions was called FA #2. FA #3 comprised all the material which was extracted at pH 7 and FA #4 was extracted at pH 10. The FA fractions were stored in dark glass bottles, at 4 °C, in the dark, under N_2, at the pH of extraction. During concentration and cleaning procedures 67% of TOC was retained by the column. The four FA fractions accounted for 98% of the TOC that had been adsorbed by the XAD-7 column.

Characteristics of the FA Fractions

Some characteristics of the four FA fractions are shown in Table II. Because of the method of extraction, these FA fractions contained a relatively high concentration of Na. In addition, a significant amount of Fe remained associated with the FA even though more than 80% of the Fe that initially was in the Lake Drummond sample was removed by the concentration and cleaning process. The remaining Fe was tightly bound (cation exchange was unsuccessful in removing the residual Fe); the molarity of bound Fe was from 1.3 to 3.1% of carboxylate normality of the FA fractions. The concentrations of Ca, Mg, Cu, Mn, K, Cl, NO_3, NO_2, PO_4 and SO_4 in the FA extracts (analyzed by standard methods) were all less than or equal to 10^{-5} *M*.

The fractions were not significantly different in molecular weight. All four FA fractions were excluded by Sephadex G-25 but were eluted from Sephadex G-50 between the exclusion and the void volumes. The eluant was tris buffer at pH 9 [24]. This behavior indicates weight averaged molecular weight between 5000 and 10,000 (the exclusion volumes for G-25 and G-50). Small-angle X-ray scattering and sedimentation equilibrium both indicated that the samples were polydisperse. From small-angle X-ray scattering the smallest particles were determined to have molecular weights of 1500 to 2500.

IR and UV spectroscopy also could not distinguish among the FA fractions [21].

Table II. Chemical Characteristics of the Four FA Fractions

Parameter	FA #1	FA #2	FA #3	FA #4
pH of extraction	4	4	7	10
TOC (mg/l)	524	501	666	271
TKN (% of organic matter)	1.1	0.9	0.8	1.0
Na (mol/l)	3.37×10^{-3}	3.41×10^{-3}	6.74×10^{-3}	4.06×10^{-3}
Ca (mol/l)	1.0×10^{-5}	3.5×10^{-6}	3.0×10^{-6}	5.5×10^{-6}
Mg (mol/l)	3.0×10^{-6}	1.5×10^{-6}	$<1.0 \times 10^{-6}$	$<1.0 \times 10^{-6}$
Fe (mol/l)	1.9×10^{-4}	1.3×10^{-4}	1.2×10^{-4}	4.2×10^{-5}

THE EXPERIMENTAL PROTOCOLS

Reaction Between H and FA Fractions

Initial Charge of the FA Materials

The charge density on FA at the beginning of an experiment was determined by measuring the concentrations of all ions that were considered to be present at significant concentrations, and using the charge balance equation to determine the initial charge of the FA. Since the residual Fe was believed to be associated with the FA, Fe was not included in the charge balance equations. The anions Cl, PO_4, SO_4, NO_3 and NO_2 were all found to be less than 1% of the Na concentration, so that the normality ($[A^-]$) of the FA solution was

$$[A^-] = [Na^+] + [H^+] \quad (1)$$

Use of the charge balance equation was preferable to other techniques such as dialysis or anion/cation exchange treatment for characterization of the absolute charge on the FA for the following reasons:

1. In preliminary experiments, TOC was lost during both dialysis through 1000-dalton dialysis tubing and anion exchange.
2. The only assurance that 100% of anions other than FA and cations have been removed from solution is to measure them analytically so that use of techniques in addition to the charge balance equation is redundant.
3. It was desirable that the FA samples not be brought too close to zero charge condition to avoid coagulation.

Errors in Potentiometric Measurements

The Orion 701 pH/mV meter has maximum accuracy specifications of ±0.1 mV. This error is unavoidable for a single titration with this equipment. Other errors are uncorrected junction potential (E_j) and meter or electrode drift. Errors in calculation of $[A^-]$ (charged equivalents per liter) as a percentage of A_T (maximum charged equivalents per liter) for an instrumental error of 0.1 mV are listed in Table III. The most trustworthy values for $[A^-]$ are those that have been derived from a system having high A_T and low $[H^+]$.

Errors in measurement of potential (E) due to junction potential (E_j) can be much larger than 0.1 mV. This error was minimized by use of an equitransferrent ionic medium. For work at pH values less than 4, the electrodes were calibrated by a titration of indifferent electrolyte with strong acid.

Table III. Effect of 0.1 mV Error on $\Delta[A^-]/[FA_T]$ Where $[A^-]$ is Calculated by Equation 1

	pH = 4	pH = 3	pH = 2
$FA_T = 10^{-2}$	0.004%	0.04%	0.4%
$FA_T = 10^{-3}$	0.04%	0.4%	4%
$FA_T = 10^{-4}$	0.4%	4%	40%

The error analysis shown in Table III and the discussion above lead to two discussions with regard to experimental protocol. The greater the concentration of FA used the better, as long as there are no experimental anomalies due to high concentration, and calculations of the charge of FA at pH less than 3 should be viewed with skepticism.

Experimental Apparatus and Protocol for H-titration of FA

Titration of FA by H was performed in a 30-ml methylation reactor with ports for pH electrode, double junction Ag/AgCl reference electrode, and N_2 bubbling tube. The N_2 bubbling tube was removed for injection of 0.1 *N* NaOH or HCl by Eppendorf micropipet or by Gilson 2-ml vernier microburet. The experimental sequence for a single acid titration was as follows:

1. To 25 ml of distilled deionized water add sufficient 3.5 *N* KCl to achieve I = 0.01, 0.1 or 1.0, deaerate with N_2 and titrate with 0.1 *N* HCl.
2. To 25 ml of FA sample add sufficient 3.5 *N* KCl to achieve I = 0.01, 0.10 or 1.0 (mindful of initial $[Na^+]$), deaerate with N_2 and titrate with 0.1 *N* HCl.
3. Repeat step 1.

The average of the values of the two calibration curves was used to calculate $[H^+]$. For FA #3 and FA #4, a known amount of HCl was added to the 25-ml aliquot of FA to bring the pH to about 4, and the solution was allowed to equilibrate overnight before acid titration. The pH values did not decrease with time in the acid regions (pH less than 7) of the titration. The experimental sequence for a single base titration was as follows:

1. Calibrate the pH electrodes with conventional pH 7 and pH 10 buffers.
2. To 25 ml of FA sample add sufficient 3.5 *N* KCl to achieve I = 0.01, 0.1 or 1.0 (mindful of initial $[Na^+]$), deaerate with N_2 and titrate with 0.1 *N* NaOH.
3. Check the calibration of electrodes using pH 7 and 10 buffers.

Activity standards were converted to concentration values using activity coefficients taken from Stumm and Morgan [25]. The range of temperatures for these experiments was 24 ± 1°C.

In the base titrations there was "hysteresis," i.e., back titration did not agree with forward titration. In these experiments, if FA was rapidly titrated to a slightly alkaline pH, allowed to stand for 2 hr, and then titrated to the initial slightly alkaline pH, the amount of base required to titrate to the initial pH was similar to the normality of Fe in the system. Martin and Reeve [26] reported similar results for the titration of a soil HA that contained Al. According to Shuman and Cromer [27] the dissociation of Cu from FA occurs in two steps, with $k_1 \sim 0.1$/sec ($t_{1/2} \sim 10$ sec) and k_2 much smaller ($t_{1/2} \sim 1$ hr). Since the easily extractable Fe was removed during extraction and cleaning of FA, and since Fe should form stronger bonds with FA than Cu, then it seems reasonable to hypothesize slower release of Fe than Cu. It was decided that the base titrations should be carried out as rapidly as possible to avoid change in pH due to the presence of complexed Fe. As a result, FA sites that are bonded to Fe are not titrated by this protocol.

Two synthetic polymers were titrated according to the procedures outlined above. Within 1 min after addition of pH titrant, the pH was stable to ±0.01 pH units for at least 5 hr. This result confirms the integrity of the experimental apparatus against CO_2 intrusion. Additionally, the charge densities which were calculated for the polymers were independent of the polymer concentration except for very acidic or basic pH regions, where disagreement was expected because of analytical error.

Reaction Between Ca and FA Fractions

The Ca–FA experiments were performed at I = 0.1 and from pH 5 to 9. The typical Ca_T concentration was from 10^{-3} to 10^{-2} *M*.

Errors in Measurement of Free Calcium

Analytical error was a greater problem for the Ca–FA system than for the H–FA system. The increased error was due to (1) the divalence of Ca, which magnifies the errors shown in Table III by a factor of four, (2) the experimental concentration of free calcium (Ca_f) typically was greater than 10^{-3} *M*, and (3) potential drift between pre- and postexperimental calibration of the Ca electrode was considerably greater than 0.1 mV. Therefore, analysis of the Ca-FA reaction was performed using techniques that allowed more frequent calibration of electrodes.

Experimental Apparatus and Protocol for the Analysis of the Ca–FA Reaction

Ca and FA were combined in a 100-ml beaker enclosed in an airtight vessel with ports for pH combination electrode, Ca electrode, double-junction reference electrode and N_2 bubbling tube, and a port which could be opened for addition of reagents. The experimental sequence for a *single data point* (i.e., only one FA_T and Ca_T) was as follows:

1. To 50 ml of FA sample add sufficient 3.5 *N* KCl to achieve I = 0.1 (mindful of initial $[Na^+]$), add sufficient 0.1 *M* $CaCl_2$ to achieve the desired Ca_T, deaerate and adjust the pH to 5.
2. To two beakers, each containing 50 ml of distilled deionized water, add sufficient 3.5 *N* KCl to achieve I = 0.1, add 0.1 *M* $CaCl_2$ to bracket the Ca_T in the FA sample, deaerate and adjust pH to 5.
3. Repetitively measure the potential of the first concentration standard, second concentration standard and FA sample to determine Ca_f.
4. Raise the pH in each beaker to 6 and repeat steps 1 through 3, then do the same at pH 7, 8 and 9.

Typically step 3 was repeated until repetitive calculations for Ca_f differed by less than 1%. In all cases, the average of the several Ca calculations was used for further data analysis. There was no consistent directional trend to the mV drift, the drift alternately going toward more positive or negative mV readings.

In addition to the extent of association between FA and Ca, it was necessary that the concomitant release of H be measured so that calculation of the charge density of the FA material could be made. This was done by titration of FA with Ca while simultaneously monitoring the equivalents of base required to keep the pH constant. The protocol previously outlined for the H–FA reaction was used for such measurements.

RESULTS AND DISCUSSION

Reactions Between H and FA

Each FA fraction was titrated at I = 0.01, 0.10 and 1.0. Titration curves for FA #2 at I = 0.01 and 1.0 are shown in Figure 1. Since FA is negatively charged over the entire pH range of the titration (see below), the addition of indifferent electrolyte causes the pH of the solution to decrease.

The charge density on the FA (meq/g of TOC) as a function of pH was calculated using the charge balance equation. The charge densities of FA #1 and FA #3 are shown in Figures 2 and 3. Charge densities of FA #2 and FA #4 are very similar to FA #3.

A notable feature of these figures is the large anionic charge which each of the FA fractions retains at pH less than 3. Between 35 and 40% of the total charge formed at pH 10 is already present at pH 3. There were no monomeric organic acids present, as shown by GPC and by high-performance liquid chromatography (HPLC) [21]. The very acidic nature of these FA materials is consistent with the expected acidity of the degradation products of humic materials [22,28]. For example, phthalic acid would be 28%, 1,2,4-benzenetricarboxylic acid would be 30%, and 1,2,4,5-benzenetetracarboxylic acid would be 37% ionized at pH 3 and I = 0.01 [29].

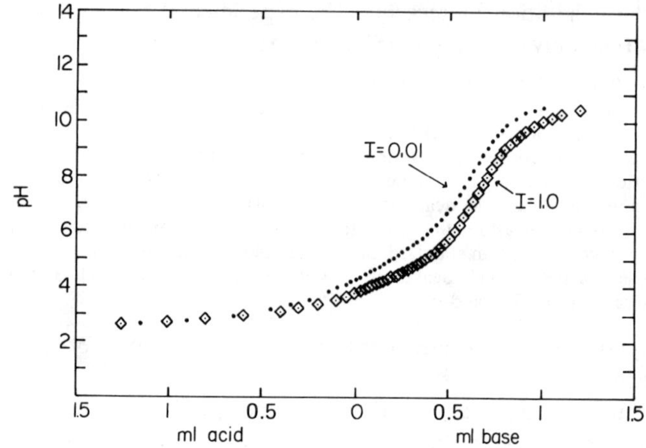

Figure 1. Effect of ionic strength on the titration of FA.

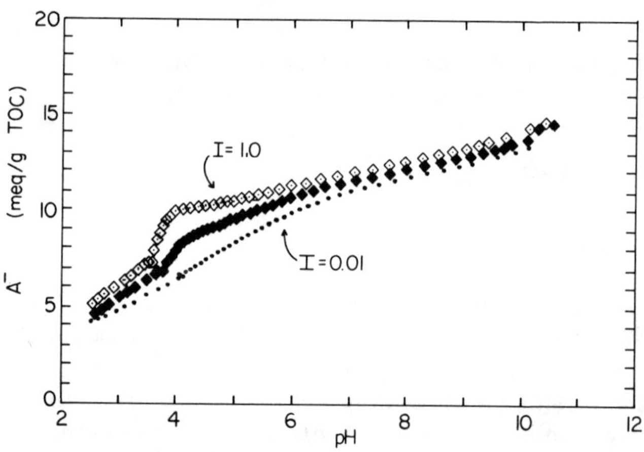

Figure 2. FA#1: Charge density as a function of pH.

FA #1 is different than the other fractions in that the charge density increases rapidly from pH 3.6 to 4.0 when I = 0.10 or 1.0. The data do not permit an unequivocal explanation of this peak. Possible explanations are (1) FA #1 may contain a large fraction of a particular acidic group and the titration is sharpened at high I due to inhibition of electrostatic interaction, (2) the space charge density (eq/volume of macromolecule)

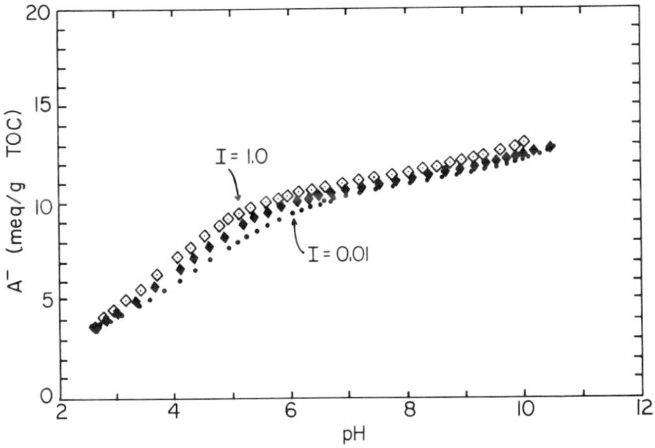

Figure 3. FA#3: Charge density as a function of pH.

reaches a level that cannot be exceeded and "charge condensation" [30] occurs until the macromolecule unfolds, or (3) rapid configurational change upon addition of base to pH 3.6 disrupts H-bonds, thus increasing the acidity of the H-donors.

The four FA fractions do not exhibit different reactivities toward the proton at I = 0.01, and only FA #1 differs from other FA fractions at higher I. In Figure 4 the charge per gram of TOC is plotted as a function of pH for all four FA fractions at I = 0.01. The similarity among these curves was unexpected. It had been anticipated that FA #1 would be more acidic than FA #4 at all pH values. The similarity in these curves, in the face of other evidence that indicates that these FA fractions are quite different, deserves further investigation.

The extent of reaction between FA and H was invariant with respect to concentration of FA or time of storage of FA. The extent of titration of FA #3 at I = 0.01 and at three concentrations of FA spanning more than one order of magnitude is shown in Figure 5. The superposition of the three curves is excellent, considering the sensitivity to error in mV measurement when the concentration of FA is low. These titrations would, of course, not be superimposed if plotted in the fashion of Figure 1. Rossotti and Rossotti [31] have shown that if the number of protons associated with a ligand is independent of the concentration of the ligand when the concentration of the ligand is not negligible, then the interaction must be mononuclear with respect to the ligand. That is, in the system studied here, there appears to be no intermolecular H-bonding. The importance of this result is discussed below.

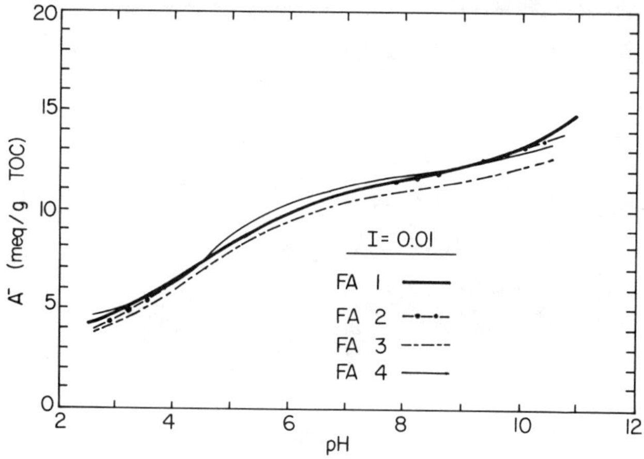

Figure 4. Charge density vs pH compared for four FA fractions at I = 0.01.

Similarly, results were identical when FA #3 was titrated at 4-month intervals. This result indicates that the storage procedure prevented degradation of FA, which would have changed the H-titration curve.

Reactions Between Ca and FA

The association between Ca and FA #3 at pH 7 is shown in Figure 6. In this and subsequent figures, the extent of reaction is expressed as meq of Ca that have associated per gram of TOC as a function of Ca_f, free calcium (at constant pH and I). The bars in Figure 6 represent the error due to ±0.2 mV deviation in potential during the analysis of Ca_f. The lines through the data points are hand-drawn to aid the reader and to assist in interpolation when data must be extracted from the figures. The data points for the two concentrations of FA overlap; similar overlap occurs at pH 5 and 9. This result enables experiments to be performed under conditions (high concentration of FA) where absolute error is minimized and permits extrapolation of results to conditions which are environmentally significant.

Misinterpretation of the nature of the reaction can easily occur if analytical difficulties are overlooked. Data which were derived from continuous titration experiments are shown in Figure 7. There was +2 mV drift during the course of these experiments. Similar anomalies were noted for the reaction between citrate and Ca. It is possible that the trend shown in Figure 7 is due to some discriminate error.

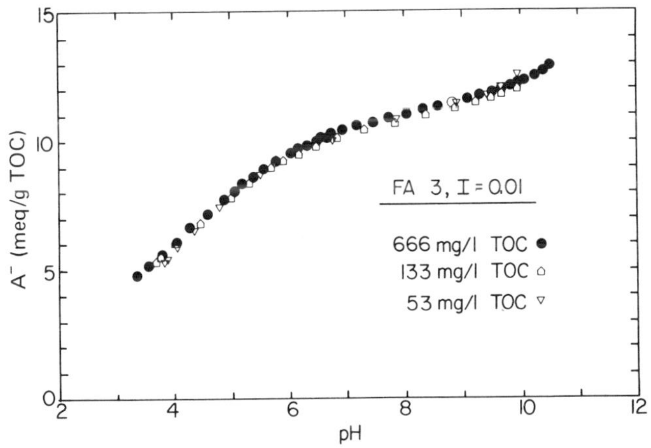

Figure 5. Independence of the H–FA reaction with respect to concentration of FA.

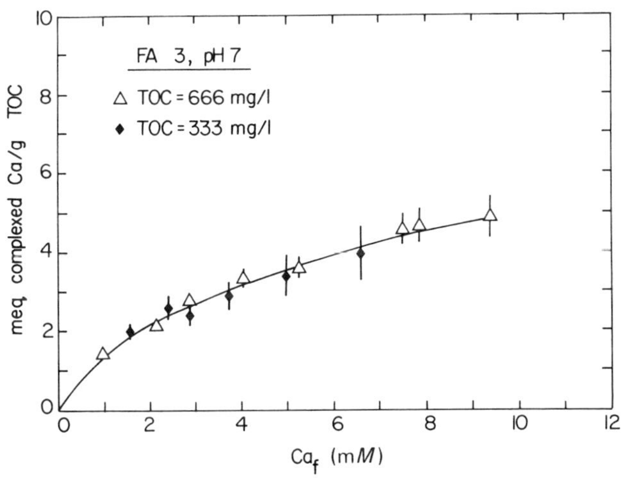

Figure 6. Association isotherms for Ca and two concentrations of FA.

The independence of the Ca–FA reaction with respect to the concentration of FA indicates that the reaction is mononuclear in FA. Either Ca is bound to only one reactive site *or* it is bound by several sites on the same FA molecule. Therefore, chelation should be considered when modeling the Ca–FA reaction and multiple-ligand complexation should not be considered.

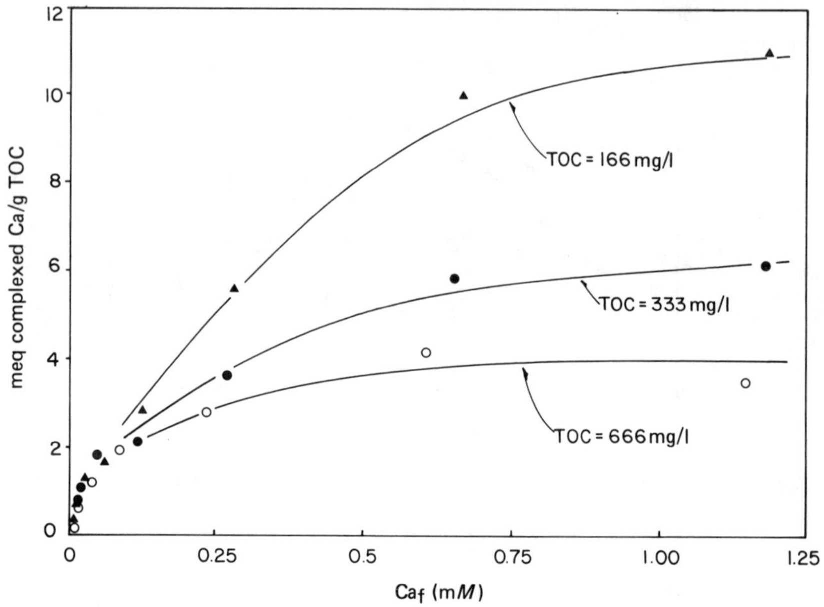

Figure 7. Bad data from titration experiments before elimination of error.

The extent of association between Ca and each of the four FA fractions as a function of Ca_f and pH is shown in Figures 8 through 11. These figures show association isotherms at pH 5, 7 and 9. As before, lines drawn through the data points are not based on any formula for minimization of variance. For FA #1 through #3 the extent of reaction increases with increasing pH. For FA #4 the reaction is strongest at pH 5 and there is little change in absolute reaction from pH 6 to pH 9. The overall range of reactivity as a function of pH decreases in order FA #1 > FA #2 > FA #3 > FA #4. The shapes of the isotherms are also different, with the slope of the isotherms at high Ca_f increasing in the order FA #1 < FA #2 < FA #3 < FA #4. It is apparent from these figures that there are considerable differences among the four FA fractions in terms of reactivity with Ca, despite the similarities in reaction with H.

Proton Loss During Ca–FA Reaction

Ca and H compete for the same sites on FA, and it is expected that some H will be released from FA upon addition of Ca [32], although some investigators have not seen significant release of H upon addition of Ca to HA [33]. In this investigation, the release of H was dependent on FA

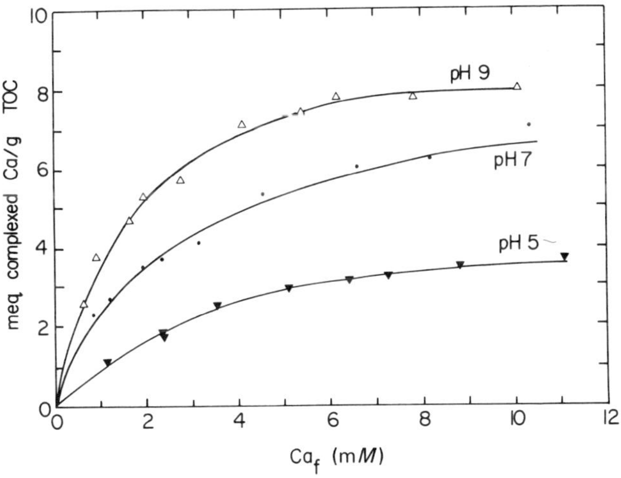

Figure 8. FA#1: Association with FA at three pH values.

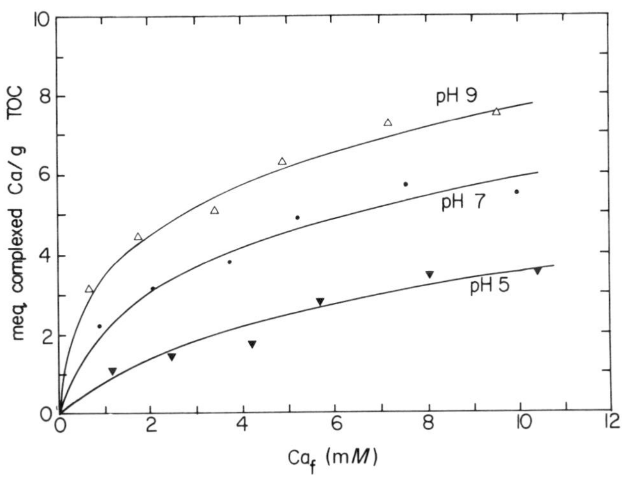

Figure 9. FA#2: Association with FA at three pH values.

fraction and pH, as shown in Table IV. Within the sensitivity and extent of the measurements made, H released/Ca complexed was invariant with respect to Ca_T at constant pH. The difference in H release with respect to FA fraction, especially at pH 5 to 7, has a significant impact on extent of charge neutralization of FA by Ca.

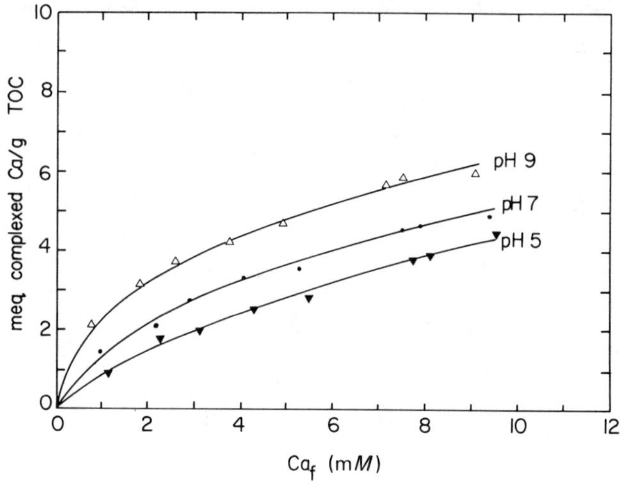

Figure 10. FA#3: Association with FA at three pH values.

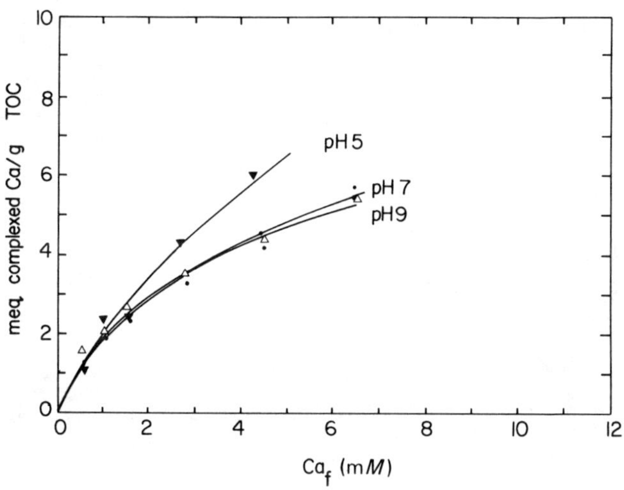

Figure 11. FA#4: Association with FA at three pH values.

Charge Neutralization of FA by Reaction with Ca

The residual charge on the FA molecules as a function of pH and Ca_f is calculated as follows:

Table IV. Equivalents H^+ Released per Equivalent of Ca Complexed for Different FA at Various pH

pH	meq H released/meq Ca complexed			
	FA #1	FA #2	FA #3	FA #4
5	0.22	0.19	0.17	0.06
6	0.21	0.14	0.16	0.07
7	0.13	0.10	0.09	0.07
8	0.12	0.13	0.11	0.07
9	0.12	0.13	0.13	0.13

$$A^- = A' - \left(1 - \frac{\text{eq H lost}}{\text{eq Ca complexed}}\right)(q_e) \qquad (2)$$

where A′ represents the anionic charge on the FA at I = 0.1 in the absence of Ca at the given pH (see Figures 2 and 3)
q_e is taken from Figures 8 through 11
H/Ca comes from Table IV

Greater than 50% of the initial charge on FA is neutralized by Ca in some cases, and the effect of this neutralization on adsorption of FA can be extensive.

The four FA fractions exhibited different charge neutralization behaviors. The residual charge on FA #1 decreases with increasing pH and the residual charge on FA #4 increases with increasing pH at constant Ca_f. FA #2 and FA #3 are intermediate in behavior. This is illustrated in Figure 12. Because of these trends, coagulation or adsorption of FA #4 should occur more readily at low pH in the presence of Ca, while FA #1 would be expected to coagulate or adsorb better at high pH when Ca_f is greater than 3×10^{-3} *M*. Theoretically, the results suggest that different reactive mechanisms or stoichiometries are relevant for the different FA reactions.

The Effect of Charge Neutralization on the Speciation of FA

Addition of Ca salts to HA typically causes extensive coagulation [14,19, 34]. This was not the case for the FA described in this chapter. For example, when 5×10^{-3} *M* Ca_T was added to FA #3 (666 mg/l TOC) at pH 5, only 7% of TOC was removed by filtration (0.2-μ pore) after 48 hr. For the same conditions, but TOC of 32 mg/l, less than 1% of TOC was removed.

Addition of Ca salts does, however, increase the adsorption of these FA on several solids. For example, FA does not adsorb to a significant extent on δ-MnO_x(s) in the absence of Ca except at acidic pH values, but addition

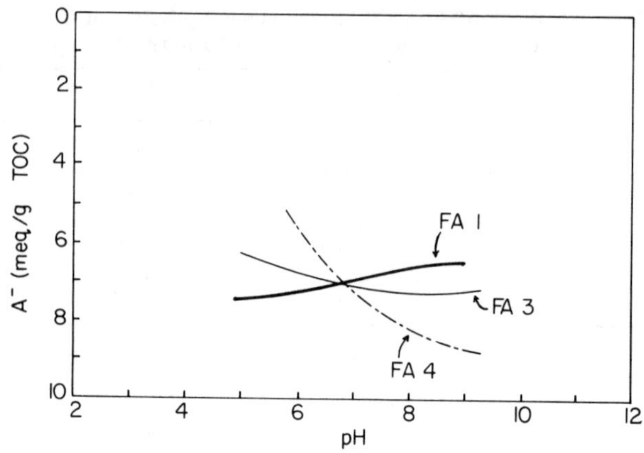

Figure 12. Charge density of three FA fractions as a function of pH and in the presence of 5×10^{-3} *M* Ca_T at I = 0.1. The y axis is inverted.

of 5×10^{-3} *M* Ca_T induces a dramatic increase in adsorption. This is illustrated in Figure 13. The most hydrophilic of the FA adsorb most strongly (FA #1 > FA #2 > FA #3 > FA #4) on δ-$MnO_X(s)$ (a hydrophilic solid). The effect of pH on the adsorption of the FA agrees qualitatively with the data shown in Figure 12; the slope of extent of reaction *vs* pH decreases in the order FA #1 > FA #2 > FA #3 > FA #4. Likewise, addition of Ca increases the extent of adsorption of FA on $Fe(OH)_3(am)$ and Ca $CO_3(s)$. $MnO_x(s)$ is typically too dilute to cause significant removal or fractionation of FA. $Fe(OH)_3(am)$ and $CaCO_3(s)$, however, can cause significant removal and fractionation of FA in natural and engineered systems [35].

Enhancement of the adsorption of polyelectrolytes by the addition of oppositely charged ions has been explained by several hypothetical mechanisms, which typically depend on *specific reaction* between the adsorbate and the counter-ion [36–38].

Independence of Reactions with Respect to Concentration of FA

The H–FA and Ca–FA reactions have been reported herein to be independent of the concentration of FA. This result is important, for it means that the homogeneous reactions of these FA are rational and predictable.

Only a limited amount of work has been done to test this result on other humic materials. Saar and Weber [39] have shown that the reaction between

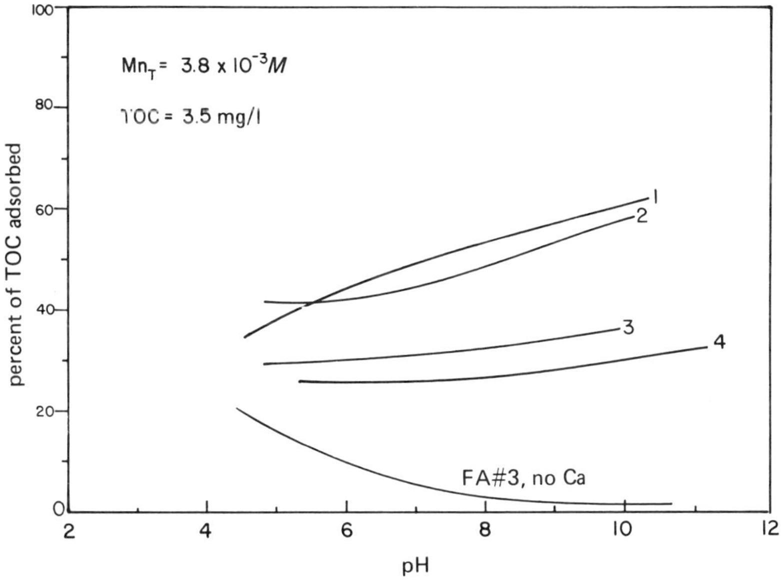

Figure 13. TOC removal by adsorption: 1 = FA#1, 2 = FA#2, 3 = FA#3, 4 = FA#4, all in presence of 5×10^{-3} *M* Ca_T.

Cu and aquatic FA is independent of the concentration of FA. They have suggested that the reaction between Cd and FA is dependent on the concentration of FA [40], but inclusion of data from their Table III and Figure 1 with data in their Figure 3 indicates that there is no consistent trend in extent of association with respect to FA concentration. Burch et al. [41] found that the H–FA reaction is independent of concentration for the two soil FA which they studied (see their Figure 1). Sposito and Holtzclaw [42] and Sposito et al. [43] reported that the number of ionizable groups per gram of sewage sludge-derived FA decreased with increasing concentration of FA [43], but this result apparently differed from earlier work on the same material [42] (see their Figures 3 and 4). Sposito et al. [44] reported that the reaction between Ca and sewage sludge-derived FA was dependent on the concentration of the FA, but this result should be viewed with caution. A maximum net potential drift of up to 3 mV was tolerated during these experiments; the maximum error in measurement of Ca_f (and therefore associated Ca) was greater than the amount of Ca which "associated" with the FA. Also, Sposito reported up to 1.3 Ca ions associated with every FA functional group (their Figure 5), which would appear to be excessively high.

MODELING THE REACTIONS

Modeling the H–FA Reaction

The information contained in Figures 2 and 3 is sufficient to allow calculation of the average charge on the FA molecule for a given pH and I. Nevertheless, it is useful to describe the H–FA reaction mathematically so that different FA fractions or different chemical conditions can be compared.

Scatchard Plots

The simplest model for reaction between H and FA is the single-site model with no interaction. This type of reaction is represented by the straight line diagonal in the Scatchard plot shown in Figure 14. The fraction of total sites associated with H is designated α_H. The middle curve in Figure 14 represents two sites, no interaction, with $\Delta pK_a = 2$. The middle curve consists of the graphical addition of two straight lines, one representing the weaker acid and the other representing the stronger acid. The bottom curve in Figure 14 shows the Scatchard plot for FA #3 at I = 0.01. This plot is qualitatively

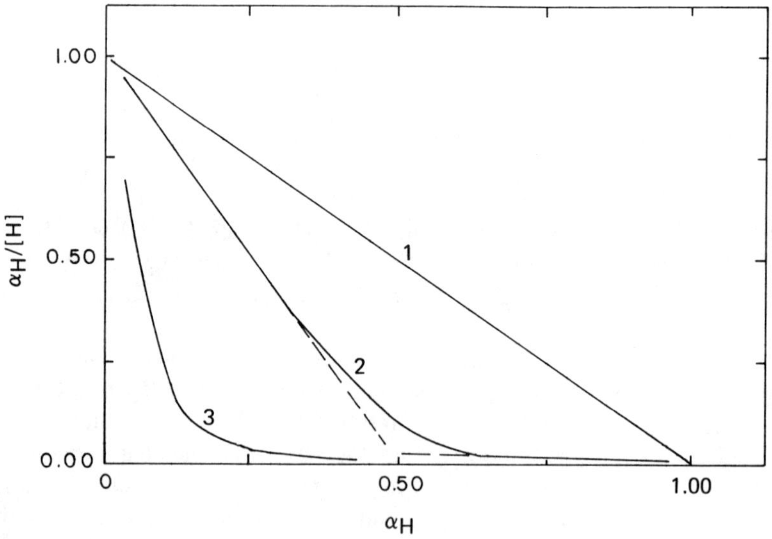

Figure 14. Scatchard plot as an analysis tool for the H–FA reaction.

similar to plots for all other titrations of FA samples in that there is continuous change in the slope over the entire length of the curve. If the curve is enlarged to examine the smaller sections the same continuous change in slope is seen. This indicates that there are many different sites *or* there is interaction between sites—*or* that both are true. The last explanation is the most reasonable. Scatchard plots are useful for determination of modeling parameters only if the nature and number of sites or the interactive parameter is known, as is sometimes the case for proteins, nucleic acids and other well characterized macromolecules [45]. Scatchard plots are of little use for the analysis of the H–FA reaction, other than to illustrate that the behavior of FA is complex.

Electrostatic Interaction

If one assumes that there are only a few types of sites and that interaction between sites is directly proportional to the charge generated, then plots of log K′ vs (1 − α) (i.e., apparent association constant vs degree of dissociation) will yield a family of straight lines obeying the equation

$$\log K' = \log K^{int} + 0.868\omega n(1 - \alpha) \quad (3)$$

where log K^{int} (association constant at (1 − α) = 0) and ω (electrostatic interaction term) are constant in the absence of configurational changes. To calculate (1 − α) one must know A_t, the total number of carboxyl sites per gram of TOC. A_t was selected as the point at which the rate of change of pH with added base was maximum. The resulting values of A_t and the pH values at which the carboxylic sites appear to be completely ionized are listed in Table V. These values of A_t should be compared to the Perdue et al. [12] value of 10.3 meq/g TOC for Satilla River humic substances. Aquatic FA described by Weber and Wilson [10] had 13.3, 17.7 and 18.2 meq of

Table V. A_t and pH When (1 − α) = 1.00

		pH		
FA	A_t	I = 0.01	I = 0.1	I = 1.0
#1	11.4 meq/g	7.55		
#2	11.3 meq/g	7.55	6.96	6.44
#3	10.75 meq/g	7.54	7.01	6.41
#4	11.6 meq/g	7.51	6.76	6.46

carboxylic groups per gram of TOC, but these values are probably too high due to analytical problems which have been discussed by Perdue et al. [12].

Plots of log K′ vs $(1 - \alpha)$ for FA #2 at three values of I are shown in Figure 15. The lines drawn through the data points are from linear regression by the method of least squares for $0.40 < (1 - \alpha) < 0.95$. Values for the intercepts (log K^{int}) and slopes (0.868 ωn) for the four FA fractions are shown in Table VI. For I = 0.01, log K^{int} increases from FA #1 to FA #4, although the differences are small and of questionable statistical significance. The slope for any FA fraction decreases with increase in I.

It is remarkable that such good linear regressions could be derived from the titration of a material which is expected to contain many different carboxyl groups. Wilson [46] modeled the H–FA reaction in this manner and he also found a very linear response. The simplest (albeit improbable) explanation is that all the carboxyl groups in a given FA fraction are identical and the slope is entirely due to electrostatic interaction.

The observed slopes can be compared to the theoretical slopes as follows. The Hermans–Overbeek equation for flexible linear polyelectrolytes [45] has been used to calculate the electrostatic interaction term (0.868 ωn) for different values of I and equivalent radius (R_e) where n is the number of carboxyl groups on an average FA molecule. The results are shown in Table VII. To put these values into perspective, if the density of the area defined by R_e is 1.05 g/cm^3, and the mass within this area is 50% water, then a spherical, hydrated FA molecule having a radius of 10^{-7} cm would contain 1350 amu of FA and thus would have 15 carboxylic acid sites. For

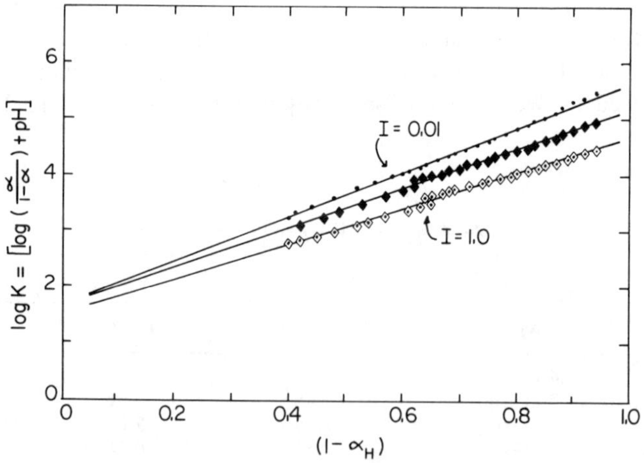

Figure 15. FA#2: apparent association constant (H–FA) vs fraction of type 1 sites which are ionized for three values of I.

Table VI. Values for a (log K^{int}) and b (0.868 ωn) in the Equation log K′ = a + b (1 − α) for "Carboxyl" Groups of Some FA Solutions

FA	I	a	b	r^2	No. of Points
#1	0.01	1.39	4.41	0.997	32
#2	0.01	1.66	4.00	0.997	28
	0.10	1.65	3.53	0.996	27
	1.00	1.50	3.18	0.993	30
#3	0.01	1.72	3.90	0.996	18
	0.10	1.91	3.16	0.998	13
	1.00	2.00	2.52	0.997	14
#4	0.01	1.91	3.56	0.995	19
	0.10	1.79	3.16	0.994	21
	1.00	1.80	2.81	0.995	23
Aldrich HA	0.10	4.30	2.17		
Smith Lake	0.10	4.23	2.72		

$R_e = 10^{-7}$ cm and n = 15, the calculated values for the slope would be 4.19, 2.49 and 0.72, respectively at I = 0.01, 0.10 and 1.0. These values are of the same order of magnitude as the experimentally observed slopes (b in Table VI). Additionally, increasing I should decrease R_e due to the inhibition of electrostatic interaction, and this would have the effect of bringing the calculated ratios for b closer to the observed ratios. Thus observations and calculations of interaction can be made to agree quantitatively (absolute value) and qualitatively (relative ratios of slopes at different I values). These results are more thought-provoking than useful at this time. Precise knowledge of molecular weight and R_e as a function of I and pH would be necessary to make conclusions about the diversity of the functional groups, and vice versa. The applicability of the Hermans–Overbeek equation is uncertain. Additionally, each of the four FA fractions contains a distribution of molecular sizes, complicating the comparison between observed and calculated values for the electrostatic interaction.

Henderson–Hasselbalch Plots

Katchalsky and Spitnik [47] and Katchalsky and Gillis [48] have shown that for flexible linear polymeric acids at low I, Equation 3 can be manipulated to derive the generalized Henderson–Hasselbalch equation

$$pH = pK_{HH} - n \log \left(\frac{\alpha}{1-\alpha}\right) \tag{4}$$

Table VII. Values for (0.868 ωn) as Calculated by the Hermans–Overbeek Equation for Flexible Linear Polyelectrolytes for Different Equivalent Radius at Various I [45]

I	$R_e = 5 \times 10^{-8}$ cm	$R_e = 1 \times 10^{-7}$ cm	$R_e = 2 \times 10^{-7}$ cm	$R_e = 4 \times 10^{-7}$ cm
0.01	0.635 n	0.279 n	0.108 n	0.034 n
0.1	0.479 n	0.166 n	0.044 n	0.0085 n
1.0	0.224 n	0.048 n	0.0080 n	0.0011 n

The K_{HH} in Equation 4 is a *dissociation* constant and represents the pH at which A_t is half titrated. Equation 4 is of less theoretical value than Equation 3, but it is very useful for the empirical description of the H–FA reaction. Values for K_{HH} and n from this work and other works are shown in Table VIII and Henderson–Hasselbalch plots of FA #2 at three values of I are illustrated in Figure 16. It should be noted in Table VIII that the three soil HA investigated by Stevenson were less acidic and had lower n values than the FA investigated in this research. Gamble [50] interpreted the

Table VIII. Values of pK_{HH} and n for "Carboxyl" Groups of Some Humic Substances

Sample	I	pK_{HH}	n	Reference
FA #1	0.01	3.71	2.77	this work
FA #2	0.01	3.76	2.66	this work
	0.10	3.52	2.44	this work
	1.00	3.17	2.32	this work
FA #3	0.01	3.74	2.59	this work
	0.10	3.56	2.29	this work
	1.00	3.29	2.06	this work
FA #4	0.01	3.83	2.35	this work
	0.10	3.50	2.20	this work
	1.00	3.33	2.07	this work
Soil HA	0.10	4.82	1.86	49
Peat HA	0.10	4.86	1.90	49
Leonardite HA	0.10	5.05	1.93	49
Type I COOH	0.10	1.88	1.29	50
Type II COOH	0.10	5.4	1.2–4.1	50
Type I & Type II	0.10	3.7	3.5	12
SRHS	0.10	3.7	3.5	12

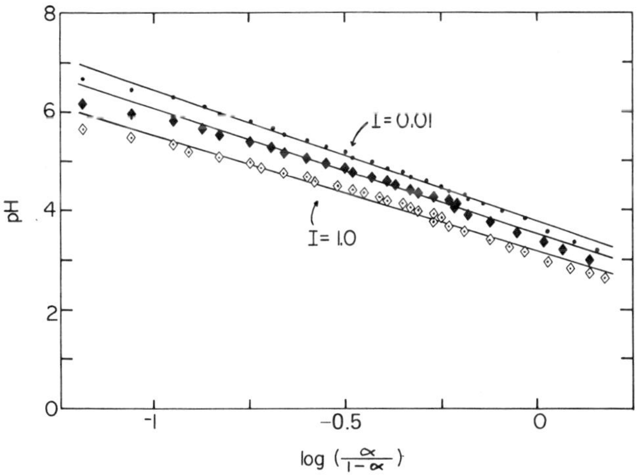

Figure 16. Henderson–Hasselbalch plots for FA#2 at three values of I.

titration of soil FA by invoking a very acidic group and a less acidic group, but Perdue et al. [12] suggested that Gamble's data were more meaningful if they were not divided into two groups. The sample titled "Type I and Type II" represents a reanalysis of Gamble's data by Perdue et al. [12]. This sample and the sample of Satilla River Humic Substances (SRHS) both had K values similar to the K values for FA #1 through #4. The n values from other studies bracket the n values for FA #1 through #4. For a monomeric acid n equals one. Katchalsky and Spitnik [47] found that n has a maximum value of 2 for a polymer with only one kind of site. The higher n values reported here and elsewhere for FA are probably due to the presence of several different acidic sites.

Finally (for modeling of the H–FA reaction) it is common to see the titration of FA reported by a plot of pK_i as a function of pH, where $K_i = (1 - \alpha)(H)/(\alpha)$. The relationship between pK_i and pH can be expressed in terms of the Henderson–Hasselbalch parameters as shown in the following equation

$$pK_i = \frac{1}{n} pK_{HH} + \frac{n-1}{n} pH \tag{5}$$

where K_i is a dissociation constant. The value of pK_{HH} and n for the last two samples listed in Table VIII were derived using Equation 5.

Modeling the Ca-FA Reaction

Apparent Association Constants

The reaction between Ca and FA appears to be mononuclear with respect to FA. Therefore, apparent association constants for the reaction have been calculated by the equation

$$K = \frac{(\text{meq Ca complexed})}{(\text{meq open sites})\ (\text{Ca}_f)} \tag{6}$$

where the numerator is derived from Figures 8 to 11 and the term (meq open sites) comes from Equation 2. These results are listed in Table IX. In all cases except FA #4 at pH 5, K decreases when Ca_f increases. Trends which have been reported in previous sections, such as the range of reactive strength as a function of pH decreasing from FA #1 to FA #4, are apparent when viewing the K values in Table IX.

Some association constants which other investigators have derived for the reaction between Ca and humic substances are shown in Table X. The various K values cannot be compared directly. For example, Schnitzer and Hansen [51] used a soil FA rather than aquatic FA materials. They used molarity rather than normality to express the concentration of FA and evaluated the association at pH 3 and 5 rather than the pH range of 5 to 9 used to generate the data shown in Table X. Additionally, the maximum Ca_f used by Schnitzer and Hansen was estimated to be <0.5 m*M* while the K values in Table X are for Ca_f in the range 1 to 7 m*M*. Mantoura et al. [52] used aquatic humic substances and pH of 8.0 but low Ca_f concentration. The log K values reported by Mantoura et al. were derived from tangents to

Table IX. Association Constants for the Reaction Between Ca and the Four FA Fractions at pH 5, 7 and 9. Calculated Using Equation 6

	FA #1			FA #2			FA #3			FA #4		
Ca_f (m*M*)	pH 5	pH 7	pH 9	pH 5	pH 7	pH 9	pH 5	pH 7	pH 9	pH 5	pH 7	pH 9
1	105	245	357	94	228	353	107	143	230	267	191	167
2	100	207	313	90	173	251	102	122	175	274	162	142
3	93	181	279	83	153	212	96	113	152	291	146	128
4	88	164	252	77	138	186	93	104	137	321	137	117
5	81	154	227	73	126	169	92	99	126		134	109
6	73	144	208	68	118	156	92	94	119			
7	67	137	190	66	112	146	93	93	113			

Table X. Association Constants Reported in the Literature for the Reaction Between Ca and Humic Substances

Maximum Ca_f (m*M*)	Ligand	Ligand Concentration Units	Log K	pH	Reference
<0.5	Soil FA	Molarity	2.6	3.0	51
<0.5	Soil FA	Molarity	3.3	5.0	51
~0.1	Lake Humics	Normality	3.3–4.1	8.0	52
10	Sludge FA	Normality	None	7.0	44
(Not reported)	Sludge FA	(Not reported)	3.12–1.23	5.0	53
7	Aquatic FA	Normality	1.83–2.55	5.0–9.0	this work

Scatchard plots, and therefore those values represent only a portion of the possible reactive sites. It is estimated that Mantoura et al. have used A_t of 1 to 3 meq/g TOC compared to typical values of 10 to 15 meq/g TOC of potential sites. Use of a higher number for A_t could decrease K by more than two orders of magnitude when the concentration of complexed Ca approaches the original A_t value.

Scatchard Plots

Scatchard plots for the Ca–FA reactions are not shown here, but some general statements regarding such plots should be made. Scatchard plots can sometimes be used to determine the total number of sites available for complexing and the strength of those sites. Because the measurement of association is less precise at high Ca_f, the total number of sites available for binding Ca can only be determined by extrapolation from data points, which represent lower values of Ca_f. However, the Scatchard plots are continuously curved, since the tangent to the plot equals the negative of K at any point. This curvature makes accurate extrapolation of the curve impossible; it also indicates multiple sites and/or that interaction among sites is occurring.

Electrostatic Interaction

A competitive interactive model for complexation of a polyelectrolyte by two cations [45,46,54] was tested to see if the association of Ca and H with FA could be expressed by two intrinsic constants and a single electrostatic term. Recall that the simple electrostatic model successfully described the H–FA reaction in both a qualitative and semiquantitative fashion. Wilson and Kinney [54] used the competitive interactive model successfully for the system trace metal-H-soluble natural organic material. For the data presented in this chapter, however, the competitive electrostatic model could not achieve the desired goals (stated above), even with estimates of total monodentate sites ranging from 0.5 to 1.5 of those shown in Table V.

Complexation and Chelation

It has been shown that the reactions between Ca and FA appear to be mononuclear with respect to FA. It is possible that more than one functional group per FA molecule reacts with each Ca ion. Traditionally it has been proposed that these chelation reactions involved contiguous functional groups, as in salicylate or phthalate [55]. However, Fettes [56] has reported that chelate rings of greater than 30 members are sometimes stable for macromolecules.

The effect of hypothetical chelation constants on apparent association constants (K′) is shown in Table XI. For the first case, it is assumed that the chelating ligand is similar to EDTA or NTA in that the chelation reactions become stronger with increasing protolysis of the chelating ligand. For the second case, it is assumed that the chelation reactions become weaker with increasing protolysis of the chelating ligand, perhaps due to the unfolding of the molecule and separation of cooperating functional groups. For both cases, it is assumed that the successive pK_a values of FA are 1.0, 3.0, 4.7, 6.5 and 10.3, so that the ligand Y is pentavalent.

Case 1 has reproduced the trends shown in Table IX for FA #1 through #3. K′ increases with increasing pH and decreases with increasing Ca_f. Case 2 has reproduced most of the trends shown in Table IX for FA #4. K′ decreases with increasing Ca_f at pH 7 and 9. For FA #4 at pH 5, the increase in K′ with increasing Ca_f has not been reproduced, and explanation of this anomaly requires further speculation. If the macromolecule is conformationally stabilized by successive chelation, then it is conceivable that association might

Table XI. Apparent Association Constants from Hypothetical Reactions[a]

Ca_f (m*M*)	Case 1 pH = 5	Case 1 pH = 7	Case 1 pH = 9	Case 2 pH 5	Case 2 pH 7	Case 2 pH 9
1	107	207	241	325	265	225
3	66	136	159	244	183	155
5	48	101	119	195	139	118
7	38	80	95	163	113	95

[a]For case 1, K_{CaH_2Y} = 300, K_{CaHY} = 600 and K_{CaY} = 1800; for case 2, K_{CaH_2Y} = 800, K_{CaHY} = 600 and K_{CaY} = 400.

increase with increasing Ca_f due to enforced proximity of cooperating functional groups. This last hypothesis is consistent with the low release of H upon addition of Ca to FA #4; if the macromolecule is constricted by the addition of Ca, then the increased space charge density might inhibit the release of protons. This hypothesis is also consistent with the observation that the surface activity of the FA increased in the order FA #4 > FA #3 > FA #2 > FA #1. Surface activity, in the case of FA, suggests that the molecules are flexible enough to present hydrophobic sites to the gas phase and hydrophilic sites to the aqueous phase. In fact, Curtis and Rogers [57] have suggested that hydrophobicity, or the ability to segregate hydrophilic moiety from hydrophobic moiety, is an important factor in the retention of humic materials by XAD resin. Following this line of reasoning, FA #4 would indeed have greater flexibility than the other FA fractions. It must be emphasized that the mechanisms proposed are consistent with the data, but are not necessarily unique explanations.

CONCLUSIONS

1. Four aquatic FA fractions were produced using a preparative scale fractionation technique based on solubility of FA as a function of pH. Reactivity with Ca and adsorptivity on inorganic solids varied among the four fractions, while the fractions were similar in molecular weight and reactivity with H.
2. Reactions of FA with H or Ca showed no dependence on FA concentration. The independence with respect to FA concentration indicates that the reactions are mononuclear. Several models were mathematically suitable for the description of the H–FA reaction but the conceptual suitability of these models is uncertain. Reactions between Ca and FA #1, #2 and #3 may be explained by simple complexation and chelation models. It is suggested that configurational changes in FA #4 at high [H] or [Ca] lead to increased association between Ca and FA #4.
3. Ca can neutralize over 50% of the anionic charge on FA and thus can affect the speciation of FA in natural and engineered environments. The extent of neutralization as a function of pH differed for the four FA fractions. In the presence of Ca, the most soluble FA fractions exhibited decreasing charge densities with increasing pH and the least soluble FA fractions had increasing charge densities with increasing pH.
4. Reaction of Ca with FA leads to adsorption of negatively charged FA on negatively charged $MnO_x(s)$. Such adsorption does not occur, at neutral pH, in the absence of Ca. It is plausible that aquatic FA are removed from freshwaters in systems containing

both divalent metal ions and inorganic solids; the mechanism of this removal is adsorption. In the absence of one or both of these reactants, FA remains in solution for long periods of time.

5. Calibration of Ca electrodes by a titration procedure is not adequate. Analytical imprecision is sometimes propagated in the form of incorrect hypotheses. Special methods have been reported in this chapter for the minimization of the analytical error associated with the measurement of the Ca–FA reaction.
6. The study of humic materials must be somewhat empirical in nature, but in this research an attempt has been made at every stage to generalize and to conceptualize the findings. Generalizations such as the independence of the association reactions with respect to FA concentration should be confirmed or discounted using other sources of FA.

ACKNOWLEDGMENTS

The research of Brian Dempsey was funded by the U.S.G.S. thesis support program. Additional funding was provided by the Office of Research and Development, Environmental Protection Agency, under Grant #R-808104. Dr. R. Wershaw provided small-angle X-ray scattering and IR analyses. The authors gratefully acknowledge advice from Phil Singer, Robert Baker and Mike Thurman.

REFERENCES

1. Yost, E. C. "Characterization of Organic Matter and Trace Metals in the Dissolved Fraction of the Upper Mississippi River," MS Thesis, Univ. of Minn. (December, 1980).
2. Buffle, J., P. Deladoey and W. Haerdi. "The Use of Ultrafiltration for the Speciation and Fractionation of Organic Ligands in Fresh Waters," *Anal. Chim. Acta 101*:339 (1978).
3. Smith, R. G., Jr. "Evaluation of Combined Applications of Ultrafiltration and Complexation Capacity Techniques to Natural Waters," *Anal. Chem. 48*:74 (1976).
4. Jewett, M. "Trihalomethane Yields of Fulvic Acids Fractionated by Sephadex Gel-Permeation Chromatography," MS Thesis, University of N.H., Durham (May, 1978).
5. Snoeyink, V. L., J. J. McCreary and C. J. Murin. "Activated Carbon Adsorption of Trace Organic Compounds," EPA-600/2-77-223 (December, 1977).
6. Davis, J. A., and R. Gloor. "Adsorption of Dissolved Organics in Lakewater by Aluminum Oxide: Effect of Molecular Weight," *Environ. Sci. Technol. 15*(10):1223 (1981).

7. Mantoura, R. F. C., and J. P. Riley. "The Analytical Concentration of Humic Substances from Natural Waters," *Anal. Chim. Acta 76*:97 (1975).
8. MacCarthy, P., M. J. Peterson, R. L. Malcolm and E. M. Thurman. "Separation of Humic Substances by pH Gradient Desorption from a Hydrophobic Resin," *Anal. Chem. 51*(12):2041 (1979).
9. Schnitzer, M., and S. U. Khan, Eds. *Soil Organic Matter* (New York: Elsevier, 1978).
10. Weber, J. H., and S. A. Wilson. "The Isolation and Characterization of Fulvic Acid and Humic Acid from River Water," *Water Res. 9*:1079 (1975).
11. Livingstone, A. D. "Chemical Components of Rivers and Lakes," U.S. Geological Survey Paper No. 440G (1963).
12. Perdue, E. M., J. H. Reuter and M. Ghosal. "The Operational Nature of Acidic Functional Group Analyses and Its Impact on Mathematical Descriptions of Acid-Base Equilibria in Humic Substances," *Geochim. Cosmochim. Acta 44*:1841 (1980).
13. Rashid, M. A. "Role of Humic Acids of Marine Origin and Their Different Molecular Weight Fractions in Complexing Di- and Tri-Valent Metals," *Soil Sci. 111*:298 (1971).
14. Sipos, S., et al. "Biopolymer-Metal Complex Systems. II. Physical Properties of Humic Substances and Their Metal Complexes," *Acta. Agron. Acad. Scient. Hung. 27*(1-2):31 (1978).
15. Sholkovitz, E. R., et al. "The Removal of Dissolved Humic Acids and Iron During Estuarine Mixing," *Earth and Planetary Science Letters 40*:130 (1978).
16. Kavanaugh, M. C. "Modified Coagulation for Improved Removal of Trihalomethane Precursors," *J. Am. Water Works Assoc. 70*:613 (1978).
17. Glazer, H. T., and J. K. Edzwald. "Coagulation and Direct Filtration of Humic Substances with Polyethylenimine," *Environ. Sci. Technol. 13*(3): 299 (1979).
18. Narkis, N., and M. Rebhun. "Stoichiometric Relationship Between Humic and Fulvic Acids and Flocculants," *J. Am. Water Works Assoc. 69*(6):325 (1977).
19. Eckert, J. M., and E. R. Sholkovitz. "The Flocculation of Iron, Aluminum, and Humates from River Water by Electrolytes," *Geochim. Cosmochim. Acta 40*:847 (1976).
20. Aiken, G. R., et al. "Comparison of XAD Macroporous Resins for the Concentration of Fulvic Acid from Aqueous Solution," *Anal. Chem. 51*(11):1979 (1979).
21. Dempsey, B. A. "The Protonation, Calcium Complexation, and Adsorption of a Fractionated Aquatic Fulvic Acid," PhD Dissertation, UNC at Chapel Hill (1981).
22. Liao, W. "Characterization of Aquatic Humic Substances," PhD Dissertation, UNC at Chapel Hill (1981).
23. Leenheer, J. A. "Comprehensive Approach to Preparative Isolation and Fractionation of Dissolved Organic Carbon from Natural Waters and Wastewaters," *Environ. Sci. Technol. 15*(5):578 (1981).
24. Swift, R. S., and A. M. Posner. "Gel Chromatography of Humic Acid," *J. Soil Sci. 22*(2):237 (1971).
25. Stumm, W., and J. J. Morgan. *Aquatic Chemistry* (New York: John Wiley and Sons, Inc., 1981), pp. 134 ff.

26. Martin, A. E., and R. Reeve. "The Extraction of Organic Matter from Podzolic B-Horizons with Organic Reagents," *Chem. Ind. 1955*:356 (1955).
27. Shuman, M. S., and J. L. Cromer. "Copper Association with Aquatic Fulvic and Humic Acids. Estimation of Conditional Formation Constants with a Titrimetric Anodic Stripping Voltammetry Procedure," *Environ. Sci. Technol. 13*(5):543 (1979).
28. Schnitzer, M., and S. U. Khan. *Humic Substances in the Environment* (New York: Marcel Dekker, Inc., 1972).
29. Purdie, N., M. B. Tomson and N. Riemann. "The Thermodynamics of Ionization of Polycarboxylic Acids," *J. Solution Chem.* 1(5):465 (1972).
30. Selegny, E. *Polyelectrolytes* (Boston: D. Reidel Publishing Co., 1974).
31. Rossotti, F. J. C., and H. Rossotti. *The Determination of Stability Constants* (New York: McGraw-Hill Book Company, 1961).
32. Van Dijk, H. "Cation Binding of Humic Acids," *Geoderma 5*:53 (1971).
33. Flaig, W., H. Beutelspacher and E. Rietz. "Chemical Composition and Physical Properties of Humic Substances." Ch. I in *Soil Components, Vol. I, Organic Components*, J. E. Gieserking, Ed. (New York: Springer-Verlag, 1975).
34. Ong, H. L., and R. E. Bisque. "Coagulation of Humic Colloids by Metal Ions," *Soil Sci. 106*:220 (1968).
35. Dempsey, B. A., and C. R. O'Melia. "Effects of Calcium on the Speciation, Adsorption, and Removal of Fulvic Acids," paper presented at the Conference of the AWWA, June 10, 1981.
36. Smellie, R. H., and V. K. LaMer. "Flocculation, Subsidence, and Filtration of Phosphate Slimes. VI. A Quantitative Theory of Filtration of Flocculated Suspensions," *J. Coll. Sci. 13*:589 (1958).
37. Sommerauer, A., D. L. Sussman and W. Stumm. "The Role of Complex Formation in the Flocculation of Negatively Charged Sols with Anionic Polyelectrolytes," *Kolloid-Zeit. und Zeit. fur Poly. 225*:147 (1968).
38. Miller, I. R., and D. C. Grahame. "Polyelectrolyte Adsorption on Mercury Surfaces. Differential Capacity in the Presence of Poly 2- and 4-Vinyl Pyridines," *J. Coll. Sci. 16*:23 (1961).
39. Saar, R. A., and J. H. Weber. "Lead(II) Complexation by Fulvic Acid: How It Differs from Fulvic Acid Complexation of Copper (II) and Cadmium(II)," *Geochim. Cosmochim. Acta* 44:1382-1384 (1980).
40. Saar, R. A., and J. H. Weber. "Complexation of Cadmium(II) with Water- and Soil-Derived Fulvic Acids: Effect of pH and Fulvic Acid Concentration," *Can. J. Chem. 57*:1263 (1979).
41. Burch, R. D., C. H. Langford and D. S. Gamble. "Methods for the Comparison of Fulvic Acid Samples: The Effects of Origin and Concentration on Acidic Properties," *Can. J. Chem. 56*:1196 (1978).
42. Sposito, G., and K. M. Holtzclaw. "Titration Studies on the Polynuclear Polyacidic Nature of Fulvic Acid Extracted from Sewage Sludge-Soil Mixtures," *Soil Sci. Soc. Am. J. 41*:330 (1977a).
43. Sposito, G., K. M. Holtzclaw and D. A. Keech. "Proton Binding in Fulvic Acid Extracted from Sewage Sludge-Soil Mixtures," *Soil Sci. Soc. Am. J. 41*:1119 (1977b).
44. Sposito, G., K. M. Holtzclaw and C. S. LeVesque-Madore. "Calcium Ion Complexation by Fulvic Acid Extracted from Sewage Sludge-Soil Mixtures," *Soil Sci. Soc. Am. J. 42*:600 (1978).

45. Tanford, C. *Physical Chemistry of Macromolecules* (New York: John Wiley and Sons, Inc., 1961).
46. Wilson, D. E. "Studies of the Metal Ion Binding Characteristics of Soluble, Natural Organic Materials," PhD Dissertation, Univ. of Alaska (1974).
47. Katchalsky, A., and P. Spitnik. "Potentiometric Titrations of Polymethacrylic Acid," *J. Polym. Sci. 2*:432 (1947).
48. Katchalsky, A., and J. Gillis. "Theory of the Potentiometric Titration of Polymeric Acids," *Rec. Trav. Chim. 68*:879 (1949).
49. Stevenson, F. J. "Stability Constants of Cu(II), Pb(II), and Cd(II) Complexes with Humic Acids," *Soil Sci. Soc. Am. 40*:665 (1976).
50. Gamble, D. S. "Titration Curves of Fulvic Acid: The Analytical Chemistry of a Weak Acid Polyelectrolyte," *Can. J. Chem. 48*(17):2662 (1970).
51. Schnitzer, M., and E. M. Hansen. "Organo-Metallic Interactions in Soils: 8. An Evaluation of Methods for the Determination of Stability Constants of Metal-Fulvic Acid Complexes," *Soil Sci. 109*(6):333 (1970).
52. Mantoura, R. F. C., A. Dickson and J. P. Riley. "The Complexation of Metals with Humic Materials in Natural Waters," *Est. Coast Mar. Sci. 6*: 387 (1978).
53. Sposito, G. "Trace Metals in Contaminated Waters," *Environ. Sci. Technol. 15*(4):396-403 (1981).
54. Wilson, D. E., and P. Kinney. "Effects of Polymeric Charge Variations on the Proton-Metal Equilibria of Humic Materials," *Limnol. Oceanog. 22*(2):281 (1977).
55. Gamble, D. S., M. Schnitzer and I. Hoffman. "Cu^{2+}-Fulvic Acid Chelation Equilibrium in 0.1 *M* KCl at 25.0°C," *Can. J. Chem. 48*:3197-3204 (1970).
56. Fettes, E. M. *Chemical Reactions of Polymers* (New York: Interscience Publishers, 1964), p. 14.
57. Curtis, M. A., and L. B. Rogers. "Effect of Molecular Size, Ionic Strength, and pH on Retentions of Aromatic Acids on XAD-8 Resins," *Anal. Chem.* 53(14):2347 (1981).

CHAPTER 13

COMPLEXATION OF HEAVY METALS BY AQUATIC HUMUS: A COMPARATIVE STUDY OF FIVE ANALYTICAL METHODS

John R. Tuschall, Jr.
Aquatic Chemistry Section
Illinois State Water Survey
Champaign, Illinois

Patrick L. Brezonik*
Department of Civil and Mineral Engineering
University of Minnesota
Minneapolis, Minnesota

ABSTRACT

Heavy metal binding with surface water humics from two swamps in Florida was examined by three published methods and two new methods developed in our laboratory. Complexometric titrations were performed using anodic stripping voltammetry (ASV), ion-selective electrode and fluorescence spectroscopy to determine the available metal binding ability (C_L) of the humic substances and the conditional stability constants (β') of the metal-humic complexes. Additionally, two new methods–continuous-flow ultrafiltration and differential spectroscopy using a competing ligand–were developed and compared to the other techniques. Copper was found to bind strongly to the surface water humics at pH 6.25 with β' values ranging from $10^{4.9}$ to $10^{8.11}$. Cadmium and zinc were less effective than copper in complexing the humic material. Values for both C_L and β' were similar

*Address all correspondence to this author.

among four of the five methods, with the ASV procedure yielding results that were considerably lower than the others. The ASV-reducible nature of the metal-humic complexes caused the ASV results to differ.

INTRODUCTION

Complexation of heavy metals by aquatic and terrestrial humus has been an important topic of research for more than a decade. The topic is of considerable practical interest because of the release of various heavy metals to aquatic systems by numerous human activities. Metal-humate complexation may affect the transport of metals through aquatic systems and also may affect biological uptake and the toxicity of the metals to aquatic organisms.

In evaluating the quantitative importance of heavy metal complexation by humic substances, at least four aspects should be considered: (1) the overall capacity of humic material to bind heavy metals; (2) the thermodynamic stability (or binding strength) of such complexes; (3) the lability of the complexes, i.e., the rate at which they dissociate; and (4) the rates of formation of metal-humate complexes. Binding capacities frequently are measured in terms of concentrations of binding (or ligand) sites, expressed as C_L, which can be determined by several titration methods. In general, thermodynamic stabilities of complexes are expressed quantitatively in terms of conditional stability constants (K'_{ML} or β'_{ML}); limitations and problems involved in using these constants to quantify metal-humate complexation are discussed below. No general criterion exists to define complex lability, although several electrochemical and spectrophotometric methods provide estimates of this phenomenon. Rates of metal-humate complex formation have not been widely studied.

Several practical questions about the role of complexation by humic material can be formulated relative to the four quantitative characteristics listed above. Whether complexation will occur at all when a heavy metal is added to a water containing humic material (in theory) can be determined from knowledge of C_L and β'_{ML}. Whether addition of a competing metal ion (such as calcium) will release a heavy metal (such as lead) from a humate complex depends not only on the values of β'_{CaH} and β'_{PbH} (where H stands for humate ligands) and on solution conditions, but also on the lability of the PbH complex. If complexation is a reversible process, then competition by other metals is a real possibility; if complexation is at least partially irreversible, then release of a toxic heavy metal upon increasing the concentration of another metal ion is of less concern. The question of complex lability has important implications regarding the biological availability of complexed metals (e.g., whether complexation enhances metal availability to microorganisms, prevents a metal from exerting a toxic effect, or has no effect on either phenomenon). Finally, the kinetics of complex formation has important analytical implications relative to the measurement of equilibrium properties of metal-humate complexes.

Progress in understanding and quantifying the role of humic material as a complexing agent for heavy metals has been slow for two principal reasons. First, as is well known, humic material itself is very complicated and heterogeneous, and our understanding of its molecular structure is still very limited. Indeed, the idea of a "molecule" of humic or fulvic acid is simplistic. Although certain monomeric units and functional groups are common to all humic macromolecules found in a given aquatic environment (e.g., freshwater swamps), the precise structural arrangements of these units probably are different for each individual macromolecule.

Furthermore, a variety of functional groups in humic material can participate in metal complexation, and treatment of humic or fulvic acid as a single ligand species is a great oversimplification. This situation implies that description of metal-humate binding in terms of a single stability constant (for a given metal) is unrealistic. Such a constant represents the "average binding strength" for the humate macromolecule over the range of titrated ligand functional groups, but as Langford [1] pointed out elsewhere in this symposium, such average constants have little thermodynamic meaning.

The second reason for slow progress in quantifying metal-humic complexation is the inadequacy of analytical methods to measure binding capacities and strengths. Although many methods have been developed and used to measure humic complexation equilibria over the past decade or so, they suffer various deficiencies, such as (1) lack of sensitivity to make measurements under ambient conditions, (2) disturbance of equilibrium conditions during (and caused by) the measurement, or (3) lack of understanding of the chemical forms actually measured by the method.

Moreover, comparability of results obtained by different groups of researchers using different procedures is hindered by the variability of solution conditions used to obtain reported data. Most workers report binding strengths in terms of *conditional* stability constants (β'), which are (undefined) functions of pH, ionic strength, and a variety of other solution conditions that often are not even measured or reported. Our limited understanding of the nature of humic materials prevents conversion of β' values obtained under one set of solution conditions to the values that would obtain under another set of conditions. Hence, comparison of literature values of β' for different waters obtained by different methods is tenuous at best.

Resolution of all aspects of the two major problems discussed above obviously is a formidable task that is beyond immediate solution. The scope of the present work was limited to the second class of problems, i.e., development and assessment of analytical methods to measure metal-humate complexation equilibria. The principal objective of this paper is to compare binding capacities (C_L) and binding strengths (β') determined by five separate complexometric titration procedures. Consistent solution conditions and the same two humic-rich waters were used in each case to allow inter-comparisons among the generated values.

Three of the methods compared here were developed elsewhere and applied previously to metal-humate complexation equilibria. The anodic stripping voltammetry (ASV) method of Shuman and Woodward [2] and the ion-selective electrode (ISE) method described by Buffle et al. [3] have been used widely in natural water complexation studies. The fluorescence-quenching procedure developed recently by Saar and Weber [4] relies on the fact that the natural fluorescence of humic material is diminished (quenched) when the material forms complexes with paramagnetic ions such as Cu^{2+}. Because of the problems and deficiencies associated with these and other existing procedures, we developed two new methods to measure metal complexation by humic material: a continuous-flow ultrafiltration method and a competing-ligand differential spectrophotometric method [5,6]. Results for C_L and β' obtained from these methods were compared to results from the three existing methods.

EXPERIMENTAL APPROACHES FOR MEASURING C_L and β'

Samples

Water samples were obtained from two hardwood (cypress) swamps in northern Florida, which our group was studying in relation to the effects of discharging secondary sewage effluent to wetland ecosystems [7]. Basin Swamp is a mixed-hardwood flood plain swamp associated with a small creek near Jasper, Florida. Waldo Swamp is a large cypress strand (a broad, forested swamp through which water flows slowly) near Waldo, Florida. Sewage effluents from the two towns have drained into the respective swamps for over 50 years, affecting levels of nutrients and major ions in the receiving waters. In both cases, however, the surface waters in the swamps are highly colored, reflecting the importance of surface drainage as the major source of water in the swamps; chemical characteristics of the waters are summarized in Table I. For all five complexation techniques, water samples from both swamps were adjusted to a pH of 6.25 (using NaOH) and to an ionic strength of 0.10 (using KNO_3). Copper (Cu^{2+}) was used as the complexing metal in all five analytical procedures.

Table I. Summary of Water Chemistry of Swamp Waters

Parameter[a]	Basin Swamp	Waldo Swamp
pH	6.1	6.2
Color	220	450
Specific conductance	150	110
Chloride	19	10
DOC	25	50
DON	0.55	0.75
DHAAN	0.08	0.11
TSP	0.30	0.13
SRP	0.30	0.12

[a]Values in mg/l except specific conductance (μS/cm), color (chloroplatinate units) and pH. DOC = dissolved organic carbon; DON = dissolved organic nitrogen; DHAAN = dissolved hydrolyzable amino acid nitrogen; TSP = total soluble phosphate-P; SRP = soluble reactive phosphate-P.

ASV Titrations

ASV complexometric titrations were done following the procedure of Shuman and Woodward [2] except that the stripping step of ASV was performed in the differential pulse (DP) mode rather than the linear (DC) mode to enhance sensitivity. Instrumentation consisted of a PAR model 174 polarographic analyzer with Kemula-type hanging mercury drop electrode and saturated calomel reference electrode connected to the solution by a 0.1 *M* KNO_3 bridge. Instrument settings, plating and scanning conditions were as described by Tuschall and Brezonik [8]. Solution pH was maintained at 6.25 by purging with a mixture of CO_2 and N_2, and titrant (Cu^{2+}) was added in μl amounts to avoid dilution.

Copper-ASV titrations of surface water from the two swamps yielded curves similar to those generally reported by users of this method (Figure 1). Peak current (i_p) was used as a measure of uncomplexed copper, although recent studies [8,9] indicate that complexed copper is reduced at the mercury electrode and is included in the measurement of peak current. Assuming that i_p is a measure of uncomplexed copper, however, Figure 1 indicates that Waldo Swamp had a greater capacity to complex copper than did Basin Swamp.

ISE Titrations

Complexometric titrations by ion-selective electrode were performed in a manner similar to that for the ASV titrations, except that a copper-ISE

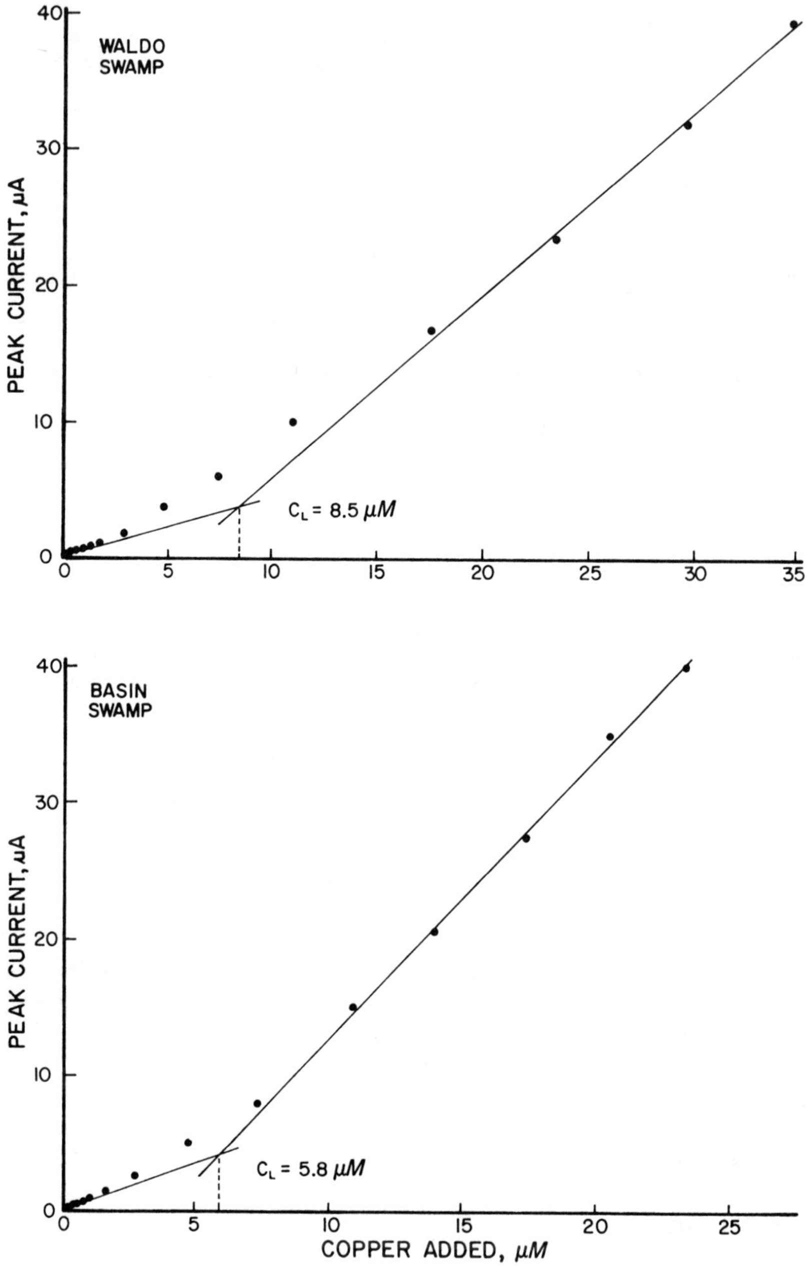

Figure 1. DPASV titration curves of two swamp water samples with copper (II) as titrant.

was used to detect uncomplexed Cu^{2+}. Samples (50 ml) were added to a polyethylene beaker maintained at 25°C in a constant-temperature bath, and potentials were measured with an Orion model 94-29 Cu-ISE and a double-junction reference electrode (SCE) containing 0.1 *M* KNO_3 in the outer reservoir. Potentials were recorded continuously with a strip-chart recorder to ensure that equilibrium had been obtained at each titration point. Other analytical details were described by Tuschall [5].

Qualitatively, the trends in titrations of the two swamp samples using the Cu-ISE to detect uncomplexed copper (Figure 2) were similar to the ASV results; i.e., Waldo Swamp had a greater complexing capacity for Cu. Moreover, for equivalent values of pCu_T, the values of Cu^{2+} were lower for Waldo Swamp than for Basin Swamp. The equivalence points of the Cu-ISE titrations (i.e., C_L) are difficult to discern from the plots in Figure 2, but plots of bound copper (Cu_b) vs total copper (Cu_T) allowed ready determination of C_L.

FLUORESCENCE MEASUREMENTS

Fluorescence-quenching (FQ) measurements followed the procedure of Saar and Weber [4]. Relative fluorescence of each sample was measured after successive additions of Cu^{2+} using an Aminco Model SPF-125 spectrofluorometer at an excitation wavelength of 350 nm and an emission wavelength of 450 nm. Titrations were continued until the concentration of copper was such that flocculation occurred. Concentrations of bound copper (Cu_b) at each titration point were calculated based on an equation of Saar and Weber:

$$(\%\ \text{quenched})/\alpha = \bar{\nu} = (Cu_b)/(L_T) \tag{1}$$

where α is an empirical constant
L_T is the total concentration of organic ligands
$\bar{\nu}$ is the ratio of bound metal to total ligand

L_T for the two swamp waters was estimated from the concentrations of dissolved organic carbon in the two waters (50 and 25 mg/l for Waldo and Basin Swamps, respectively), assuming the organic matter was 50% carbon and that it had an average molecular weight of 1000. This yielded estimated L_T values of 1×10^{-4} *M* (Waldo) and 5×10^{-5} *M* (Basin Swamp). These values also were used for comparative purposes in Scatchard analyses described later.

Saar and Weber used a value of 57 for α, based on calibration of their FQ data with ISE titration results. Application of this value to samples from the Waldo and Basin Swamps led to values of $Cu_b > Cu_T$, i.e., greater

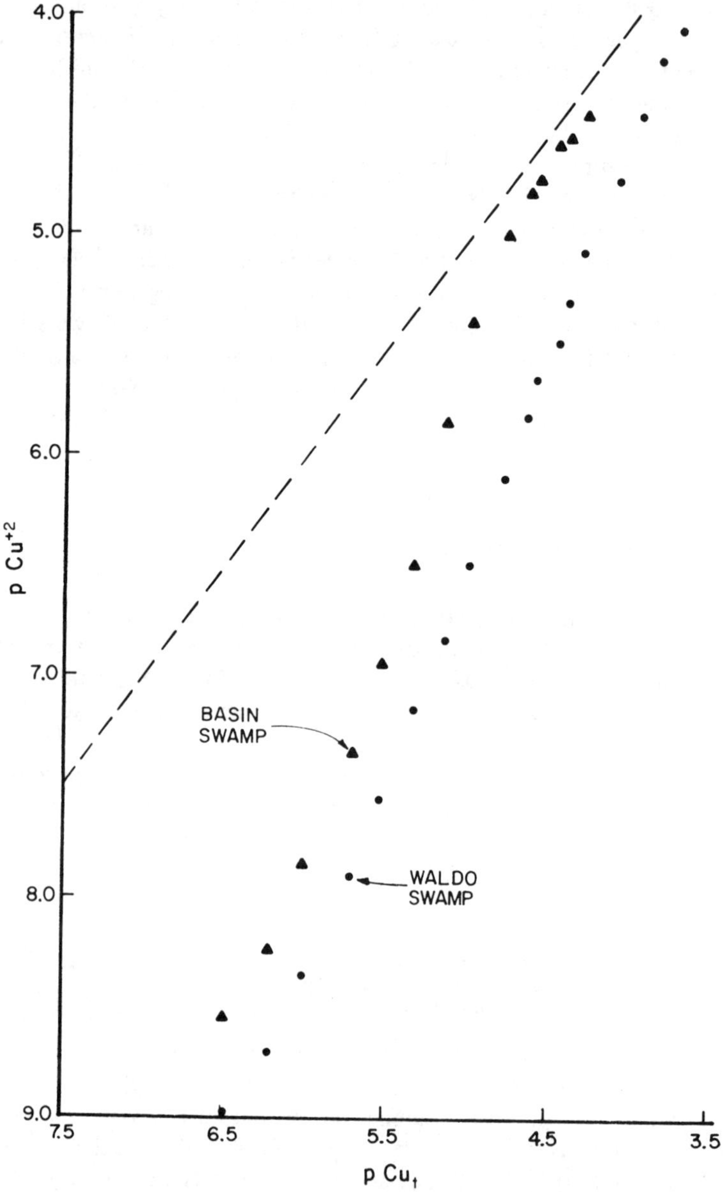

Figure 2. Copper titrations of swamp waters using a Cu-ISE to detect free Cu^{2+} Dashed line is standard curve for background electrolytes.

than the total Cu, which is physically impossible. Values of α were obtained for these waters by calibrating at two values of percentage quenched with Cu_b values obtained (at equivalent values of Cu_T) from the Cu-ISE titrations performed on the samples. For Waldo Swamp, we found $\alpha = 104$, and for Basin Swamp, $\alpha = 150$. As a result of the required calibration of the FQ data using ISE results, the FQ technique is not a completely independent method to determine C_L and β'.

The fluorescence of both swamp samples was quenched to a similar extent for copper additions up to $\sim 3 \times 10^{-5}$ *M* (30% quenched) (Figure 3). Therefore, quenching was greater in Waldo Swamp water than in Basin Swamp water. Quenching did not approach 100% for either sample, suggesting that not all humic fluorophores are involved in copper complexation. The maximum quenching achieved was 69% for Waldo and 56% for Basin, both at $Cu_T = 3.1 \times 10^{-4}$ *M*. At higher copper concentrations, a rapidly settling floc was produced. Copper binding capacities were obtained from Figure 3 and Equation 1 as the asymptotic values on a plot of pCu_b vs pCu_T. Calculated values of C_L were 7.4×10^{-5} *M* for Waldo Swamp and 2.1×10^{-5} *M* for Basin Swamp.

CONTINUOUS ULTRAFILTRATION

Ultrafiltration (UF) has been used by several workers to evaluate the extent of metal binding by high-molecular-weight organic fractions in natural waters [10–12]. The technique is based on the use of membranes that allow ions and small organic molecules to pass through them but retain larger organic molecules and any metal ions complexed with them. In previous applications of the UF technique, determination of metal distribution was done by filtering a sample at a single metal and ligand concentration. To construct a titration curve, filtration experiments were repeated at several metal concentrations–a tedious and time-consuming approach.

Based on work by Ryan and Hanna [13] on the binding of steroids and other small organic molecules to proteins, we developed a continuous UF technique to measure metal binding by macromolecular humic materials [5,6]. The method involves continuous addition of a solution of metal titrant into a stirred ultrafiltration cell containing a constant volume (70 ml) of humic-containing water, initially with no added metal. Selection of a membrane represents a compromise between decreasing retention of organic matter with increasing membrane porosity and increasing solute interaction with ionic surface sites as filter porosity decreased. An Amicon UM-10 membrane (nominal pore size of 10^4 daltons) was found to give optimum results, and >95% of the organic matter from the two swamps was retained by this membrane.

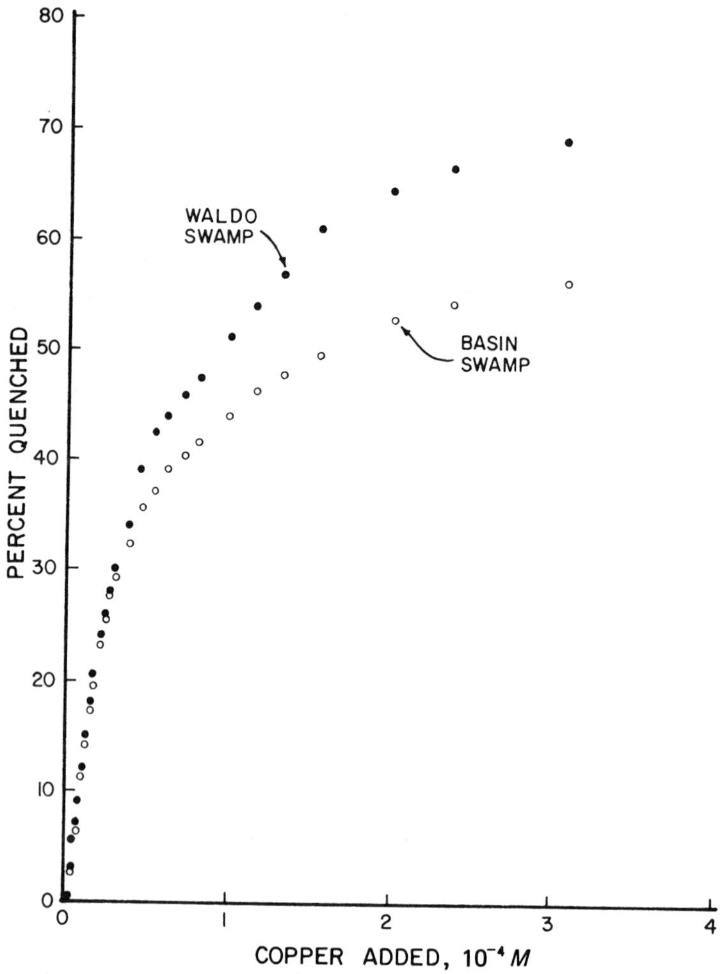

Figure 3. Percentage fluorescence quenched (λ_{em} = 450 nm) vs added copper concentration for the two swamp waters.

Filtrate from the cell was obtained at a flow rate of 0.2 ml/min and collected in discrete (3 to 7 ml) fractions using a Gilson GME automatic fraction collector. The concentration of metal in each fraction was taken to be the concentration of unbound metal in the cell when the fraction was collected. Metal analyses were performed by flameless atomic absorption spectrophotometry using a Perkin-Elmer Model 5000 at instrument settings recommended by the manufacturer.

The amount of metal bound to organic matter (M_b) in the UF cell can be calculated for each fraction by a simple materials balance:

$$M_b = V_{out}(M_i) - V_c(M_e) - M_{out} \tag{2}$$

where V_{out} is the total volume of effluent collected
V_c is the UF cell volume
(M_i) is the influent concentration of metal
(M_e) is the effluent concentration of metal
M_{out} is the total amount of metal collected in the effluent

For the jth fraction

$$M_{out} = \sum_{n=1}^{j} V_n(M_{en})$$

where V_n is the volume of each fraction
(M_{en}) is the effluent metal concentration in each fraction

A programmable calculator was used to reduce the experimental data and compute M_b for each fraction.

Plots of M_e vs V_{out} are curvilinear, even in the case where no organics are in the UF cell (Figure 4a) and where M_e is thus predictable based on simple mixing and dilution. Because of the curvilinear nature of the results, deviations from ideal behavior are not easily detected in such plots. However, for ideal behavior (i.e., no complexation by macromolecules), the results can be linearized [6] by plotting log $\{(M_i)/[(M_i) - (M_e)]\}$ vs V_{out}. As shown in Figure 4b, the copper data for water containing no organics fell on the theoretical line for simple dilution, indicating that rejection or binding free metal ions by the membrane was not occurring. Both ways of plotting the continuous UF data clearly show that copper was retained by the swamp water samples (see Figures 4a and b for the Waldo sample). Binding capacity C_L cannot be determined directly from Figure 4, but this parameter can be determined from a plot of M_b (from Equation 2) vs V_{out}. Such plots tend toward an asymptotic value that can be extrapolated to obtain C_L. Calculated values of C_L were 4.3×10^{-5} *M* for Waldo Swamp and 1.2×10^{-5} *M* for Basin Swamp.

One advantage of the continuous UF technique is that it is readily applicable to a variety of metals. For comparative purposes, we performed titrations of the two swamp waters with cadmium and zinc, as well as with copper. Computed values for C_L were 5 to 10 times lower for Cd and Zn than the C_L values we found for Cu (Table II). In all cases C_L for Basin Swamp was about 40% of that for Waldo Swamp; the C_L values for Cd and Zn in a given swamp were quite similar.

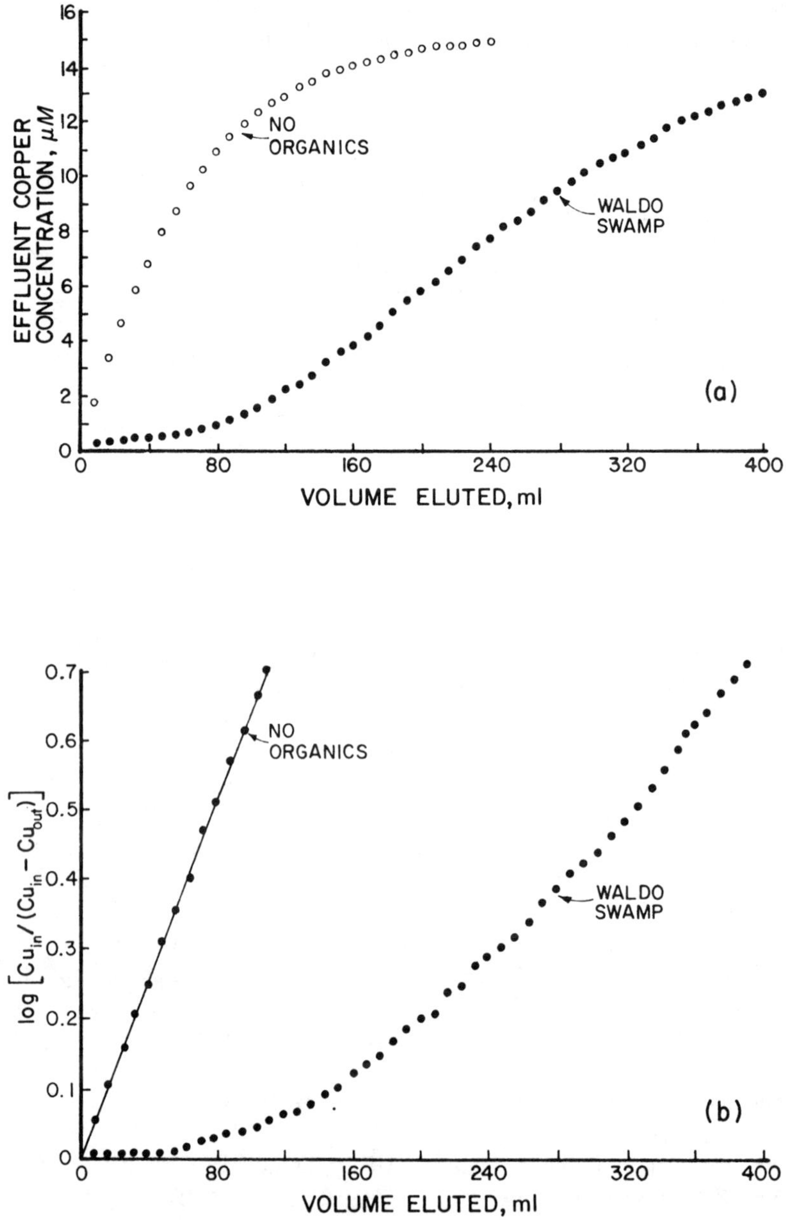

Figure 4. Elution patterns of copper for continuous ultrafiltration procedure. Solid line in lower graph represents theoretical dilution line; circles represent experimental values for background electrolyte solution (no organics) and for Waldo Swamp.

Table II. Complexing Capacities (C_L) and Average Conditional Stability Constants ($\log \beta'$) for Two Swamp Waters from Continuous Ultrafiltration Method

	C_L (M)	$\log \beta'$
Cadmium		
Waldo Swamp	5.25×10^{-6}	5.21
Basin Swamp	2.3×10^{-6}	5.36
Zinc		
Waldo Swamp	4.9×10^{-6}	4.78
Basin Swamp	1.8×10^{-6}	5.08

Competing-Ligand Differential Spectroscopy (CLDS)

This method has not been used previously in studies on metal complexation by aquatic humus, but it has been used in other environmental studies to determine conditional stability constants for iron with microbially produced siderophores using EDTA as the competing ligand [14,15]. In addition, Campbell et al. [16] used this approach to study the binding of zinc by river water organic matter. Unfortunately, many of the strong absorption bands for metal-ligand complexes are in the 200 to 300 nm range, in which natural organic matter also absorbs strongly. The key to the CLDS method is finding a ligand that (1) forms a strong complex with the metal of interest and (2) has a large bathochromatic shift in absorbance for the complexed species.

The CLDS method we developed [5,6] involves the equilibria:

$$\begin{aligned} CuL &= Cu_f + L \\ Cu_f + SAL &= CuSAL \end{aligned} \qquad (3)$$

where L is the ligand of interest (in this case natural water humic material)
Cu_f is uncomplexed copper
SAL is the competing ligand, salicylate

Salicylate absorbs strongly at 281 nm, and the addition of Cu^{2+} causes a bathochromatic shift in the absorption band to a maximum at 320 nm. The concentration of CuSAL is proportional to absorbance at 320 nm and can be measured simply by differential spectroscopy, i.e., solution with added Cu^{2+} in the sample path of a spectrophotometer and solution with no added Cu^{2+} in the reference path. By quantifying (CuSAL) and knowing β'_{CuSAL} and the analytical concentrations of copper (Cu_T), ligand (L_T) and salicylate (SAL_T), the concentrations of Cu_f and CuL can be calculated.

The concentration of free copper is determined from β'_{CuSAL}:

$$\beta'_{CuSAL} = 10^{3.5} = CuSAL/(Cu_f)(SAL) \quad (4)$$

and the mass balance equation for salicylate:

$$SAL_T = SAL + CuSAL \quad (5)$$

Subsequently, CuL is determined from the mass balance for copper:

$$Cu_T = Cu_f + CuL + CuSAL \quad (6)$$

Under the conditions we used with salicylate, the limit of detection for Cu_f was about 10^{-7} *M*; it can be shown [6] that the maximum value of β' that could be determined under these circumstances was on the order of 10^9. For natural organics with larger values of β' (that produce values of $Cu_f < 10^{-7}$ *M*), a competing ligand with a β'_{Cu} greater than that for CuSAL would need to be used.

Titrations of the swamp water samples by CLDS resulted in the data presented in Figure 5. As with the other procedures, the Waldo Swamp sample complexed more copper than did the Basin Swamp sample. The horizontal difference between the dashed line (standard curve) and each sample point is a measure of the copper bound (Cu_b) to natural organic ligands in the sample. A plot of Cu_b vs Cu_T was used to compute C_L for the samples; values of 1.3×10^{-5} and 6.8×10^{-5} were found for Basin Swamp and Waldo Swamp, respectively.

Determination of Conditional Stability Constants

A single value of β' at various levels of added metal is not observed with most natural water samples; as noted previously, aquatic humic material is heterogeneous and contains a variety of ligands with a range of binding strengths. In cases where only a few ligand types (two, or at most three) are present in a solution, stability constants for each type can be resolved from a titration curve using plots of the type developed by Scatchard [17]. In these plots of $\bar{\nu}/(M_f)$ vs $\bar{\nu}$ (where $\bar{\nu}$ is the ratio M_b/L_T), the slope (i.e., tangent) of the curve at any point is a measure of the "instantaneous β'" applicable at that value of $\bar{\nu}$.

Workers using Scatchard plots on titrations of natural water organics generally have been able to resolve the curves into two limiting linear regions at high and low values of $\bar{\nu}$, and have interpreted these results as indicating the existance of two types of ligands: a strong site (of high β') from the

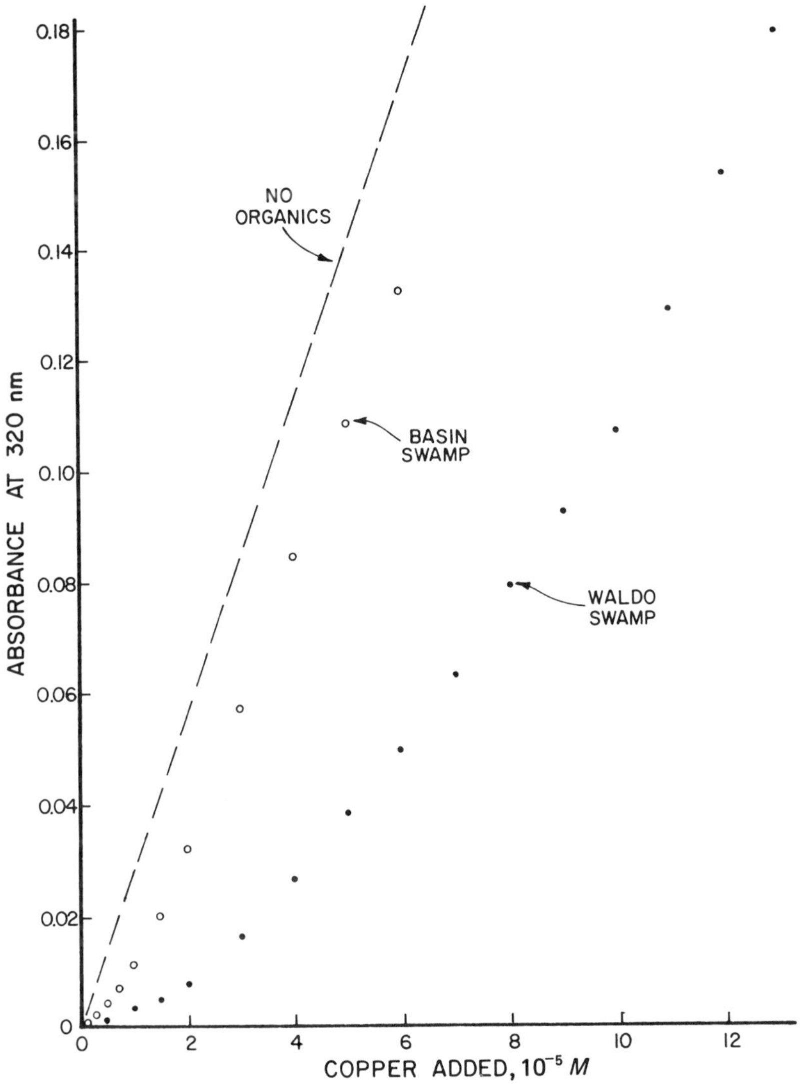

Figure 5. Copper titration curves for swamp samples by competing ligand-differential spectroscopy. Dashed line is standard curve for CuSAL formation in background electrolyte solution (absent of competing organic ligands). $SAL_T = 1 \times 10^{-3}$ *M* in all solutions.

straight line at low values of $\bar{\nu}$, and a weak site (of lower β') from the linear region at high values of $\bar{\nu}$. It is recognized that for aquatic humus, with possibly a wide variety of ligand types (and hence values of β'), Scatchard plots are still a simplification of reality. Nonetheless, the plots provide a greater measure of realism than use of a single (average) value for β'.

In this study, Scatchard plots were constructed using data from the five types of titrations, and the plots were divided into three regions (for Waldo Swamp, $\bar{\nu} < 0.02$, $0.021 < \bar{\nu} < 0.2$, and $\bar{\nu} > 0.21$; for Basin Swamp, $\bar{\nu} < 0.025$, $0.026 < \bar{\nu} < 0.125$, $\bar{\nu} > 0.126$). Values of L_T used for all the Scatchard plots were 1×10^{-4} *M* for Waldo Swamp and 5×10^{-5} *M* for Basin Swamp, and were estimated for these waters as described earlier. It should be noted that the value used for L_T in the Scatchard analysis does not affect the value of the calculated β'. Figure 6 shows an example using the ISE titration data for Waldo Swamp. The slopes of the most linear portion of each segment were used as measures of β' associated with these regions of $\bar{\nu}$ (i.e., ratios of M_b/L_T). The values of β' thus obtained are not to be taken as measures of intrinsic conditional stability constants for specific types of ligand groups, and the values are used here only so that comparisons could be made among the five analytical procedures used to determine the levels of complexed and uncomplexed metal.

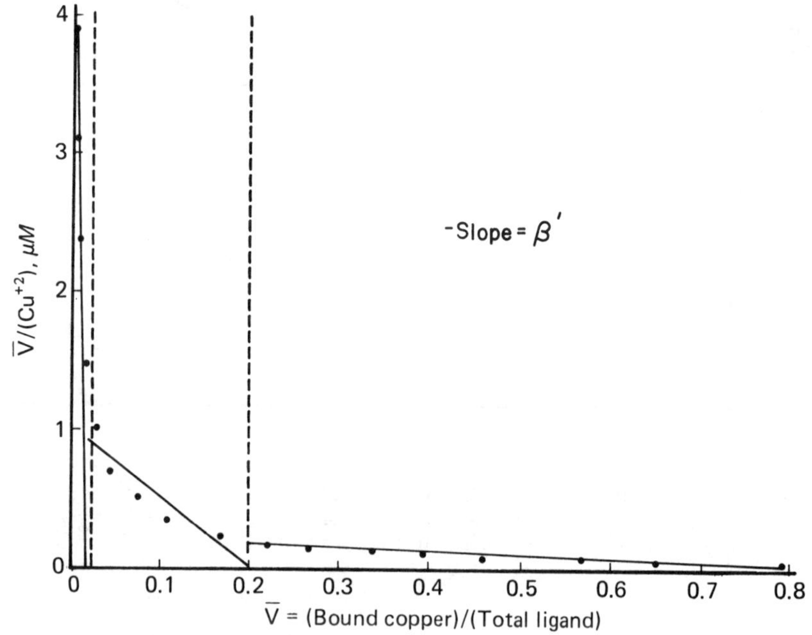

Figure 6. Scatchard plot for Cu-ISE titration of Waldo Swamp water showing the three regions in which slopes (= $-\beta'$) were obtained.

COMPARISON OF ANALYTICAL METHODS

Values of C_L determined on the two swamp samples by the five analytical methods are tabulated in Table III. For each sample, agreement among the procedures was quite good, except for the ASV method, which yielded C_L values considerably lower than those from the other four methods. For example, the range of C_L for Waldo Swamp was less than a factor of two (4.3×10^{-5} *M* to 7.9×10^{-5} *M*) for the four non-ASV methods, whereas ASV gave a value of 0.8×10^{-5} *M*. The concordance of results among the four methods is encouraging, considering the diversity of analytical measurements involved and the difficulties in obtaining reliable data with some of the methods. The closeness of the results suggests that the four (non-ASV) procedures provide fairly unbiased estimates of C_L. The lower values obtained by ASV probably reflect problems caused by the reducible nature of the copper-organic complex [5,8,9].

In all five procedures, Waldo Swamp water complexed more copper than did Basin Swamp water. Average values for C_L were 5.5×10^{-5} *M* and 1.3×10^{-5} *M*, respectively. These values are generally higher than the C_L values Truitt and Weber [18] reported for seven surface water samples in New Hampshire (range of 0.1 to 1.5×10^{-5} *M*).

Values of β' determined from the three regions of the Scatchard plots for the five methods are listed in Table IV. Three of the procedures (FQ, continuous UF, CLDS) were not sufficiently sensitive to speciate copper

Table III. Copper Binding Capacities of Swamp Waters Determined by Five Procedures

	C_L (*M*)	
Method of Analysis	Waldo Swamp	Basin Swamp
ASV	8.5×10^{-6}	5.8×10^{-6}
ISE	7.9×10^{-5}	1.4×10^{-5}
Fluorescence spectroscopy	7.4×10^{-5}	2.1×10^{-5}
Continuous ultrafiltration	4.3×10^{-5}	1.2×10^{-5}
Competing-ligand differential spectroscopy	6.8×10^{-5}	1.3×10^{-5}
Range	8.5×10^{-6} to 7.9×10^{-5}	5.8×10^{-6} to 2.1×10^{-5}
Mean	5.45×10^{-5}	1.31×10^{-5}
Standard deviation	2.9×10^{-5}	5.42×10^{-6}

Table IV. Conditional Stability Constants (log β') for Copper Determined on Swamp Waters from Scatchard Plots Using Five Analytical Methods[a]

Range of $\bar{\nu}$[b]	ASV	ISE	Flourescence	Continuous Ultrafiltration	Competing Ligand	$\overline{X}$
Waldo Swamp						
0–0.02	5.25	8.11	ND[c]	ND	ND	7.81
0.021–0.2	5.36	6.72	6.80	6.54	6.61	6.60
0.21–0.8	ND	5.34	5.17	5.70	5.48	5.47
Basin Swamp						
0–0.025	6.04	7.82	ND	ND	ND	7.53
0.026–0.125	5.30	6.85	5.70	6.67	6.72	6.55
0.126–0.4	ND	6.26	4.87	5.56	5.54	5.38

[a]All values are log β'.
[b]$\bar{\nu} = (M_b)/L_T$.
[c]ND = Not detected.

in the lowest range of $\bar{\nu}$ (i.e., $\bar{\nu} < 0.02$). For the other two ranges of $\bar{\nu}$ there was good agreement among the β' values determined by ISE, continuous UF and CLDS. The β' values computed by fluorescence-quenching for Waldo Swamp agreed favorably with values from the three procedures mentioned above, but the FQ values for Basin Swamp were lower (by factors of ~5 to 10) than the ISE, UF and CLDS results.

In all cases, the β' values computed from Scatchard plots of the ASV titrations were considerably lower than the corresponding values obtained by the other methods. In the range $\bar{\nu} < 0.02$ the β' values obtained by ISE were larger by factors of almost 10^3. The discrepancies between ASV results and those of the other methods are less pronounced in the intermediate range of $\bar{\nu}$, but β' values from ASV still were more than a factor of 10 lower than the average β' for the four other methods. It is apparent that the ASV procedure used here is not an accurate method to determine β' for copper-organic complexes.

The discrepancy between the ASV results and those of the other four methods probably is related to the fact that metal complexes with natural organics in the swamp samples are directly reducible at the mercury electrode under the conditions used for analysis in this study. Cyclic voltammograms [5] determined on aliquots of Waldo Swamp water containing 1×10^{-5} *M* Cu_T showed the presence of a single reduction peak at a potential 120 mV cathodic of the reduction potential of Cu^{2+}. This difference in potentials is not great enough to allow plating of free Cu^{2+} without reducing the complexed metal.

It is interesting to note that the other four methods yielded comparable results in spite of the fact that different species of metal are measured in the methods. For example, the ISE method measured only free Cu^{2+}, whereas the continuous UF method measured free Cu^{2+} plus the Cu present in inorganic complexes and in complexes with small organic ligands.

In summary, this comparative study indicated that four different methods (ion-selective electrode titrations, fluorescence quenching continuous ultrafiltration, and competing-ligand differential spectroscopy) produced comparable values of complexing capacity (C_L) and complex stability (β'). Of the four methods, the ISE procedure has the greatest sensitivity and can measure stability constants at low values of $\bar{\nu}$ using a Scatchard plot.

ACKNOWLEDGMENTS

Support for this research was provided by grants from the National Science Foundation (Grant Number PFR 78-19199) and the U.S. Environmental Protection Agency (Grant Number EPA-R-806286020).

REFERENCES

1. Langford, C. "Interaction of Metal Ions with a Well Characterized Fulvic Acid," Chapter 11, This Volume.
2. Shuman, M. S., and G. P. Woodward, Jr. "Chemical Constants of Metal Complexes from a Complexometric Titration Followed with Anodic Stripping Voltammetry," *Anal. Chem.* 45:2032-2035 (1973).
3. Buffle, J., F. L. Greter and W. Haerdi. "Measurement of Complexation Properties of Humic and Fulvic Acids in Natural Waters with Lead and Copper Ion-Selective Electrodes," *Anal. Chem.* 49:216-222 (1977).
4. Saar, R. A., and J. H. Weber. "Comparison of Spectrofluorometry and Ion-Selective Electrode Potentiometry for Determination of Complexes between Fulvic Acid and Heavy-Metal Ions," *Anal. Chem.* 52:2095-2100 (1980).
5. Tuschall, J. R., Jr. "Heavy Metal Complexation with Naturally Occurring Organic Ligands in Wetland Ecosystems," PhD Thesis, University of Florida, Gainesville, 1981.
6. Tuschall, J. R., Jr., and P. L. Brezonik. "Application of a Continuous-Flow Ultrafiltration Procedure and Competing Ligand Differential Spectroscopy to Measurement of Heavy Metal Complexation by Natural Water Organic Matter," Submitted to *Anal. Chem.* (1981).
7. Tuschall, J. R., Jr., K. C. Ewel and P. L. Brezonik. "Tertiary Treatment of Wastewater Using Flow-Through Wetland Systems." *Proc. 1981 Nat. Conf. Environmental Engineering*, Atlanta, GA. (New York: Amer. Soc. Civil Eng., 1981), pp. 558-565.
8. Tuschall, J. R., Jr., and P. L. Brezonik. "Evaluation of the Copper Anodic Stripping Voltammetry Complexometric Titration for Complexing Capacities and Conditional Stability Constants," *Anal. Chem.* 53:1986-1989 (1981).

9. Greter, F. L., J. Buffle and W. Haerdi. "Voltammetric Study of Humic and Fulvic Acid, Part 1. Study of the Factors Influencing the Measurement of their Complexing Properties with Lead," *J. Electroanal. Chem.* 101:211-229 (1979).
10. Gjessing, E. T. "Ultrafiltration of Aquatic Humus," *Environ. Sci. Technol.* 4:437-438 (1970).
11. Guy, R. D., and C. L. Chakrabarti. "Studies of Metal-Organic Interactions in Model Systems Pertaining to Natural Waters," *Can. J. Chem.* 54:2600-2611 (1976).
12. Hoffman, M. R., E. C. Yost, S. J. Eisenreich and W. J. Maier. Characterization of Soluble and Colloidal-Phase Metal Complexes in River Water by Ultrafiltration: A Mass Balance Approach," *Environ. Sci. Technol.* 15:655-661 (1981).
13. Ryan, M. T., and N. S. Hanna. "Investigation of Equilibrium Ultrafiltration as a Means of Measuring Steroid-Protein Binding Parameters," *Anal. Biochem.* 40:364-379 (1971).
14. Anderegg, G., L. L'Eplattenier and G. Schwarzenbach. "Hydroxamatkomplexe III. Eisen (III): Austauch Zwischen Sideraminen and Komplexonen. Diskussion der Bildungskonstanten der Hydroxamatkomplexe," *Helv. Chim. Acta* 46:1409-1422 (1963).
15. Harris, W. R., C. J. Carrano, S. R. Cooper, S. R. Sofen, A. E. Avdeef, J. V. McArdle and K. N. Raymond. "Coordination Chemistry of Microbial Iron Transport Compounds, 19: Stability Constants and Electrochemical Behavior of Ferric Enterobactin and Model Complexes," *J. Am. Chem. Soc.* 101:6097-6104 (1979).
16. Campbell, P. G. C., M. Bisson, R. Gagne and A. Tessier. "Critical Evaluation of the Copper (II) Solubilization Method for the Determination of the Complexation Capacity of Natural Waters," *Anal. Chem.* 49:2358-2362 (1977).
17. Scatchard, G. "The Attractions of Proteins for Small Molecules and Ions," *Anal. N. Y. Acad. Sci.* 51:660-672 (1949).
18. Truitt, R. E., and J. H. Weber. "Copper (II)–and Cadmium(II)–Binding Abilities of Some New Hampshire Freshwaters Determined by Dialysis Titration," *Environ. Sci. Technol.* 15:1204-1208 (1981).

CHAPTER 14

A CRITICAL EXAMINATION OF METAL-LIGAND COMPLEXATION MODELS: APPLICATION TO DEFINED MULTILIGAND MIXTURES

Edward Michael Perdue and Charles Russell Lytle

Environmental Sciences/Chemistry Department
Portland State University
Portland, Oregon

ABSTRACT

From the application of the two-component Scatchard equation to metal binding in complex, but defined, multiligand mixtures, it has been demonstrated that the resulting fitting parameters are totally empirical and bear no relation to the composition of the ligand mixtures. This observation casts considerable doubt on the applicability of this model to metal-humic substances complexation. The two multiligand mixtures tested were: (1) a hypothetical continuum of log K' distributed normally about an average log K' of 2.00 with a deviation of ±3.00; and (2) a set of 23 ligands chosen from the MINEQL simulation program. A statistical-chemical approach is proposed, with the Gaussian distribution used as a convenient example. Such a distribution model shows promise for predicting the extent of metal complexation at environmentally significant metal concentrations from laboratory results obtained at much higher metal concentrations.

INTRODUCTION

Humic substances are ubiquitous in soil and aquatic environments, being formed through random reactions that presumably occur during microbial degradation of biomass. Despite the fact that elemental analyses, functional group analyses, number-average molecular weights, etc. indicate that some bulk properties of humic substances are relatively invariant, all attempts to fractionate humic substances into simpler subfractions have been unsuccessful. For example, Reuter and Perdue [1] have found that the infrared spectra of molecular size fractions of aquatic humus do not decrease in complexity with decreasing molecular weight. Curtis et al. [2] have found that pH gradient chromatography of humic substances on XAD-8 resins yields essentially a continuous elution profile. Further evidence for the tremendous complexity of humic substances is found in numerous studies of their acid-base properties [3-6].

The fact that metal complexation by humic substances results in the release of protons indicates that the highly diverse acidic functional groups of humic substances are directly involved in metal complexation reactions. For complexation of a metal ion M by the ith ligand or binding site at constant pH,

$$M + L_i \leftrightarrows ML_i \tag{1}$$

and

$$K_i = \frac{[ML_i]}{[M][L_i]} \tag{2}$$

Because of competition by H^+ and the possibility that the ith metal binding ligand may bind several protons, conditional stability constants are commonly used to describe metal complexation at constant pH.

$$K_i' = \frac{[ML_i]}{[M][H_{x_i}L_i]} \tag{3}$$

where $H_{x_i}L_i$ is the concentration of *all* ligand species that are not bound to M

The total stoichiometric concentration of the ith ligand can thus be given by

$$C_i = [ML_i] + [H_{x_i}L_i] \tag{4}$$

Finally, it should be noted that

$$K_i' = K_i\left(\frac{[L_i]}{[H_{x_i}L_i]}\right) \tag{5}$$

In other words, the conditional stability constant (K_i') equals K_i times the fraction of $H_{x_i}L_i$ that is not protonated. This fraction is constant at constant pH.

In a complex mixture of ligands, an average stability "constant" ($\bar{K}$) and a *conditional* average stability "constant" ($\bar{K}'$) can be defined, the latter being readily calculated from experimental data.

$$\bar{K} = \frac{\sum_i [ML_i]}{[M]\sum_i [L_i]} = \frac{\sum_i K_i[L_i]}{\sum_i [L_i]} \tag{6}$$

$$\bar{K}' = \frac{\sum_i [ML_i]}{[M]\sum_i [H_{x_i}L_i]} = \frac{\sum_i K_i'[H_{x_i}L_i]}{\sum_i [H_{x_i}L_i]} = \frac{(C_M - [M])}{[M](C_L - C_M + [M])} \tag{7}$$

The conditional average stability "constant" equals the thermodynamic average stability "constant" times the fraction of uncomplexed ligand that is not protonated.

$$\bar{K}' = \bar{K}\left(\frac{\sum_i [L_i]}{\sum_i [H_{x_i}L_i]}\right) \tag{8}$$

If the numerator and denominator of Equation 7 are divided by the activity of all uncomplexed forms of an arbitrarily selected reference ligand ($H_{x_r}L_r$), Equation 9 is obtained.

$$\bar{K}' = \frac{\sum_i K_i'\left[\frac{[H_{x_i}L_i]}{[H_{x_r}L_r]}\right]}{\sum_i \left[\frac{[H_{x_i}L_i]}{[H_{x_r}L_r]}\right]} \tag{9}$$

Because, in the most general case, $H_{x_i}L_i$ and $H_{x_r}L_r$ will have different affinities for the metal (M), their ratio will vary as the overall ligand/metal ratio in solution changes. Accordingly, $\bar{K}'$ must also vary with solution composition and *cannot be regarded as a constant at all.* This inescapable fact has been clearly stated by previous workers [7,8]. Although Equations 6 through 9 were derived for 1:1 complexes, analogous equations are obtained for higher-order complexes (e.g., 1:2, 1:3). Again, the average "constants" are clearly a function of solution composition [7].

The analytical methods employed in metal complexation studies often require C_M values in excess of 10^{-6} *M* (e.g., ion-selective electrode measurements), while concentrations of 10^{-8} *M* are probably more characteristic of environmental levels of trace metals such as Cu^{+2}, Cd^{+2}, Pb^{+2}, etc. The experimental results of several researchers seem to indicate that $\bar{K}'$ values increase asymptotically as C_M/C_L approaches zero [8-11]. The fact that $\bar{K}'$ is a function of C_L and C_M rather than a constant immediately raises some important questions. Can $\bar{K}'$ values determined at one set of C_L and C_M values be extrapolated to other experimental conditions? If so, what type of chemical model most effectively describes the concentration-dependence of $\bar{K}'$? It is the purpose of this chapter to examine currently used models and other potentially useful models of metal complexation to determine the most appropriate formalism for describing metal complexation by humic substances.

Because of the inherent difficulty in chemically characterizing humic substances and the possible structural variabilities that make comparisons of results from different laboratories rather tenuous, a more rigorous comparison of multiligand complexation models can be made in complex, but well defined, ligand mixtures. In this chapter, two such ligand mixtures will be examined, with the expectation that the conclusions reached in this study will be relevant to the study of metal complexation by humic substances.

DISCRETE MULTILIGAND MODELS

Multiligand models or the analogous multiple binding site models usually assume 1:1 stoichiometry for all metal-ligand complexes. It is convenient to combine Equation 3 with Equation 4 to obtain

$$\nu_i \equiv \frac{[ML_i]}{C_i} = \frac{K_i'[M]}{1 + K_i'[M]} \qquad (10)$$

When summed for all ligands at a total ligand concentration of C_L

$$\bar{\nu} = \frac{\sum_i \nu_i C_i}{\sum_i C_i} = \sum_i \left(\frac{K_i'[M]}{1 + K_i'[M]}\right)\left(\frac{C_i}{C_L}\right) = \frac{C_M - [M]}{C_L} \qquad (11)$$

As indicated in Equation 11, experimental $\bar{\nu}$ values are directly calculable from C_M, C_L and [M]. Substitution from Equation 11 into Equation 7 yields $\bar{K}'$ values.

$$\bar{K}' = \frac{1}{[M]}\left(\frac{\bar{\nu}}{1-\bar{\nu}}\right) \tag{12}$$

As indicated earlier, however, $\bar{K}'$ values are not constant in multiligand mixtures. The only real constants in such systems are the K_i' values of individual ligands or binding sites, and then only at constant pH. While some researchers have elected to present metal complexation by humic substances simply as the experimental $\bar{K}'$ values [8], others have used Equation 11, generally truncated to two terms, to try to resolve K_i' and C_i/C_L values from experimental data [9-11].

When attempting to analyze experimental metal complexation data, it is frequently *initially* assumed that humic substances have only one type of ligand or binding site, so that Equation 11 can be written as

$$\bar{\nu} = \left(\frac{K_1'[M]}{1 + K_1'[M]}\right)\left(\frac{C_1}{C_L}\right) \tag{13}$$

which is easily rearranged to yield the Scatchard equation [12].

$$\frac{\bar{\nu}}{[M]} = K_1'\left(\frac{C_1}{C_L}\right) - K_1\bar{\nu} \tag{14}$$

This simple equation, which predicts a linear relationship between $\bar{\nu}/[M]$ and $\bar{\nu}$, never adequately describes the curvilinear plots of $\bar{\nu}/[M]$ vs $\bar{\nu}$ that are always obtained in studies of metal-humus complexation equilibria [9-11]. If one accepts the description of humic substances as a highly complex mixture of ligands, the failure of Equation 14, which assumes only one ligand or binding site, is not unexpected.

Almost without exception, those researchers who use the Scatchard equation to analyze experimental data respond to the failure of the linear model by including the second term from Equation 11, yielding

$$\bar{\nu} = \left(\frac{K_1'[M]}{1 + K_1'[M]}\right)\left(\frac{C_1}{C_L}\right) + \left(\frac{K_2'[M]}{1 + K_2'[M]}\right)\left(\frac{C_2}{C_L}\right) \tag{15}$$

The addition of the second term (which increases the number of curve-fitting parameters from two to four) generally results in a greatly improved fit of the experimental data, leading to the erroneous conclusion that humic substances, a highly complex mixture of substances extracted from soils by meteoric waters, contain only two types of metal-binding ligands. Alternatively, it has been suggested that the K'_1, C_1/C_L, K'_2 and C_2/C_L values are actually *average* values for two distinct *classes* of ligands, but this cannot possibly be correct. In a typical experiment, C_L/C_M ratios are varied from one to two orders of magnitude to generate data ($\bar{\nu}$ and [M]). When Equation 15 is used to fit such data, it is implicitly assumed that K'_1 and K'_2 are constants, independent of C_L/C_M. The mathematical developments leading to Equation 9 are certainly applicable to two classes of ligands, each class containing a large number of similar but nonidentical ligands. The conclusion remains that average "constants" are not, in fact, constant, but must vary with C_L/C_M. It therefore follows that, even if the ligands of humic substances do fall into two distinct classes, the average equilibrium "constant" for each class of ligands would be a function of $\bar{\nu}$. Thus, the whole idea of extracting *average* equilibrium *constants* from the two-component Scatchard equation is erroneous. If humic substances contained only two nonidentical binding sites, rather than a very large number of such ligands, then K'_1, C_1/C_L, K'_2 and C_2/C_L would be chemically meaningful parameters. As it is, however, the four constants that are obtained from the two-component Scatchard equation must be regarded as nothing more than mathematical curve-fitting parameters that have no chemical significance. Statements to the effect that humic substances contain two classes of metal-complexing ligands are therefore potentially misleading and any speculation about the chemical structures of these presumed ligands is clearly unwarranted. The limitations of the Scatchard analysis in complex systems are well documented, particularly in biochemical literature [13-16].

Even though the four parameters (K'_1, C_1/C_L, K'_2 and C_2/C_L) are empirical constants, they *can* be used in the two-component Scatchard equation to accurately predict the degree of complexation of a metal by humic substances in the experimental range of C_M or C_L values. It will subsequently be shown that the two-component Scatchard equation is not readily extrapolated to C_M or C_L values outside the experimental concentration range.

The conclusions given in the preceding paragraph are not intended as a criticism of the Scatchard equation per se, but rather of the manner in which it is applied to humic stubstances. The Scatchard equation has been widely and successfully applied to biomolecules such as proteins and nucleic acids. It is not at all unreasonable to expect that a purified protein molecule might have only one or two distinct sites for binding of a metal ion or other substrate.

Metal complexation equilibria can also be described by other linear equations. For example, if the one-component Scatchard equation (Equation 14) is inverted and rearranged, Equation 16 is obtained.

$$\frac{1}{\bar{\nu}} = \frac{C_L}{C_1} + \frac{C_L}{C_1}\left(\frac{1}{K_1'[M]}\right) \tag{16}$$

By plotting $1/\bar{\nu}$ vs $1/[M]$, C_1/C_L and K_1' can be obtained for a system that contains only one type of ligand. This equation has been in use for as long as the Scatchard equation for describing equilibria in biochemical systems [17]. More recently, Equation 16 has been used in an attempt to describe metal binding by humic substances [18]. Because Equation 16 is simply a rearranged form of the Scatchard equation, it is only applicable to single-ligand systems. Plots of $1/\bar{\nu}$ vs $1/[M]$ will necessarily be curvilinear for complex multiligand systems such as humic substances.

Bufflé et al. [18], having observed the failure of Equation 16 to describe metal complexation by humic substances, suggested that the deviation from linearity was attributable to the formation of both 1:1 and 1:2 metal-ligand complexes. Indeed, just as the addition of a second term to the Scatchard equation (Equation 15) resulted in a much better mathematical description of metal complexation by humic substances [9–11], the same effect is achieved by inclusion of both 1:1 and 1:2 stoichiometries. As in the former case, however, the calculated "constants" for 1:1 and 1:2 complexes cannot be regarded as anything more than curve-fitting parameters with no chemical significance. While it is indeed possible that 1:1 and 1:2 complexes are formed, it is not possible to determine the relevant average equilibrium "constants" because they are *not* constant in multiligand systems. Like their two-ligand model counterparts, the principal utility of two-stoichiometry models lies in their ability to accurately predict overall metal complexation by humic substances within the experimental range of C_M and C_L values.

CONTINUOUS MULTILIGAND MODELS

The discussion in the preceding section may have given the impression that there is no rigorous mathematical method by which metal complexation by humic substances can be efficiently described (i.e., some process other than tabulation of experimental results [8]). However, MacCarthy and Smith [7] have shown that, under severely limiting conditions of excess ligand that may not be attainable in laboratory studies or in the environment, the average stability "constants" for 1:1, 1:2, and higher complexes in multiligand mixtures approach constant-limiting values. While this conclusion is of theoretical interest, it is of little practical importance, given the analytically imposed limitation on C_M and C_L values in laboratory studies. The development of a mathematical model that can be calibrated under laboratory conditions and extrapolated accurately to environmental concentrations of metal and ligand is clearly warranted.

Proton binding by humic substances is obviously analogous to metal binding, the same complex ligand mixture being involved in both reactions. Posner [19,20] reported that proton binding by humic substances was efficiently described by a continuous multiligand distribution model, the relative concentration of each ligand being normally distributed relative to the pK_i of the ligand. In other words, most ligands had "average" pK_i values, with fewer very strong or very weak ligands. In terms of a Gaussian distribution

$$\frac{C_i}{C_L} = \frac{1}{\sigma\sqrt{2\pi}} e^{-\frac{1}{2}\left(\frac{\mu - pK_i}{\sigma}\right)^2} dpK \qquad (17)$$

where C_i/C_L is the mole fraction of ligands in the interval dpK whose acid dissociation constant is expressed as a negative logarithm (pK_i)
σ is the standard deviation for the distribution of pK_i values about the mean pK value (μ)

Similar models have been used to describe association equilibria in complex biochemical systems [17,21,22] and the adsorption of gases on catalytic surfaces [23]. Currently, there are no data to suggest that a Gaussian distribution of metal-binding ligands occurs in humic substances. Nevertheless, some sort of distribution is intuitively expected to exist. The general features of distribution models can be illustrated by the Gaussian distribution model. The Gaussian distribution model (Equation 17) is easily combined with the previously discussed multiligand model for 1:1 complexes (Equation 11) and converted from a summation to a continuous integral.

$$\bar{\nu}_{calc} = \frac{1}{\sigma\sqrt{2\pi}} \int_{-\infty}^{\infty} \left(\frac{[M]10^{\log K'}}{1 + [M]10^{\log K'}}\right) e^{-\frac{1}{2}\left(\frac{\mu - \log K'}{\sigma}\right)^2} d\log K' \qquad (18)$$

Given a set of μ and σ values, Equation 18 can be evaluated by numerical integration (e.g., Simpson's Method) at any [M] value. Finite lower and upper limits of ($\mu - 4\sigma$) and ($\mu + 4\sigma$) for log K′ will yield sufficiently accurate integrals while minimizing computational time. To model experimental $\bar{\nu}$ data [$\bar{\nu}_{exp} = (C_M - [M])/C_L$], $\bar{\nu}_{calc}$ values are calculated at each value of [M] using an initial set of μ and σ values. The residual sum of squares (RSS) is calculated via Equation 19,

$$(RSS) = \sum_{j=i}^{n} (\bar{\nu}_{calc} - \bar{\nu}_{exp})^2/\bar{\nu}_{exp} \tag{19}$$

where n is the number of data points ($\bar{\nu}_{exp}$, [M]) used in the experiment

An iterative nonlinear regression technique is used to determine the set of μ and σ values for which RSS is minimized. This general approach has been recently used to describe proton binding by aquatic humic substances [24].

The Gaussian distribution model has several distinct advantages and disadvantages relative to the discrete two-ligand model or the two-stoichiometry model. First of all, it has to be a more realistic chemical model of humic substances, even if the actual distribution of ligands is non-Gaussian. The symmetry of the Gaussian function makes it theoretically possible to define the shape of the curve at higher C_M values where $\bar{\nu} \geqslant 0.5$ and extrapolate to lower C_M levels where $\bar{\nu}$ approaches zero. By expressing C_i/C_L as an explicit function of K_i' and by utilizing K_i' as the variable of integration (d log K′), the overall behavior of a complex mixture of ligands can be described without knowing any actual K_i' or C_i values. Secondly, from a purely empirical point of view, the Gaussian distribution model with two curve-fitting parameters (μ and σ) is less constrained than the two-component Scatchard equation, which has four curve-fitting parameters (K_1', C_1/C_L, K_2', C_2/C_L). On the negative side, the determination of optimum μ and σ values and the use of Equation 18 to calculate $\bar{\nu}$ values are both complicated by the necessity of using numerical integration techniques, although this criticism is less severe today due to the general availability of laboratory computers. One additional and somewhat nebulous perceived negative aspect of the continuous distribution model is the inability of the model to reduce multiligand chemical equilibrium systems to a table of thermodynamically defined equilibrium constants. It is somehow psychologically less than satisfactory to have to describe metal complexation equilibria with what is essentially a statistical model. The demonstrated successes of two-ligand models and two-stoichiometry models in very simple chemical systems have unfortunately led numerous researchers to try to apply these fundamentally inappropriate models to humic substances. This alternative is worse than the statistical approach, in our view, because the curve-fitting parameters are often mistakenly assumed to have chemical significance, when, in fact, these models are also unable to extract thermodynamically defined equilibrium constants for multiligand systems. It is our hope that the following section of the chapter will adequately demonstrate the overall superiority of the continuous distribution model in describing metal complexation in multiligand systems.

APPLICATIONS

Gaussian Ligand Set

The two-component Scatchard equation (Equation 15) will be used to analyze a hypothetical ligand mixture in which the ligand concentrations (C_i) are, in fact, normally distributed (Figure 1). By definition, the Gaussian distribution will perfectly describe this ligand mixture. The point of the exercise is to examine the sensitivity of the Scatchard analysis to the range of C_L, C_M and [M] values used to obtain the four curve-fitting parameters of the Scatchard equation (K'_1, C_1/C_L, K'_2, C_1/C_L). Using Equation 18 and Equation 12 with $\mu = 2.00$ and $\sigma = 3.00$, $\bar{\nu}$ and $\overline{K}'$ values for this hypothetical ligand mixture were calculated for $-11.0 \leqslant \log [M] \leqslant -3.0$ in increments of 0.1 log units for a constant C_L value of 2.0×10^{-4} *M*. The corresponding C_M value at each [M] value is

$$C_M = [M] + \bar{\nu} C_L \tag{20}$$

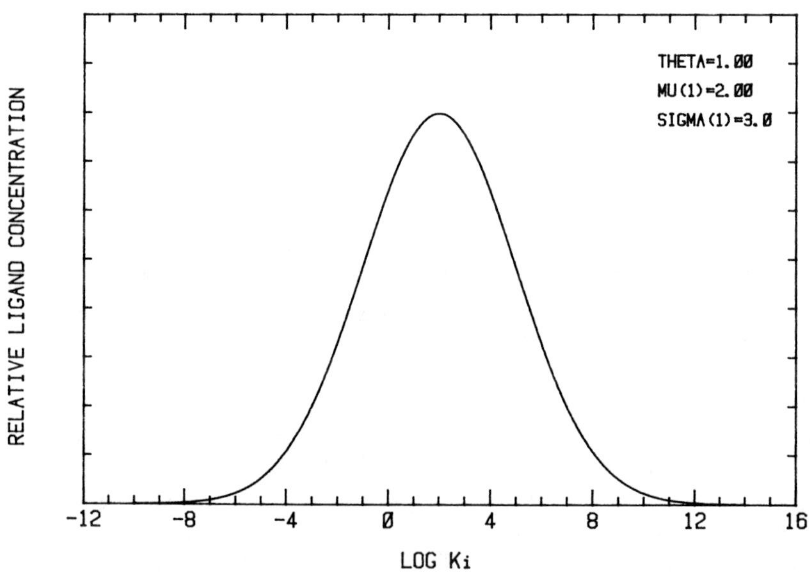

Figure 1. Normally distributed ligand concentrations (C_i) used for the Gaussian ligand set. $\mu = 2.00$, $\sigma = 3.00$.

To simulate experimental limits of detection of [M], three plots of $\bar{\nu}/[M]$ vs $\bar{\nu}$ were constructed assuming that the analytical detection limits for [M] are 1.26×10^{-8} *M* (Fit 1), 1.26×10^{-7} *M* (Fit 2), and 1.26×10^{-6} *M* (Fit 3). (See Figure 2.) The corresponding C_M values are 5.72 μM, 11.5 μM, and 22.1 μM.

Using all data points from the assumed "detection limit" to the highest value of (M), a weighted nonlinear regression procedure was used to fit data to the two-component Scatchard equation (Equation 15), yielding the "best" values of K_1', C_1/C_L, K_2' and C_2/C_L. The calculated results are

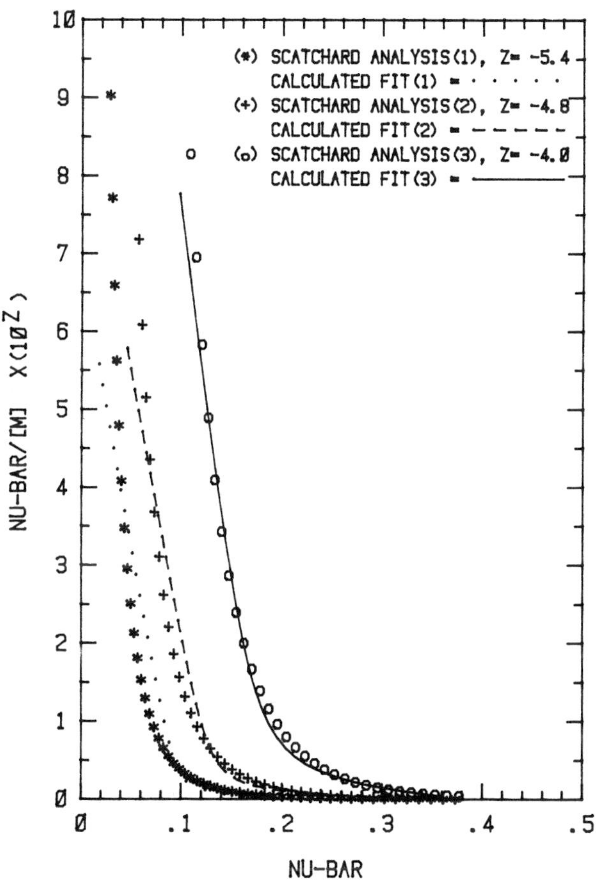

Figure 2. Experimental $\bar{\nu}$ and calculated fits for the Gaussian ligand set using the two-component Scatchard equation for three titration ranges. Lowest C_M of 5.72 μM = Fit 1; 11.5 μM = Fit 2; 22.1 μM = Fit. 3.

tabulated in Table I, and the calculated curves are superimposed on the data in Figure 2.

From the results in Table I, it is apparent that the values of K'_1 and C_1/C_L are particularly sensitive to the "detection limit," a fourfold change in C_M leading to a 20-fold change in K'_1. The other parameters are relatively less sensitive to changes in C_M because the region of the data set that mainly determines K'_2 and C_2/C_L (high $\bar{\nu}$ values) is unchanged in the three test cases. It is also clearly the case that K'_1, C_1/C_L, K'_2 and C_2/C_L have no obvious chemical significance because, rather than consisting of two ligands or even two classes of ligands, the data were generated assuming a continuous Gaussian distribution of ligands with an average log K of 2.00.

Accepting the empirical nature of the Scatchard parameters, it remains to be determined whether experimental results can be extrapolated to very low $\bar{\nu}$ values that typify natural waters. In Figure 3, a plot of log $\bar{K}'$ vs log (C_L/C_M) is given. The points from the Gaussian model are still rising fairly steeply at log (C_L/C_M) values greater than two. (In natural waters containing 10 mg C/L of aquatic humus and a metal concentration of $C_M = 10^{-8}$ *M*, log (C_L/C_M) is approximately three.) In contrast, all fitted Scatchard curves level off at log $\bar{K}'$ values that are several orders of magnitude

Table I. Sensitivity of Scatchard Equation K′ to Titration Range for the Gaussian and MINEQL Ligand Sets, and Gaussian and Scatchard Parameters for the Calculated Fits in Figures 1 through 4

Gaussian Ligand Set						
Lowest C_M (μM) Used in Fit	Gaussian Fit[a]		Scatchard Fit			
	μ	σ	Log K'_1	C_1/C_L	Log K'_2	C_2/C_L
5.72	2.00	3.00	7.26	0.094	4.52	0.243
11.5	2.00	3.00	6.65	0.126	4.21	0.234
22.1	2.00	3.00	6.03	0.166	3.89	0.220
MINEQL Ligand Set						
Lowest C_M (μM) Used in Fit	Gaussian Fit		Scatchard Fit			
	μ	σ	Log K'_1	C_1/C_L	Log K'_2	C_2/C_L
6.31	3.28	3.41	8.27	0.095	3.97	0.657
12.6	3.64	1.77	6.88	0.130	3.82	0.674
20.0	3.71	1.42	6.86	0.138	3.79	0.675

[a]The Gaussian model with μ = 2.00 and σ = 3.00 was used to generate the data set, so the fit is perfect by definition.

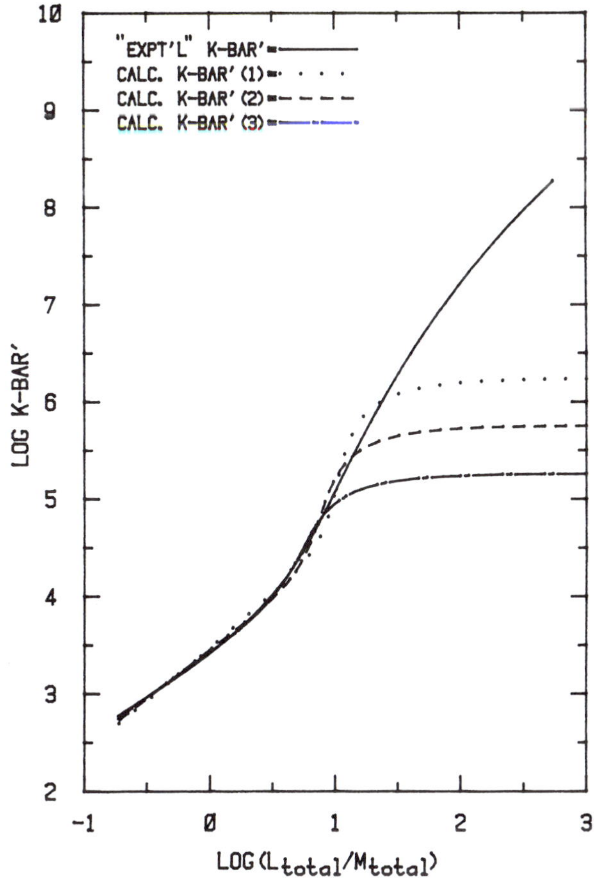

Figure 3. "Experimental" log $\overline{K}'$ values and calculated fits for the Gaussian ligand set using the two-component Scatchard equation. Lowest C_M of 5.72 μM = Fit 1; 11.5 μM = Fit 2; 22.1 μM = Fit. 3.

too low. As a result, the concentration of uncomplexed metal (M) at a log (C_L/C_M) value of three (2.5×10^{-12} *M*) is not accurately predicted by the Scatchard fits. The best predicted [M] value (1×10^{-9} *M*), which is obtained for Fit 1 with the lowest "detection limit," is still more than two orders of magnitude too high. In that fit, a departure from the gradual increase of log $\overline{K}'$ of the Gaussian set is seen. In fact, if the "detection limit" is lowered even further, the Scatchard curve will rise vertically near log (C_L/C_M) = 1 and level off at even higher values of $\overline{K}'$. Thus, the gradual change of $\overline{K}'$ that characterizes the Gaussian set cannot be modeled by the two-component Scatchard equation at high C_L/C_M ratios typical of natural waters.

MINEQL Ligand Set

Moving from a purely mathematical ligand set to a real, defined multiligand system, the titration of a 23-ligand mixture ($C_L = 2.0 \times 10^{-4}$ *M*) at pH 5.0 and I = 0.1 with Cu(II) was simulated using MINEQL [25]. Total Cu(II) values ranged from 10^{-8} *M* to 10^{-3} *M*. The individual ligands, K_i' values, and C_i values are summarized in Table II. The choice of ligands and concentrations was designed to produce a gradual titration curve and to avoid complexes of greater than 1:1 stoichiometry, hopefully eliminating ambiguity about the choice of model for fitting the data. The end result is a ligand mixture in which only 1:1 complexes are formed in significant concentrations. The actual concentration of every chemical species is, of course, given in the MINEQL output, so the system is essentially defined and is easily reproduced by other researchers with access to MINEQL.

Table II. Concentrations and Log K′ Values for Copper Complexation for the 23 Ligands Used in the MINEQL Computer-Simulated Titration

Ligand	Log K′[a]	Conc (μM)
Acetate	1.65	10.45
Alanine	3.58	11.01
Arginine	3.58	11.01
Aspartate	3.98	10.83
Citrate	6.37	9.05
DCTA	14.66	1.87
EDTA	14.35	5.41
Ethylenediamine	3.78	9.48
Glutamate	3.41	11.01
Glycine	3.58	11.01
Histidine	6.52	7.93
Isoleucine	3.68	10.92
Leucine	3.68	10.92
Norcardamine	3.10	6.72
NTA	8.34	3.64
Ornithine	−2.42	0.09
Oxalate	6.14	8.67
Phenylalanine	3.68	10.92
Phthalate	2.99	11.01
Salicylate	2.26	10.45
Sulfosalicylate	3.44	11.01
Tartrate	3.37	10.55
Valine	3.48	11.01

[a]K′ = the conditional stability constant at pH = 5.00 and I = 0.1 *M*.

Given this ligand mixture and the calculated equilibrium data at each level of added Cu(II), how well can the results be described by the two-component Scatchard equation (Equation 15) and by the Gaussian equation (Equation 18)? Scatchard plots are given for this data set at C_M "detection limits" of 6.31 μM (Fit 1), 12.6 μM (Fit 2), and 20.0 μM (Fit 3) in Figure 4. The K'_1, C_1/C_L, K'_2 and C_2/C_L values for each fit are given in Table I, and the calculated curves are superimposed on the data points in Figure 4. As in the case of the Gaussian ligand set, K'_1 and C_1/C_L are sensitive to the

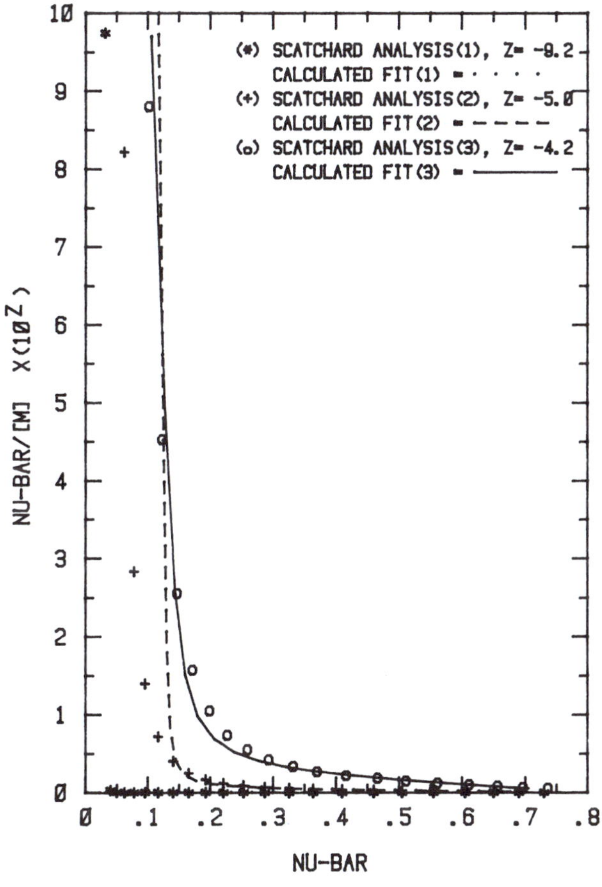

Figure 4. Experimental $\bar{\nu}$ and calculated fits for the MINEQL ligand set using the two-component Scatchard equation for three titration ranges. Lowest C_M of 6.31 μM = Fit 1; 12.6 μM = Fit 2; 20.0 μM = Fit 3.

C_M "detection limit," a threefold change in C_M leading to an 80-fold change in K'_1.

The predictions of the Scatchard model and the Gaussian model for the MINEQL data set were compared using that portion of the data set for which the lowest C_M = 6.31 μM. The best values of the respective curve-fitting parameters are given in Table I. The calculated variation of log $\bar{K}'$ with log (C_L/C_M) for each model is given in Figure 5. Neither model is very effective at describing the experimental curve at both the low C_L/C_M values found in most laboratory studies and the high C_L/C_M values that typify natural waters. Interestingly, the Gaussian model is more successful

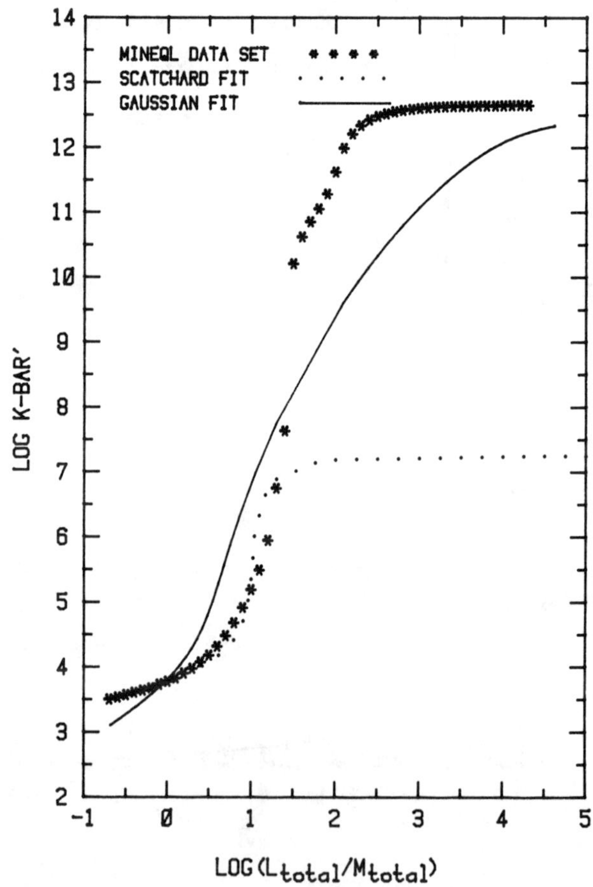

Figure 5. "Experimental" log $\bar{K}'$ values and calculated fits for the MINEQL ligand set using the Scatchard model and the Gaussian model. Lowest C_M = 6.31 μM.

in extrapolating laboratory data to environmental C_L/C_M values, even though it contains only half as many curve-fitting parameters as the two-component Scatchard model.

The above comparison is all the more interesting in view of the fact that the MINEQL data set is non-Gaussian and is definitely not a continuous ligand set. The flat region of log $\bar{K}'$ above a log (C_L/C_M) of three is due to the titration of the strongest ligand in the data set.

Humic substances are almost certainly intermediate in complexity between the discrete, 23-ligand MINEQL set and a continuous distribution exemplified by the Gaussian ligand set. Intuitively, however, a distribution model is probably a better representation of humic substances, even if the actual ligand distribution in humus is non-Gaussian. It is therefore anticipated, and will hopefully soon be demonstrated, that metal complexation by humic substances is realistically and efficiently described by a continuous distribution model.

CONCLUSIONS

Average stability "constants" for metal complexation in a multiligand mixture, whether for the total ligand mixture or for a subclass of similar, but nonidentical ligands, are not constant at all, but vary with solution composition. This observation bears directly on the two-component Scatchard equation, commonly used to analyze metal complexation by humic substances. That equation, while being rigorous for two-component systems, becomes simply an empirical curve-fitting equation with four adjustable parameters when applied to multiligand systems that contain three or more ligands; this is analogous to using a cubic equation to curve-fit data. The derived constants have little or no chemical significance. The large number of adjustable parameters does make it possible to curve-fit data with the two-component Scatchard equation, even in complex ligand mixtures like humic substances.

Given the empirical nature of the K_1', C_1/C_L, K_2' and C_2/C_L values obtained by fitting complex ligand mixtures with the two-component Scatchard equation, their values would be expected to depend on the range of concentrations of ligand and metal used to obtain a data set. This expectation was confirmed by analysis of two defined multiligand mixtures. The K_1' and C_1/C_L values in particular were quite sensitive to the lowest C_M values used in the analysis.

For complex ligand mixtures such as humic substances, rigorous discrete complexation models can be approximated by continuous models. One such model, a Gaussian distribution model, shows promise for describing complex multiligand systems with only two curve-fitting parameters, the mean log $\bar{K}'$

for the ligand mixture and the standard distribution of ligand log $\bar{K}'$ values about the mean.

The Gaussian distribution model is superior to the two-component Scatchard equation in its ability to predict log $\bar{K}'$ values at C_M values that typify natural waters from experimental data at much higher C_M values.

ACKNOWLEDGMENT

The authors thank Dr. R. Parrish for writing the two Fortran programs used for the nonlinear least-squares analyses and for the determination of optimum μ and σ values.

REFERENCES

1. Reuter, J. H., and E. M. Perdue. "Calculation of Molecular Weights of Humic Substances from Colligative Data: Application to Aquatic Humus and its Molecular Size Fractions," *Geochim. Cosmochim. Acta* 45: 2017-2022 (1981).
2. Curtis, M. A., A. F. Witt, S. B. Schram and L. B. Rogers. "Humic Acid Fractionation Using a Nearly Linear pH Gradient," *Anal. Chem.* 53:1195-1199 (1981).
3. Gamble, D. S. "Titration Curves of Fulvic Acid: The Analytical Chemistry of a Weak Acid Polyelectrolyte," *Can. J. Chem.* 48:2662-2669 (1970).
4. Gamble, D. S. "Potentiometric Titration of Fulvic Acid: Equivalence Point Calculations and Acidic Functional Groups," *Can. J. Chem.* 50:2680-2690 (1972).
5. Burch, D. R., C. H. Langford and D. S. Gamble. "Methods of Comparison of Fulvic Acid Samples: The Effects of Origin and Concentration on Acidic Properties," *Can. J. Chem.* 56:1196-1201 (1978).
6. Perdue, E. M., J. H. Reuter and M. Ghosal. "The Operational Nature of Acidic Functional Group Analyses and Its Impact on Mathematical Descriptions of Acid-Base Equilibria in Humic Substances," *Geochim. Cosmochim. Acta* 44:1841-1851 (1980).
7. MacCarthy, P., and G. C. Smith. In: *Chemical Modeling in Aqueous Systems*, E. A. Jenne, Ed. (Washington, DC, American Chemical Society–Symposium Series No. 93, 1979), p. 201.
8. Gamble, D. S., A. W. Underdown and C. H. Langford, "Copper (II) Titration of Fulvic Acid Ligand Sites with Theoretical, Potentiometric, and Spectrophotometric Analysis," *Anal. Chem.* 52:1901-1908 (1980).
9. Mantoura, R. F. C., and J. P. Riley. "The Use of Gel Filtration in the Study of Metal Binding by Humic Acids and Related Compounds," *Anal. Chim. Acta* 78:193-200 (1975).
10. Bresnahan, W. T., C. L. Grant and J. H. Weber. "Stability Constants for Complexation of Copper (II) Ions with Water and Soil Fulvic Acids Measured by an Ion Selective Electrode," *Anal. Chem.* 50:1675-1679 (1978).

11. Sposito, G., K. M. Holtzclaw and C. S. LeVesque-Madore. "Cupric Ion Complexation by Fulvic Acid Extracted from Sewage Sludge-Soil Mixtures," *Soil Sci. Soc. Amer. J.* 43:1148-1155 (1979).
12. Scatchard, G., J. S. Coleman and A. L. Shen. "Physical Chemistry of Protein Solutions VII. The Binding of Some Small Anions to Serum Albumin," *J. Am. Chem. Soc.* 79:12-20 (1957).
13. Klotz, I. M., and D. L. Hunston. "Properties of Graphical Representations of Multiple Classes of Binding Sites," *Biochem.* 10(16):3065-3069 (1971).
14. Hunston, D. L. "Two Techniques for Evaluating Small Molecule-Macromolecule Binding in Complex Systems," *Anal. Biochem.* 63:99-109 (1975).
15. Thakur, A. K., P. J. Munson, D. L. Hunston and D. Rodbard. "Characterization of Ligand-Binding Systems by Continuous Affinity Distributions of Arbitrary Shape," *Anal. Biochem.* 103:240-254 (1980).
16. Scheinberg, I. H. "Scatchard Plots," *Science.* 215:312-313 (1982).
17. Karush, F., and M. Sonenberg. "Interaction of Homologous Alkyl Sulfates with Bovine Serum Albumin," *J. Am. Chem. Soc.* 71:1369-1376 (1949).
18. Bufflé, J., F. L. Greter and W. Haerdi. "Measurement of Complexation Properties of Humic and Fulvic Acids in Natural Waters with Lead and Copper Ion-Selective Electrodes," *Anal. Chem.* 49:216-222 (1977).
19. Posner, A. M. In: *Proceedings of the 8th International Congress of Soil Science. Part II.* (Bucharest, Romania, 1964), p. 161.
20. Posner, A. M. "The Humic Acids Extracted by Various Reagents from a Soil. Part I. Yield, Inorganic Components, and Titration Curves," *J. Soil Sci.* 17:65-78 (1966).
21. Pauling, L., D. Pressman and A. L. Grossberg. "The Serological Properties of Simple Substances. VII. A Quantitative Theory of the Inhibition by Haptens of the Precipitation of Heterogeneous Antisera with Antigens, and Comparison with Experimental Results for Polyhapenic Simple Substances and for Azoproteins," *J. Am. Chem. Soc.* 66:784-792 (1944).
22. Nisonoff, A., and D. Pressman. "The Annular Nitrogen of Pyridine as a Determinant of Immunologic Specificity," *J. Am. Chem. Soc.* 79:5565-5572 (1957).
23. Sips, R. "On the Structure of a Catalytic Surface," *J. Chem. Phys.* 16:490-495 (1948).
24. Parrish, R. S., E. M. Perdue, N. L. Wolfe and J. H. Reuter. "On Estimation of Pearson Mixing Distributions with an Application to Titration," *Technometrics*, submitted for publication.
25. Westall, J. C., J. L. Zachary and F. M. M. Morel. "MINEQL. A Computer Program for the Calculation of Chemical Equilibrium Composition of Aqueous Systems," *Techn. Note No. 18*, Mass. Inst. Techn. (1976).

CHAPTER 15

METAL ION SPECIATION STUDIES IN THE PRESENCE OF HUMIC MATERIALS

James H. Weber

Department of Chemistry
University of New Hampshire
Durham, New Hampshire

ABSTRACT

Researchers do metal ion speciation studies in the presence of humic matter by a variety of techniques. The complexity of humic matter makes it advantageous to compare the results of at least two types of experiments. This chapter describes speciation studies using equilibrium dialysis (ED), differential pulse anodic stripping voltammetry (DPASV) and fluorescence spectrometry (FS). Emphasis will be on the implicit and explicit assumptions involved in the above approaches, and their advantages and disadvantages. The following points exemplify the scope of the discussion. (1) FS and DPASV are nonseparation techniques that are useful only for limited metal ions. In contrast, the ED separation approach is appropriate for any metal ion determinable by atomic absorption spectrophotometry. (2) DPASV and ED determine free metal ion concentrations, while only FS determines free humic matter concentrations. (3) The use of DPASV requires critical evaluation of the problems of kinetic current corrections and adsorption of humic matter on mercury electrodes. (4) A model for metal ion binding by FS yields the titration parameters of total ligand concentration and conditional stability constant.

INTRODUCTION

Many scientists are interested in metal ion speciation in natural waters because biological and physiochemical properties of metal ions are changed by complexation. First, certain free metal ions are more toxic than metal ions complexed to organic ligands. Second, geochemical transport of metal ions depends on the extent of complexation. In addition, speciation of metal ions in natural water matrices is a challenging problem that can lead to widely applicable advances in analytical methodology. Although considerable expertise is available for determining total metal ions [1–4], their speciation is more difficult and controversial [5]. Methods espoused by various research groups are attacked by others.

Our earlier emphasis on solving the speciation problem has been on experiments with fulvic acids isolated from soil (SFA) and water (WFA) [6]. We differentiated total added and free metal ions by several techniques: ion selective electrodes (ISE) for Cu^{2+} [7], Cd^{2+} [8] and Pb^{2+} [9,10], electron paramagnetic resonance for VO^{2+} [11], and differential pulse anodic stripping voltammetry (DPASV) on the glassy carbon electrode for Cu^{2+} [12].

Our research group is problem-oriented, not technique-oriented, when studying metal ion–humic matter binding. That is, we test and evaluate any potentially appropriate technique, and are not limited to a specific approach such as ISE or DPASV. Comparative experiments are advantageous for isolated or in situ humic matter systems because of the diversity of their ligands. This chapter describes our results with the equilibrium dialysis (ED), DPASV and fluorescence spectrometry (FS) techniques. I will discuss the advantages and disadvantages of ED, DPASV and FS in the next three sections of the chapter.

DESCRIPTION OF TECHNIQUES

Equilibrium Dialysis

The principle behind the ED experiment is that free metal ions, but not complexed metal ions, can permeate the dialysis membrane. Thus one can measure free metal ion concentrations inside the membrane and total metal ion concentrations outside it. Figure 1 exemplifies the resulting data. The x-axis value of the intersection of the upper and lower branches of the curve is the total ligand concentration (C_L). C_L is the maximum amount of metal ion that a humic matter mixture can bind in the presence of excess metal ion.

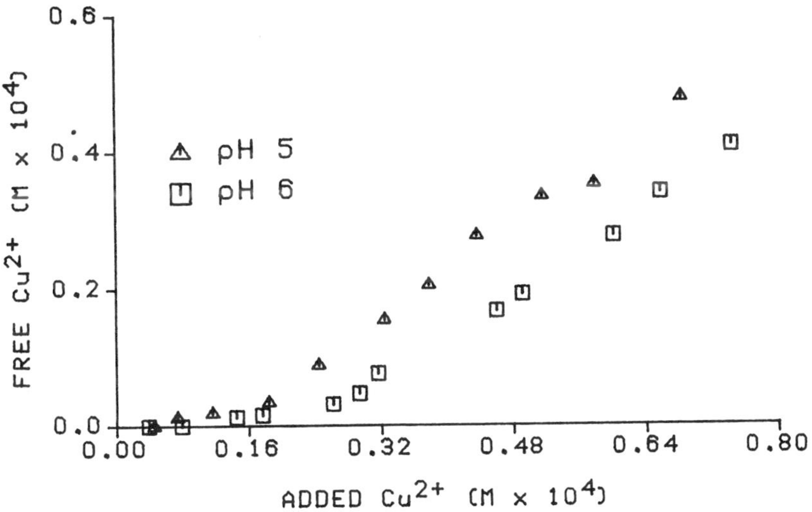

Figure 1. Equilibrium dialysis titration curves for 10 mg/l SFA. Sample titrations by Cu^{2+} at pH 5 and 6. These are not pooled data and C_L values are different than in Table I (reference 17).

The ED technique is the only example of a separation approach in this chapter. The ED method, therefore, has one major advantage over the DPASV and FS approaches, because nearly any metal ion of interest can be determined by atomic absorption spectrometry (AAS). In addition, it requires no modification of natural water samples by addition of electrolytes, buffers, etc. There are potential problems with ED experiments. We have already minimized the amount of organic matter that permeates the dialysis membrane and learned how to purify it [13]. Adsorption of low concentration metal ions on the dialysis membrane is an important potential problem. However, adsorption of metal ions is minimized by membrane purification [13] and using a medium of appreciable ionic strength. For example, Hart and Davies [14,15] proved with estuarine samples that spikes of Cd^{2+}, Cu^{2+}, Pb^{2+} and Zn^{2+} in the sub- μM range could be completely recovered. In addition, we determine total metal ion concentration and do not rely on the amount added. It is very unlikely that we can completely prevent adsorption of organic matter on the membranes, but we believe the problem will be minimized by the ionic strength of natural water and fulvic acid samples.

In a recent paper [13] we described the ED determination of C_L values of 10 mg/l SFA solutions for Cu^{2+} and Cd^{2+} in 0.001 M KNO_3, which is similar to the ionic strength of freshwater samples. We titrated in Cu^{2+} or

Cd^{2+} and waited 2 days for equilibrium between additions. That paper, which contains a detailed experimental section, demonstrated the validity of the ED method. First, very little of the complexing capacity of SFA permeated the dialysis membrane over the 30-day experiment. Second, Cu^{2+} and Cd^{2+} readily diffuse through the membrane. Third, C_L values determined for 6.25 μM ethylenediaminetetraacetic acid (EDTA) with Cu^{2+} and Cd^{2+} are within 2.4% of the theoretical value. The latter experiment demonstrates that the membrane is impermeable to the EDTA complexes. Table I contains the C_L values for titrations of 10 mg/l SFA with Cu^{2+} and Cd^{2+} at pH 5, 6 and 7. The results agree with our past ISE results [7,8,16]. C_L values for SFA are greater for Cu^{2+} than for Cd^{2+} at the same pH, and increase for both metal ions as pH increases.

Subsequently we did a further ED study of metal ion complexation by 10 mg/l SFA in 0.001 M KNO_3 [17]. This study differs from the previous 30-day ED study [13] in two important respects. First, we set up about 15 different solutions containing different metal ion concentrations for each experiment, and the total dialysis time was 48 hr for each sample. The permeation of organic matter is less than in the 30-day experiments. Second, we studied Mn^{2+}, Ni^{2+} and Zn^{2+} as well as Cu^{2+} and Cd^{2+} in the pH 5 to 7 range. Table I contains the resulting C_L values.

Most C_L results (Table I) demonstrate three major trends. First, C_L values for each M^{2+} increase with pH increases. Second, C_L values for Mn^{2+}, Ni^{2+}, Zn^{2+} and Cd^{2+} at each pH are similar. Third, at any pH C_L values for Cu^{2+} are significantly higher than for the other four M^{2+}. One could rationalize the C_L trends on the basis of SFA conditional stability constants, which increase as pH increases [7-9]. In addition, Cu^{2+} binds more strongly to SFA than Cd^{2+} or Ni^{2+} under the same experimental conditions [7,8,16], and probably shows the same trend with Mn^{2+} and Zn^{2+}. We presently have insufficient understanding to explain deviations from the trends, or anomalous results such as $C_L = 0$ for Zn^{2+} at pH 5.

Table I. Complexing Capacity (C_L) Values for 10-mg/l Soil Fulvic Acid Solutions

	C_L (μM)				
pH	Cu^{2+}	Cd^{2+}	Mn^{2+} [b]	Ni^{2+} [b]	Zn^{2+} [b]
5	16.2,[a] 10.8[b]	8.0,[a] 5.5[b]	5.2	5.2	0
6	24.1,[a] 20.3[b]	15.7,[a] 10.6[b]	4.8	4.8	10.2
7	28.7,[a] 14.2[b,c]	19.3,[a] 9.7[b]	11.6	9.4	11.8

[a]From Truitt and Weber [13].
[b]From Rainville and Weber [17].
[c]100% error in C_L.

Because of the nature of the experiments, C_L values should not change as the metal ion or pH changes. If, for example, we did similar experiments with a model ligand such as 10 μM salicylic acid, the resulting C_L values should be independent of M^{2+} and pH. The same result is expected for a mixture of salicylic acid and phthalic acid (a common fulvic acid model), or the complex fulvic acid mixture itself. Thus, the solution chemistry of mixtures does not completely explain the wide range of C_L values.

We conclude that the observations result from metal ion-induced aggregation of SFA. We [9,10] and other research groups [18,19] previously suggested such a phenomenon. We contend that C_L values are higher for Cu^{2+} than the other M^{2+} of this study because of the ability of Cu^{2+} to aggregate SFA via formation of inner sphere complexes. A nuclear magnetic resonance study [20] demonstrated that Cu^{2+} forms inner sphere complexes with fulvic acid, but Mn^{2+} forms outer sphere ones. These results confirm our conclusion that Cu^{2+} forms different kinds of complexes with SFA than does Mn^{2+}, and probably Cd^{2+}, Ni^{2+} and Zn^{2+} as well. Increased C_L values at higher pH are also in agreement with an aggregation model and might be caused by formation of mixed M(SFA)(OH) complexes. The difficulty in measuring C_L values at pH 7 for Cu^{2+} (Table I) might result from nonreproducible aggregation.

We also applied the ED technique to seven southeastern New Hampshire freshwater samples [21]. Because of the ED approach we were not required to add chemicals to the samples, and other than filtering through 0.45-μm polycarbonate membranes they were unmodified. The study of unchanged samples is a major advantage of the ED approach. We titrated all samples with Cu^{2+} and Cd^{2+} while maintaining the original pH. The measured C_L values (Table II) were generally larger for Cu^{2+} (1 to

Table II. Complexing Capacity (C_L) Values of Some New Hampshire Freshwaters

	C_L (μM)	
Sample	Cu^{2+}	Cd^{2+}
Portsmouth Reservoir	8.6	0
Lamprey River	10.7	0.4
Oyster River	1.1	b
Exeter River	2.1[a]	4.3
Durham Reservoir	5.0	3.1
Drew Pond	11.9	9.7
Barrington Swamp	15.1	2.0

[a]Incomplete titration due to malachite formation.
[b]Microbes coated dialysis membrane.

15 μM) than Cd^{2+} (0 to 10 μM) as expected on the basis of previous SFA experiments [7,8,13]. The only exception (Exeter River) occurred because the Cu^{2+} titration could not be completed due to the formation of $Cu_2(OH)_2CO_3$ (malachite) resulting from the high pH and alkalinity of that sample.

Determination of the water properties of pH, dissolved organic carbon (DOC), hardness, color, alkalinity, conductance, and concentrations of iron, copper and cadmium allowed us to do a correlation study of all measured water properties at the 90% confidence level. We found strong positive correlations among major inorganic properties such as pH, hardness, alkalinity and conductance. However, the DOC and C_L values show poor correlation with all other properties, with the exception of a DOC-iron concentration correlation. Thus the Cu^{2+} and Cd^{2+} C_L values showed no correlation at the 90% level with the DOC and color organic matter parameters of the samples.

Differential Pulse Anodic Stripping Voltammetry

DPASV experiments have been used extensively to determine metal ion speciation in natural water samples because researchers can measure very low (~10^{-9} M) metal ion concentrations [5,22,23]. There are two disadvantages to DPASV experiments in common with other techniques. DPASV experiments, like ISE ones, have an ionic strength requirement, and DPASV, in common with other nonseparation techniques such as ISE and FS, has the disadvantage of responding to only a limited number of metal ions. This list of metal ions of interest easily determined by DPASV includes only Pb^{2+}, Cu^{2+}, Cd^{2+} and Zn^{2+}. DPASV has two additional major drawbacks that limit its effectiveness in natural water matrices. First, DPASV is the only nonseparation technique that disturbs the equilibrium between free and complexed metal ions; therefore, measured stripping currents (i_s) must be corrected for kinetic current contributions. Second, humic matter adsorbs on mercury, making the results from the most commonly used electrode questionable [24]. The adsorption makes calculations of free metal ion based on peak potential shifts invalid [25].

Research groups [26-31] that determined C_L values for natural water samples by ASV did not make corrections for kinetic current or adsorption of organic matter on the mercury electrode. As we show below, corrections to raw i_s data significantly change C_L values. It is the need for these corrections that makes ASV so difficult to use for metal ion speciation studies in the presence of humic matter, despite its capability of measuring low trace metal ion concentrations.

We titrated 10, 20 and 40 mg/l SFA solutions in 0.1 M KNO_3 by Cu^{2+} under N_2 at pH 5 and 6 while measuring i_s by DPASV [32]. Figure 2

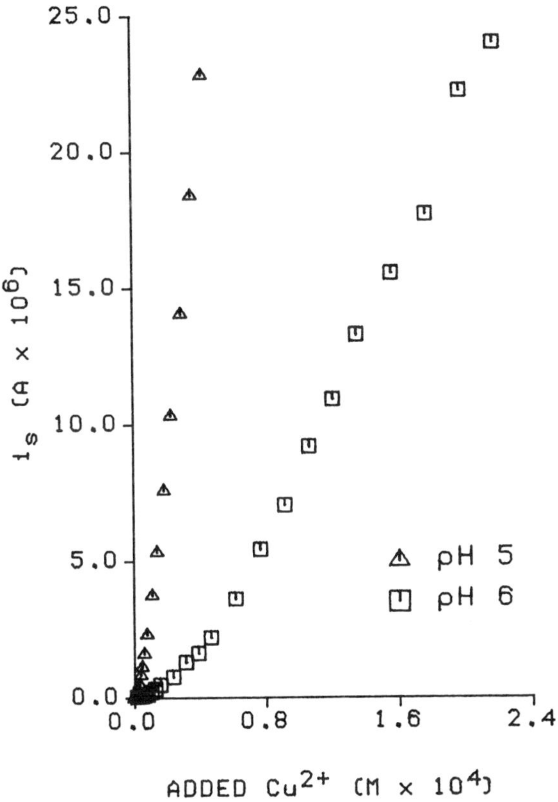

Figure 2. Stripping current vs total $[Cu^{2+}]$ added to 40 mg/l SFA at pH 5 and 6. Only initial data plotted at pH 6 to emphasize breaks in the titration curves.

shows typical raw data for 40 mg/l SFA at pH 5 and 6. Uncorrected i_s values cannot be directly converted into free $[Cu^{2+}]$ before making kinetic current corrections. The i_s and C_M (total metal ion added) data were first fitted to Shuman's equation [33] by a nonlinear regression program:

$$i_s = \frac{S_u}{2}\left[C_M - C_L - \frac{1}{K} + \sqrt{\left(C_L - C_M + \frac{1}{K}\right)^2 + \frac{4C_M}{K}}\right] \tag{1}$$

The program fits the data and refines the estimated C_L (total ligand concentration), S_u (upper slope; slope of i_s vs C_M at high C_M values), and K (conditional stability constant) values. Our computer program provides stripping currents corrected for kinetic current contributions as well as corrected C_L, S_u, K and free [Cu^{2+}] values. The corrected S_u and i_s values allow calculation of free [Cu^{2+}]. Table III gives the C_L, S_u and K data at pH 5 and 6 during various stages of the calculations.

Table III shows that the kinetic current corrections have a large effect on the titration parameters. For every set of experimental conditions, C_L values increase from untreated data to fitted data to kinetic current corrected data. The pH 5 data in the presence of 20 mg/l SFA exemplifies the trend. The untreated 4.8 *M* C_L value yields a 25% higher 6.0 μM value for fitted data and the kinetic current correction increases the value another 45% to 8.7 μM. We expected kinetic current corrections to yield higher C_L values, because subtraction of the kinetic current from each i_s value decreases the currents at constant C_M. The average K at pH 5 also increases as the data are refined. The enhanced K value is expected because uncorrected i_s values are due to free and labile [Cu^{2+}] and consequently underestimate complexed [Cu^{2+}]. S_u values decrease during kinetic current corrections for the same reason.

The trends in pH 6 data are similar to those of pH 5, but there are some differences (Table III). The values for C_L are much higher at pH 6 than at pH 5 for the same SFA concentrations. Nonlinear regression treatment of the unfitted data increases C_L values by about 5 to 8% and kinetic current correction further increases them by 2 to 10%. S_u values are usually unchanged by fitting the data, but they decrease with kinetic current correction as in the pH 5 experiments.

The data treatment described above eliminates the problem due to dissociation of the complex during the plating step, and avoids the uncertainty caused by adsorption of humic matter on the mercury electrode. The upper slopes (Table III) decrease with increases in [SFA] at all three stages of the data treatment, indicating that for the same free metal ion concentration the stripping current decreases with increasing [SFA]. The decrease in i_s values reflected by decreased upper slopes is probably caused by an increasing inhibition of electron transfer rate due to increasing amounts of SFA adsorption. This variation in upper slopes with [SFA] and with pH changes demonstrates the need for in situ calibration curves. Calibration curves in the absence of SFA are inappropriate. Corrected S_u and i_s values allow calculation of free [Cu^{2+}]:

$$i_s^{corr} = S_u^{corr}[Cu^{2+}] \qquad (2)$$

The free [Cu^{2+}] can now be used in any desired data treatment.

The major conclusion from this DPASV work is that voltammetric techniques for measurement of free metal ion concentration in the presence

Table III. Complexing Capacity (C_L), Upper Slope (S_u), and Conditional Stability Constants (K) Determined for Soil Fulvic Acid Solution by DPASV

Status of Data	pH 5				pH 6		
	SFA (mg/l)	C_L (μM)	S_u ($\mu A/\mu M$)	$K \times 10^{-5}$	C_L (μM)	S_u ($\mu A/\mu M$)	$K \times 10^{-5}$
Unfitted	10	2.8	0.90		17.4	0.18	
	20	4.8	0.80		24.6	0.16	
	40	6.7	0.67		37.4	0.13	
Nonreg fitted	10	5.2	0.96	3	18.2	0.18	1
	20	6.0	0.82	6	26.7	0.16	2
	40	7.9	0.67	3	40.5	0.13	7
Average				4			3
Kinetic current corrected	10	7.4	0.85	6	19.4[a]	0.14	5
	20	8.7	0.75	8	29.9[a]	0.12	3
	40	13.9	0.62	5	41.2[a]	0.09	1
Average				6			3

[a]Calculated from first iteration because of nonconvergence.

of humic matter are feasible when two precautions are taken. First, it is necessary to correct measured i_s values for metal ion caused by dissociation of the complex during the plating step of the process. Second, researchers should be aware of adsorption of humic matter on the mercury electrode, and minimize its effects by calculating free metal ion concentration from in situ i_s vs C_M calibration curves in which the current has been corrected for complex dissociation.

Fluorescence Spectrometry

Humic matter and natural water samples fluoresce [13,16,34,35], but the fluorescence of organic ligands is generally quenched by complexation to paramagnetic metal ions [36,37], i.e., those with unpaired electrons. The FS experiment has certain advantages over most other methods. First, fluorescence can be measured in a very dilute solution such as 0.6 mg/l SFA, which has lower DOC than many natural waters. Second, there is no ionic strength requirement as in DPASV or ISE. Third, unlike the DPASV or ED methods, one measures the decrease in free fluorescing organic matter, not the free metal in concentration. The third advantage results in a direct measurement of C_L values. The major disadvantage of the FS technique is that it is generally most effective with paramagnetic metal ions, especially Cu^{2+}.

SFA and WFA fluorescence spectra exhibit a broad, featureless emission peak at 445 to 450 nm after excitation at 350 nm [16]. The emission intensity depends on pH; it increases from pH 2 to 5 (maximum value), and decreases slightly to pH 8. Two requirements must be met before using the fluorescence property of FA for metal binding studies. Uncomplexed metal ions must not quench fulvic acid fluorescence, and metal ion quenching of fluorescence must be proportional to metal ion complexation. In initial studies [16] we proved the first point by observing that at pH 1.4, where metal ions are not complexed to FA, its fluorescence is essentially unquenched. Second, we demonstrated with the model ligand salicyclic acid (pH 6) that % Cu^{2+} bound (calculated from known stability constants) and percent fluorescence quenched are equal during titration by Cu^{2+}. The two preliminary experiments encouraged us to begin metal ion complexation studies of FA using fluorescence as a probe.

We titrated 32 mg/l solutions of SFA and water-derived fulvic acid (WFA) with Cd^{2+}, Cu^{2+}, Pb^{2+}, Co^{2+} and Ni^{2+} at various pH values [16]. For the same metal ion/FA ratio the percent fluorescence quenched for each metal ion increases as pH increases. This trend agrees with past complexation studies done by ISE [7-10] that demonstrated increased K values at higher pH values. The effectiveness of metal ions in quenching FA fluorescence under similar experimental conditions is: $Cu^{2+} > Pb^{2+} >$

$Co^{2+} \sim Ni^{2+} > Cd^{2+}$. Cu^{2+} and Pb^{2+} form strong FA complexes with similar K values, but paramagnetic Cu^{2+} is more effective than diamagnetic Pb^{2+} in quenching fluorescence, as expected. Similarly, among weakly bound Co^{2+}, Ni^{2+} and Cd^{2+}, paramagnetic Co^{2+} and Ni^{2+} are more effective fluorescence quenchers than diamagnetic Cd^{2+}, which has no effect on fluorescence.

Still to be proven was that the complexation of FA by Cu^{2+} is proportional to its fluorescence quenching. A series of Cu^{2+} ISE and FS quenching experiments at pH 3, 4, 5 and 6 demonstrated this requirement. We plotted the results as percentage fluorescence quenched or bound Cu^{2+} (ISE result) as y-axis variables and total Cu^{2+}/total FA as the x-axis variable. The latter variable is a measure of the progress of either type of titration. The FS and ISE results demonstrate that Cu^{2+} binding to FA is proportional to fluorescence quenching. A similar set of experiments demonstrated the same correlation of ISE and fluorescence quenching for diamagnetic Pb^{2+}. However, diamagnetic Pb^{2+} is a less effective fluorescence quencher than paramagnetic Cu^{2+} at the same level of FA binding.

Our recent results [38] extend the earlier work by describing a quantitative method for determining the micromolar complexing capacities or total ligand concentrations (C_L) of fluorescing ligands for a metal ion using a titration procedure. We titrated solutions of 36 μM L-tyrosine in 0.1 M $KClO_4$ or 10 mg/l SFA in 0.1 M KNO_3 by Cu^{2+} at various constant pH values (see Table IV). The fluorescence excitation (emission) wavelength was 277 nm (296 nm) for L-tyrosine and 350 nm (445 nm) for SFA. Scattering measurements at 400 nm or 500 nm were done simultaneously with fluorescence measurements. Fluorescence measurements made after scattering became double its initial value in the absence of Cu^{2+} were discarded.

We treated data on the basis of our derived equation:

$$\frac{[ML]}{C_L} = \frac{I_L - I}{I_L - I_{ML}} = \frac{1}{2KC_L}\left[(KC_L + KC_M + 1) - \sqrt{(KC_L + KC_M + 1)^2 - 4K^2C_LC_M}\right] \tag{3}$$

where I is the fluorescence intensity measured during titration by metal ion
I_L, which is set to 100 units, is the fluorescence intensity of *free* organic matter before any metal ion is added
I_{ML} is the residual fluorescence intensity after the titration endpoint

As with the DPASV method, the experimental variables are fitted by nonlinear regression analysis. The I and C_M experimental variables are fitted to yield the parameters C_L, K and I_{ML}.

Titrations of the model compound L-tyrosine with Cu^{2+} tested the proposed theory and the ability of the curve-fitting program. The average

Table IV. Average Fitted Parameters for Fluorescence Titrations with Cu^{2+} [a]

Sample	pH	Experimental Concentration	C_L (μM)	$K \times 10^{-5}$	I_{ML} (%)[d]
L-tyrosine	6	35.7 μM	33.0	0.58[c]	2.6
SFA	5	10.0 mg/l	22.1[b]	0.48[b]	40.2[b]
SFA	6	10.0 mg/l	19.7	1.08	22.4
SFA	7	10.0 mg/l	19.6	2.84	20.7

[a] Average of at least three replicates.
[b] Only one experiment at pH 5.
[c] Value calculated from thermodynamic stability and acid dissociation constants is 0.59×10^5.
[d] Based on I_L = 100%.

values determined for the C_L, K and I_L constants for triplicate experiments are given in Table IV. The measured 33.0 μM value for C_L is quite close to the experimental 35.7 μM concentration. The average K value obtained also agrees well with the first conditional stability constant of 0.59×10^5 calculated from the literature values for the thermodynamic stability constant and acid dissociation constant [39]. The I_{ML} value is very low at 2.6% (±1.7% s.d.) and is essentially zero, indicating that the complex is considerably lower in fluorescence efficiency than the free ligand.

Figure 3 gives the results of Cu^{2+} titrations of 10 mg/l SFA at pH 5, 6 and 7 in 0.1 M KNO_3. At least three replicates are shown for each pH. The curves show the expected trends with a higher degree of quenching as pH increases due to increased binding. The curve for pH 7 shows the sharpest break with broader curves for pH 6 and 5. The nonlinear regression results in Table IV are averaged results for three experiments at each pH except for pH 5, which is discussed below. The C_L values for pH 6 and 7 are about 20 μM and the K values increase with a pH increase. The final fluorescence value (I_{ML}) is approximately 21% for pH 7 and 22% for pH 6, as shown in Figure 3. This is markedly different from the L-tyrosine experiments and can be interpreted in two ways. First, fluorescence efficiency of the complex may be about one-fifth that of the free ligand, giving rise to the significant I_{ML} values after all the ligand is bound. Second, the residual fluorescence might be due to nonbinding fluorescent molecules in the SFA mixture. In this case the *binding material* is completely or almost completely quenched, as in the case of L-tyrosine.

The experiments conducted at pH 5 gave reproducible data (Figure 3), but did not give reasonable titration parameters when analyzed by nonlinear regression. I_{ML}, K and C_L values for four experiments were highly scattered and often gave negative C_L values. The reason for this nonreproducibility is probably the very weak binding at pH 5. The only experiment

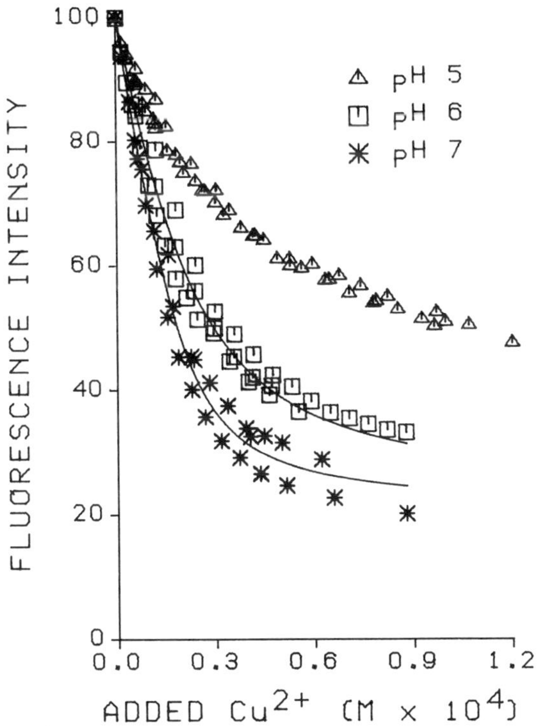

Figure 3. Fluorescence quenching curves for 10 mg/l SFA by Cu^{2+} at pH 5, 6 and 7. The line at pH 6 and 7 is the fitted curve.

at pH 5 we could fit gave K and C_L values that follow the trends for pH 6 and 7 (Table IV). The I_{ML} value however, is 40.2%, which is higher than those at pH 6 and 7 and may be erroneous.

Metal ions can precipitate FA under certain conditions [9,10,40]. This phenomenon has been generally ignored in the context of complexing capacity titrations even though precipitation is very likely near the end of the titrations. As metal ion is added and binds to SFA, scattering increases. Although the increase is gradual at first, it eventually undergoes a rapid increase and then seems to level off. These trends are similar to those observed by Pagenkopf and Whitworth [40] for humic acid with Cu^{2+} at pH 7. Increased scattering with increased added metal ion means that metal ions are removed by a process other than simple complexation. After insoluble solids form in the sample its properties are no longer explainable by solution chemistry. Interactions between ions in solutions and solid surfaces may become the dominant process. One possible result is removal of metal ion from solution by adsorption onto the surface of particles

[9,10]. As mentioned above, previous work on complexing capacities has relied on determining free metal ion in equilibrium with complex and ligand. Any process that removes free metal ion from solution will be interpreted as complexation, but many past studies probably had aggregates near the end of the titrations.

We conclude that fluorescence spectrometry has the required sensitivity to determine micromolar complexing capacities of natural organic matter. Combined with nonlinear regression analysis it gives results that are comparable to techniques such as DPASV, ISE and dialysis/atomic absorption. The fact that fluorescence differentiates free and bound ligand makes it an excellent complement to other complexing capacity techniques which differentiate free and bound metal ion. Additional advantages of FS are that it is fairly rapid and requires no supporting electrolyte. Solution scattering can also be measured with the same instrumentation, giving valuable information on precipitate formation. The usual goal of complexing capacity determinations is an understanding of solution equilibria. By carefully considering both fluorescence intensities and scattering, we can distinguish complexation from adsorption of metal ions on particulates. This approach gives a more complete picture of processes related to metal ion binding for the complex humic materials system. Continuing work in our laboratory is the application of the FS technique to natural water samples.

CONCLUSIONS

Researchers should be extremely wary when analyzing metal ion speciation data in the presence of isolated humic material or in natural water samples. There are advantages and disadvantages with separation techniques such as dialysis, gel filtration chromatography, or ultrafiltration, and with nonseparation approaches such as ion-selective electrodes, voltammetry, electron paramagnetic resonance and fluorescence spectrometry. We should ask questions such as the following when evaluating the various techniques. Does the measurement disturb equilibria? To what extent do humic materials and trace metal ions adsorb on electrodes and membranes? Are all data from solution processes or has aggregation occurred? To what extent must we modify natural water samples for the experiment?

ACKNOWLEDGMENTS

This work was partially supported by Office of Water Resources Technology grant B004-NH and National Science Foundation grants OCE 79-10571 and CHE 79-08399 for the Cary 219. I am grateful to the students and postdoctoral associates mentioned in the references for providing the creative ideas, critical insights and experimental skills that made this work possible.

REFERENCES

1. Smits, J., J. Nelissen and R. Van Grieken. "Comparison of Preconcentration Procedures for Trace Metals in Natural Waters," *Anal. Chim. Acta* 111:215-226 (1979).
2. Windom, H. L., and R. G. Smith, Jr. "Copper Concentrations in Surface Waters off the Southeastern Atlantic Coast, U.S.A.," *Mar. Chem.* 7:157-163 (1979).
3. Sturgeon, R. E., S. S. Berman, J. A. H. Desaulniers, A. P. Mykytivk, J. W. McLaren and D. S. Russell. "Comparison of Methods for the Determination of Trace Elements in Seawater," *Anal. Chem.* 52:1585-1588 (1980).
4. Bruland, K. W., R. P. Franks, G. A. Knauer and J. H. Martin. "Sampling and Analytical Methods for the Determination of Copper, Cadmium, Zinc, and Nickel at the Nanogram per Liter Level in Sea Water," *Anal. Chim. Acta* 105:233-245 (1979).
5. Florence, T. M., and G. E. Batley. "Chemical Speciation in Natural Waters," *C.R.C. Crit. Rev. Anal. Chem.* 9:219-296 (1980).
6. Weber, J. H., and S. A. Wilson. "The Isolation and Characterization of Fulvic Acid from River Water," *Water Res.* 9:1079-1084 (1975).
7. Bresnahan, W. T., C. L. Grant and J. H. Weber. "Stability Constants for the Complexation of Copper(II) Ions with Water and Soil Fulvic Acids Measured by an Ion Selective Electrode," *Anal. Chem.* 50:1675-1679 (1978).
8. Saar, R. A., and J. H. Weber. "Complexation of Cadmium(II) with Water- and Soil-Derived Fulvic Acids: Effect of pH and Fulvic Acid Concentration," *Can. J. Chem.* 57:1263-1268 (1979).
9. Saar, R. A., and J. H. Weber. "Conditional Stability Constants, Solubility, and Implications for Lead(II) Mobility," *Environ. Sci. Technol.* 14:877-880 (1980).
10. Saar, R. A., and J. H. Weber. "Lead(II) Complexation by Fulvic Acid: How it Differs from Fulvic Acid Complexation of Copper(II) and Cadmium(II)," *Geochim. Cosmochim. Acta* 44:1381-1384 (1980).
11. Wilson, S. A., and J. H. Weber. "An EPR Study of the Reduction of Vanadium(V) to Vanadium(IV) by Fulvic Acid," *Chem. Geol.* 26:345-354 (1979).
12. Smart, R. B., and J. H. Weber. "Differential Pulse Anodic Stripping Voltammetry of Copper(II) at the Glassy Carbon Electrode," *Anal. Chim. Acta* 115:331-336 (1980).
13. Truitt, R. E., and J. H. Weber. "Determination of Complexing Capacity of Fulvic Acid for Copper(II) and Cadmium(II) by Dialysis Titration," *Anal. Chem.* 53:337-342 (1981).
14. Hart, B. T., and S. H. R. Davies. "A New Dialysis-Ion Exchange Technique for Determining the Forms of Trace Metals in Water," *Aust. J. Mar. Freshwater Res.* 28:105-112 (1977).

15. Hart, B. T., and S. H. R. Davies. "A Batch Method for the Determination of the Ion-Exchangeable Trace Metals in Natural Waters," *Aust. J. Mar. Freshwater Res.* 28:397-402 (1977).
16. Saar, R. A., and J. H. Weber. "Comparison of Spectrofluorometry and Ion-Selective Electrode Potentiometry for Determination of Complexes Between Fulvic Acid and Heavy-Metal Ions," *Anal. Chem.* 52:2095-2100 (1980).
17. Rainville, D. P., and J. H. Weber. "Complexing Capacity of Soil Fulvic Acid for Cu^{2+}, Cd^{2+}, Mn^{2+}, Ni^{2+}, and Zn^{2+} Measured by Dialysis Titration: A Model Based on Soil Fulvic Acid Aggregation," *Can. J. Chem.* 60:1-5 (1982).
18. Gamble, D. S., A. W. Underdown and C. H. Langford. "Copper Titration of Fulvic Acid Ligand Sites with Theoretical, Potentiometric, and Spectrophotometric Analysis," *Anal. Chem.* 52:1901-1908 (1980).
19. Templeton, G. D., III, and N. D. Chasteen. "Vanadium-Fulvic Acid Chemistry: Conformational and Binding Studies by Electron Spin Probe Techniques," *Geochim. Cosmochim. Acta* 44:741-752 (1980).
20. Deczky, K., and C. H. Langford, "Application of Nuclear Magnetic Resonance Relaxation Times to Study of Metal Complexes of the Soluble Soil Organic Fraction Fulvic Acid," *Can. J. Chem.* 56:1947-1951 (1978).
21. Truitt, R. E., and J. H. Weber. "Copper(II)- and Cadmium(II)-Binding Abilities of Some New Hampshire Freshwater Determined by Dialysis Titration," *Environ. Sci. Technol.* 15:1204-1208 (1981).
22. Gilain, G., G. Duyckaerts and A. Disteche. "Direct and Simultaneous Determination of Zn, Cd, Pb, Cu, Sb, and Bi Dissolved in Seawater by Differential Pulse ASV with HMDE," *Anal. Chim. Acta* 106:23-37 (1979).
23. Zirino, A., and S. P. Kounaves. "Stripping Polarography and the Reduction of Copper(II) in Sea Water at the Hanging Mercury Drop Electrode," *Anal. Chim. Acta* 113:79-90 (1980).
24. Cominoli, A., J. Buffle and W. Haerdi. "Voltammetric Studies of Humic and Fulvic Substances—Part III. Comparison of the Capabilities of Various Polarographic Techniques for the Analysis of Humic and Fulvic Substances," *J. Electroanal. Chem.* 110:259-275 (1980).
25. Buffle, J., and F.-L. Greter. "Voltammetric Study of Humic and Fulvic Substances. Part II. Mechanism of Reaction of the Pb-Fulvic Complexes on the Mercury Electrode," *J. Electroanal. Chem.* 101:231-251 (1979).
26. Srna, R. F., K. S. Garrett, S. M. Miller and A. B. Thum. "Copper Complexation Capacity of Marine Water Samples from Southern California," *Environ. Sci. Technol.* 14:1482-1486 (1980).
27. Plavšić, M. M., S. Kozar, D. Krznarić, H. Bilinski and M. Branica. "The Influence of Organics on the Adsorption of Copper(II) and β-Al_2O_3 in Seawater. Model Studies with EDTA," *Mar. Chem.* 9:175-182 (1980).
28. Shuman, M. S., and G. P. Woodward, Jr. "Stability Constants of Copper-Organic Chelates in Aquatic Samples," *Environ. Sci. Technol.* 11:809-813 (1977).
29. Blutstein, H., and R. F. Shaw. "Characterization of Copper Binding Capacity in Lake Water," *Environ. Sci. Technol.* 15:1100-1102 (1981).
30. Baccini, P., and U. Suter. "Melimex and Experimental Heavy Metal Pollution Study: Chemical Speciation and Biological Availability of Copper in Lake Water," *Schweiz. Z. Hydrol.* 41:291-314 (1979).

31. Tuschall, J. R., Jr., and P. L. Brezonik. "Characterization of Organic Nitrogen in Natural Waters: Its Molecular Size, Protein Content, and Interaction with Heavy Metals," *Limnol. Oceanogr.* 25:495-504 (1980).
32. Bhat, G. A., R. A. Saar, R. B. Smart and J. H. Weber. "Titration of Soil-Derived Fulvic Acid by Copper(II) and Measurement of Free Copper(II) by Anodic Stripping Voltammetry and Copper(II) Selective Electrode," *Anal. Chem.* 53:2275-2280 (1981).
33. Shuman, M. S., and J. L. Cromer. "Copper Association with Aquatic Fulvic and Humic Acids. Estimation of Conditional Stability Constants with a Titrimetric Anodic Stripping Voltammetry Procedure," *Environ. Sci. Technol.* 13:543-545 (1979).
34. Gagosian, R. B., and D. H. Stuermer. "The Cycling of Biogenic Compounds and their Biogenetically Transformed Products in Seawater," *Mar. Chem.* 5:605-632 (1977).
35. Stewart, A. J., and R. G. Wetzel. "Fluorescence: Adsorbance Ratios–A Molecular Weight Tracer of Dissolved Organic Matter," *Limnol. Oceanog.* 25:559-564 (1980).
36. Chen, R. F. in: *Biochemical Fluorescence: Concepts,* R. F. Chen, H. Edelhoch, Eds., (New York: Marcel Dekker, Inc., 1976), Vol. 2, Chap. 13.
37. Parker, C. A. *Photoluminescence of Solutions with Applications to Photochemistry and Analytical Chemistry*, (Amsterdam: Elsevier, 1968).
38. Ryan, D. K., and J. H. Weber. "Fluorescence Quenching Titration for Determination of Complexing Capacities and Stability Constants of Fulvic Acid," *Anal. Chem.* 54:986-990 (1982).
39. Smith, R. M., and A. E. Martell, *Critical Stability Constants. Volume 1: Amino Acids* (New York: Plenum Press, 1974), p. 31.
40. Pagenkopf, G. K., and C. J. Whitworth. "Precipitation of Metal-Humate Complexes," *J. Inorg. Nucl. Chem.* 43:1219-1222 (1981).

CHAPTER 16

CONDITIONAL STABILITY CONSTANTS OF TRACE METALS AND NATURALLY OCCURRING HUMIC MATERIALS: APPLICATION IN EQUILIBRIUM MODELS AND VERIFICATION WITH FIELD DATA

James J. Alberts

Savannah River Ecology Laboratory
Aiken, South Carolina

John P. Giesy

Pesticide Research Center and
Department of Fisheries and Wildlife
Michigan State University
East Lansing, Michigan

ABSTRACT

Ion-selective electrode techniques and Scatchard plots were employed to determine the conditional stability constants and number of binding sites for the trace elements Cu, Cd and Pb with naturally occurring organic matter isolated from several streams of the southeastern United States. The data obtained from these studies were used to (1) investigate the degree of binding of oxy-functional groups exhibited by these organic ligands, and (2) determine the usefulness of the values for stability constants and binding sites when employed in thermodynamic chemical equilibrium models. Field data describing the chemical speciation of Ca, Cd, Cu and Fe were used to verify model predictions and to estimate the conditional constants of both Ca and Fe with these organic ligands. The results of these studies demonstrate that large-molecular-weight organic matter plays a significant role in the complexing of metals and that the conditional stability constants determined for organic matter in the southeastern U.S. have applicability over a wide geographical range.

INTRODUCTION

A major goal of environmental biogeochemistry is to gain a sufficient understanding of the transport and fate of materials introduced into natural systems to allow the prediction of the short- and long-term impacts of these materials. Knowledge of the chemical form of an element in natural waters is essential to understanding and predicting an element's distribution between the various biotic and abiotic components of the system. Considerable progress has been made in determining the chemical form of trace metals in natural waters where inorganic ligands dominate the anionic component of the system. The formulation of thermodynamic equilibrium models using formation constants for these inorganic complexes has provided several potentially powerful tools for understanding the chemical form of elements in water [1]. However, while chemical equilibrium models have much potential, they do have limitations, not the least of which is their present inability to successfully predict the chemical forms of many trace metals in systems where naturally occurring organic compounds comprise a significant portion of the anions present [2]. This natural organic matter is often dominated by humic and fulvic acids [3] whose complex structures still continue to perplex investigators.

While the absolute structure of humic acids may evade investigators for some time to come, the role of these compounds in complexing metals is well known [4,5] and advances have been made in incorporating terms into equilibrium models to address natural organic matter/trace metal complexation [6]. Of particular interest is the quasiparticle model, which simplifies the complex metal-organic interactions of surface waters by replacing the actual assembly of organic compounds present with an assembly of hypothetical, identical macromolecules whose mole mass is the number average mole mass of the actual mixture and whose metal-complexation reactions closely mimic those of the real system [6]. This model allows for the incorporation of overall, conditional formation constants and stoichiometries determined by methods such as Scatchard plot analysis, while providing a generally applicable conceptualization of metal-organic binding which may be of value to predicting chemical speciation of metals in natural waters.

It is the purpose of this chapter to describe some of our work aimed at quantifying the binding of metals by naturally occurring organic matter, and to determine the applicability of the data for systems in the Southeast to other regions of the country. To this end, we have used many techniques of measurement, data analysis and chemical modeling that will not be detailed in this chapter, but which are or will soon be described in the literature. Our purpose is not to detail the minutiae of currently available analytical techniques, even though we feel that the importance of proper analytical applications cannot be overstated; rather, we examine how current techniques may allow us to predict the chemical form of trace elements over a wide geographical area, for waters which contain significant concentrations of naturally occurring organic ligands.

METHODS

Water samples (50 litres) were collected from eight sites in the Southeast. The sites included: four rivers (the Caloosahatchee, Peace, and Suwannee rivers in Florida, and the Satilla River in Georgia), Banks Lake (a blackwater lake in south Georgia), and the Okefenokee Swamp. The limnological parameters of these waters have been discussed previously [7-9] and are summarized in Table I. The waters represent a wide range of pH, conductivity and dissolved organic carbon values.

The water samples were filtered through 0.45-μm membrane filters (Millipore Corp.) and organic matter was concentrated successively with two ultrafilters (Diaflo Hollow Fiber, Amicon Corp.). The ultrafilters used thus provided two organic size fraction concentrates, namely, 0.45 μm > fraction I > 50 Å and 50 Å > fraction II > 13 Å. Ultrafilters also may separate by charge characteristics of the molecules, so these fractions must be considered as operationally defined.

Each fraction was then analyzed for total exchangeable proton concentration and the concentration of carboxylic and phenolic protons per gram of organic matter was estimated [4]. In addition, the complexing capacity of each fraction was determined for Cd, Cu and Pb with ion-selective electrodes (Orion Corp.) [10]. Data from these titrations were analyzed after the method of Scatchard [5,11] to determine the conditional stability constants (pH = 5.0) of the metal-organic complexes as well as the number of sites active in binding metals with a specific stability constant.

All calculations of elemental distributions were conducted using the thermodynamic equilibrium models REDEQL2 [12] or GEOCHEM [13].

Table I. Ranges of Chemical Characteristics of Southeastern Waters Studied

pH	4.0–8.3
Conductivity ($\mu\Omega$)	36–1950
Alkalinity (mg/l)	0–3.0
DOC (mg/l)	11.4–47.3
% DOC removed by ultrafiltration (>5000 daltons)	44.5 ± 18.5 (n = 8)

Finally, to determine the effect of competing ions on binding of these metals, the binding capacity measurements were carried out as before, except that the organic fraction being tested was placed into solutions containing the competing metal (Al, Ca or Mn) prior to titration with the metals of interest (Cd, Cu or Pb) [14]. Again ion-selective electrodes were employed to measure the metal activities.

RESULTS AND DISCUSSION

DOC Distribution and Nature

In Table I, it can be seen that there are a wide range of dissolved organic carbon concentrations (DOC, defined as that material which passes through a 0.45-μm filter) in the waters of this study. Also of interest is the fact that approximately 45% of the DOC is removed by ultrafiltration, indicating a large molecular size or colloidal form for this material. This observation is in agreement with others [3,15] who have shown that DOC may be to a large extent humic or fulvic material. However, 40.9% (±19.2) of the total DOC is removed by the first ultrafilter, indicating a size range of 50 to 4500 Å, while only 3.8% (±2.9) of the total DOC is removed by the succeeding ultrafilter (size range, 13 to 50 Å). These data, representing eight samples taken from six water bodies, agree with other data from the Southeast [7-10] which show that DOC can be fractionated by ultrafiltration techniques and that these fractions may represent different proportions of the total DOC depending on season and water type. These data emphasize that while DOC should be thought of as a continuum of organic material, it must be remembered that the distribution of carbon is not equal across the size spectrum.

Our primary concern in this chapter is determining the extent of the binding of metals to organic matter. To this end, a discussion of the distribution of carbon within organic fractions is of limited use as metals are bound to the carbon skeleton through heteroatoms such as N, O or S. Of these elements, S is in low concentrations in natural organic compounds and little is known of metal binding to organic compounds via this element in natural systems. In contrast, N is present in amounts adequate to make a significant impact on metal binding and has been implicated in metal binding in recent work [16]. However, the majority of the evidence at this point argues for oxygen binding to metals via carboxylic and phenolic groups [4,17].

We determined the total number of exchangeable protons in these organic fractions by means of the $Ba(OH)_2$ precipitation technique [4] and found the values to range from 3.9 to 11.9 (meq/g ash-free dry wt of organic matter). Furthermore, using the calcium acetate precipitation method [4], we observed that the exchangeable protons were equally divided between the carboxylic and phenolic functional groups (% carboxylic protons = 47.6 ± 15.9 for 16 samples). While Perdue [17] has correctly pointed out the limitations of the calcium acetate method, it is interesting that the percentage carboxylic nature of these organics is in such close agreement when the range of physical parameters observed for these waters is considered. Furthermore, the errors attributable to the calcium acetate method have little effect on the conclusions drawn for metal-to-organic binding.

Organic Binding of Cd, Cu and Pb

Using ion-selective electrodes [10] and the method of Scatchard [5,11], we approximated the conditional stability constants of the organic fractions with Cd, Cu and Pb. The Scatchard plot analysis was employed because it was developed to determine both stability constants (K) and number of binding sites (n) for metal-protein complexes. As these complexes often have the same difficulties in analysis (primarily undetermined molecular weights of the organics and stoichiometry of the complex), it was hoped that this method, which has been used for humic-metal complexes [18], would allow us to produce data which were comparable across a wide range of water types. Of particular interest was the prospect of determining the number of binding sites (n) per gram of DOC, as this value has eluded investigators in the past. The results of our analyses are given in Table II as the average log K values and average −log n values determined for the organic matter isolated by ultrafiltration from southeastern waters. The two fractions from each sample location have been combined so that the data reported in Table II are representative of the entire organic fraction

Table II. Conditional Stability Constants (log K) and Concentrations of Binding Sites (−log [n]) for Natural Organic Matter Isolated from Southeastern United States Waters[a]

	Cd^{2+}	Cu^{2+}	Pb^{2+}
K_1	5.83 ± 0.35 (8)	6.52 ± 0.45 (8)	5.09 ± 0.46 (8)
n_1	5.96 ± 0.56 (8)	5.62 ± 0.37 (8)	5.63 ± 0.63 (8)
K_2	3.13 ± 2.45 (6)	4.89 ± 0.82 (8)	2.00 ± 3.36 (7)
n_2	4.04 ± 1.66 (6)	4.94 ± 1.72 (8)	3.29 ± 1.26 (7)

[a]Values in ( ) = number of samples.

removed. Two points are apparent from the data: (1) for all three metals, there appear to be two sites for organic binding (K_1, n_1 and K_2, n_2, respectively) at most of the locations sampled; (2) while some variation occurs, the agreement of the values between locations was quite good.

The presence of two binding sites is in agreement with previous work, which indicates phthalic and salicylic binding sites as the primary binding types for metals with humic acids [4,19]; however, some recent studies postulate that the two sites represent steric hindrance rather than differing functional groups [18]. The magnitude of the stability constants in Table II is in reasonable agreement with previously reported data [18,20,21]. Of interest is the fact that K_2 values for Cd and Pb are low, and that in some cases K_2 and n_2 values could not be measured. This observation indicates that the binding of Cd and Pb to organic matter may occur primarily through the stronger sites indicated by the K_1 and n_1 values, while Cu binding appears to occur at both sites in all cases studied.

At the outset of the study, we were concerned about the variation in n values, particularly with respect to different metals binding at similar sites, i.e., n_1 for Cd, Cu and Pb. This concern arose from our wish to employ predictive models. To accomplish this end, the n values for a given site would have to be similar for different metals. Only one n value per site may be entered in the computations, whereas any stability constant may be entered for a specific metal-organic complex. As can be seen from Table II, the agreement for n_1 values is quite good. While the agreement for n_2 values is not as good as for n_1 values, they are still usable since the variability indicates an overlap of the possible values. More importantly than allowing the use of this data in further computations is the good agreement among both n_1 and n_2 when the range of waters is considered. The agreement of the data, as noted by the relatively low standard deviations of the values, indicates that given a wide range of conditions and possible organic carbon sources, the organic fractions isolated by ultrafiltration do not vary greatly with respect to stability constants or number of binding sites available for metal binding. Thus, it may be possible to derive some general expressions for metal-organic binding on a regional basis.

An additional calculation may be made with the data derived for milliequivalents of carboxylic and phenolic acidities and the n_1 and n_2 values, which are expressed in μg atm/l. From these data and the mg C/l data, it is possible to calculate the percentage of carboxylic and phenolic sites that may be active in binding metals if one assumes that n_1 represents only carboxylic bonding of the phthalate type and n_2 represents bonding that incorporates phenolic bonds. With these calculations, it is found that 0.8 to 26% of the carboxylic groups may participate in n_1-type binding, and 0.8 to >100% of the phenolic groups may participate in n_2-type binding. The ranges are generated by consideration of the standard deviations of the n values and hence the >100% values. It must be pointed out that while these calculations are interesting, they do not take into consideration the stoichiometry of the possible complexes, nor the degree of dissociation of the oxy-functional groups. However, the calculations do point out the internal consistency of the measurements and indicate the relative magnitudes of n_1- and n_2-type binding.

Equilibrium Models and Field Verification

As previously described, a valid test of the data is their applicability to various systems. In an attempt to determine the generality of our data, we incorporated our organic stability constant values into the chemical equilibrium, thermodynamic model GEOCHEM [13] and attempted to predict the chemical speciation of Cd and Cu from waters in Maine [10]. The field data were chosen because the pH and DOC values were within the range of values studied in the Southeast. In addition, the investigators had passed samples of centrifuged waters through anion and cation resins and analyzed the eluants so that the percent of anionic and cationic metals had been determined. Two lakes (Sunken Pond and Bog Pond) were chosen from these studies based on their pH and DOC values. Table III shows the pertinent chemical data from the ponds, and the percentage anionic and cationic character of four elements studied are reproduced in Table IV.

Table III. Chemical Characteristics of Two Ponds from Maine

	Bog Pond	Sunken Pond
pH	4.6	4.6
Alkalinity (mg/l)	1	1
DOC (mg/l)	19.3	13.7
Na (mg/l)	3.2	3.2
K (mg/l)	0.4	0.97
Ca (mg/l)	2.3	0.73
Cd (μg/l)	0.40	0.32
Cu (μg/l)	22.5	27.9
Fe (μg/l)	0.18	0.11
CO_3^{2-} (*M*)	4.5×10^{-13}	4.5×10^{-13}
SO_4^{2-} (*M*)	5.9×10^{-5}	8.8×10^{-6}

Table IV. Percent Ca, Cd, Cu and Fe Removed from Two Pond Surface Waters by Ion Exchange Resins

	Bog Pond		Sunken Pond	
	% Cation	% Anion	% Cation	% Anion
Ca	>99	<1	>99	<1
Cd	65	37	56	55
Cu	<1	98	5	96
Fe	10	90	19	81

The model GEOCHEM was employed because it has provisions for fulvic acid ligands with stability constants, at least for K_1-type binding with Cu and Cd. For this reason, GEOCHEM was used to predict the distribution of these two metals for Sunken Pond and Bog Pond and then compared those predictions to calculations in which the K_1 and K_2 values determined in southeastern waters had been inserted for the values in GEOCHEM. This latter model was called GEOPLAY. The n_1 and n_2 values from the Southeast were adjusted to the DOC of the ponds and used in both models as the concentrations of available sites. In addition, since the type I and type II binding sites are usually considered to be phthalate- and salicylate-type bonds respectively, the model REDEQL2 was also used to predict the chemical forms of Cd and Cu in the two Maine ponds. REDEQL2 is an earlier version of GEOCHEM that has a smaller thermodynamic data base and does not have fulvic acid-type ligands but does have phthalic acid and salicylic acid data. Hence REDEQL2 was used without alteration of the thermodynamic data but with the same corrected n_1 and n_2 values used in GEOCHEM and GEOPLAY as the concentration of phthalic and salicylic acids, respectively.

The results of the computations can be seen in Table V. For both metals, in the two ponds, the models with unaltered thermodynamic data predicted that >95% of the metals should exist as cations. The model GEOPLAY, which had had the K_1 and K_2 values from Table II inserted in the thermodynamic data base of the model, predicted that approximately 40% of the Cd in both ponds would be cationic, as would about 9% of the Cu. Comparison of the predictions from the models with the empirical data gathered in the field (Table IV) clearly shows that the predictions using the stability constants determined for organic matter from the Southeast give a much better agreement with measured distributions than do those of either of the other models.

The reason for the discrepancies can be seen with an examination

Table V. Percentage Cd and Cu Predicted by Chemical Models to be in Anion or Cation Forms in Two Maine Surface Waters

	Bog Pond			
	Cd		Cu	
	% Cation	% Anion	% Cation	% Anion
REDEQL2	99.0	<1	98.6	<1
GEOCHEM	97.7	0.8	95.9	2.9
GEOPLAY	41.1	58.3	9.9	90.0
	Sunken Pond			
	Cd		Cu	
	% Cation	% Anion	% Cation	% Anion
REDEQL2	99.8	<1	99.3	<1
GEOCHEM	98.9	1.1	95.5	3.5
GEOPLAY	40.1	59.8	9.2	90.8

of the stability constant data within the GEOCHEM thermodynamic data base. The data base has log K_1 and log K_2 values of 3.5 and 2.7, 4.3 and 2.5 for Cd and Cu, respectively, with fulvic material. Comparing these values with those from this study (log K_1 = 5.8, log K_2 = 3.1, for Cd; log K_1 = 6.5, log K_2 = 4.9, for Cu; Table II) shows a two log unit discrepancy for the K_1 values for both metals and a similar discrepancy for the log K_2 value of Cu. Since models such as GEOCHEM are designed to solve sets of simultaneous, competitive equations by successive iterations, it is apparent that the two orders of magnitude difference in the value of a stability constant can have significant effects on the results, especially when many of the stability constants are similar. In the case of the thermodynamic data deck of GEOCHEM, of 36 metals (including H^+) only 9 metals have log K values listed. For log K_1 the 9 values range from 3.1 for Mg^{+2} to 4.7 for Pb^{+2}, a range of only 1.5 log units for all complexes considered. When this information is considered, it is easy to understand how the differences in predictions arise.

Influence of Ca and Fe

The results of the computations shown in Table V would indicate that the stability constants for Cd and Cu with natural organics from the Southeast are a relatively good estimate of the values for similar complexes from the Northeast. However, as has been noted, the models solve competitive equations. It has been pointed out previously that major ions such as Ca will be present in such quantities as to effectively compete for binding sites with the trace metals [22,23]. This is particularly true if the stability constants are similar, as we have seen in existing models. However, the stability constant values which best predict the field data for Cd and Cu may be as much as two orders of magnitude greater than those used for Ca-fulvic complexes in GEOCHEM. To investigate the effect of changing the stability constants for the Ca-fulvic complex, we again employed GEOCHEM and ran a series of calculations in which the Ca-fulvic stability constant values were increased to determine the magnitude of the log K_1 and log K_2 values which would be required to cause a deviation in the predicted value from that determined in the field (Table IV). The results of this study are seen in Figure 1.

In both cases, significant differences were noted at log $K_1 \sim 5$ and log $K_2 \sim 3$. These values represent approximately one log unit higher values than have been reported for the stability constants of Ca-humic compounds [21]. Values in excess of the literature data do not appear warranted in the case of calcium due to the observed cationic character of Ca in field studies (Table IV). Hence the values of log $K_1 = 3.6$ and log $K_2 = 1.7$ used in the model's data base were used in further calculations.

A similar exercise is warranted with respect to Fe. The field data indicate that, in both ponds, Fe is primarily anionic (Table IV). Examination of the thermodynamic data reveals that no stability constants exist for Fe^{+3}-fulvic complexes, and that Fe^{+2}-fulvic complexes are reported as log $K_1 = 4.4$ and log $K_2 = 2.7$, which appear to be too low for the measured distribution. Hence, the model GEOCHEM was again employed to approximate realistic stability constants for Fe^{+3}-fulvic complexes. As before, the n_1 and n_2 values for southeastern systems were used as the concentrations of binding sites with K_1 and K_2 values, respectively. The stability constants were changed to higher values in successive runs of the model and the percentage cationic Fe was plotted against K values (Figure 2). For both systems, log $K_1 = 9.6$ and log $K_2 = 8.6$ values gave the best estimate of the field data (Table IV). Thus, it would appear that a divalent alkaline earth such as Ca has relatively low stability constants relative to metals Cd and Cu (3.5 and 1.5 vs ~6 and ~4 for Cd and Cu), and the trivalent Fe has stability constants higher than the divalent species (~9.5 and ~8.5). This would mean that under equilibrium conditions, the trivalent Fe would be expected to bind to organic matter more strongly than the divalent metals, which in turn would bind more strongly than the alkaline earths. Some support is provided for this argument by the studies of competitive binding of metals by organic matter.

Natural organic matter that had been isolated from southeastern waters

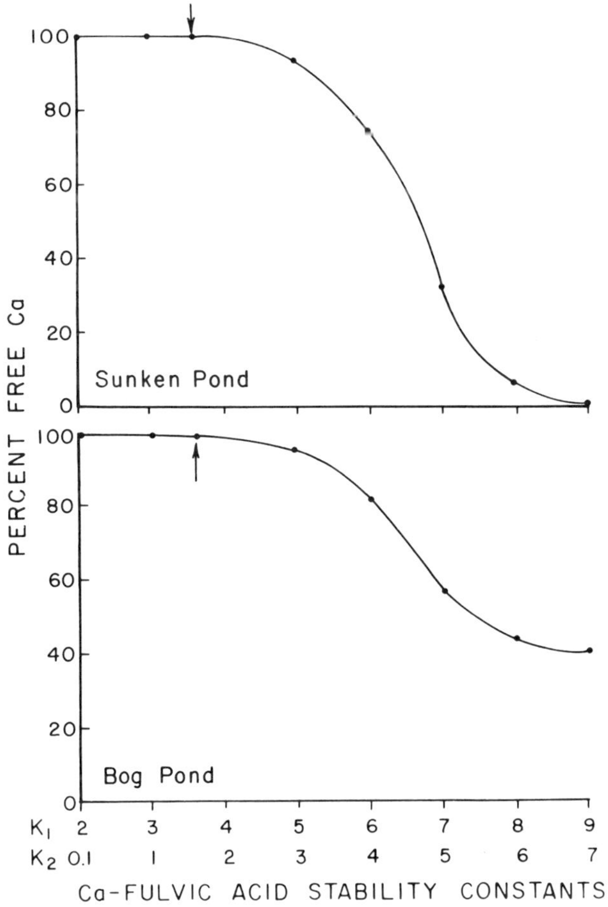

Figure 1. Percentage cationic Ca^{2+} predicted as a function of the two stability constants (K_1 and K_2) of the Ca–fulvic acid complex. Arrows represent values currently in the thermodynamic data base of the equilibrium model "GEOCHEM."

by ultrafiltration was allowed to react with various elements including Al^{+3}, Ca^{+2} and Mn^{+2} [14]. The binding capacities of these organics were then determined by titration with Cd^{+2}, Cu^{+2} and Pb^{+2}. The binding capacities determined in this manner were then compared to the binding capacities of separate samples of the organic matter which had not been allowed to react with the competing ions Al^{+3}, Ca^{+2}, Mn^{+2}. The results of a portion of this work are shown in Table VI. As can be seen, Ca^{+2} did not

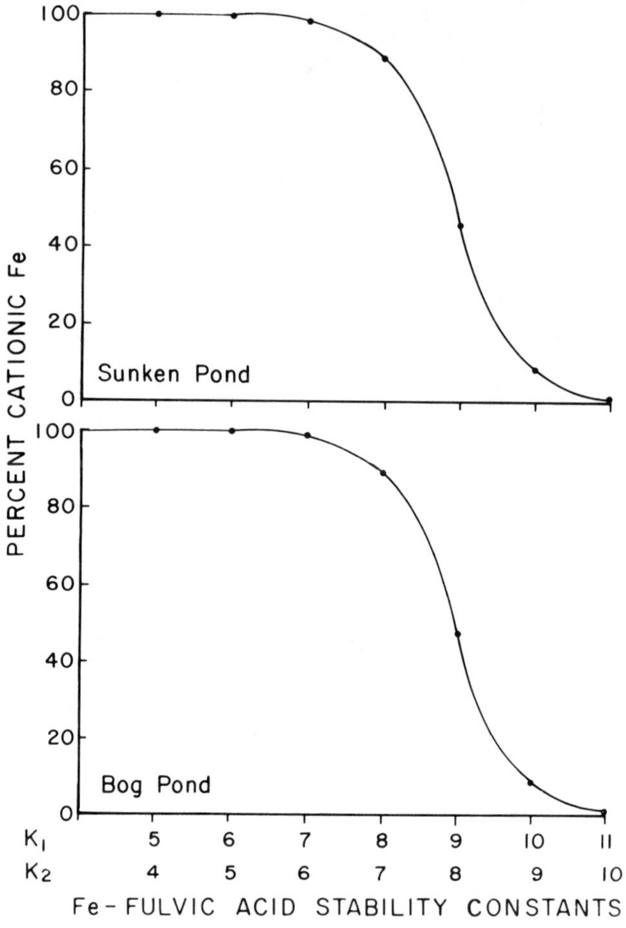

Figure 2. Percentage cationic Fe^{3+} predicted as a function of the two stability constants (K_1 and K_2) of the Fe–fulvic acid complex.

compete well with any of the divalent metals, whereas the trivalent Al was very effective at blocking the binding sites of the organic from reacting with divalent metals. The divalent Mn ion was intermediate in its ability to block sites, but appeared to be more effective against Cd and Pb, which have lower stability constants with organics (Table II) than does Cu. Recent studies [24] have arrived at similar results, although some discrepancies in order of reactivity do arise, particularly as a function

Table VI. Effect of Competing Ions on Cd, Cu and Pb Binding Capacities of Surface Water Organic Matter

Competing Ion	% of Original Binding Capacity Remaining in the Presence of Competing Ions		
	Cd^{2+}	Cu^{2+}	Pb^{2+}
Al^{3+}	3.2 ± 5.4[a]	ND[b]	3.6 ± 5.4
Ca^{2+}	95.7 ± 4.8	89.3 ± 8.0	94.7 ± 4.6[a]
Mn^{2+}	25.6 ± 16.5[a]	51.3 ± 20.2	24.8 ± 19.4

[a]n = 9, all others n = 12.
[b]ND = not determined.

of pH. While there are always difficulties in interpretating competition studies, the general trend supports the lower organic-metal stability constants of Ca^{+2} and the high values of Fe^{+3} relative to Cd and Cu.

Since as we have noted the model solves competitive reaction equations, we again used the model GEOPLAY to predict the distribution of chemical species of Maine lakes. In this effort, we incorporated the stability constants we had derived for Ca and Fe along with those for Cd and Cu and again used n_1 and n_2 as described above. The results of those calculations are compared to the field observations in Table VII. The agreement between calculated and observed distribution was quite good. The calculated values for Cu are less accurate with the incorporation of the Fe and Ca values; but, considering the cumulative errors possible in the data gathering and calculations, they must still be considered in good agreement. In addition, it must be remembered that other metal-organic

Table VII. Percentage Cationic Ca, Cd, Cu and Fe Predicted by Chemical Models and Found by Ion Exchange Resin in Maine Ponds

	Bog Pond		Sunken Pond	
	Found	Pred.	Found	Pred.
Ca^{2+}	>99	98.7	>99	99.4
Cd^{2+}	65	67.7	56	62.6
Cu^{2+}	<1	19.0	5	16.5
Fe^{3+}	10	19.5	19	18.1

interactions not considered in this treatment (i.e., Al^{+3}, Mg^{+2}) are required before fine tuning of these predictions can be considered.

SUMMARY

The work reported here shows that approximately 50% of the DOC of six southeastern waters can be extracted by ultrafilters designed to remove particles >13 Å in diameter. In addition, despite a wide range of physical and chemical differences in these waters and possible differences in source terms, this organic matter has narrow ranges of stability constants and numbers of binding sites available for complexing with Cu, Cd and Pb. The number of exchangeable protons associated with this organic matter ranges from 3.9 to 11.9 meq/g ash-free dry wt organic matter, with approximately 50% of those protons being carboxylic and 50% being phenolic. However, it appears that only a relatively small portion of those proton exchange sites may be active in metal binding.

Computer model predictions in conjunction with field observations and laboratory data presented in this work indicate that trivalent ions have higher stability constants for binding with natural organic matter than do divalent heavy metals, which, in turn, form stronger bonds than do alkaline earths. More importantly, it appears that data gathered relative to metal-organic binding in the southeastern United States is applicable on a broad regional basis and may well be applicable across regions of the U.S. Hence, it may be possible to obtain accurate predictions of trace metal speciation in highly organic waters without the requirement of determining stability constants for each location, as has been suggested [21].

In conclusion, the quasiparticulate model is not a preferred or even theoretically appealing alternative to the complete description of the metal-ligand interactions, but may be the only pragmatic method of describing complex natural systems in a manner that allows the prediction of the free metal concentrations. Thus, it would appear that natural organic-metal interactions are not intractable, and many useful measurements and predictions of chemical speciation of metals in high organic concentration waters may be made using current techniques and knowledge.

ACKNOWLEDGMENT

This research was supported by contract EY-76-C-09-0819 between the University of Georgia and the U.S. Department of Energy.

REFERENCES

1. Nordstrom, D. K., L. N. Plummer, T. M. L. Wigley, T. J. Wolery, J. W. Ball, E. A. Jenne, R. L. Bassett, D. A. Crerar, T. M. Florence, B. Fritz, M. Hoffman, G. R. Holdren, Jr., G. M. Lafon, S. V. Mattigod, R. E. McDuff, F. Morel, M. M. Reddy, G. Sposito and J. Thrailkill. "A Comparison of Computerized Chemical Models for Equilibrium Calculations in Aqueous Systems," in: *Chemical Modeling in Aqueous Systems*, E. A. Jenne, Ed. (Washington, DC: American Chemical Society, Symp. Ser. 93, 1979), p. 857.
2. Jenne, E. A. "Chemical Modeling–Goals, Problems, Approaches, and Priorities," in *Chemical Modeling in Aqueous Systems*, E. A. Jenne, Ed. (Washington, DC: American Chemical Society, Symp. Ser. 93, 1979), p. 3.
3. Reuter, J. H., and E. M. Perdue. "Importance of Heavy Metal–Organic Matter Interactions in Natural Waters," *Geochim. Cosmochim. Acta* 41:325-334 (1977).
4. Schnitzer, M., and S. U. Khan. *Humic Substances in the Environment* (New York: Marcel Dekker, Inc., 1972), p. 203.
5. Giesy, J. P., and J. J. Alberts. "Trace Metal Speciation: The Interaction of Metals with Organic Constituents of Surface Waters," in prep.
6. Sposito, G. "Trace Metals in Contaminated Waters," *Environ. Sci. Technol.* 15(4):396-403 (1981).
7. Alberts, J. J., and D. W. Evans. "Comparative Chemistry of Submicron Sized Organic Fractions in Four Rivers of the Southeastern United States," *Prog. Phys. Chem. of the Earth* 13:463-468 (1979).
8. Schindler, J. E., and J. J. Alberts. "Analysis of Organic–Inorganic Associations of Four Georgia Reservoirs," *Arch. Hydrobiol.* 74:429-440 (1974).
9. Giesy, Jr., J. P., and L. A. Briese. "Metals Associated with Organic Carbon Extracted from Okefenokee Swamp Water," *Chem. Geol.* 20:109-120 (1977).
10. Giesy, Jr., J. P., L. A. Briese and G. J. Leversee. "Metal Binding Capacity of Selected Maine Surface Waters," *Environ. Geol.* 2(5):257-268 (1978).
11. Scatchard, G. "The Attractions of Proteins for Small Molecules and Ions," *N.Y. Acad. Sci.* 51:660-672 (1949).
12. McDuff, R. E., and F. M. Morel. "Technical Report EQ-73-02: Description and Use of the Chemical Equilibrium Program REDEQL2," W. M. Keck Lab of Environmental Engineering Science, California Institute of Technology, Pasadena (updated July 1975).
13. Sposito, G., and S. V. Mattigod. "Geochem: A Computer Program for the Calculation of Chemical Equilibria in Soil Solutions and Other Natural Water Systems," Kearney Foundation of Soil Science, University of California, Riverside (1979).
14. Giesy, Jr., J. P., and L. A. Briese. "Metal Binding Capacity of Soft, Acid Organic Waters," in prep.
15. Beck, K. C., J. H. Reuter and E. M. Perdue. "Organic and Inorganic Geochemistry of Some Coastal Plain Rivers of the Southeastern United States," *Geochim. Cosmochim. Acta* 38:341-364 (1974).

16. Tuschall, Jr., J. R., and P. L. Brezonik. "Characterization of Organic Nitrogen in Natural Waters: Its Molecular Size, Protein Content, and Interactions with Heavy Metals," *Limnol. Oceanog.* 25(3):495-504 (1980).
17. Perdue, E. M. "Solution Thermochemistry of Humic Substances. Acid-Base Equilibria of River Water Humic Substances," in: *Chemical Modeling in Aqueous Systems*, E. A. Jenne, Ed. (Washington, DC: American Chemical Society, Symp. Ser. 93, 1979), p. 99.
18. Bresnahan, W. T., C. L. Grant and J. H. Weber. "Stability Constants for the Complexation of Copper (II) Ions with Water and Soil Fulvic Acids Measured by an Ion Selective Electrode," *Anal. Chem.* 50(12):1675-1679 (1978).
19. Martell, A. E. "The Influence of Natural and Synthetic Ligands on the Transport and Function of Metal Ions in the Environment," *Pure and Appl. Chem.* 44(1):81-113 (1975).
20. Shuman, M. S., and G. P. Woodward, Jr. "Stability Constants of Copper-Organic Chelates in Aquatic Samples," *Environ. Sci. Technol.* 11(8):809-813 (1977).
21. Mantoura, R. F. C., A. Dickson and J. P. Riley. "The Complexation of Metals with Humic Materials in Natural Waters," *Estuar. Coastal Mar. Sci.* 6:387-408 (1978).
22. Wolf, A., K. Bunzl, F. Dietl and W. F. Schmidt. "Effect of Ca^{2+}-Ions on the Absorption of Pb^{2+}, Cu^{2+}, Cd^{2+} and Zn^{2+} by Humic Substances," *Chemosphere* 5:207-213 (1977).
23. O'Shea, T. A., and K. H. Mancy. "The Effect of pH and Hardness of Metal Ions on the Competitive Interaction Between Trace Metal Ions and Inorganic and Organic Complexing Agents Found in Natural Waters," *Water Res.* 12:703-711 (1978).
24. Kerndorff, H., and M. Schnitzer. "Sorption of Metals on Humic Acid," *Geochim. Cosmochim. Acta* 44:1701-1708 (1980).

CHAPTER 17

DISTRIBUTION OF STABILITY CONSTANTS AND DISSOCIATION RATE CONSTANTS AMONG BINDING SITES ON ESTUARINE COPPER-ORGANIC COMPLEXES: ROTATED DISK ELECTRODE STUDIES AND AN AFFINITY SPECTRUM ANALYSIS OF ION-SELECTIVE ELECTRODE AND PHOTOMETRIC DATA

Mark S. Shuman, Bradley J. Collins, Paul J. Fitzgerald and Dean L. Olson

Department of Environmental Sciences and Engineering
School of Public Health
University of North Carolina
Chapel Hill, North Carolina

ABSTRACT

An "affinity spectrum" technique is introduced to evaluate metal binding to humic substances. The model on which it is based considers binding to occur at a large number of nonequivalent sites distributed randomly on the macromolecule and possessing binding constants with an arbitrary distribution of values. An evaluation of Cu binding to estuarine organic materials collected from the Ogeechee River estuary near Savannah, Georgia was carried out on samples concentrated and fractionated in the laboratory by ultrafiltration. Estimates of stability constants and dissociation rate constants were obtained using equilibrium data from a cupric ion-selective electrode and kinetic data from a photometric method. A rotated disk electrode technique was also used to estimate rate constants. Binding constants and rate constants were found to have numerical values distributed within two or three regions separated by about 1.5 orders of magnitude. Most of the Cu was weakly bound, with less than half dissociating slowly enough to distinguish bound Cu kinetically from unbound Cu.

INTRODUCTION

The distribution of chemical forms or the speciation of metals is important to investigations of metal transport and toxicity in freshwater and marine environments. Various attempts to model metal speciation mathematically using estimates of complex stability constants have been successful only to the extent that the set of constants employed are complete and accurate. Current research aims to provide reliable estimates of stability constants for complexes of metals with natural organic matter. Thermodynamic stability data alone however, address but one aspect of the biogeochemistry of metals. Another aspect, the kinetic dissociation of metal complexes as it relates to metal bioavailability and toxicity, must also be considered.

Dissociation Kinetics

Metal toxicity to aquatic organisms cannot be gauged simply by measuring total metal concentration. Recent investigations suggest that ionic or uncomplexed metal is the toxic form [1] and the organic material through complexation masks metal toxicity in the oceans [2-4] and in rivers and lakes [5,6]. The molecular size of organic chelators appears to be an important component of metal availability. Giesy et al. [7] found that Cd toxicity of *Simocephalus serrulatus* decreased with increasing molecular size of pond organic material, and Petruzzelli and Guidi [8] observed that the highest-molecular-weight fractions of soil humics lowered Cu availability to wheat. This molecular size-dependence could arise from complex stability (the larger molecules may have a greater affinity for metals), or it could arise from kinetics (the larger metal complexes may dissociate at a very slow rate).

The detailed biological mechanism for metal uptake by organisms is not known, but both active and passive transport have been suggested [9]. Passive transport includes adsorption, ion exchange and diffusion, whereas active transport is thought to involve formation of an enzyme-substrate complex on one side of the cell membrane and translocation and release on the other side. This process is often expressed in oversimplified fashion by a Michaelis-Menten equation

$$E + S \rightleftarrows ES \rightarrow E + P \qquad (1)$$

where E is an enzyme
P is a product that incorporates the metal
S, the substrate, is a metal, M

If all reactions indicated in Equation 1 are rapid, the process would be controlled by the rate of availability of substrate to the enzyme, that availability expressed by

$$ML \underset{k_f}{\overset{k_d}{\rightleftarrows}} M + L \qquad (2)$$

where ML is a metal complex
L is a ligand
k_d and k_f are the rate constants of dissociation and formation

This mechanism is not unusual in biochemical processes; in catalytic schemes involving ternary enzyme-metal ion-substrate complexes, there are several dissociation steps [10] and the rate of metal chelate ring opening can be the slowest step. On the other hand, metal complex stability may be important to both active and passive transport; competition for metal between enzyme or adsorption sites within the cell and chelators in the surrounding media could control uptake, the more stable chelates preventing uptake.

The question of whether there is a kinetic component to toxicity is not merely academic. In estuaries, regions may exist with very low tolerance for added trace metals, and this possibility needs to be investigated. These regions are where uncomplexed metal concentration is greatest, where the largest organic molecules have precipitated and where dilution has not greatly decreased metal concentration. The equilibrium models of Sibley and Morgan [11] show that ionic concentrations of Cu, Ni and Pb reach a maximum between about 2 to 15‰, principally because of desorption from fluvial clays. Further, it is over this same range of salinity that the largest organics, those that appear to be most efficient in masking the toxicity of ionic metal, coagulate and precipitate [12].

Very few estimates of metal-complex dissociation rates appear in the literature. Langford and Khan [13] estimated a Fe-fulvic acid dissociation rate constant of about 10^{-2}/sec at pH 1.5 from values of the stability constants and the kinetic formation rate constant obtained photometrically by stopped flow. The fulvic acid in this case was derived from a Prince Edward Island podzol soil. Shuman and Michael [14] using a rotated disk electrode (RDE) estimated the dissociation rate constants of Cu-organic complexes in marine coastal samples to be about 2/sec and Shuman and Cromer [15] estimated the dissociation rate constant of Cu complexes of humic and fulvic acid isolated from a North Carolina lake to be about 0.1 and 2.7/sec respectively.

Complex Stability Constants

Stability constants of metal-humic complexes are estimated from experimental data of complexed and uncomplexed metal concentrations obtained at equilibrium under conditions of varying humic and metal concentrations. The "binding curve," i.e., the relationship between bound and unbound metal, is given by

$$\bar{v} = \frac{nK[M]}{1 + K[M]} \tag{3}$$

where $\bar{v}$ is the molar ratio of bound metal to total ligand
[M] is the molar concentration of uncomplexed metal
K is the site binding constant
n is the number of equivalent binding sites

Several linear transformations of this equation are used; the one suggested by Scatchard [16]

$$\frac{\bar{v}}{[M]} = nK - \bar{v}K \tag{4}$$

leads to a graphical solution, a "Scatchard plot" in which $\bar{v}/[M]$ is plotted against $\bar{v}$, and K is estimated from the slope and n from the intercept.

The above equations apply to a single binding site. For multiple sites, as would be encountered with humic materials, Equation 3 can be rewritten [17]

$$\bar{v} = \sum_{i=1}^{m} \frac{n_i K_i [M]}{1 + K_i [M]} \tag{5}$$

where i represents the ith class with n_i sites and with an independent site binding constant, K_i

The total number of sites is n_0, obtained by summing over m classes of sites

$$n_0 = \sum_{i=1}^{m} n_i \tag{6}$$

Graphical representations of multiple binding sites using transformations of Equation 5 are nonlinear and rather featureless. The limitations in their use for estimating K_i and n_i are discussed by Klotz and Hunston [17], who give expressions for limiting slopes and intercepts in terms of the sums and products of K_i and n_i. A unique solution from which individual binding constants can be abstracted is possible only for two independent classes of sites. Relationships for a system with site interaction have also been given [18]. Thus, Scatchard plots or similar transformations of Equation 5 are useful for estimating formation constants only when it can be assumed that there exist at most two classes of binding sites. This seems an oversimplification for metal binding with humic materials, although Scatchard plots and similar transformations have been used extensively in studies of metal-humic complexation [19-23]. One thing is clear from this literature: all graphical representations of equilibrium data are nonlinear and indicate multiple binding sites.

These approaches to analyzing complex equilibria data have been criticized by Gamble et al. [24] and by MacCarthy and Smith [25] who demonstrate that stability "constants" obtained with them are not thermodynamic constants but vary with organic and metal concentrations. Nonlinear Scatchard plots provide a good example of this variability and can be interpreted as reflecting a large number of nonequivalent binding sites.

Objectives and Scope

This chapter describes a photometric and a RDE method for estimating Cu-organic complex dissociation rate constants, and an ion-selective electrode method for estimating stability constants. It presents preliminary results of the application of these methods to organics that have been concentrated and separated according to molecular size by ultrafiltration. It is also the first report in the literature where an affinity spectrum technique is employed to analyze equilibrium and kinetic data obtained from metal-natural organic complex systems. The work described is part of an effort to use these methods to reveal the nature of binding sites on natural organic matter, to determine if molecular size affects kinetic dissociation rate constants and to investigate how metal bioavailability as reflected by kinetic dissociation varies along the salinity gradient of an estuary.

PROCEDURES AND DATA ANALYSIS TECHNIQUES

Ultrafiltration/Diafiltration

Surface samples of 18 liters each were collected by peristaltic pump from nine stations along the Ogeechee River, Georgia during a cruise of the R/V *Blue Fin*, (Skidaway Institute of Oceanography). The samples for ultrafiltration were immediately filtered with a 0.45-μ polycarbonate cartridge filter prewashed with HNO_3 and rinsed with sample before filtration, placed in acid-washed 5-gal carboys, spiked with sodium azide to 1 mg/l and refrigerated at 8 °C upon return to Chapel Hill. Six liters were freeze-concentrated to about 2 liters prior to sequential ultrafiltration, from largest to smallest pore diameter, using Amicon membranes, XM-50 (pore diameter 3.1 nm), UM-2 (1.3 nm) and UM-05 (1.04 nm). The UM-2 and UM-05, which have a residual negative charge, were presoaked with 5% NaCl to minimize ionic charge effects. Ultrafiltration proceeded until the retentate was about 50 ml, whereupon about 250 ml of distilled water was filtered at constant retentate volume. This "diafiltration" was necessary to desalt the sample. Diafiltration with 5X the retentate volume left <1% of the original solutes smaller than the pore diameter in the retentate. The first ultrafiltrate was then ultrafiltered and diafiltered again with the next smallest membrane. The three fractions, with nominal molecular diameters >3.1 nm, 1.3 to 3.1 nm and 1.04 to 1.3 nm, were refrigerated until used. Further details concerning these procedures, including an organic carbon balance, are presented elsewhere [26].

Rotated Electrode Technique

The utility of a mercury-plated RDE for measuring dissociation kinetics of metal complexes in natural waters was first suggested by work with model systems of Cd-NTA and Cd-EDTA solutions [27] and first demonstrated with copper complexes in marine coastal samples [14]. Anodic stripping voltammetry (ASV) is combined with the RDE, which allows kinetic measurements at a very low metal concentration. Electrolysis of the uncomplexed metal (the complex itself is not reduced at the applied potential) is described by the overall reaction

$$\mathrm{ML} \underset{k_f}{\overset{k_d}{\rightleftarrows}} \mathrm{M(II)} + \mathrm{L} \quad \text{(in solution)} \tag{7}$$

$$\mathrm{M(II)} + 2e^- \rightarrow \mathrm{M(0)} \quad \text{(at the electrode)}$$

and is carried out at constant potential for a period of 5 to 10 min for a series of electrode rotation rates, ω, in the range of about 50 to 3000 rpm. The coulombs of electricity passed during electrolysis, Q_k, is measured by integrating the oxidation (stripping) current and is a function of the kinetic dissociation constant of the complex, k_d, the ratio of uncomplexed to complexed metal, K, and the rotation rate. In general, Q_k increases as k_d and K increase (increased availability of M to the electrode) and decreases as the rotation rate is increased (less time for dissociation). The data are plotted as the ratio Q_k/Q_0, or its inverse, as a function of rotation rate where Q_0 is the coulombs of electricity that would be passed during electrolysis if all the metal present were reducible. When Q_k/Q_0 is plotted against ω or $\omega^{1/2}$, $Q_k/Q_0 \rightarrow 1.0$ as $\omega \rightarrow 0$ because at very slow rotation rates all complexes have time to dissociate. At the limit of high rotation rates, there is no time for dissociation, and only uncomplexed metal is reduced at the disk. Therefore $Q_k/Q_0 \rightarrow \frac{[M]}{[M] + [ML]} = \frac{K}{K + 1}$ as $\omega \rightarrow \infty$. Kinetic information is obtained from data between these two limits, either by comparing Q_k/Q_0 vs $\omega^{1/2}$ with the theoretical equations of Hale [28] or by using a linearized approximation given by Levich [29] from which rate constants are obtained from the slope of Q_0/Q_k vs $\omega^{1/2}$. Data analysis is not quite this straightforward if more than one complex exist together in solution. However, it is readily shown theoretically that the contribution of each complex to the total current is additive and that a simple graphical procedure can be used to separate the contribution of the slower from the faster dissociating complexes. Figure 1, constructed using Hale's theory, illustrates how this is accomplished with two complexes, one for which $K_s = 5.5$, $k = 0.01$/sec and another $K_F = 5.0$, $k = 10.0$/sec. Note that Q_0/Q_k vs $\omega^{1/2}$ for the slower dissociating complex increases sharply at very low rotation rates and becomes independent of rotation rate at about 40 rpm whereas Q_0/Q_k for the faster dissociating complex is nearly linear over the entire range. The overall Q_0/Q_k is a simple sum, minus 1.0, of the contributions of the two complexes, and the extrapolation of the linear portion back to the origin gives the ratio $(K_s + 1)/K_s$ for the slower complex, which in this case has a value of 1.18. This same intercept, when determined for experimental data, is used to correct it point by point for the presence of the slow complex (see below). Q_k/Q_0 is the corrected Q_k/Q_0, which is then fitted to Hale's theory with a nonlinear curve-fitting routine. The range of rate constants that can be measured by this technique depends somewhat on K, but is typically 10^{-3} to 10^2/sec [30].

Ion-Selective Electrode Titrations

Samples with 20 mg/l DOC (25 ml) were titrated using a Fisher piston buret and 10^{-3} *M* Cu. Uncomplexed Cu was determined with an Orion cupric ion-selective electrode (ISE) and double-junction reference electrode. The pH was maintained at 6.00 throughout the titration by adjusting P_{CO_2}.

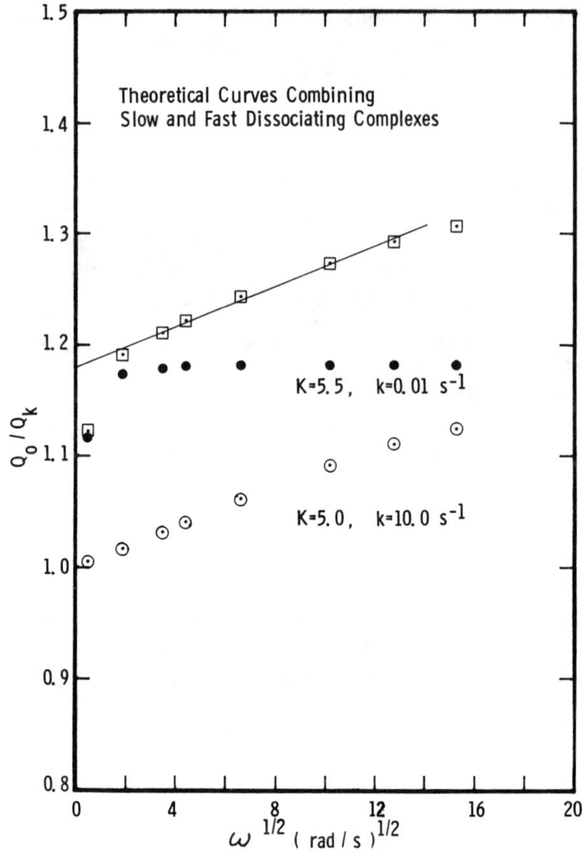

Figure 1. Theoretical RDE Q_0/Q_k curves for two complexes with widely different rate constants. ⊙ K = 5.0, k = 10/sec; ● K = 5.5, k = 0.01/sec; ⊡ combined Q_0/Q_k (see text). The line is drawn to intercepts at 1.18.

Constant ionic strength of 0.1 *M* was maintained with $NaNO_3$. For acid titrations, potassium salts were used.

Titration data were first transformed to give a Gran's plot. This transformation is appropriate to obtain accurate estimates of the endpoint for titrations involving weak metal-ligand interactions such as those encountered here [31].

The Gran's plot endpoint and data in the linear portion after the endpoint were used to establish the Cu standard curve applicable to the titration solution. The standard curve obtained is used to determine uncomplexed Cu concentrations from mV readings and these are subtracted from total Cu to give bound Cu. This and total ligand concentration expressed in terms of Cu concentration at the Gran's plot endpoint (the "complexation capacity") were used to compute v. Estimates of the endpoint were affected by uncertainties in the electrode response value used to compute the Gran's function. The binding curve of $\bar{v}$ vs log[Cu] was smoothed with a cubic spline and used to calculate N(K). N(K) is plotted vs −log[Cu] to give the affinity spectrum.

The acid titration of K_3PO_4 followed with a pH electrode was used as a model system to test the procedure. This system was chosen because acidity constants are well known, and have widely separated values. The pH electrode also gives a greater response and is linear over a much wider concentration range than the Cu electrode. The Cu electrode was used to titrate the natural organics that had been concentrated and isolated by ultrafiltration from samples obtained from the Ogeechee River estuary.

Photometric Measurements

Photometric measurements of kinetic dissociation were based on the appearance of a complex of Cu with 4-(2-pyridylazo)resorcinol (PAR). PAR was found by the method of continuous variations to form both 1:1 and 1:2 Cu complexes at pH 7.5 with an absorption peak at 508 nm. The complex obeys Beer's law in the range 4.0 to 40.0 × 10^{-6} *M* and has a molar absorptivity of 46,996/*M*-cm at pH 7.5 and 0.1 *M* ionic strength. A mixture of Cu and humic material with humics in excess as judged from the complexation capacity was allowed to react with a large molar excess of PAR. The overall reaction can be written

$$\mathrm{CuHM} \underset{k_2}{\overset{k_{-2}}{\rightleftarrows}} \mathrm{Cu + HM}$$

$$\mathrm{Cu + PAR} \underset{k_{-1}}{\overset{k_1}{\rightleftarrows}} \mathrm{CuPAR} \qquad (8)$$

where HM is the humic material. The rate constant k_1 is quite large as determined by stopped flow (see below). Thus, the rate of formation of CuPAR in the above reaction is determined by the dissociation of CuHM and the rate expression is simply

$$\frac{d(\mathrm{CuPAR})}{dt} = k_{-2}(\mathrm{CuHM}) \qquad (9)$$

provided that PAR is in sufficient excess of HM so that k_1 (PAR) $\gg k_2$ (HM), i.e., provided that the reaction is essentially irreversible.

The CuHM and the PAR solutions were injected with syringes into a 2-cm absorption cell and the CuPAR appearance at 508 nm followed with a Cary Model 219 spectrophotometer from 30 sec to 40 min. Normally, data are analyzed by plotting log ΔA vs t where $\Delta A = A_\infty - A_t$, the difference between the measured absorbance at infinity and the absorbance at time t. The slope of this plot gives the microscopic rate constant k_{-2} and the zero time intercept is the logarithm of the initial CuHM concentration in terms of Cu. Rate constants and site concentrations for multiple sites can be estimated graphically from these plots [32] if rate constants are sufficiently separated. However, the alternative method described below circumvents the problem of estimating regions of linearity within what is usually a monotonically decreasing curve.

The rate constants k_{-1} and k_1 of Reaction 8 were estimated by stopped flow under pseudo-first-order conditions with apparatus consisting of a Beckman DU spectrophotometer and an Aminco-Morrow stopped flow system. Data acquisition and analysis were made with the aid of a Charles River Data System Inc., Model MF-211, LSI-11 based microcomputer system and a 12-bit A/D converter. Experiments with 10 μM Cu and 100 to 130 μM PAR at pH 7.5 gave $k_1 = 2.58 \times 10^6\ s^{-1} M^{-1}$, $k_{-1} = 85.2$/sec.

Affinity Spectrum Technique

An alternative approach to linear transformation of Equation 5 was suggested by Hunston [18] for estimating binding constants of proteins with small molecules. This approach is based on replacing the summation with a continuous function

$$\bar{v} = \int_{-\infty}^{\infty} \frac{N(K)K[M]}{1 + K[M]}\, d(\log K) \tag{10}$$

where

$$n_0 = \int_{-\infty}^{\infty} N(K)\, d(\log K) \tag{11}$$

The integral of Equation 10 cannot be solved in closed form but a finite difference, numerical approximation technique, developed by Ferry [33] for the dynamic response of viscoelastic materials, can be used to solve for N(K) given experimental values of [M] and $\bar{v}$. This approximation is of the form,

$$N(K) = \left| \frac{\bar{v}(a[M]) - \bar{v}([M]/a)}{2 \log a} - \frac{a}{(a-1)^2} \times \frac{[\bar{v}(a^2[M]) - \bar{v}([M]/a^2)] - 2[\bar{v}(a[M]) \quad \bar{v}([M]/a]}{2 \log a} \right|_{[M] = \frac{1}{K}} \tag{12}$$

where log a is chosen to be 0.2. Equation 12 is solved using a computer program which first smooths the experimental binding curve and then uses values of $\bar{v}$ at [M] in a procedure to give N(K). The result is an "affinity spectrum" of N(K) vs log K reflecting a distribution of K and an area under each peak proportional to $\bar{v}$. This data analysis views metal binding in a way fundamentally different from that expressed by Equation 5 in which discrete independent sites are envisioned. Rather, this approach views bonding represented by a distribution of sites where "N(K) d(log K)/n_0 represents the probability of finding a binding site with a binding constant between log(K) + d(log K)" [18] on the macromolecule. Further details about the numerical approximation are available in Thakur et al. [34].

A similar approach to kinetic data analysis is possible. The integrated form of Equation 9 is given by

$$C(t) = C_0 e^{-kt} \tag{13}$$

where
k = dissociation rate constant
$C(t)$ = CuHM concentration at time t = $\Delta A/(\epsilon_{CuPAR})(b)$
C_0 = CuHM concentration at t = 0
ϵ_{CuPAR} = molar absorptivity of CuPAR; b = path length

For a mixture of n complexes which all dissociate to form CuPAR

$$C(t) = \sum_{i=1}^{n} C_{0_i} e^{-k_i t} \tag{14}$$

As above, a continuous function H(k,t) can be substituted for the summation where H(k,t)d(ln k) is now the probability of finding a site with a dissociation rate constant in the region ln k, ln k + d(ln k), and k is treated as a variable. The integral takes the form [33]

$$C(k,t) = S \int_{-\infty}^{\infty} H(k,t) e^{-kt} d(\ln k) \tag{15}$$

A solution can be obtained using the first and second derivatives of C(k,t) [35] resulting in

$$H(k,t) = \frac{\partial^2 C(k,t)}{\partial(\ln t)^2} - \frac{\partial C(k,t)}{\partial(\ln t)} \tag{16}$$

and

$$S = \frac{4C(k,t)}{\sum_{i=1}^{n} C_{0_i}} \tag{17}$$

H(k,t) and S originate from the approximate inversion of a Laplace transform expression equivalent to Equation 15 where the integral is taken with respect to k instead of ln k. The value of S depends on whether the 1st, 2nd or n^{th} derivative is taken in the approximation. Any computed value of H(k,t) corresponds not to k = t, but k = 2/t. When H(k,t) is plotted vs ln k = ln(2/t) it gives the distribution of rate constants about a peak value of ln k with an area under the peak proportional to the concentration of complexes since

$$\int_{-\infty}^{\infty} H(k,t)\, d(\ln k) = \sum_{i=1}^{n} C_{0_i} \tag{18}$$

RESULTS AND DISCUSSION

Rotated Disk Electrode

A typical RDE experiment for which both Q_k/Q_0 and Q_0/Q_k are plotted against $\omega^{1/2}$ is shown in Figure 2. Extrapolation of Q_0/Q_k to $\omega = 0$ results in an intercept >1.0 which is interpreted as indicating the presence of a slowly dissociating complex. This intercept is used to transform the data in the following way. For a mixture of two complexes, one which dissociates slowly, CuL_s, and another which dissociates much more rapidly, CuL_F,

$$(Q_0/Q_k)_{\omega=0} = \frac{Cu_T}{[Cu] + [CuL_F]} \tag{19}$$

and

$$(Q_k/Q_0)_{\omega\to\infty} = \frac{[Cu]}{[Cu] + [CuL_F] + [CuL_s]} = \frac{[Cu]}{Cu_T} \tag{20}$$

where Cu_T is the analytical concentration of copper

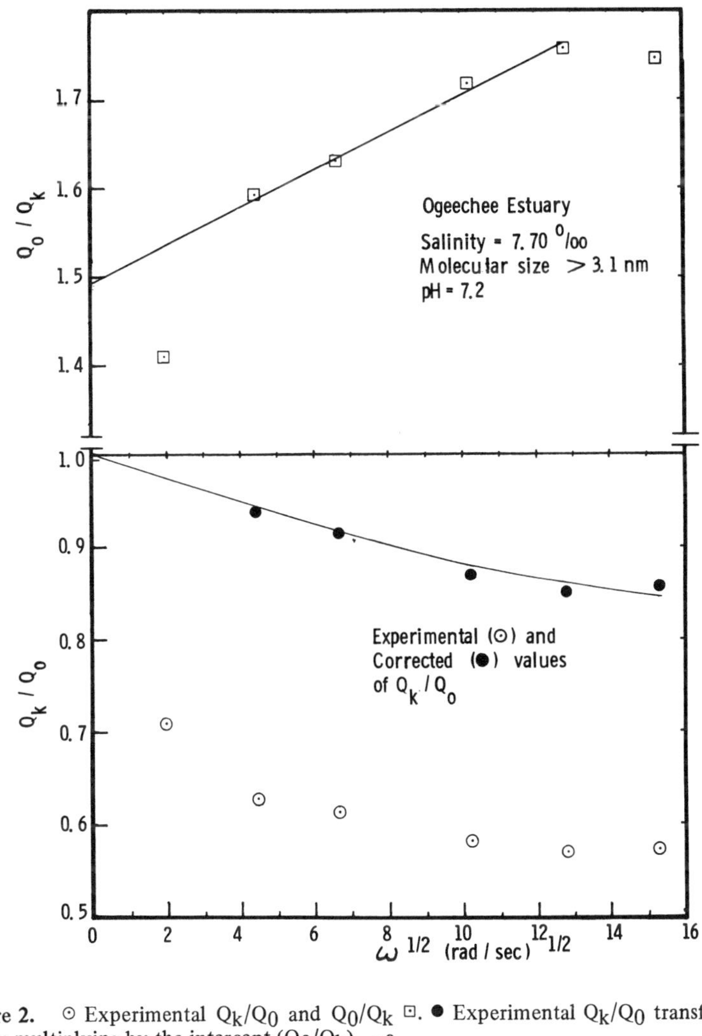

Figure 2. ⊙ Experimental Q_k/Q_0 and Q_0/Q_k ⊡. ● Experimental Q_k/Q_0 transformed by multiplying by the intercept $(Q_0/Q_k)_{\omega=0}$.

The intercept and the limit at large rotation rates are easily obtained from the data and their product is

$$\frac{[Cu]}{[Cu] + [CuL_F]} = \frac{K}{K+1} \qquad (21)$$

where $K = \dfrac{[Cu]}{[CuL_F]}$

Further,

$$[(Q_0/Q_k)_{\omega=0} - 1](K + 1) = \frac{[CuL_S]}{[CuL_F]} \tag{22}$$

Multiplying Q_k/Q_0 at any ω by $(Q_0/Q_k)_{\omega=0}$ approximates its value for the situation where CuL_F is present alone. How good an approximation this is depends on the rate constant and K for CuL_S, improving substantially with decreasing rate constants. Examination of Figure 1 verifies that the approximation also improves with increasing ω. These data are fitted to Hale's theory to estimate the rate constant and K for CuL_F and the untransformed data are likewise fitted to estimate the corresponding constants for CuL_S. Further iterations are possible but are not warranted because of the rather crude nature of the approximation.

Results for the RDE are given in Table I. The percentage of total Cu that dissociates slowly enough to be detected by the RDE is only 35 to 65% for these solutions, in which the Cu concentration is 17% of the complexation capacity. Although most of the Cu is associated with the organic matter at this percentage saturation, a large fraction of it is rapidly released and indistinguishable from uncomplexed Cu. Of the complexes detected, the slower dissociating ones predominate. An increase in the rate constants of the more rapidly dissociating complexes is observed with decreasing molecular size for 1.98‰. This trend is repeated at 7.7‰. The rate constants are much larger for samples from the two higher salinities, a trend that reflects a change with salinity in the distribution and nature of the complexes formed.

Table I. Rotating Disk Electrode Estimates of Concentration Quotients and Kinetic Rate Constants of Metal-Organic Complexes (13.8 mg/l Organic Carbon, 6.0 μM Cu, 2 mM CO_3^{2-}, pH = 6.8 to 7.35)

Salinity, ‰ of Original Sample	Size Fraction and % of Total Carbon	[Slow]/[Fast]	k_f, sec^{-1} Fast	k_f, sec^{-1} Slow	% Cu Complexed
1.98	>3.1 nm, 9%	8.3	0.25	0.02	65
	1.3–3.1 nm, 62%	1.1	0.56	0.15	35
	1.04–1.3 nm, 5%	3.7	1.8	0.04	42
7.70	>3.1 nm, 5%	1.1	6.0	0.12	34
	1.04–1.3 nm, 5%	1.4	13	0.25	38
19.2	1.3–3.1 nm, 58%	3.0	6.6	0.03	59

A larger sampling will be necessary to establish the characteristics of these changes.

Ion Selective Electrode Titrations

The protonation of tribasic phosphate is depicted by the binding curve in Figure 3. This model was chosen because its widely separated acidity constants accentuate the features common to all data analyses by the affinity spectrum technique. The value of $\bar{v}$ was computed by dividing the concentration of protonated phosphate at each point along the titration by the total concentration of phosphate. It was anticipated for this simple chemical system that $\bar{v}$ should vary through three plateau regions at $\bar{v}$ = 1.0, 2.0 and 3.0. The first plateau was in reality 0.82, indicating some protonation at the beginning of the titration. The difference between the first and second plateaus was 1.0, but the third plateau reached only about 57% of saturation of the monobasic phosphate before the difference between the concentration of free protons estimated with the pH electrode and the concentration of total acid added became

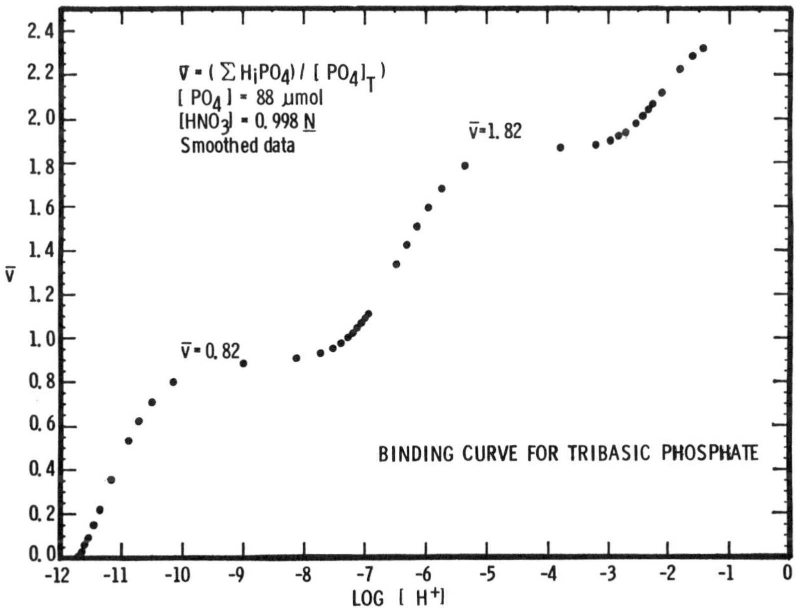

Figure 3. Binding curve for the titration of 88 μM tribasic phosphate with HNO_3.

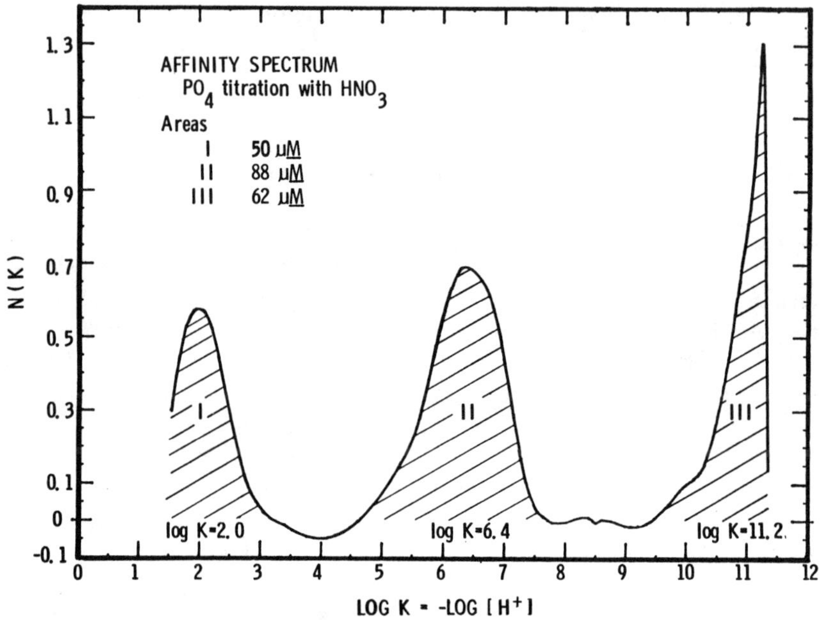

Figure 4. Affinity spectrum of the titration of tribasic phosphate. Shading indicates areas for each region I-III.

so small that estimates of $\bar{v}$ were highly imprecise. The affinity spectrum in Figure 4 has peaks corresponding to the inflection points of the binding curve at values of log K of 2.0, 6.4 and 11.2. These compare with values given by Smith and Martell [36] of 2.0, 5.7 and 11.7. The areas under each peak reflect the concentration of each of the three phosphate forms, for example, 88 μM under region II in Figure 4, or 100% of the dibasic phosphate.

The Cu binding curve and affinity spectrum for Ogeechee River estuary organic material is shown in Figures 5 and 6 respectively. As with the tribasic phosphate, computation of $\bar{v}$ becomes imprecise at large titrant concentrations so that any binding sites with log K < 4 are not included in the spectrum. The area under the curve, however, indicates that these are less than 5% of the total Cu. At the other extreme, log K values > 7 are also not included because of the loss of sensitivity of the electrode below a *total* Cu concentration of about $10^{-6.5}$ and also because only a very small fraction of the total binding capacity consists of sites with log K >7.

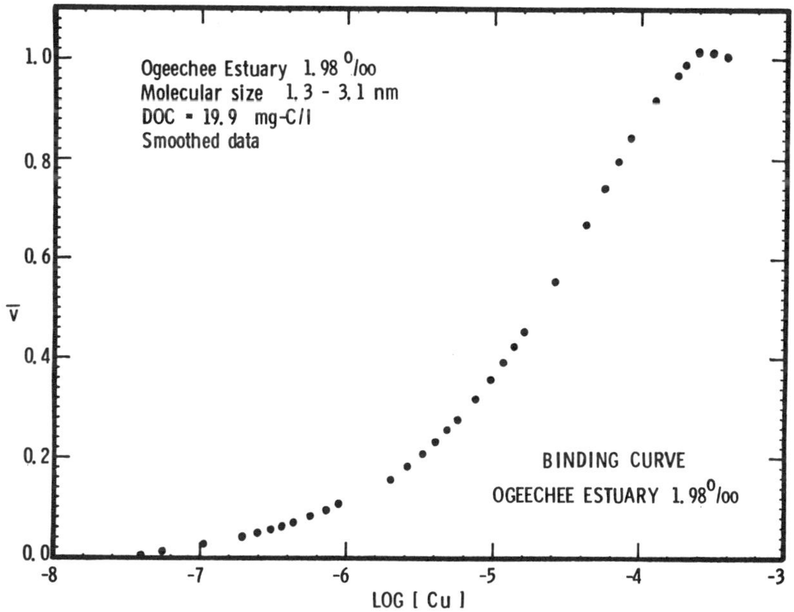

Figure 5. Binding curve for Ogeechee River estuary organics, 19.9 mg/1 DOC, titrated with Cu.

The affinity spectrum shows a distribution of complexes in three well defined regions designated I, II and III centered around log K = 4.4, 5.3 and 6.2 respectively. Region I, the weakest complexes, contains the largest concentration, about 68% of the total complexation capacity; region II has 27% and region III about 5%. Only a small percentage of the complexation capacity forms strong complexes with Cu.

Photometric Rate Data

Organics from the 1.98‰ sample, the 1.3 to 3.1 nm size fraction, were used. Two solutions with 21 mg/l DOC, one containing 7.1 μM Cu and the other 34.5 μM Cu, were reacted with 100 μM PAR at pH 7.5, ionic strength 0.1 M and 25 °C. Absorbance changes were followed 50 msec to 30 sec with the stopped-flow apparatus and at longer times with a Cary Model 219 spectrophotometer. Experiments carried out at 1000 μM PAR gave results indistinguishable from the lower PAR concentration. Data analysis of the 7.1 μM Cu solution is shown in Figure 7.

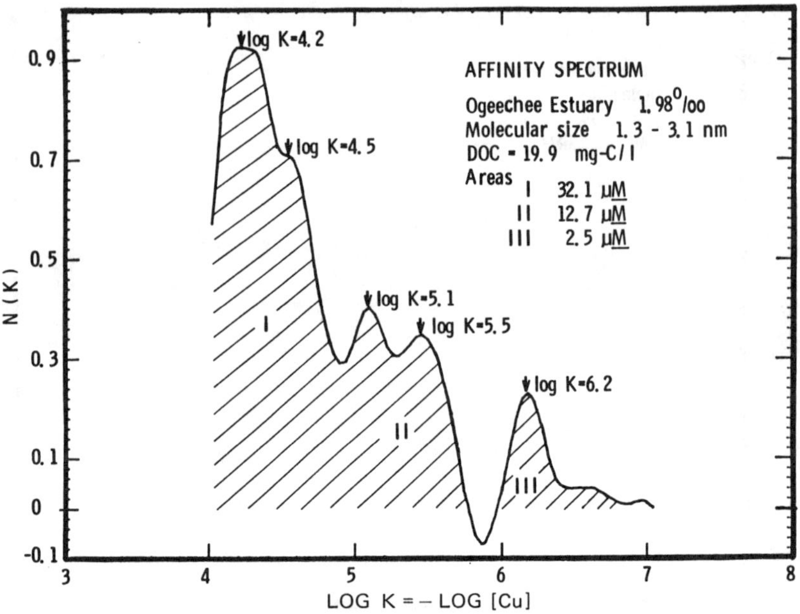

Figure 6. Affinity spectrum of the titration of Ogeechee River estuary organics. Shading indicates areas under each region, I-III.

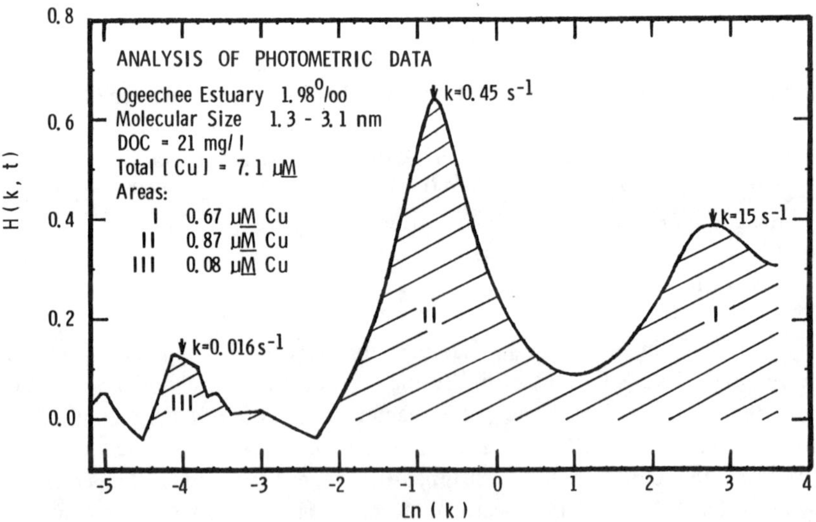

Figure 7. Analysis of photometric data. Ln(k) = ln(2/t).

Three distinct regions, I, II and III, are seen where rate constants are distributed around 15, 0.45 and 0.016 sec, respectively. Characteristics of this spectrum and the spectrum for the 34.5 μM Cu solution are summarized in Table II. Only about 23% of the 7.1 μM Cu solution is associated with complexes that dissociate slowly enough to be observed by the stopped flow, those with rate constants smaller than about 50 sec. With 34.5 μM Cu, this percentage drops to 10.6%. These results compare quite well with the RDE technique, which was carried out at slightly lower pH (pH 7.1). It is evident that a significant fraction of the total copper associated with the organic material dissociates very rapidly. It also appears that the distribution of Cu is a function of metal saturation of the organic material. The rate constant distribution is remarkably similar to the stability constant distribution shown in Figure 6, although the titration was carried out at a lower pH, which shifts the concentration distribution to weaker complexes.

CONCLUSIONS

Stability constants and dissociation rate constants for Cu complexed with estuarine organic macromolecules are distributed in two or three distinct regions separated by 1 to 1.5 orders of magnitude. Less than half of the total Cu dissociates slowly enough to be distinguished kinetically from uncomplexed Cu. Dissociation rate constants estimated with the RDE decrease with the salinity of the samples from which the organics were obtained and decrease also with molecular size. The distribution of rate constants varies with the degree of saturation of the organics

Table II. Results of Affinity Spectrum Analysis of Photometric Data

Total Cu Concentration (μM)	Rate Constants at Peaks (sec^{-1})		Complex Concentration (μM Cu)	% of Total Cu
7.1	0.016		0.080	1.1
	0.45		0.87	12.2
	15.1		0.67	9.4
		Totals	1.62	22.7
34.5	0.49		0.83	2.4
	34.0		2.8	8.2
		Totals	3.63	10.6

by Cu. This work demonstrates that the affinity spectrum technique and the RDE are appropriate and effective for analyzing kinetic and equilibrium experiments in which Cu is associated with natural organic materials.

REFERENCES

1. Sunda, W., and R. R. L. Guillard. "The Relationship Between Cupric Ion Activity and the Toxicity of Copper to Phytoplankton," *Water Res.* 34:513-529 (1976).
2. Barber, R. T., and J. H. Ryther. "Organic Chelators: Factors Affecting Primary Production in the Cromwell Current Upwelling," *J. Exp. Mar. Biol. Ecol.* 3:191-199 (1969).
3. Prakash, A., and M. A. Rashid. "Influence of Humic Substances on the Growth of Marine Phytoplankton Dinoflagellates," *Limnol. Oceano.* 13:598-606 (1968).
4. Davey, E. W. "Potential Roles of Metal Ligands in the Marine Environment," in *Toxicity to Biota of Metal Forms in Natural Water*, R. W. Andrew, P. V. Hodson, D. E. Konasewich, Eds. (International Joint Commission, Windsor, Ont., 1975), pp. 197-209.
5. Gachter, R., K. Lum-Shue-Chan and Y. K. Chau. "Complexing Capacity of the Nutrient Medium and its Relation to Inhibition of Algal Photosynthesis by Copper," *Schweiz. Z. Hydrol.* 35:253-261 (1973).
6. Stokes, P., and T. C. Hutchinson. "Copper Toxicity to Phytoplankton, as Affected by Organic Ligands, Other Cations and Inherent Tolerance of Algae to Copper," in *Toxicity to Biota of Metal Forms in Natural Water*, R. W. Andrew, P. V. Hodson, D. E. Konasewich, Eds. (International Joint Commission, Windsor, Ont., 1975), pp. 159-185.
7. Giesy, J. P., Jr., G. J. Leversee and D. R. Williams. "Effects of Naturally Occurring Aquatic Organic Fractions on Cadmium Toxicity to *Simocephalus Serrulatus* (Daphnidae) and *Gambusia Affinis* (Poeciliidae)," *Water Res.* 11:1013-1020 (1978).
8. Petruzzelli, G., and G. Guidi. "Influence of Soil Organic Matter on Copper Availability to Plants," *Z. Pflanzenernaehr. Bodenkd* 6:679-684 (1976).
9. Gertz, S., and I. Suffet. "The Biological Fate of Radionuclides in Aquatic Environments," in *Fate of Pollutants in the Air and Water Environments*, J. N. Pitts and M. L. Metcalf, Eds. (New York: John Wiley & Sons, Inc., 1977).
10. Frey, C. M., and J. Stuehr. "Kinetics of Metal Ion Interactions with Nucleotides and Base-Free Phosphates," in *Metal Ions in Biological Systems, Vol. 1, Simple Complexes*, H. Sigel, Ed. (New York: Marcel Dekker, Inc., 1974), pp. 51-116.
11. Sibley, T. H., and J. J. Morgan. "Equilibrium Speciation of Trace Metals in Freshwater:Seawater Mixtures," in *Proceedings of the International Conference on Heavy Metals in the Environment,* Toronto, Ont., 1975), pp. 319-338.

12. Sholkovitz, E. R. "Flocculation of Dissolved Organic and Inorganic Matter During the Mixing of River Water with Seawater," *Geochim. Cosmochim. Acta* 40:831-845 (1976).
13. Langford, C. H., and T. R. Khan. "Kinetics and Equilibrium of Binding of Fe^{3+} by a Fulvic Acid: A Study by Stopped Flow Methods," *Can. J. Chem.* 53:2979-2984 (1975).
14. Shuman, M. S., and L. C. Michael. "Application of Rotating Disk Electrode Techniques to the Measurement of Copper-Organic Complex Dissociation Rate Constants in Marine Coastal Samples," *Environ. Sci. Technol.* 12:1069-1072 (1978).
15. Shuman, M. S., and J. L. Cromer. "Copper Association with Aquatic Fulvic and Humic Acids. Estimations of Conditional Formation Constants with a Titrimetric Anodic Stripping Procedure," *Environ. Sci. Technol.* 13:543-545 (1979).
16. Scatchard, G. "The Attraction of Proteins for Small Molecules and Ions," *Ann. N. Y. Acad. Sci.* 51:600-672 (1949).
17. Klotz, I. M., and D. L. Hunston. "Properties of Graphical Representations of Multiple Classes of Binding Sites," *Biochem.* 10:3065-3069 (1971).
18. Hunston, D. L. "Two Techniques for Evaluating Small Molecule-Macromolecule Binding in Complex Systems," *Anal. Biochem.* 63:99-109 (1975).
19. Buffle, J., F. Greter and W. Haerdi. "Measurement of Complexation Properties of Humic and Fulvic Acids in Natural Waters with Lead and Copper Ion-Selective Electrodes," *Anal. Chem.* 49:216-222 (1977).
20. Bresnahan, W. T., C. L. Grant and J. M. Weber. "Stability Constants for the Complexation of Copper (II) Ions with Water and Soil Fulvic Acids Measured by an Ion-Selective Electrode," *Anal. Chem.* 50:1675-1679 (1979).
21. Brady, B., and G. K. Pagenkopf. "Cadmium Complexation by Soil Fulvic Acid," *Can. J. Chem.* 56:2331-2336 (1978).
22. Mantoura, R. F. C., and J. P. Riley. "The Use of Gel Filtration in the Study of Metal Binding by Humic Acids and Related Compounds," *Anal. Chim. Acta* 78:193-200 (1975).
23. Van den Berg, C. M. G., and J. R. Kramer. "Determination of Complexing Capacities of Ligands in Natural Waters and Conditional Stability Constants of the Copper Complexes by Means of Manganese Dioxide," *Anal. Chim. Acta* 106:113-120 (1979).
24. Gamble, D. S., A. W. Underdown and C. H. Langford. "Copper (II) Titration of Fulvic Acid Ligand Sites with Theoretical, Potentiometric, and Spectrophotometric Analysis," *Anal. Chem.* 52:1901-1908 (1980).
25. MacCarthy, P., and G. C. Smith. "Stability Surface Concept: A Quantitation Model for Complexation in Multiligand Mixtures," in *Chemical Modeling in Aqueous Systems*, E. A. Jenne, Ed. (Washington, D.C.: ACS Symposium Series 93, ACS, 1979), pp. 201-222.
26. Fitzgerald, P. J. MSPH Technical Report, Department of Environmental Sciences & Engineering, University of North Carolina, Chapel Hill, NC (1981).
27. Shuman, M. S., and L. C. Michael. "Ring-Disk Techniques for Studying Kinetics of Metal Complex Formation and Dissociation in Natural

Waters," in *Proceedings of the International Conference on Heavy Metals in the Environment*, (Toronto, Ont., 1975), pp. 227-248.

28. Hale, J. M. "Transients in Convective Systems II. Limiting Current and Kinetically Complicated Galvanostatic Transients at Rotating Disc Electrodes," *J. Electroanal. Chem. Interfacial Electrochem.* 8:332-349 (1964).
29. Levich, V. G. *Physicochemical Hydrodynamics* (Englewood, NJ: Prentice-Hall, 1962).
30. Albery, W. J., and M. L. Hitchman. *Ring-Disc Electrodes* (Oxford: Clarendon Press, 1971).
31. Rossotti, F. J. C., and H. Rossotti. "Potentiometric Titrations Using Gran's Plots–A Textbook Omission," *J. Chem. Ed.* 42:375-378 (1965).
32. Mark, H. B., G. A. Rechnitz and R. A. Grienke. *Kinetics in Analytical Chemistry* (New York: Wiley-Interscience, 1968).
33. Ferry, J. D. *Viscoelastic Properties of Polymers*, 2nd ed. (New York: John Wiley & Sons, Inc., 1970), Chapter 4.
34. Thakur, A. K., P. J. Munson, D. L. Hunston and D. Rodbard. "Characterization of Ligand-Binding Systems by Continuous Affinity Distributions of Arbitrary Shape," *Anal. Biochem.* 103:240-254 (1980).
35. Schwarzl, F., and A. J. Staverman. "Higher Approximations of Relaxation Spectra," *Physica* 18:791-799 (1952).
36. Smith, R. M., and A. E. Martell. *Critical Stability Constants, Vol. 4* (New York: Plenum Press, 1976).

CHAPTER 18

PHENOLIC ALDEHYDES AS INDICATORS OF THE ORIGIN OF HUMIC SUBSTANCES IN MARINE ENVIRONMENTS

Janusz Pempkowiak*

Institute of Oceanology
Polish Academy of Sciences
Sopot, Poland

Roger Pocklington

Atlantic Oceanographic Laboratory
Bedford Institute of Oceanography
Dartmouth, Nova Scotia, Canada

ABSTRACT

Nitrobenzene oxidation of humic substances isolated from soils, waters and sediments from the Baltic Sea and the Vistula River provides a number of phenolic aldehydes. The total yield of these varies from 0.4% for soil humics to 0.1% for marine dissolved humics. Based on the yield of phenolic aldehydes from the oxidation reactions, the percentage of land-derived humics in Baltic Sea environments is about 50%. A similar result is obtained from stable carbon isotope ratio measurements. Seasonal changes were noted in the concentration of humics in both marine and river waters.

*Present address: Atlantic Oceanographic Laboratory, Bedford Institute of Oceanography, Dartmouth, NS, Canada.

INTRODUCTION

Organic matter present in soils is essential to plant growth and plays an important role in the mobilization and transport of metals [1]. Particularly important constituents of organic matter in these respects are humic substances. Fractions of the humic matter in soils become dissolved in water and leached into freshwater streams [2]. As conditions within the freshwater regime change, the humic substances become further segregated. However, they bind, in the form of complexes, a significant portion of the trace elements dissolved in water [2], and have a considerable influence on the growth of organisms [3].

In the mixing of fresh and saline waters, a fraction of the dissolved humic substances–in particular, the higher-molecular-weight fraction–flocculates and eventually precipitates to the bottom sediments in the coastal zone [4]. As a result, organic matter in the open ocean tends to be primarily of marine provenance, whereas in coastal environments and landlocked seas the proportion of land-derived organic matter can be substantial [5]. Because of the important role it plays in a variety of processes (e.g., the formation of oil and gas), determining the sources of organic matter found in the marine environment is one of the most important problems in marine chemistry. Since organic matter of terrestrial origin includes lignin and its derivatives, land-derived organic matter may be distinguished from that of marine origin by the occurrence of aromatic structures which are difficult to derive from purely marine sources. Thermogravimetric methods have been used to show differences between fractions of humics on the basis of changes in aromaticity [1], but these methods do not provide sufficient resolution to allow for quantitative interpretation. Oxidation of organic matter within marine sediments and suspensions has proven to be a more successful means of evaluating the source of organic matter in sedimentary environments [6]. Phenolic aldehydes and acids present in the products of the oxidation provide quantitative information as useful as that obtained from isotopic ratios [7]. Since lignins are one of the precursors of humics on land, phenolic aldehydes produced by the mild oxidation of soil humics may be used as indicators of the origin and the composition of humics present in marine environments. This chapter, which concerns only humic substances, is the continuation of earlier work on the use of phenolic aldehydes as indicators of the origin of organic matter in the marine environment [6,8,9]. We have used this oxidation approach to determine the proportion of land-derived humic substances in the Baltic Sea off Poland.

MATERIALS AND METHODS

Sampling Sites and Procedures

Sampling sites are shown in Figure 1. The area under study represents a region marginal to the world ocean, yet interesting from the perspective of the processes occurring there. The average salinity of surface waters in the study area, both coastal and open sea, is 7‰, with variations not exceeding 0.2‰. The salinity of the bottom layer equals 10‰ and is also constant. Surface water is strongly influenced by river runoff, and bottom water is influenced by inflow of water from the North Sea. However, the exchange of water between the Baltic and the North Sea is very limited since it does not exceed 3% of the total volume of water in the Baltic per year. Open sea water samples of 100 to 200 liters were collected in a 12-liter plastic sampler in November 1979, and February and June of 1980. Coastal water samples of 23 liters were collected 20 times a month over the period of June 1978 through December 1979. River water samples of 100 to 150 liters were collected once a week over the period of January 1978 through June 1980. Sediment samples were collected in 1979 using a grab sampler of the Van Graf type. The sampler collected the upper (0 to 10 cm) layer of sediment. Four soil samples were collected from a depth of 30 to 70 cm at four different points located 2 to 15 km from the river and mixed subsequent to analysis. Suspended matter from seawater was collected by means of a sediment trap from a depth of 70 m at the location shown in Figure 1 during the period of September 1978 to July 1979. Suspended matter from river water was collected by decantation and centrifugation of the samples collected in the manner described above.

Two different species of phytoplankton that are predominant in the Baltic (*Chlorella vulgaris* and *Scenedesmus microspina*) were grown in Praft solution in the laboratory as axenic cultures. Cells were harvested by decantation and centrifugation once per week.

Preparation of the Amberlite XAD-2 Column

Amberlite XAD-2 resin (70 to 80 g) (Serva, West Germany and Rohm and Haas, USA) was extracted with acetone for about 40 hr and subsequently with methanol for the same period. A 60-cm glass column was filled with the resin to form a 2- X 30-cm resin bed. The resin was washed with 0.2 *M* HCl solution (16 hr) and finally with 1.0 *M* NH_4OH solution in a 50:50 mixture of double-distilled water and methanol. Prior to use, the resin was washed with double-distilled water.

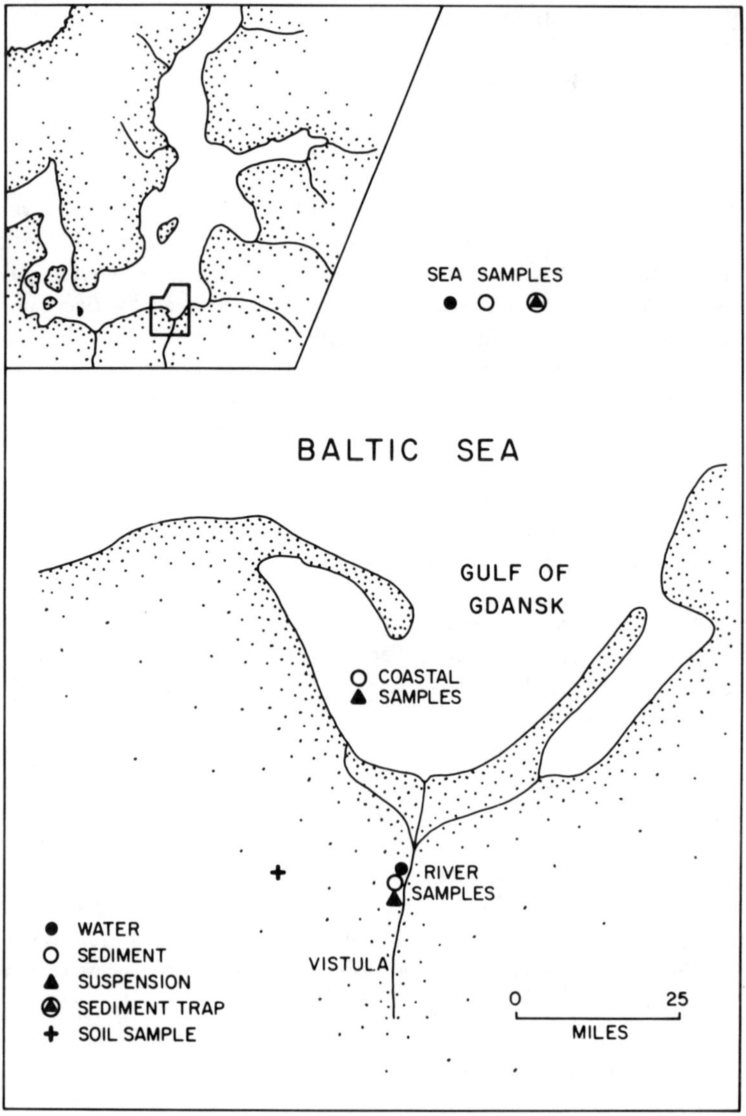

Figure 1. Map of the southern Baltic showing sampling sites.

Isolation of Humics

Water samples were filtered through glass fiber filters (GF/F; Whatman, U.K.) within 1 to 5 hr of collection and acidified to pH 1.7 with concentrated HCl. The acidified water was passed through the Amberlite

XAD-2 column at an average flow rate of 1.0 to 1.2 l/hr (≈0.2 bed volume per minute). After passing 200 to 300 liters, the resin was washed with distilled water until chloride ions could no longer be detected by precipitation of AgCl. Sorbed substances were then eluted with 1 *M* NH_4OH solution in a 50:50 mixture of water and methanol over 150 hr until the final volume of effluent reached 250 ml.

Sediment samples, soil and river particulate matter were washed with 0.1 *M* HCl and distilled water and then extracted with 0.2 *M* NaOH solution (1:25 solid:solution ratio) in glass chambers under nitrogen with continuous mixing. Several extractions at 60 °C were necessary to complete the removal of humic substances. The samples were centrifuged at 9000 g, and the supernatants combined and acidified to pH 1.7 with concentrated HCl. The precipited humic acids (HA) were separated; the solution containing fulvic acids (FA) was passed through the Amberlite XAD-2 column and the FA recovered in the way described for humics from water.

Oxidation of Humics

Portions of the humics (50 to 100 mg) were transferred to glass reaction vials and 1 ml of nitrobenzene and 5 ml of 2 *M* NaOH solution were added. The vials were placed in a constant-temperature block and held for 4 hr at 170 °C. The contents of each vial were washed into a separatory funnel (100 ml) using an equal volume of 2 *M* NaOH, the solids were resuspended, recentrifuged, and the supernatant was added to the contents of the separatory funnel. The solution was extracted twice with methylene chloride (2 X 5 ml) and the CH_2Cl_2 layer discarded. The aqueous layer was acidified to pH 1 with 5 ml of 6 *M* HCl and the phenolic aldehydes were extracted into methylene chloride (2 X 5 ml). The CH_2Cl_2 extract was dried over $CaSO_4$ (Drierite®), then filtered and washed (10 ml CH_2Cl_2) in a Buchner funnel on a glass-fibre filter. The filtrate was reduced in volume to 500 μl under N_2. An aliquot (2 μl) of this solution was injected into the gas chromatograph (Model 5731-A; Hewlett Packard, USA). The following conditions of separation were used: injector in splitless mode; injection port temperature 250 °C; column temperature 150 °C for 2 min increased to 240 °C at a rate of 4 °C/min. Using a capillary column (SP2100, 25 m long) the average retention times of phenolic aldehydes of interest were: vanillin, 5.2 min; p-hydroxybenzaldehyde, 8 min; syringaldehyde, 8.8 min. An example of a gas chromatogram of the oxidation products of river humic substances is shown in Figure 2. None of the blank peaks interfere with the compounds of interest and this facilitates quantitative interpretation of chromatograms. A minimum of 3 ng/μl of the aldehydes injected gives a recognizable signal. The precision of the method was tested with samples

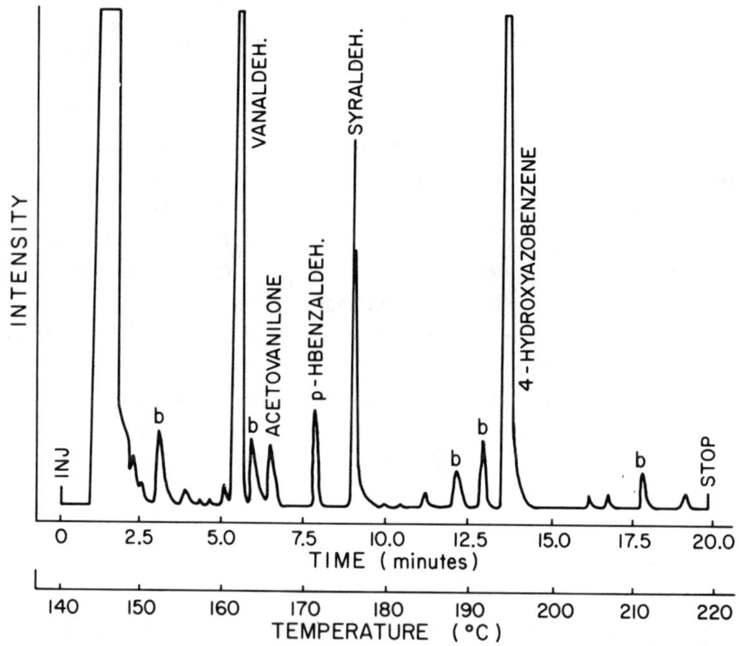

Figure 2. A typical gas chromatogram of the oxidation products of river dissolved humic substances (b marks peaks which appear in the blanks).

of river humic substances. The coefficient of variation of seven replicate oxidations was 6.2%. A linear relationship between concentration and integrated area in the range 0 to 500 µg/ml of each aldehyde was established by injection of standard solutions of aldehydes. The actual amount of aldehydes in the injected subsample was found from the calibration curves and then recalculated on the basis of the substance that had been taken for analysis and expressed as µg/mg.

Stable Carbon Isotopes Ratio Measurement

Samples of humics were acidified with 10% HCl for a minimum 5 hr at 50 °C and then transferred to precombusted (500 °C) porcelain boats. The humic materials were combusted and the resultant CO_2 gas was purified in a separate vacuum line containing Cu and MnO_2 at 470 °C to remove oxides of nitrogen and sulfur. Finally the $^{13}C/^{12}C$ ratios were measured on a 15 cm, 60^0 sector ratio mass spectrometer (Nuclide Corp., USA). The $^{13}C/^{12}C$ ratio is expressed in $\delta^{13}C$ notation defined by:

$$\delta^{13}C(‰) = \frac{(^{13}C/^{12}C)\text{ Sample} - (^{13}C/^{12}C)\text{ Standard}}{(^{13}C/^{12}C)\text{ Standard}} \times 1000$$

The standard used was PDB Cretaceous Belemnite and $\delta^{13}C$ is expressed in units of parts per thousand (‰). The accuracy as well as precision of the determination was given in the recent paper by Tan and Strain [10]. In their study NBS-21 standard graphite was subjected to the same oxidation and purification procedure used for these samples. The coefficient of variation was less than 0.03% and their results compared well quantitatively with those of various other workers [10].

RESULTS

The average concentration of humics isolated from different environments as well as the yields of phenolic aldehydes are presented in Table I. Vanillin constituted 70 to 80% of the total phenolic aldehydes. Syringaldehyde constituted 15 to 25%, where p-hydroxybenzaldehyde was usually 0 to 5% of total aldehydes. There was significant correlation between the yield of vanillin and the yield of syringealdehyde ($r = 0.96$). Other than in soil, p-hydroxybenzaldehyde rarely occurred in amounts allowing quantitative determination.

Seasonal changes of phenolic aldehydes presented in Figure 3 refer to changes in the content of the sum of phenolic aldehydes (ΣPA) in the oxidation products of humics isolated from river water and coastal water. The content of ΣPA in river humics ranges from 1.85 μg/mg in July and 1.96 μg/mg in June to 2.81 μg/mg in October and 2.76 μg/mg in December. Variation of ΣPA in coastal waters is more distinct and amounts to 2.14 μg/mg in February and 0.73 μg/mg in April and 0.62 μg/mg in September. In both cases, systematically higher values occur in fall and winter 1979 than in spring and summer 1979. Concentrations of ΣPA in humics isolated from open sea water are close to those in coastal waters and amount to 0.98 μg/mg, 1.50 μg/mg and 1.37 μg/mg in June, November and February respectively.

The percentages of terrestrial humics in various environments of the Baltic Sea are presented in Table II. In order to calculate the percentages in seawater the average value of ΣPA equal to 2.44 μg/mg was taken as characteristic of dissolved riverine humics. The ΣPA value characteristic of marine derived humics is difficult to obtain since no organic matter can be isolated from sea water or sediments and assumed to be of purely marine origin. Therefore organic matter isolated from two cultures of phytoplankton characteristic of Baltic water was tested

Table I. Average Yield of Phenolic Aldehydes (PA) in Oxidation Products of Humic Substances (HS) Isolated from Various Environments of the Baltic Sea

Origin and Fraction of Humic Substances	Number of Samples	Concentration of HS (mg/l)	Yield of PA (μg/mg)			
			Vanillin	Syringaldehyde	p-Hydroxy-benzaldehyde	ΣPA
Soil						
Humic acids	1[a]		3.74	0.86	0.31	4.91
Fulvic acids	1[a]		2.23	0.63	0.44	3.30
Vistula River						
Dissolved HS	17	7.9	1.92	0.46	0.06	2.44
Suspended HS	12	2.0	2.51	0.70	0.12	3.33
Coast waters						
Dissolved HS	22	2.5	1.21	0.18	0.00	1.39
Open sea						
Dissolved HS	3	1.6	1.09	0.15	0.00	1.24
Marine suspended matter	1[b]		0.18	0.00	0.00	0.18
Sediments						
Humic acids	4		1.31	0.13	0.02	1.46
Fulvic acids	4		1.63	0.32	0.09	1.97
Phytoplankton[c]	2		0.00	0.00	0.00	0.00

[a]Combined sample.
[b]Sample collected over period of 10 months.
[c]Axenic strains of the Baltic algae *Chlorella vulgaris* A1-76 and *Scenedesmus microspina* B2-76.

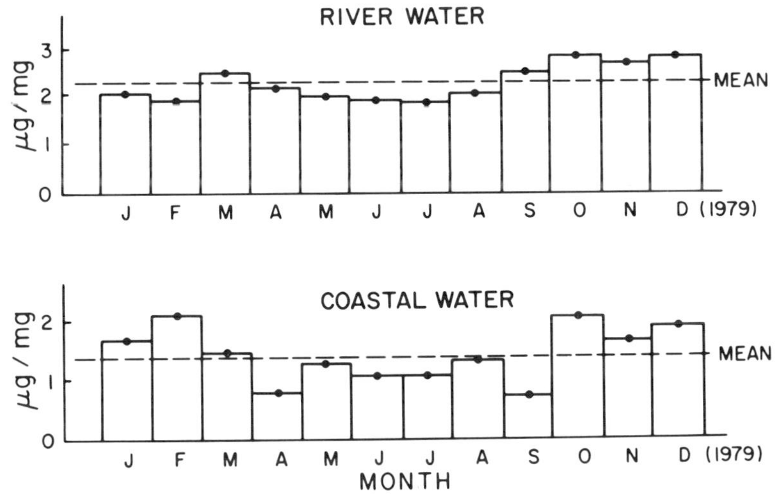

Figure 3. Seasonal changes in the sum of phenolic aldehydes in dissolved humic substances isolated from coastal and river water.

for phenolic aldehydes. ΣPA in the phytoplankton, and therefore in open sea humics, equals zero. It is assumed that values of ΣPA are proportional to the content of river humics in marine dissolved humics. The percentages of terrestrial humics in marine suspension and sediments were calculated on the basis of the average value of ΣPA in river suspended humics equal to 3.33 μg/mg and the assumption of proportional changes between ΣPA and the percentage of river humics in marine suspension and sediments. Again the assumption was made that ΣPA in marine end-members equals zero. These percentages were also calculated on the basis of $\delta^{13}C$ values, according to the following formula:

$$F_t(\%) = \frac{\delta^{13}Cs - \delta^{13}Cm}{\delta^{13}Cr - \delta^{13}Cm} \times 100$$

where $\delta^{13}C_s$, C_m and C_r represent values of $\delta^{13}C$ in sample, marine and river humics (Table II)

$\delta^{13}C$ values of terrestrial end-members are −28.36‰ for river dissolved humics and −29.02‰ for river suspended humics. The $\delta^{13}C$ value characteristic of marine derived humics is −23.73‰.

Table II. Percentage of Terrestrial Humic Substances (HS) in Various Environments of the Baltic Sea, Calculated on the Basis of the Sum of Phenolic Aldehydes (ΣPA) and the Ratio of Stable Carbon Isotopes ($\delta^{13}C$) in Average Samples

Environment	Average ΣPA (μg/mg)	Terrestrial HS Based on ΣPA (%)	Average $\delta^{13}C$ (‰)	Terrestrial HS Based on $\delta^{13}C$ (%)
Coastal dissolved humics	1.39	57	−26.46	59
Open sea dissolved humics	1.24	50	−26.13	52
Marine suspension	0.18	5.4	−23.73(−23.5)[a]	6.5 (0.0)
Sedimentary humics	1.42	42(54)[b]	−26.51	60
River HS				
Water	2.44	100	−28.36	100
Suspension	3.33	100	−29.02	100

[a]Assuming 6% contribution of land-derived organic matter to marine suspended organics.
[b]The value obtained after river dissolved humic substances were considered.

DISCUSSION

The first problem which had to be faced was that of the isolation of humics from sediments and water. Isolation of humics from soil and sediments was carried out by a standard procedure of alkaline extraction. Several consecutive extractions resulted in quantitative removal of humics. Isolation of humics from water was accomplished by using absorption on Amberlite XAD-2 resin. This method has been reported to give an excellent yield for both the sorption and desorption stages [11]. Our own laboratory studies indicated that on passing acidified water through the resin, up to 85% of those organic substances which absorb in the ultraviolet were removed. Usually not more than 85% of sorbed substances can be desorbed even when an alkaline solution of methanol is used [5]. Using our method, about 40% of organic carbon from river water and 25% of organic carbon from seawater could be removed [5]. All the analyses were performed on organic matter isolated from water, and it should be noted that all numbers refer to isolated substances.

Under these conditions of separation, a proportion of low-molecular-weight substances is isolated which could hardly be included within the humic fraction, namely fatty acids, aldehydes and polycyclic aromatic compounds [12]. The molecular weight distribution investigated by gel filtration shows that some 20% of dissolved humic substances exhibit molecular weights below 500 [5]. Accounting for the remaining yellow color of this fraction, further analysis of dissolved humics was performed on unfractionated material. Actually the results do not refer strictly to humics alone but rather to that fraction of organic matter isolated by means of sorption on Amberlite XAD-2. However, considering the low concentrations of fatty acids and other compounds in comparison to humics, we believe that this subtlety can be ignored.

As shown in Table I, the concentration of humics in seawater is much lower than that in river water, probably because of lower concentrations of organic matter in seawater as compared to river water, and the higher percentage of humic substances in organic matter dissolved in river water [5]. The fact that humic substances of high molecular weight flocculate in the mixing zone has also some influence on the differences in concentration between the concentrations of dissolved humics between river and sea [4,13].

Isolated humic substances were oxidized under mild conditions to give aldehydes and ketones as major products [1,9]. The yield of the three major phenolic aldehydes is listed in Table I. The proportion of aldehydes in the oxidation products of humics is much lower than those in wood [1] or even in organic matter from lignin-rich sediments [7]. Therefore, amounts of the order of 50 mg of humics must be oxidized in order to achieve sufficient precision.

Even though Baltic water is rich in humic substances [5,14], some 50 liters must be passed through the resin to provide sufficient amounts of substances for analysis. Applying the same method to ocean water would require the passage of hundreds of liters of water [15].

The total amount of phenolic compounds in oxidation products of organic matter is usually related to the plant material from which it came [6,9]. Phenolic aldehydes are a valuable source of information on the nature of the source material [7]. In our work, a different approach was taken. In calculating the amount of organic matter of different provenance it was assumed that dissolved humics from river water constitute part of the dissolved humics from seawater. Furthermore, it was assumed that humics originating in the marine environment do not give phenolic aldehydes upon nitrobenzene oxidation. Based on these assumptions, a simple method for estimating the percentage of land-derived humics was developed using the following formula:

$$F_t(\%) = \frac{\Sigma PA_s - \Sigma PA_m}{\Sigma PA_t - \Sigma PA_m} \times 100 = \frac{\Sigma PA_s}{\Sigma PA_t} \times 100$$

where F_t is the percentage of land-derived humics
ΣPA_s is the sum of phenolic aldehydes in the sample
ΣPA_t is the sum of phenolic aldehydes characteristic of rivers
ΣPA_m is the sum of phenolic aldehydes characteristic of the sea (which equals zero)

The highest proportion (57%) of terrestrial humics is found in coastal water. This is most reasonably related to the influence of river discharge. Surprisingly, however, the percentage of terrestrial humics in open sea water is only slightly lower (50%) than that in coastal water. The substantial exchange of water between the Gulf of Gdansk and the Baltic Sea, which exceeds fourfold the volume of water in the Gulf, as well as the system of currents giving a prevailing anticlockwise movement of the water masses in the Gulf [16], may be responsible for this effect. The total volume of water in the Gulf of Gdansk is some 700 km^3. Measurements in 1975 and 1976 indicated that between 2500 and 3000 km^3/yr of open sea water enters the Gulf.

The presence of small amounts of phenolic aldehydes in marine suspension indicates that they contain some terrestrial organic matter which, because it is more concentrated in the adjacent open sea sediments, must be more resistant to degradation than is marine organic matter. Considering that the sample was collected in the open sea, it is presumed that physical forces transport small particles of terrestrial organic matter to that site [17]. Such particles are generated, most probably, in the process of flocculation and are subsequently transported to the open sea.

Sedimentary humics of open sea samples consist of only 43% of terrestrial organic matter; the rest is of marine origin. Since the proportion of terrestrial humics in seawater is some 50%, one might expect the same proportion in marine sediments. The 43% found for terrestrial humics in sediments suggests that there must be another source of marine humics in sediments besides water. The most likely source is labile marine organic matter which, in the course of transformation, is converted into humic substances, naturally, of marine origin. This indicates that biochemical processes occurring in the surface layers of bottom sediments are capable of producing humics from organic matter of marine origin. Chemical and physical properties of humics isolated from bottom sediments of the Baltic Sea indicate changes in composition of these substances [18,19].

It should be emphasized that the percentages of terrestrial humics in the Baltic Sea are surprisingly high when compared with other coastal regions, for example, the Gulf of St. Lawrence [9]. A study of the organic carbon present in the Baltic Sea indicated that about 25% of dissolved organic substances are riverborne [18]. Recent results from Nyquist [14] indicated that as much as 20 to 30% of the so-called "humic substances" present in the Baltic Sea might be composed of lignin sulfonates produced in paper mills and discharged directly, or via rivers, to the sea. These compounds exhibit properties similar to humic substances and it is highly probable that they are isolated from seawater together with the latter [16].

Hedges and Mann [7] showed that the total amount of phenolic compounds in oxidation products of marine sedimentary organic matter can be correlated with stable carbon isotope ratios. The same method can be used to verify the percentages of terrestrial humics in the Baltic Sea environment. The $\delta^{13}C$ values, listed in Table II, suggest that there is a difference between substances isolated from land and marine environments. The difference can be expressed quantitatively assuming that humic substances in the Baltic Sea environment with different values of $\delta^{13}C$ comprise both land-derived ($\delta^{13}C_t$ = −29.02‰ for sedimentary and −28.36‰ for dissolved humics) and marine-derived ($\delta^{13}Cm$ = −23.73‰) humic substances, and the formula earlier defined is used [10].

Despite the differences between the methods used, similar results are obtained for dissolved humics isolated from both coastal and offshore waters. The percentage of terrestrial humics in marine sediments is much higher when calculated on the basis of $\delta^{13}C$. It can be attributed to the differences in ΣPA in dissolved humics and suspended humics from river water. If the average ΣPA of suspended humics and dissolved humics from river water were used, which is justified by the high stability of humics, the percentage of terrestrial humics calculated on the basis of ΣPA would be 54%, a value closer to that obtained on the basis of $\delta^{13}C$.

The $\delta^{13}C$ values of humics (in Table II) are low, as is to be expected of refractory organic matter [13]. The $\delta^{13}C$ values of river organic matter have been reported to range between −28.5 and −29.4 [20], whereas the resistant fraction of planktonic organic matter lies between −22.5 and −23.5 [21]. Humic substances isolated from Sargasso Sea water exhibit $\delta^{13}C$ values close to those of the resistant fraction [21], which justifies using the $\delta^{13}C$ of marine suspensions as a value characteristic of humic substances originating in the Baltic Sea. According to Skopintsev [13], refractory organic matter comprising lignins or/and other biopolymers such as cellulose is richer in the heavier isotopes in comparison to lower-molecular-weight amino acids and sugars. The $\delta^{13}C$ value of the labile fraction of planktonic organic matter is reported to range from −16.7 to −19.2 [21].

The agreement between results obtained by ΣPA and $\delta^{13}C$ methods is excellent. Hedges and Mann [17] reported a good correlation between phenolic compounds and $\delta^{13}C$ in their study and yet the calculated percentages of the terrestrial fraction of organic matter differed by as much as 5 to 10% between the two methods.

Seasonal changes of ΣPA in HS isolated from river water are presented in Figure 2. The higher values are observed during spring and fall and the lower ones in summer and winter. No correlation was found between water discharge and ΣPA. Correlation between the concentration of dissolved humic substances and ΣPA is also very poor. This would appear to compare to periods of enhanced input of terrigenous material: in spring after snowmelt and in fall with input of leaves and other debris from the previous growing season.

Much bigger seasonal changes of ΣPA were observed in dissolved humic substances from coastal waters. Lower values recorded in summer were more than 50% below the mean concentration and much lower than those observed in winter. This is probably due to a substantial inflow of dissolved organic matter during the period of increased primary productivity. These substances undergo biochemical degradation and produce substances which behave like humic substances but which are much more labile [13]. In summer their concentration remains constant mainly because of a steady influx of new material. In fall and winter, labile humics formed from marine organic matter are stabilized, and since there is no other organic matter available for their replacement, the concentration decreases.

The changes in concentration of dissolved humics are accompanied by changes in composition. The concentration of terrestrial humics and ΣPA in seawater remains constant, but the percentage of terrestrial humics in seawater has to change because the concentration of dissolved humics produced from marine organisms changes considerably [22]. The same effect was observed in a marsh lake in the Netherlands where concentration of dissolved humics changed from 3 mg/l to 14 mg/l mostly because of the inflow of fresh humic substances formed from precursors in the water, which subsequently were incorporated by the process of humification [23].

CONCLUSIONS

Phenolic aldehydes produced by mild oxidation of humic substances with nitrobenzene proved to be useful indicators of the origin of humic substances in the marine environment. However, the low yield of the oxidation, giving only some 0.1% aldehydes, requires oxidation of 50 to 100 mg of humic substances. This limits the applicability of the method to environments with fairly high concentrations of humic substances.

The percentage of terrestrial humic substances in the Baltic Sea calculated on the basis of the sum of aromatic aldehydes in oxidation products show that some 50% are land-derived. This percentage agrees within a few percent, with values calculated on the basis of stable carbon isotope ratio. Over twofold changes in the yield of phenolic aldehydes, recorded in the coastal zone of the Baltic Sea, are attributed to the changes in the composition of aquatic humus caused mostly by inflow of fresh marine humic substances during periods of increased primary productivity.

ACKNOWLEDGMENTS

We wish to thank Dr. F. C. Tan and his technical staff for performing the $\delta^{13}C$ determinations. Dr. Pempkowiak's work in Canada was supported by NRC of Canada postdoctoral fellowship program.

REFERENCES

1. Schnitzer, M., and S. Khan, Eds. *Humic Substances in the Environment* (New York: Marcel Dekker, Inc., 1972).
2. Gjessing, E., Ed. *Physical and Chemical Characteristics of Aquatic Humus* (Ann Arbor, MI: Ann Arbor Science Publishers, Inc., 1976).
3. Christman, R. F., and M. Ghassemi. "Chemical Nature of Organic Color in Water," *J. Am. Water Works Assoc.* 58:723-741 (1966).
4. Sholkovitz, E. "Flocculation of Dissolved Organic and Inorganic Matter during the Mixing of River Water and Seawater," *Geochim. Cosmochim. Acta* 40:831-846 (1976).
5. Pempkowiak, J., and G. Kupryszewski. "The Input of Organic Matter to the Baltic from the Vistula River," *Oceanology* 12:79-98 (1980).
6. Hedges, J. "Chemical Indicators of Organic Carbon Sources in Rivers and Estuaries," in *Flux of Organic Carbon by Rivers to the Oceans*, U.S. Department of Energy, Report No. CONF 8009140 (Springfield, VA: National Technical Information Service, U.S. Department of Commerce, 1980), pp. 109-141.
7. Hedges, J. L., and D. C. Mann. "The Lignin Geochemistry of Marine Sediments from the Southern Washington Coast," *Geochim. Cosmochim. Acta* 43:1809-1818 (1979).
8. Gardner, W., and D. Menzel. "Phenolic Aldehydes as Indicators of Terrestrially Derived Organic Matter in the Sea," *Geochim. Cosmochim. Acta* 38:813-822 (1974).
9. Pocklington, R., and C. MacGregor. "The Determination of Lignin in Marine Sediments and Particulate Form in Seawater," *Inter. J. Environ. Anal. Chem.* 3:81-93 (1973).
10. Tan, F., and P. Strain. "Organic Carbon Isotope Ratios in Surface Sediments in the St. Lawrence Estuary," *Estuar. Coast. Mar. Sci.* 8:213-225 (1979).
11. Mantoura, R., and J. P. Riley. "The Analytical Concentration of Humic Substances from Natural Waters," *Anal. Chim. Acta* 76:97-105 (1975).
12. Leonard, J., and N. Crewe. Personal communication (1980).
13. Skopintesev, B. A. "Refractory Organic Matter in the Marine Environments," in *Marine Organic Chemistry*, E. K. Dursma and R. Dawson, Eds. (Amsterdam: Elsevier Corp., 1981), pp. 125-177.

14. Nyquist, G. "Fluorescence Measurements in the Baltic Proper," *Proc. 11th Conf. Baltic Oceanog.* (Rostock, German Democratic Republic, 1978), pp. 101-118.
15. Stuermer, D. H., and G. H. Harvey. "Humic Substances from Sea Water," *Nature* 250:480-481 (1974).
16. Kowalik, Z. Personal communication (1978).
17. Emelyanov, E. M., and O. S. Pustelnikow. "Baltic Sea Coasts and Their Role in Sediment Supplying," in *Geology of the Baltic Sea*, V. Gudelis and E. Emelyanov, Eds. (Vilnus: Mokslas Publishers, 1976), pp. 141-159.
18. Bojanowski, R., and J. Pempkowiak. "Humic Substances in the Baltic Sea," in *Management of Environment*, B. Patel, Ed. (New Dehli: Wiley Eastern Ltd., 1980), pp. 246-260.
19. Pempkowiak, J. "Organic Substances in the Baltic," *Studia Materialy KBM* 17:205-230 (1977).
20. Williams, P. M., and L. J. Gordon. "Carbon 13; Carbon 12 in Dissolved and Particulate Organic Matter in the Sea," *Deep Sea Res.* 17:19-27 (1970).
21. Degens, E. T., M. Behrend, B. Gotthand and E. Reppman. "Metabolic Fractionation of Carbon Isotopes in Marine Plankton," *Deep Sea Res.* 15:11-20 (1968).
22. Pempkowiak, J., G. Kupryszewski and D. Skrobot. "Physical and Chemical Properties of Humic Substances in the Mixing Zone of the Vistula and the Gulf of Gdanisk," *Studia i Materialy KBM* 26:235-252 (1979).
23. De Haan, H. "Some Structural and Ecological Studies on Soluble Humic Compounds from Tjeukemeer," *Verh. Int. Ver. Theor. Angew. Limnol.* 18:685-695 (1972).

CHAPTER 19

INVESTIGATIONS ON HUMIC SUBSTANCES IN NATURAL WATERS

N. Plechanov, B. Josefsson, D. Dyrssen

Department of Analytical and Marine Chemistry
Chalmers University of Technology and University of Gothenburg
Göteborg, Sweden

K. Lundquist

Department of Organic Chemistry
Chalmers University of Technology and University of Gothenburg
Göteborg, Sweden

ABSTRACT

Gram quantities of water-soluble humic substances (WSHS) have been isolated from natural waters using an anion exchange resin as well as XAD-7 adsorbent. The WSHS samples were characterized by spectroscopic methods (UV, fluorescence, IR, ^{1}H- and ^{13}C-NMR) and elemental and functional group analyses. Molecular weight determinations indicated that the weight-average molecular weight of the WSHS was about 1.5×10^3; ionization of acidic groups resulted in an increase in apparent molecular weight due to electrostatic expansion. As evidenced by spectroscopic and analytical data, the samples from polluted waters contained a component which might be derived from lignosulfonic acids.

INTRODUCTION

The bulk of the organic matter in surface waters derived from natural sources consists of dark brown acidic polymers known as humic substances [1,2]. According to Reuter and Perdue [3], about 60 to 80%

of the dissolved organic carbon and particulate organic carbon are humic substances. Water-soluble humic substances (WSHS) consist of a great variety of relatively stable by-products of biological activity. WSHS are known to act as complexing agents for inorganic ions [3], to promote the solubilization of pesticides [4] and to be precursors for haloforms in chlorine treated waters [5]. According to Odén [6] and Page and Dutoit [7] WSHS can be divided into two fractions–fulvic acids (FA), soluble in water at pH 1, and humic acids (HA), precipitating from aqueous solutions with pH <1. The concentration of WSHS in natural waters is about 10 mg/l, which in most cases is too low to make it possible to investigate them directly in the aquatic environment. Therefore the examination of their physical and chemical properties requires a preconcentration step. In recent years two techniques have been widely used for this purpose, namely anion exchange and adsorption on molecular-adsorbing resins [8,9]. Acidic and negatively charged species are concentrated using anion exchange resins, while almost all types of organic compounds are concentrated on molecular-adsorbing resins. Both methods give mixtures of compounds defined as humic substances. Such an operational definition of WSHS makes it difficult to compare results obtained in different studies.

In this work, gram quantities of fulvic and humic acids were isolated from natural waters using an anion exchange resin as well as XAD-7 adsorbent. The samples obtained were subjected to a systematic investigation by physicochemical methods in order to better understand the structure and properties of water-soluble humic substances and their role in the aquatic environment.

MATERIALS AND METHODS

Extraction and Purification of WSHS

Extraction

Humic acids and fulvic acids were isolated from waters of Göta River (relatively polluted) and Finnsjön (lake, unpolluted). The principal scheme of the concentration procedures is shown in Figure 1. A series of humic acid (HA) and fulvic acid (FA) samples were obtained (Table I).

Preliminary studies of the adsorption and elution of the HA and FA (defined as in [6,7]) on XAD-7 showed that FA were completely (>95%) eluted with three bed volumes of dioxane-water (1:1) while only traces of HA were eluted with this eluent.

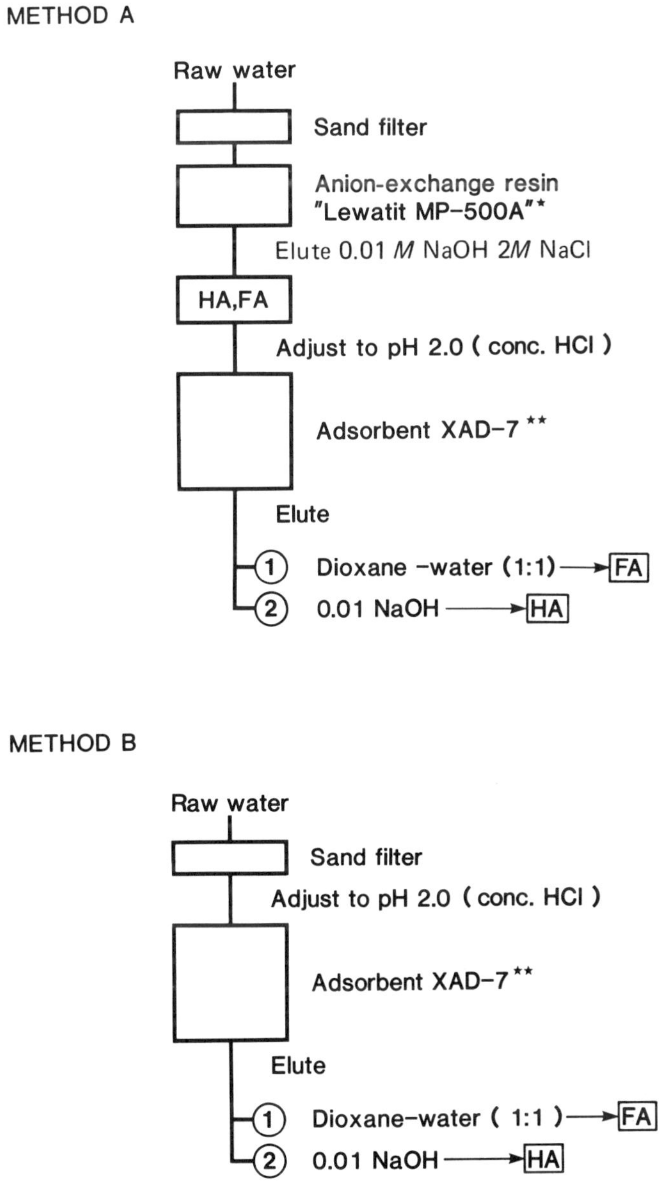

Figure 1. Concentration procedures (Methods A and B) used for isolation of humic acids (HA) and fulvic acids (FA) from natural waters. A. Using anion exchange resin Lewatit MP-500 (Bayer AG, Leverkusen Germany) as preconcentration step. B. Using adsorbent XAD-7 (Rohm and Haas Nordiska AB, Sweden).

Table I. Origin and Designation of Humic Acid and Fulvic Acid Samples Isolated

Sample	Lake/River	Concentration Procedure	Time	Yield (g)	Amount of Water Processed (ton)
FA-0	Göta River	A	May 1980	8.5	7.1
FA-1	Göta River	A	May–June 1981	15	27.3
HA-1	Göta River	A	May–June 1981	12	27.3
FA-2	Göta River	B	May–June 1981	2.1	1.7
HA-2	Göta River	B	May–June 1981	0.5	1.7
FA-3	Finnsjön	B	June 1981	0.3	1.6
HA-3	Finnsjön	B	June 1981	0.2	1.6

Sodium hydroxide solution was adopted for the elution of HA. HA were precipitated by acidification to pH 1.0 centrifuged off and dissolved in 70% ethanol. This solution and the solution of FA in dioxane-water (1:1) were used in the subsequent purification step.

Purification

The solution of FA/HA was passed several times through a strong cation exchange resin in H^+ form (Zeocarb 225*) until the concentration of sodium was less than 1 ppb. The solution was then lyophilized, dissolved in ethanol and centrifuged at 10,000 rpm in order to remove particles. In addition, a minute amount of a jelly was obtained, which tentatively has been ascribed a carbohydrate nature. The resulting FA/HA were precipitated into absolute ether in order to remove low-molecular-weight organic impurities. The precipitate was isolated by centrifugation, washed several times with absolute ether and dried in vacuo over P_2O_5. All samples were stored in a desiccator over P_2O_5.

Spectroscopic Analyses

UV spectra were recorded on a Hitachi spectrophotometer, model 100-60. Fluorescence spectra (excitation and emission spectra) were recorded on an Aminco Bowman SPF spectrofluorimeter. The procedure has been described in detail by Almgren et al. [10]. IR spectra were recorded on a Perkin Elmer model 399 instrument using KBr pellets. ^{1}H-NMR spectra were recorded with a 270-MHz instrument working in the pulse

*Permutit Co. Ltd., England

Fourier mode (Bruker WH 270). Solvents were dimethylsulfoxide-d_6, and pyridine-d_5-CD_3COOD (1:1) with tetramethyl silane (TMS) as an internal reference.

Determination of the Molecular Weight Distribution

Ultrafiltration was performed on a PM-10 Diaflo membrane (Amicon Corporation, USA) using an Amicon ultrafiltration cell, Model 402. The retention of solutes by the membrane was determined using the following organic compounds as standards for molecular weight: bacitracin (molecular weight 1400), insulin (molecular weight 5733), cytochrome-C (molecular weight 12,600) and ribonuclease (molecular weight 13,600). The cell volume was kept constant and the flushing solution had the same pH as the sample (washing technique). The retention factor R was calculated from the following equation:

$$\ln \frac{C_c}{C_0} = R \frac{V}{V_0}$$

where C_0 and C_c are the sample concentrations in the cell before and after flushing with volume V

V_0 is the constant initial volume in the cell

C_0 and C_c were determined by monitoring the absorbance at 300 nm.

Gel filtration chromatography (GFC) was performed on columns (2.5 X 35 cm) of Sephadex* G-25 and G-50, using water and water-dioxane (1:1) as eluant and on Sephadex LH-60 with dimethylformamide-acetic acid (200:1) [11] as eluant. The same compounds as in the ultrafiltration experiments were used for the calibration of the Sephadex G-50 columns with water as eluant. Tannic acid (molecular weight 1700, Fluka**) was used as reference in the experiments with Sephadex LH-60.

Potentiometric Titrations

The potentiometric titrations were carried out with the following equipment: Radiometer G 202 C glass electrode, Radiometer K 4040 calomel reference electrode, Solatron 7055 voltmeter (resolution 1 μl), thermostated titration vessel (t = 25.00 °C ± 0.05 °C). A Commodore PET 3016 computer, programmed in BASIC, controlled the voltmeter readings and the additions

---

*Pharmacia Products, S-751 04, Uppsala, Sweden.

**Fluka AG, CH-9470 Buchs, Switzerland.

from the burette. Hydrochloric acid (1.0 *M*) and sodium hydroxide (1.0 *M*) were used. All the titrations were performed in nitrogen atmosphere. The H^+ concentrations during the titration were calculated from the following equation

$$E = E_0 + 59.16 \log[H^+] + j_H[H^+]$$

where E_0 is an operationally defined standard electrode potential
$j_H[H^+]$ corrects for the change in liquid junction potential at low pH

E_0 and j_H were determined separately by titrating the medium (0.1 *M* NaCl) with solutions of 1.0 *M* HCl and 1.0 *M* NaOH.

RESULTS AND DISCUSSION

Elemental and Functional Group Analyses

The results of elemental and functional group analyses are summarized in Table II.

Table II. Elemental and Functional Group Analyses of Humic Acids and Fulvic Acids[a]

Sample	%C	%H	%N	%O	%S	%Ash	%-OCH_3	-COOH (meqv/g)[b]
FA-0	51.96	4.32	1.04	39.80	3.05	<0.02	6.5	4.25
FA-1	53.32	4.51	1.20	38.90	1.77	0.03	8.2	4.00
HA-1	56.00	4.51	1.36	37.25	0.72	0.37	6.0	4.25
FA-2	53.15	4.76	1.04	37.59	2.60	1.02	7.7	3.95
HA-2	55.29	4.53	1.26	37.08	1.78	0.25	6.1	4.15
FA-3	51.47	4.51	1.63	40.78	1.14	0.34	1.9	3.80
HA-3	55.00	4.30	2.10	n.d.[e]	0.8	n.d.	n.d.	3.33
FA-RT[c]	41.4	5.1	2.8	49.5	n.d.	1.18	n.d.	4.31
FA-OR[d]	51.1	3.62	1.13	n.d.	n.d.	1.0	n.d.	6.8
HA-OR[d]	53.4	3.73	2.10	n.d.	n.d.	4.3	n.d.	4.5

[a]Elemental, ash and methoxyl analyses were carried out by Prof. Dipl. Ing., Dr. H. Malissa, G. Reuter Analytische Laboratorien, Elbach (Germany) and by the Analytical Laboratory, Studsvik Energiteknik AB (Sweden).
[b]Determined in this work by potentiometric titration.
[c]Fulvic acids from River Tamar [12].
[d]Fulvic and humic acids from Oyster River [8].
[e]n.d. = Not determined.

It is clear from Table II that all the samples isolated in this work except FA-2 have a very low ash content. This allowed us to study the properties of the organic matter without interference from inorganic impurities. Carbon, nitrogen and hydrogen contents in our samples are close to those obtained from the Oyster River [8], concentrated using an anion exchange resin, and differ from those isolated from the River Tamar [12], concentrated using XAD 7.

All the samples in this work have relatively high methoxyl contents (1.9 to 8.2%). This may be interpreted as indicating that the samples originate from lignin. Substantial amounts of sulfur are also present in the samples (0.7 to 3.0%). The presence of methoxyl groups and sulfur may indicate the presence of lignosulfonic acids (more or less biochemically degraded). In this connection it is noteworthy that FA-3, which is obtained from a lake supposed to be unpolluted, has the lowest methoxyl content. Obviously, elemental and functional group analyses suggest that the humic materials isolated by the two methods used (A and B, Fig. 1) have similar composition; this is further supported by the other investigations made in this work.

Spectroscopic Studies

Ultraviolet and Fluorescence Spectra

The UV spectra of some HA and FA samples are shown in Figure 2. In aqueous solutions the spectra of HA and FA exhibit an absorption maximum at 205 nm and shoulders at 230 nm and 270 nm. The shoulder at 270 nm becomes more pronounced with increasing sulfur content. Since lignosulfonic acids exhibit an absorption peak at 280 nm [13], this might indicate the presence of lignosulfonic acids (or components derived from lignosulfonic acids) in the samples. However the pH-dependence of the appearance of the spectrum (see below) adds some uncertainty to this suggestion. UV spectra of FA and HA samples were also recorded in 0.1 *M* NaOH and 0.5 *M* HCl. In 0.1 *M* NaOH solution the spectra showed a maximum at 216 nm and a shoulder at 270 nm. When 0.5 *M* HCl was used as solvent the spectra exhibited a maximum at 205 nm and shoulders at 230 nm and 320 nm. Tentatively, the variation of the UV spectra with pH has been interpreted as indicating of the presence of arylconjugated acids [14]. The pH-dependence of the appearance of the spectra requires additional studies to obtain more specific structural information.

The fluorescence spectra of the HA and FA samples are shown in Figure 3. The spectra of all HA and FA samples show one broad band in the excitation spectrum (λ_{max} = 325 nm) as well as in the emission

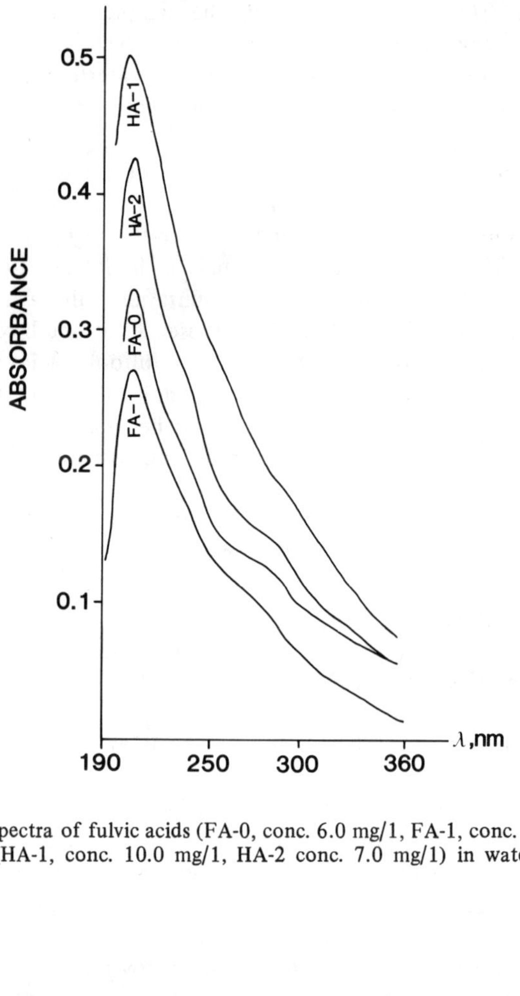

Figure 2. UV spectra of fulvic acids (FA-0, conc. 6.0 mg/1, FA-1, conc. 6.0 mg/1) and humic acids (HA-1, conc. 10.0 mg/1, HA-2 conc. 7.0 mg/1) in water. Cell length 1 cm.

spectrum (λ_{max} = 420 nm). The fluorescence spectra of the FA and HA agree with those obtained for a humic acid, isolated from the water of Göta River, using Chelex-100 loaded with iron(III) [15]. Studies of the homogeneity of the samples using the fluorescence technique described by Lundquist et al. [16] provided some evidence for the presence of a lignosulfonic acid component together with humic substances.

Infrared (IR) Spectra

The IR spectra of FA and HA studied are rather similar (Figure 4). The broad band at about 3400 cm^{-1} and the inflexion at about 2600 cm^{-1}

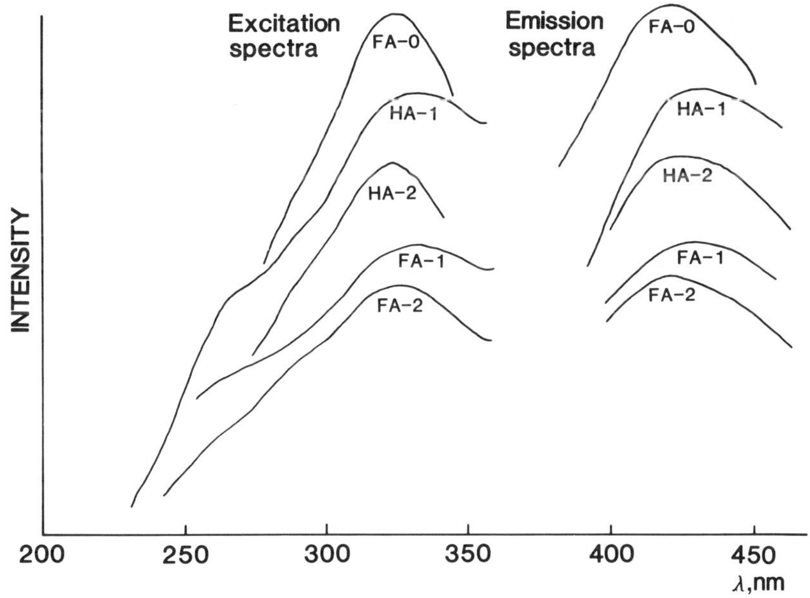

Figure 3. Qualitative fluorescence spectra of fulvic acids (FA) and humic acids (HA).

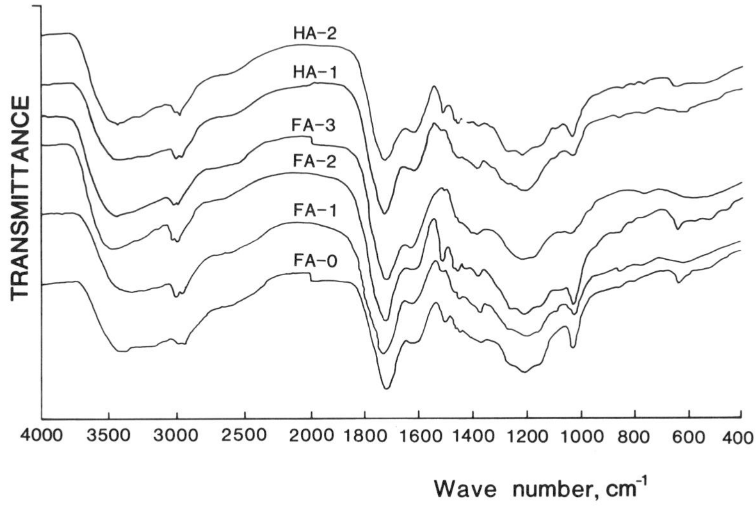

Figure 4. IR spectra of fulvic acids (FA-0,FA-1,FA-2,FA-3) and humic acids (HA-1,-HA-2).

are attributed to O-H stretching. The bands at 2990 cm^{-1} and 2970 cm^{-1} are due to C-H stretching. Carbonyl stretching gives rise to bands at 1725 cm^{-1} (unconjugated C=0) and 1620 cm^{-1} (conjugated C=0). Signals from aromatic nuclei (skeletal vibrations) probably both contribute to the latter band and are also responsible for the band at 1510 cm^{-1}. The complex band at about 1205 cm^{-1} presumably is due to C-O stretching of phenols and alcohols and O-H deformation in carboxyl groups. A fairly well defined band at 1040 cm^{-1} is primarily attributed to aromatic C-H in-plane deformation [13]. The band at 650 cm^{-1} may be due to S-O stretching as in sulfonic acids [13]. Interestingly, the sulfur content of the samples seems to match the intensity of this peak.

Finally, it should be pointed out that our IR spectra exhibited strong similarities in general appearance with those obtained by Beck and Reuter [17].

NMR Spectroscopy

A series of ^{1}H-NMR spectra of humic and fulvic acids in DMSO-d_6 solution are shown in Figure 5. All the spectra exhibit peaks at δ 1.18, 3.71, 4.03 and $\approx$7. In addition there are broad peaks in the region above δ = 7, which are due to signals from hydroxyl protons. The

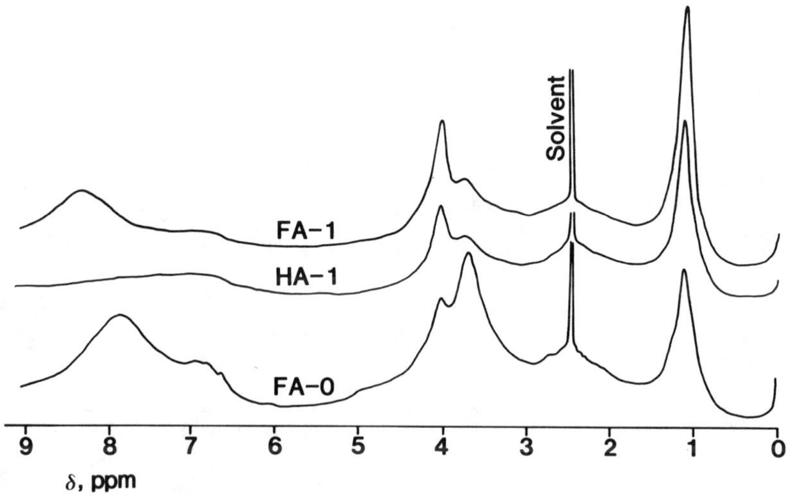

Figure 5. ^{1}H-NMR spectra of fulvic acids (FA-0) and FA-1) and a humic acid sample (HA-1). Solvent, DMSO-d_6.

peak at $\delta \approx 7$ can be attributed to aromatic protons. The presence of alkyl groups could explain the peak at $\delta = 1.18$ and the very broad peak in the region δ 1 to 3. Spectra of humic and fulvic acids show the same general appearance.

Signals from methoxyl groups ought to contribute to the 3.71 peak. The peak at δ 4.03 is most likely due to signals from >CH-O and $-CH_2$-O groupings. A series of small, equally spaced peaks (δ 6.89, 7.07 and 7.27) appeared in some of the spectra. These peaks are very small in the spectra shown in Figure 5. The peaks can be explained by signals from protons coupling with nitrogen nuclei. Such coupling is rather unusual but occurs for a spherical charge distribution on a nitrogen nucleus (e.g., amines in strong acid) [18]. A spectrum of FA-O in pyridine-d_5-CD_3COOD (1:1), shown in Figure 6, resembles those taken in DMSO solutions.

The broad peaks are consistent with complex and polymeric structures of the samples. The sharp peaks in the ^{1}H-NMR spectra of humic substances are invariably due to contaminants. Purification by precipitation into ether was found to efficiently release components which caused sharp peaks. With the exception of variations in the relative intensities of the peaks, the spectra shown in Figures 5 and 6 are very similar. As yet we have not found any satisfactory explanation for the variation in peak intensities. One possibility would be that the samples consist of two groups of substances and that fractionation during the workup procedures results in a variation of the proportions of these two groups of materials.

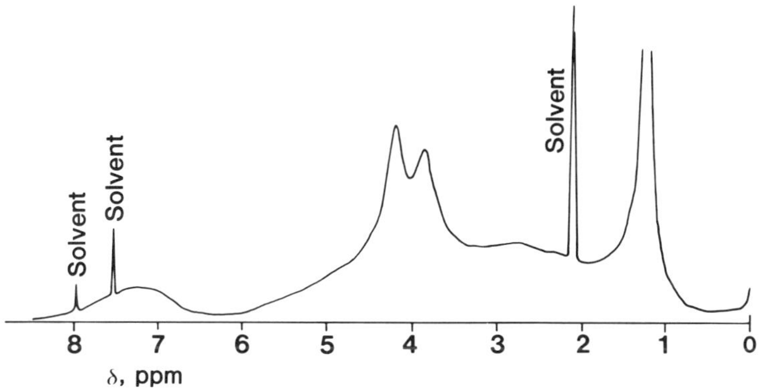

Figure 6. ^{1}H-NMR spectrum of a fulvic acid sample (FA-0) in pyridine-d_5–CD_3COOD(1:1).

^{13}C-NMR spectroscopy (DMSO solution) showed a broad peak at about δ 170 which could be attributed to carboxyl groups. In addition, peaks appeared at $\delta \approx 120$ (broad peak, aromatic carbon), δ 56.1 ($-OCH_3$), δ 51.6 ($-CH_2$-O), and $\delta \approx 30$ (very broad peak, alkyl groups).

Estimation of the Molecular Weight of WSHS by Gel Filtration Chromatography and Ultrafiltration

Gel filtration chromatography (GFC) has been used to study the molecular weight distributions of HA and FA.

There are usually two main kinds of gel-solute interactions operating in GFC experiments: coulombic interactions, caused by charged sites of the gel and solutes (marked when distilled water is used) and adsorption effects. The effect of the coulombic interactions usually can be eliminated by adding a charge-suppressing electrolyte to the elution system. Therefore, sodium chloride (0.1 *M*) was added to the eluant. The adsorption effects can be suppressed by adding a proper organic solvent to the system.

A series of GFC experiments has been carried out with Sephadex G-25. To describe the GFC properties of the solute we used a distribution coefficient, K'_D, which is defined, according to Lundquist and Wesslén [19], as

$$K'_D = (V_e - V_0)/(V_t - V_0)$$

where V_e is the elution volume of the solute
V_0 is the void volume of the column, determined by using Blue Dextran 2000
V_t is the available fractionation volume of the column, as determined by measuring the elution volume of acetone

The chromatograms of FA-1 obtained from Sephadex G-25 runs at different pH are shown in Figure 7. It is clear from Figure 7 that the elution peak of FA-1 shifts to higher K'_D values with decreasing pH. This is attributed to adsorption effects. The elution curve from the pH 7.9 run is expected to be closest to a presentation of the molecular weight distribution, since at this pH the adsorption effects are counteracted by far-reaching ionization of the molecules. This is supported by the fact that the elution curve obtained with water (0.2 *M* NaCl, 0.02 *M* HCl)-dioxane (1:1) as eluant resembles that obtained at pH 7.9.

It is clear from the Sephadex G-25 experiments that compounds with molecular weight <500 ($K'_D > 0.3$ [19]) are not present in FA-1 and this was found to be true also for the other samples examined in this work. It is notable that K'_D for FA is close to 1 at pH 2 (Figure 7),

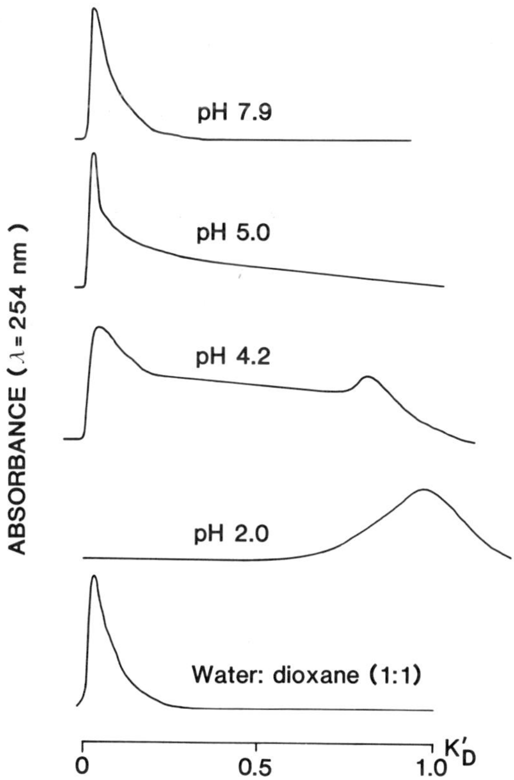

Figure 7. Gel filtration chromatograms (GFC) of a fulvic acid sample (FA-1) on Sephadex G-25. Eluants, 0.1 *M* NaCl (pH varied) and water (0.2 *M* NaCl, 0.02 *M* HCl): dioxane (1:1).

while K'_D for HA under the same conditions has been found to be $\gg$1 (pronounced adsorption). According to the results presented above, molecular weight distributions cannot be determined at low pH in aqueous solution using Sephadex gels; therefore, such investigations were made at pH 7.9 (Sephadex G-50, see below). In addition, the molecular weight distribution was examined, using the system Sephadex LH-60/dimethylformamide-acetic acid (200:1). The molecular weights of FA-0, FA-1 and HA-1 in water solutions (ionic strength 0.1, NaCl) at pH 2.0 and 7.9 were also evaluated, using ultrafiltration technique (PM-10 filter). The filter was calibrated by plotting log molecular weight against the

Table III. Average Molecular Weights of Fulvic Acids and Humic Acid Determined by Gel Filtration Chromatography and Ultrafiltration

Sample	Molecular Weight $\times 10^{-3}$	pH	Method of Determination
FA-0	1.5	2.0	UF, water, 0.1 *M* NaCl
	6.0	7.9	UF, water, 0.1 *M* NaCl
	4.5	7.9	GFC, Sephadex G-50, water, 0.1 *M* NaCl
	1.6		GFC, Sephadex LH-60, DMF-Acetic acid (200:1)
FA-1	1.4	2.0	UF, water, 0.1 *M* NaCl
	4.7	7.9	UF, water, 0.1 *M* NaCl
	4.9	7.9	GFC, Sephadex G-50, water, 0.1 *M* NaCl
HA-1	1.9	2.0	UF, water, 0.1 *M* NaCl
	4.5	7.9	UF, water, 0.1 *M* NaCl
	4.8	7.9	GFC, Sephadex G-50, water, 0.1 *M* NaCl

retention coefficient (R). The nominal molecular weight of the solute was determined from this plot.

The results of the molecular weight determinations by ultrafiltration together with those obtained by gel filtration are listed in Table III.

It appears from Table III that the average molecular weight of both FA and HA is about 1.5×10^3. At the point close to that of complete dissociation of carboxylic groups, the observed apparent average molecular weight is about 5×10^3. This indicates that an expansion of FA and HA polymeric molecules, caused by repulsion forces between groups of the same charge, takes place in water solution.

It is interesting to compare the weight average molecular weight (GFC,UF) with a number-average molecular weight. The number average molecular weight for FA-0 was determined to be 760.* The ratio weight average molecular weight/number average molecular weight is a measure of the breadth of the molecular weight distribution. For FA-0 sample this ratio was found to be close to 2, i.e., the value for the Flory most probable molecular weight distribution [20].

Potentiometric Titration Studies

Potentiometric titrations were performed for a number of FA and HA samples. A typical potentiometric titration curve is shown in Figure 8.

*Determined by vapor pressure osmometry, with dimethylformamide as solvent. Analysis was performed in Prof. Dipl.-Ing. Dr. H. Malissa, G. Reuter Analytische Laboratorien, Elbach (Germany).

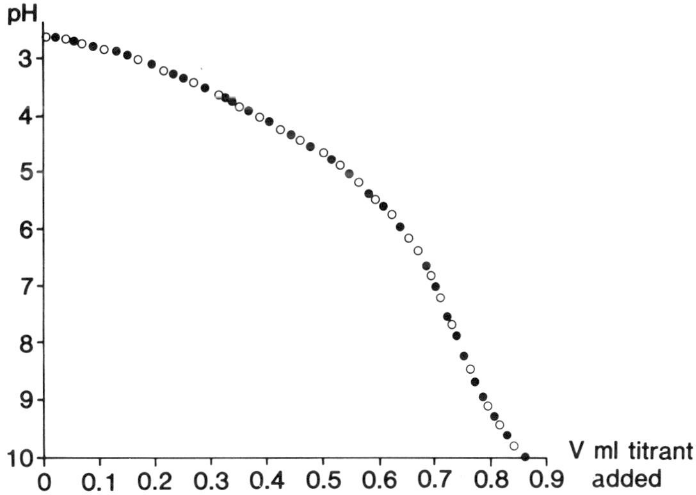

Figure 8. Titration curve of FA-1(185) mg) in water solution. Ionic strength 0.1 (NaCl). Initial volume 100 ml. Experimental points: ○-forward titration with 1.0 *M* NaOH, ●-reverse titration with 1.0 *M* HCl. V ml titrant added = (V ml NaOH added– V ml HCl added).

No hysteresis was observed between forward and reverse titrations. The equivalence point corresponding to the neutralization of the most acidic groups (mainly carboxylic) was estimated from the maximum of ΔpH/Δml titrant added. For all the FA and HA the equivalence point was around pH 7.9.

The quantitative examination of proton-binding equilibria in humic and fulvic acids requires a model taking into account their polyelectrolytic nature. Among several possible models [12] we found that the best results were obtained for the model based on a modified equation of Tanford [21]:

$$pH - \log\left(\frac{\alpha}{1-\alpha}\right) = p\overline{K}_i + n\alpha \qquad (1)$$

where α is the degree of dissociation of the polymeric material
n is a term responsible for electrostatic interactions

For chemically homogeneous polymers the value of $p\overline{K}_i$ is close to the pK of the repetitive monomeric unit (intrinsic constant). Because of the irregular and chemically heterogeneous structures of the humic and fulvic acids, it is likely that the value of $p\overline{K}_i$ represents a statistical weight average of the intrinsic constants. The value of α in every titration point was calculated according to equation:

$$\alpha = ([H^+] - K_W/[H^+] + C_{OH} - C_H)/C_t \qquad (2)$$

where K_W is the ionic product of water in 0.1 *M* NaCl solution
C_t is the concentration of the titrable groups, determined as the number of mmol of NaOH consumed before the equivalence point (forward titration)
C_{OH} is the $NaOH_{added}$ concentration in titration vessel
C_H is the HCl_{added} concentration in the titration vessel
$[H^+]$ is the concentration of protons

The value of $p\overline{K}_i$ was obtained by extrapolating $[pH - \log(\alpha/1 - \alpha)]$ to $\alpha = 0$ (Figure 9).

The parameters C_t, $p\overline{K}_i$ and n for a number of fulvic and humic acids are listed in Table IV.

It is clear from Table IV that all humic and fulvic acids have similar acid properties, i.e., the C_t, $p\overline{K}_i$ and n values are very close regardless of the isolation procedure by which they have been obtained.

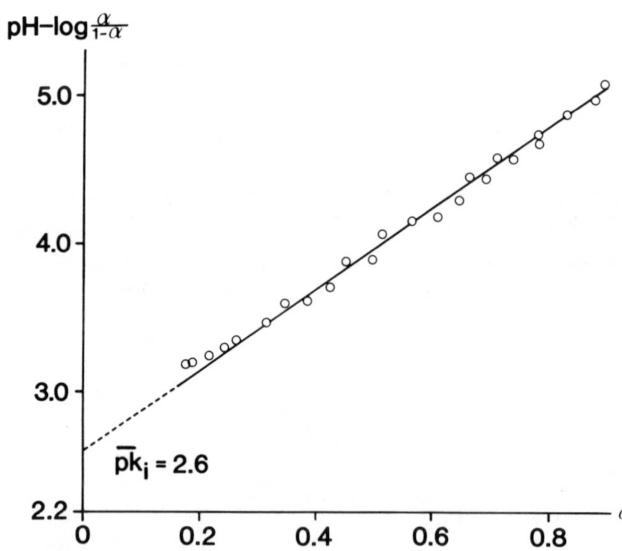

Figure 9. Plot of $[pH - \log(\alpha/1 - \alpha)]$ vs α for HA-1.

Table IV. Parameters Determined from Potentiometric Titrations of Fulvic Acid and Humic Acid[a]

Sample	C_t $\left(\frac{meq}{g}\right)$	$p\bar{K}_i$	n
FA-0	4.15	2.21	2.99
FA-1	4.00	2.30	2.80
HA-1	4.25	2.60	2.80
FA-2	3.95	2.35	2.76
HA-2	4.15	2.50	2.82

[a]Ionic strength 0.1, NaCl.

Using the parameters C_t, $p\bar{K}_i$ and n, pH in water solution (I = 0.1, NaCl) was calculated as a function of the HA concentration (Figure 10). The difference between the calculated pH values and those measured experimentally by titrating the medium (water, I = 0.1, NaCl) is ±0.005 pH units. The same agreement was observed for the other humic and fulvic acids studied. This indicates that the proposed model based on Equation 1 can be applied for a quantitative dissection of proton-binding equilibria for humic materials.

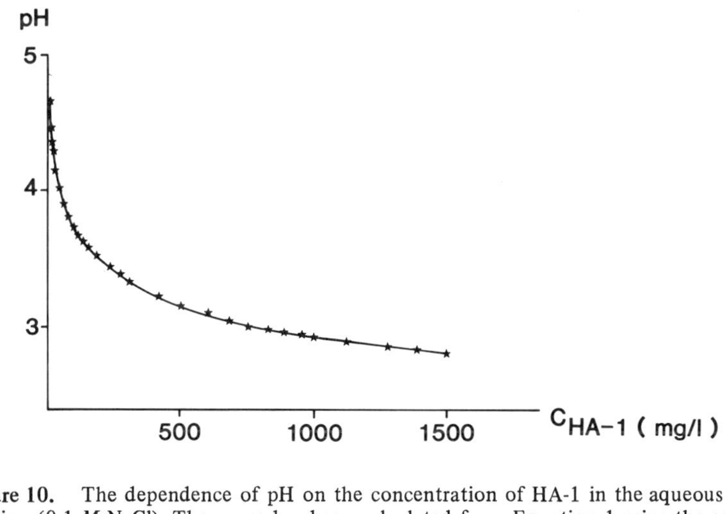

Figure 10. The dependence of pH on the concentration of HA-1 in the aqueous solution (0.1 *M* NaCl). The curve has been calculated from Equation 1 using the parameters given in Table IV, * = experimental points.

CONCLUSIONS

1. Concentration of water-soluble humic substances (WSHS) by using anion exchange resin or molecular adsorbent resulted in quantitative rather than qualitative differences in the composition of the materials isolated.
2. The composition of WSHS suggests that they originate at least in part from lignin.
3. The lignin component in WSHS from polluted waters might be derived from lignosulfonic acids.
4. ^{1}H-NMR results suggest that WSHS in addition to lignin-derived compounds contain varying amounts of a second group of materials characterized by the presence of alkyl groups.
5. The molecular weight distributions of humic acids and fulvic acids are rather similar.
6. A model based on the equation of Tanford is found to describe quantitatively the proton-binding equilibria for the water-soluble humic substances.

ACKNOWLEDGMENTS

The authors thank Mr. R. Hammarberg and Mr. Ö. Andersson (Department of Sanitary Engineering) for their help in isolation of humic substances and Dr. G. Nyquist (Department of Analytical and Marine Chemistry) for recording fluorescence spectra. The authors also thank Dr. Danielle Robert (C.E.N.G., DRF–Chimie Organique Physique, Grenoble, France) for recording a ^{13}C-NMR spectrum.

REFERENCES

1. Bordovskiy, O. K. "Accumulation and Transformation of Organic Substances in Marine Sediments," *Marine Geol.* 3:1-114 (1965).
2. Jackson, T. A. "Humic Matter in Natural Waters and Sediments," *Soil Sci.* 119(1):56-64 (1975).
3. Reuter, J. H., and E. M. Perdue. "Importance of Heavy Metal-Organic Matter Interactions in Natural Water," *Geochim. Cosmochim. Acta* 41:325-334 (1977).
4. Schnitzer, M. In: *Soil Organic Matter*, M. Schnitzer and S. U. Khan, Eds. (New York: Elsevier Scientific Publishing Company, 1978) pp. 1-64.
5. Trussel, R. R. and M. D. Umphres. "The Formation of Trihalomethanes," *J. Am. Water Works Assoc.* 70:604-612 (1978).
6. Odén, S. "Humic Acids," *Kolloidchem Beineftc.* 11:75 (1919).
7. Page, H., and M. Dutoit. "Studies on the Carbon and Nitrogen Cycles in Soil, III The Formation of Natural Humic Matter," *J. Agric. Sci.* 20:478 (1930).

8. Weber, J. H. and S. A. Wilson. "The Isolation and Characterization of Fulvic and Humic Acid from Riverwater," *Water Res.* 9:1079-1084 (1975).
9. Mantoura, R. F. C. and J. P. Riley. "The Analytical Concentration of Humic Substances from the Natural Waters," *Anal. Chim. Acta* 76:97-106 (1975).
10. Almgren, T., B. Josefsson and G. Nyquist. "A Fluorescence Method for Studies of Spent Sulfite Liquor and Humic Substances in Sea Water," *Anal. Chim. Acta* 78:411-422 (1975).
11. Lundquist, K., R. Simonson, and K. Tingsvik. "Investigation of Lignins from Soda and Soda/Antraquinone Cooking of Spruce Wood," *Paperi ja Puu* 63(11):709-712 (1981).
12. Varney, M. S., R. F. C. Mantoura, M. Whitfield, D. R. Turner, and J. P. Riley. "Potentiometric and Conformational Studies of the Acid-Base Properties of Fulvic Acid from Natural Waters," paper presented NATO Conference "Trace Metals in Sea Water," Sicily, March 30-April 3, 1981.
13. Sarkanen, K. V. and C. H. Ludvig, Eds. *Lignins, Occurrence, Formation, Structure and Reactions* (New York: John Wiley & Sons, Inc., 1971) pp. 241-293.
14. Aulin-Erdtman, G. and R. Sanden. "Spectrographic Contributions to Lignin Chemistry. IX. Absorption Properties of Some 4-Hydroxyphenyl, Guaiacyl, and 4-Hydroxy-3,5 dimethoxyphenyl Type Model Compounds for Hardwood Lignins," *Acta Chem. Scand.* 22:1187-1209 (1968).
15. Lundquist, K., B. Josefsson, and G. Nyquist. "Analysis of Lignin Products by Fluorescence Spectroscopy," *Holzforschung* 32:27-32 (1978).
16. Lundquist, K., I. Egyed, B. Josefsson, and G. Nyquist. "Lignin Products in Pulping Liquors and Their Fluorescence Properties," *Cellulose Chem. Technol.* 15:669-679 (1981).
17. Beck, K. C. and J. H. Reuter. "Organic and Inorganic Geochemistry of Some Coastal Plain Rivers of the Southeastern United States," *Geochim Cosmochim. Acta* 38:341-364 (1974).
18. Bible, R. H., Jr. *Interpretation of the NMR Spectra. An Empirical Approach.* (New York: Plenum Press, 1965) p. 66.
19. Lundquist, K. and B. Wesslén. "Gel Filtration of Lignin Model Compounds," *Acta Chem. Scand.* 25:1920-1922 (1971).
20. Yau, W. W., J. J. Kirkland, and D. D. Bly. *Modern Size-Exclusion Liquid Chromatography* (New York: John Wiley & Sons, Inc., 1979) p. 8.
21. Tanford, C. *Physical Chemistry of Macromolecules* (New York: John Wiley & Sons, Inc., 1963) p. 551.

CHAPTER 20

GEOCHEMISTRY OF HUMIC ACID DURING ESTUARINE MIXING

Lewis E. Fox*

College of Marine Studies
University of Delaware
Lewes, Delaware

ABSTRACT

Chemical analyses of the waters from various estuaries located on the east coast of the United States were made in order to examine the transport of aquatic humic substances. The analyses included determination of salinity, humic acid carbon, salt-precipated carbon, dissolved organic carbon, high-molecular-weight carbon, total amino acids, individual amino acids, elemental analyses, total carboyhdrate, total organic nitrogen, ammonia, nitrate plus nitrite and proton-NMR.

Interpretation of these analyses reveals the following:

1. Humic acids are removed from the water column in all the estuaries studied.
2. Humic acid removal affects 20% or less of the dissolved organic carbon transported through an estuary.
3. Removal of humic acid also removes from 30 to 60% of the high-molecular-weight organics in an estuary.
4. Salt-induced aggregation is not consistently responsible for humic acid removal.

---

*Current address: Center for Earth and Planetary Physics, Harvard University, Cambridge, Massachusetts.

5. The humic substances extracted by salt aggregation were compositionally different from humic substances extracted by acid precipitation although there were similarities in the protein compositions of both.

INTRODUCTION

The chemical behavior of organic matter in estuaries is not clearly understood. Ignorance in this area remains a major impediment to the determination of the quantity and chemical character of terrestrial organic carbon exported to the oceans. It appears that some terrestrially derived organic carbon undergoes aggregation and precipitation within estuaries [1–8] but the proportion of the total estuarine organic carbon this removal encompasses and the uniformity of that proportion among different estuaries is unknown.

A major component of the high-molecular-weight organic matter in lake and river waters is composed of humic substances [4,10–16] and the current opinion is that these substances are removed from estuarine waters [1,4,5,7,8,17]. Much of the evidence for the removal of terrestrially derived humics is based on spectrophotometric determinations of humic concentrations from various watersheds. Unfortunately, these methods of analysis could be subject to fundamental inaccuracies when applied to humic substances.

Essentially, spectrophotometric determinations of humic concentrations are done two ways: (1) by measurement of the absorbances of water samples at the wavelength of maximum humic absorption; or (2) by the preparation of solutions from humics extracted from water samples followed by measurements of absorbances of these solutions at an appropriate wavelength. Absorbances are generally converted to humic concentrations by calibration with a standard solution prepared from either aquatic or soil humic extracts.

The primary weakness of this procedure lies in the uncertainty that an equivalent amount of humic substance will always have an equivalent absorption value. Humic substances are composed of heterogeneous mixtures of organic compounds [18,19], thereby presenting the possibility that mixtures may vary from sample to sample. Since the total absorbance depends on the absorbances of the various components, the intensity may vary among samples or between standards and samples. Theng et al. [20] reported appreciable differences in visible absorptions between different preparations of soil humic acids. Similarly, Levesque [21] notes that fluorescence characteristics of humic compounds vary with molecular weights. Christman and Ghassemi [22] found that intensities of humic compound color and fluorescence were affected by pH. Shapiro [9] observed that humate solutions would darken with the addition of iron.

Changes in organic structure that reduce absorption cannot be distinguished from actual organic removal by spectral detection methods even if calibrated with organic matter extracted from the river under study. The natural occurrence of such structural changes may explain the observation that changes in aquatic humic acid color do not represent equal changes in organic carbon content [16].

In spite of the weakness of spectrophotometric analysis, it is unlikely that the observance of humic acid removal in estuaries is merely an analytical artifact. In fact, the removal of humic acids may be an important sink for terrestrial organic carbon for two reasons. First, the terrestrial abundance and physiochemical properties of humic acids suggests that their contribution to total carbon removed in an estuary is large. Second, the reported aggregation of humic acid in response to increases in ionic strength [6,23] tends to suggest a mechanism for their observed behavior. However, humic acids refer to an enormously complex class of organic compounds which have similar physiochemical characteristics but are structurally very diverse [18,19]. Better understanding of their role in the estuarine transport of terrestrial organic matter requires a systematic analytical approach supported by a better method of humic acid quantification.

The analytical approach taken in this study is based on comparisons of both the behavior and composition of organic removal products from estuaries with substantially different drainage basin environments. The approach is implemented by chemical analyses of the organic matter removed in selected estuaries, laboratory mixing experiments and field studies, and features a method for humic acid determination based on carbon analysis. The method combines extraction of humic acid via functional definition with total carbon analysis. The greatest advantage to this approach lies in its amenability to absolute standardization and the comparability of its results to other organic determinations such as dissolved organic carbon (DOC).

This chapter will demonstrate that (1) humic acid removal is a common estuarine phenomenon, but it is rarely caused by salt-induced aggregation, (2) humic acid removal has little effect on the quantity of carbon transported through an estuary although it has a substantial effect on the size distribution of estuarine organic matter, (3) the proportion of dissolved organic carbon composed of humic acid fluctuates with time and geographic location.

METHODS

Study Area

During the period from July 1979 to July 1980 the Delaware Bay, the Broadkill, Murderkill, St. Jones, Mispillion and Leipsic Rivers, which are tributaries to the Delaware Bay, and the Mullica River estuary were analyzed for dissolved humic acid (Figure 1). These watersheds were selected because they

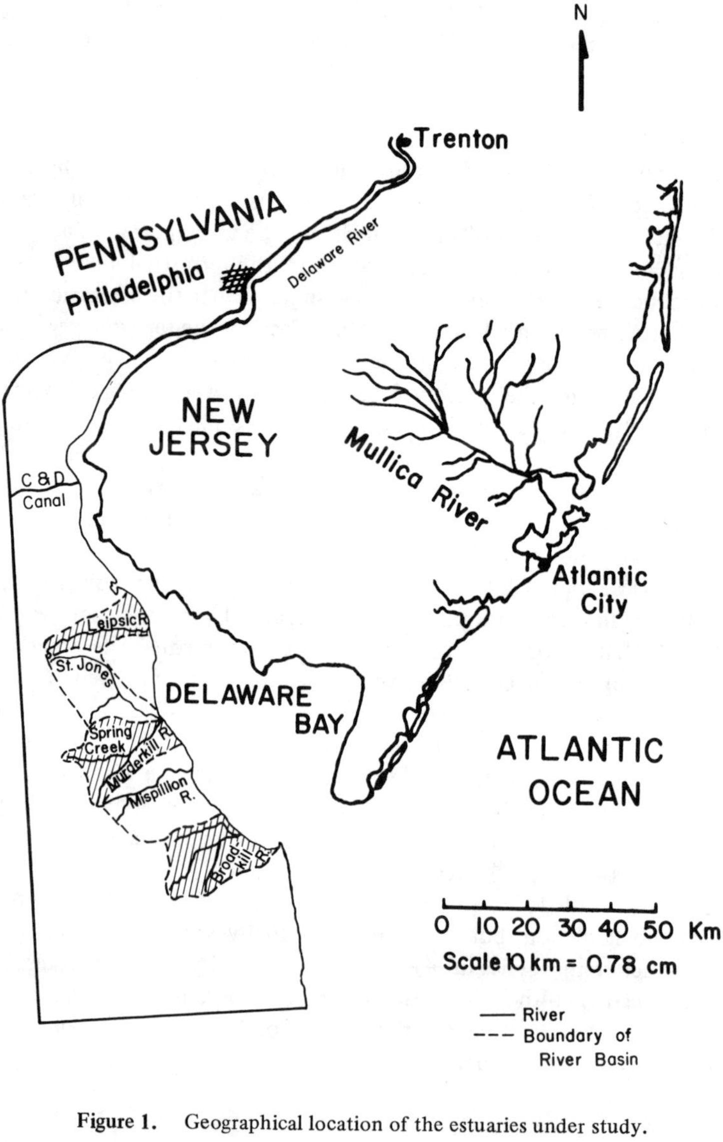

Figure 1. Geographical location of the estuaries under study.

offer a variety of soil and vegetative characteristics. The Mullica estuary was especially desirable for study because it has been reported to contain high concentrations of humic acid [24,25]. Details about the physical characteristics of these watersheds are given elsewhere [26].

Procedure

Field samples were collected in 2-liter polypropylene bottles with care to avoid the collection of surface material. Samples were immediately poisoned with mercuric chloride (1 ml/l).

All samples were filtered through Whatman glass fiber GFC filters (functional pore size 1.2 m) [27] to remove particulates immediately before humic acid analysis. All analyses were done within 48 hr of collection.

Dissolved humic acid was also measured during laboratory mixing experiments. Specific volumes of less than 0.5 ‰ (salinity), unfiltered river water were mixed with enough GFC-filtered seawater (salinity 33.4 ‰) to make solutions of desired salinities. Samples of 50 ml each were periodically taken from these solutions over a 3-hr period, filtered through GFC filters and measured for humic acid. This experiment was performed on the Broadkill, St. Jones and Mullica estuaries and repeated at six different salinities.

Salinities were determined with a Model RA-7A, Industrial Instrument induction salinometer. Analysis of dissolved organic carbon (DOC) was done by persulfate oxidation [28].

Both field and laboratory measurements of humic acid carbon were made using the following procedure. Duplicate 100-ml water samples were filtered through GFC filters. To 100 ml of filtrate, 0.8 ml of 1 *M* H_2SO_4 was added to aggregate humic acid. This lowered the pH of the sample to between 1.9 and 2.0. The acidified samples were then filtered through prebaked (450 to 500 °C, 2 hr) GFC filters immediately after acid addition. After filtration, the filters were dried in a vacuum desiccator and subsequently analyzed for carbon by automated Dumas method [29]. Machine filter blanks were determined from prebaked GFC filters wetted with 0.8 ml of deionized water.

Background carbon (carbon that adsorbs to the glass fiber filters without acid-induced aggregation) was determined from the y-intercept of a graph of humic acid carbon as a function of five different volumes of the same sample. This measurement was made separately for each estuary studied. Details concerning the quality control of the method are given elsewhere [26].

The extraction of organic matter proceeds as follows. River water was vacuum-filtered through a sandwich of three 12.5-cm-diameter Whatman #1 filters alternated with three Reeve Angle 93AH filters. The resulting filtrate was then refiltered through 4.25-cm Whatman GFC filters. The filter residue was composed of riverine particulates and discarded.

For extraction of humic acid (HA), the filtered sample was acidified to a pH of between 1.9 and 2.0 with concentrated H_2SO_4 and filtered through a Whatman GFC filter.

Humic acid residue was extracted from the filters by placing them in a 500-ml beaker with 100 ml 0.01 *M* NaOH solution and grinding with a magnetic stirrer. The resulting filter NaOH slurry was filtered through a wire mesh filter holder to remove ground filter particles. The resulting filtrate was washed with 0.01 *M* NaOH until the color was removed.

The humic acid solution was then pressure-filtered through a GFC filter in an ultrafiltration apparatus under 40 psi N_2 to remove pieces of filter which passed the wire mesh. Filtration was followed by acidification to pH 2.00 with 1 *M* H_2SO_4. The humic solution was dialyzed in 1000-molecular-weight dialysis tubing against EDTA for 12 hr and then against deionized water for 12 hr and then lyophilized.

The aquatic organic matter aggregated by the addition of seawater is called salt filter residue (SFR). Extraction of SFR follows the same procedure as extraction of humic acid (HA) except the acidification step is replaced by the filtration of a 50-50 filtered sample-filtered seawater mix.

The high-molecular-weight fraction of aquatic organic matter is called ultrafilter residue (UFR). For extraction of UFR, filtered river water was passed through a TRHK Millipore filter membrane ($>10^5$ molecular weight retention) under 40 psi N_2. Following filtration, the filter residue was washed from the ultrafilter membrane into 0.01 *M* NaOH solution rather than grinding. Occasionally the NaOH concentration had to be increased to dissolve more filter residue. The subsequent steps were identical to those of the other extractions.

The following chemical analyses were performed on the extractants: (1) total organic carbon; (2) total organic nitrogen; (3) total nitrogen; (4) nitrite plus nitrate; (5) ammonia; (6) total amino acid; and (8) proton-NMR (Figure 2).

Extracts were prepared for proton-NMR (^{1}H-NMR) in the following manner. Extracts were washed with dry ethyl ether five times. The washed extract was lyophilized and resuspended in 0.01 *M* NaOH solution, and then lyophilized and resuspended in high-grade D_2O (99.96% D). Samples were run on a Bruker WM 250, 250-MHz nuclear magnetic resonance instrument equipped with Fourier transfer.

Dried extracts were analyzed for carbon, nitrogen and hydrogen by automated Dumas on an elemental analyzer. The proportion C, N, H and S + O were computed on an ash-free basis from the elemental analysis. Sulfur plus oxygen was calculated by difference.

Total amino acid was determined by fluorometric analysis through fluorescamine amine derivatization after acid hydrolysis [26]. Total carbohydrate was determined by acid hydrolysis of polysaccharides to monosaccharides followed by reduction to their corresponding aldol and the subsequent cleavage of equivalent amounts of formaldehyde with periodic acid. Formaldehyde is subsequently determined colorimetrically with 3-methyl-2-benzothiazolinone hydrozonehydrochloride (MBTH) [30,31].

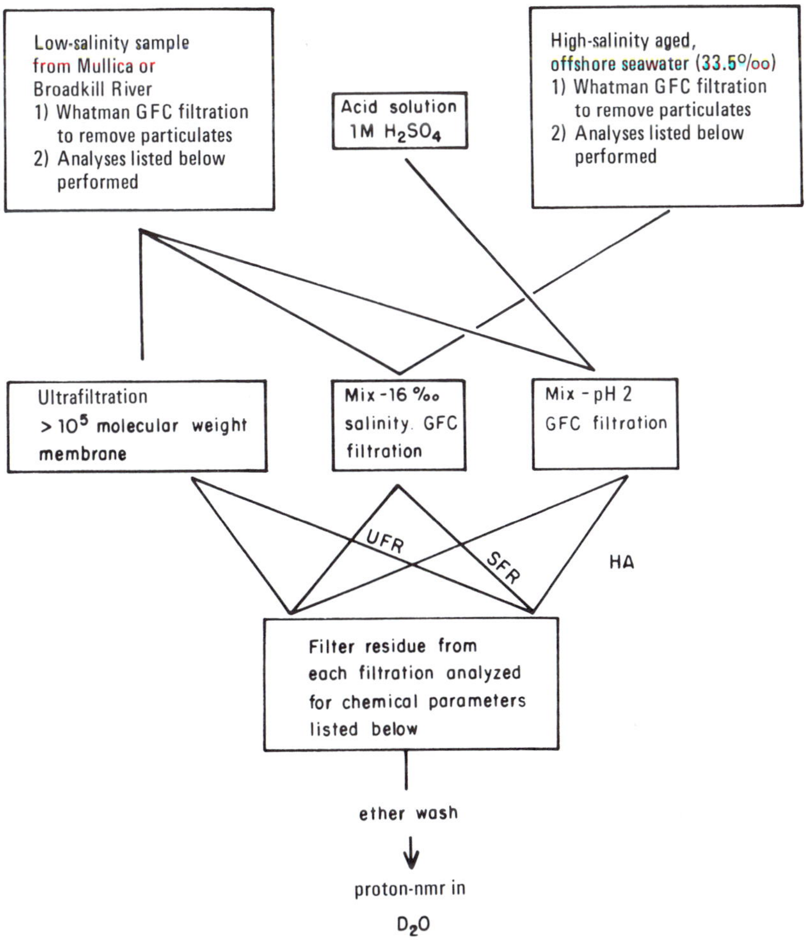

1. Total organic carbon
2. Total organic nitrogen
3. Total nitrogen
4. Nitrite plus nitrate
5. Ammonia
6. Total carbohydrate (hydrolyzable)
7. Total amino acid

Figure 2. The experimental flow diagram.

Analysis of total organic nitrogen was done by persulfate oxidation of organic matter followed by NO_3^- + NO_2^- quantification after Cu-Cd reduction [32].

Nitrite plus nitrate was determined by Cu-Cd reduction [33]. Ammonia was determined colorimetrically by the indophenol blue method [33].

RESULTS AND DISCUSSION

Field measurements of dissolved humic acid carbon are plotted as a function of salinity and illustrated in Figure 3. The negative deviation of data points from linearity suggests humic acid carbon is undergoing removal from all the estuaries studied. Such an interpretation is based on the conditions that the estuaries under study have only one major freshwater and saltwater source [34,35], and that they are in steady-state with respect to humic acid and salt inputs relative to the residence time of their waters [36]. These results correspond to observations of humic acid removal in other estuaries [5,7].

Graphs of dissolved organic carbon (DOC) as a function of salinity are presented in Figure 4. In spite of scatter, these plots are essentially linear, suggesting the conservative nature of DOC. Scatter probably results from small variable fluxes of carbon from upland drainage, surrounding marshes, in situ productivity and sediments. Such variations are common in environments of abundant organic matter such as estuaries.

Comparison of Figures 3 and 4 reveals that humic acid carbon is undergoing removal at salinities where DOC is either conservative or being added. Since humic acid carbon is part of the DOC, humic acid removal should theoretically be visible in the DOC vs salinity diagrams. The fact that it is not has two possible explanations. Either the humic acid carbon removed is too small to be resolved by DOC measurements, or humic acid carbon is not actually removed.

Table I gives humic acid carbon as a percentage DOC at the highest humic acid concentrations in various estuaries. It is evident from the table that humic acid carbon comprises less than 20% of the DOC. Such a percentage is smaller than the variability of DOC concentrations along the axis of the estuaries and is probably the most reasonable explanation for the inability to observe humic acid removal from DOC measurements.

Humic acid may be removed when no absolute loss of carbon occurs in the estuary. Since the measurement of humic acid depends on its aggregation at low pH, any chemical alteration in the molecular structure which would render it insensitive to pH change will appear as humic acid removal. In a sense, humic acid is removed since it no longer fits its operational definition,

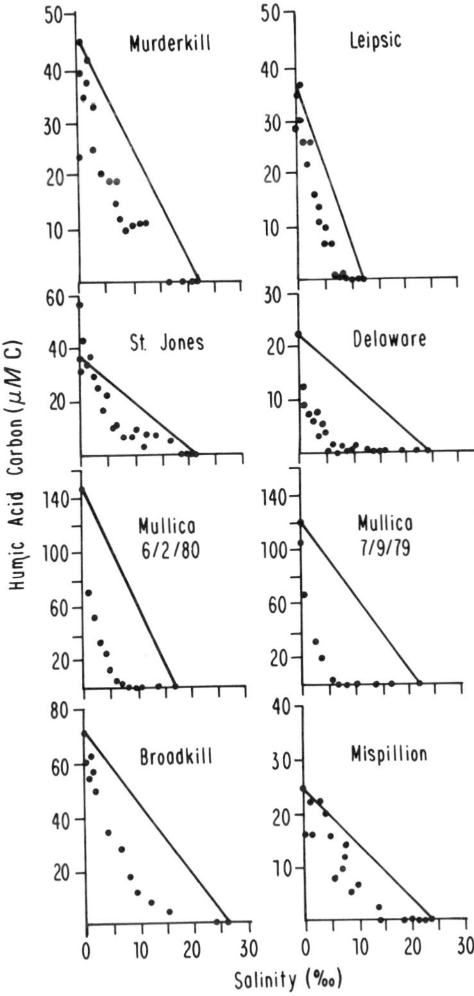

Figure 3. Humic acid carbon as a function of salinity. The standard error of the mean is 5%.

but the dissolved organic carbon composition of the estuary will remain conservative. Such molecular alterations could be caused by biological or mechanical interactions and would probably tend toward smaller molecules which are less likely to aggregate.

Two important conclusions can be drawn from the data at this point in spite of the uncertainty that an actual loss of terrestrial carbon is occurring

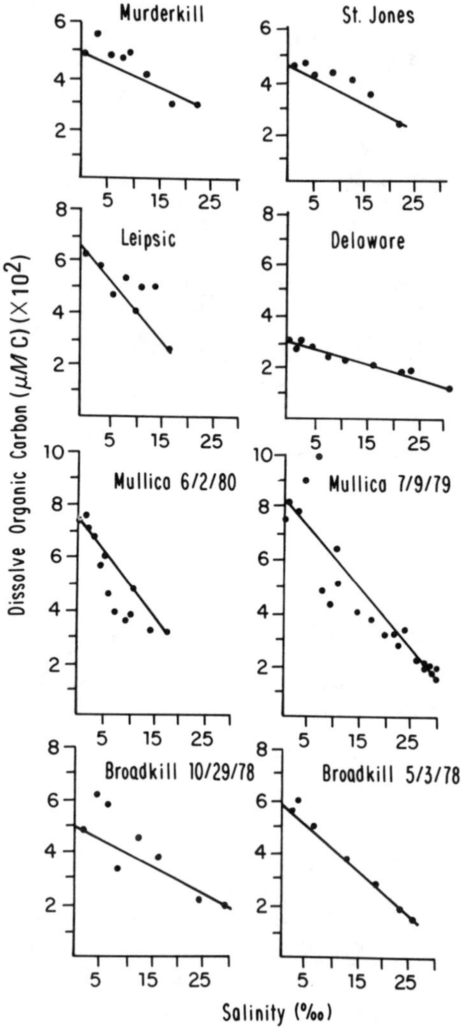

Figure 4. Dissolved organic carbon as a function of salinity.

within an estuary. First, the maximum loss of humic acid carbon in the estuaries under study is 20% or less. This indicates that the maximum loss of humic acid has little influence on the magnitude of carbon transported through an estuary. This does not imply the loss of humic acid is unimportant to the character of estuarine organic matter, however. Data that will be discussed later indicate that the removal of humic acid has a substantial effect

Table I. Range of DOC Values and Comparison to HAC

River	DOC Range[a] (μM C/l)	% HAC[b]
Mullica 6/2/80	750–340	20
Mullica 7/9/79	1000–150	14
Broadkill 10/24/79	620–210	14
Broadkill 7/80		18
Murderkill	570–290	9
St. Jones	460–230	13
Leipsic	540–270	5
Delaware	310–120	8
Amazon	1000–170	3–6 [7]

[a]Four Scottish rivers 1125-340 [5]; world average 420 [7].
[b]Determined at the maximum HAC concentration.

on the size distribution of organic matter within an estuary. Second, humic acid carbon is clearly not a constant proportion of the estuarine DOC. The proportions of organic constituents that compose the DOC of a particular watershed are probably functions of the biological characteristics of the drainage basin of that watershed. Such environments are highly variable with respect to location and season.

The kinetics of the aggregation of humic acid in response to increasing salinities were investigated by mixing experiments performed in the laboratory. Surprisingly, in light of the reported behavior of humic acid [6,37], humic acid concentrations measured at timed intervals over a range of salinities illustrated no consistent removal patterns with either variable. Iron, on the other hand, which was determined simultaneously, was consistently removed by a second-order kinetic mechanism [26,38]. The behavior of iron suggested that the variable behavior of humic acid was not an experimental artifact.

Plots of humic acid carbon as a function of salinity generated from the mixing experiments are presented in Figure 5. Clearly, humic acid is not removed (within 10% analytical error) in response to increasing salinity from two of three estuarine samples involved in the experiment. The results of these experiments indicate that salt-induced aggregation is not a universal removal mechanism for estuarine humic acid. It is altogether possible that humic acid is lost by salting out in one estuary and by biological or chemical degradation in another. Which is which could depend on the chemical composition of the humic acids, which in turn depends on the environmental characteristics of the watershed.

The geographic variability of the organic composition of estuarine humic substances has been further investigated by a suite of analyses performed on

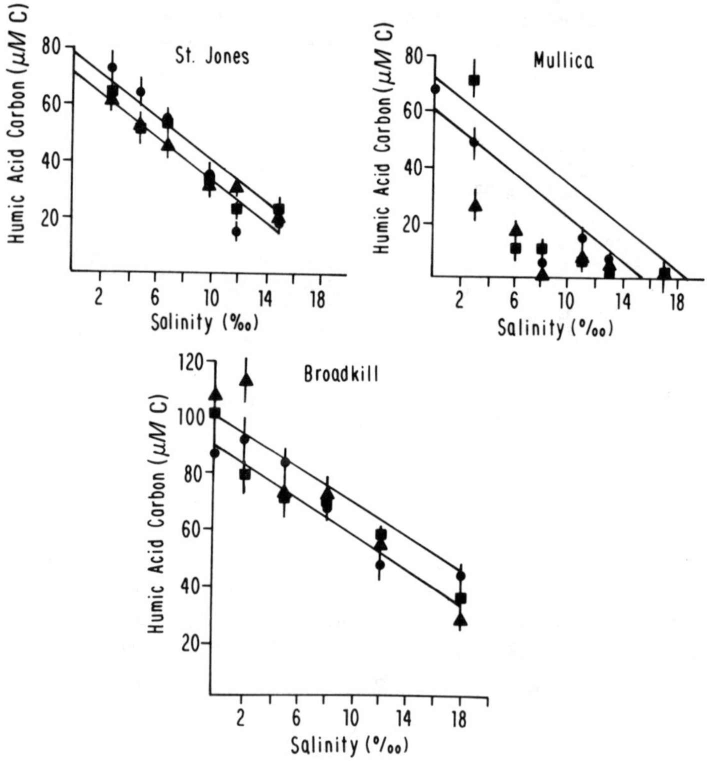

Figure 5. Humic acid carbon is plotted as a function of salinity. Humic acid measurements taken at 60 (●), 120 (▲) and 180 (■) min after mixing river water with seawater. The solid lines represent the expected boundaries of humic acid carbon that would result from dilution of the zero salinity sample.

humic acid and other functionally defined organics extracted from the Broadkill and Mullica Rivers (Figure 2). The results of these analyses are summarized in Tables II through IV and Figure 6.

Measurements of humic acid carbon, nitrogen, carbohydrate and protein are given as percentages of the total respective riverine constituents in Table II. Proportions of the riverine carbon, nitrogen and protein composed of humic acid are similar in both rivers. However, the proportions of DOC composed of humic and carbon (Table I) is temporally variable in both estuaries. Whether these proportions remain seasonally constant is unknown, but such consistency is unlikely in estuaries of vastly different climate.

Elemental composition data (Table III) indicate Broadkill humic acid to be more saturated with respect to hydrogen than humic acid from the Mullica.

Table II. Extract Carbon, Nitrogen, Carbohydrate and Amino Acid as a Percentage of Total Respective Riverine Constituent[a]

River Extract	% Organic C	% Organic N	% Carbohydrate	% Protein
Mullica HA	19.2 ± 2.5	24.1 ± 5.6	29.1 ± 8.9	42.9 ± 6.3
Broadkill HA	18.0 ± 3.2	19.8 ± 2.5	14.4 ± 1.3	53.3 ± 7.6
Mullica SFR	16.4 ± 3.9	8.0 ± 5.4	6.8 ± 1.7	19.4 ± 2.3
Broadkill SFR	2.8 ± 0.3	3.6 ± 4.5	B.D.[b]	B. D.
Mullica UFR	33.2 ± 4.2	51.7 ± 14.7	11.2 ± 2.9	27.6 ± 8.3
Broadkill UFR	25.1 ± 3.9	26.2 ± 7.3	18.2 ± 2.1	69.2 ± 7.5

[a]± Confidence interval at 90% probability level.
[b]B. D. = Below detection.

Table III. The Elemental Composition of Riverine Extracts and Various Humic Substances[a] (These values are the mean proportions of extract elements that form volatile oxides presented on an ash-free basis. The percentage of oxygen plus sulfur was determined by the difference between the mass of C + H + N and the ash-free mass of the sample. From the literature, sulfur generally represents less than 1%.)

Extract	% C	% H	% N	% O + S
Mullica HA	38.0 ± 1.8	3.8 ± 0.4	1.6 ± 0.6	56.6 ± 1.9
Mullica SFR	42.4 ± 0.9	5.1 ± 0.5	6.1 ± 0.6	46.4 ± 1.2
Broadkill HA	40.5 ± 4.6	5.8 ± 0.5	4.0 ± 0.4	49.7 ± 4.7
Broadkill SFR	12.5 ± 0.8	3.6 ± 1.0	2.5 ± 0.3	81.4 ± 1.3

[a]± standard deviation.

Table IV. The Extract Proportions of Amino Acids and Carbohydrate on a Carbon and Nitrogen Basis[a] (Amino acids were determined on an amine nitrogen basis and converted to carbon values by using the average amino acid C/N ratio of 3.37 as established from a suite of amino acids of known C/N ratio. Carbohydrate was measured, following acid hydrolysis, on a glucose basis and converted to moles of glucose carbon.)

River Extract	Amino Acids N as % Extract N	Amino Acids C as % Extract C	Carbohydrate C as % Extract C
Mullica HA	70.5 ± 18.5	8.6 ± 1.4	8.2 ± 1.4
Mullica SFR	70.3 ± 21.7	4.7 ± 0.9	2.3 ± 0.6
Broadkill HA	50.7 ± 4.2	13.2 ± 2.2	10.2 ± 1.1
Broadkill SFR	B.D.[b]	B.D.	B.D.

[a]± standard deviation.
[b]B.D. = below detection.

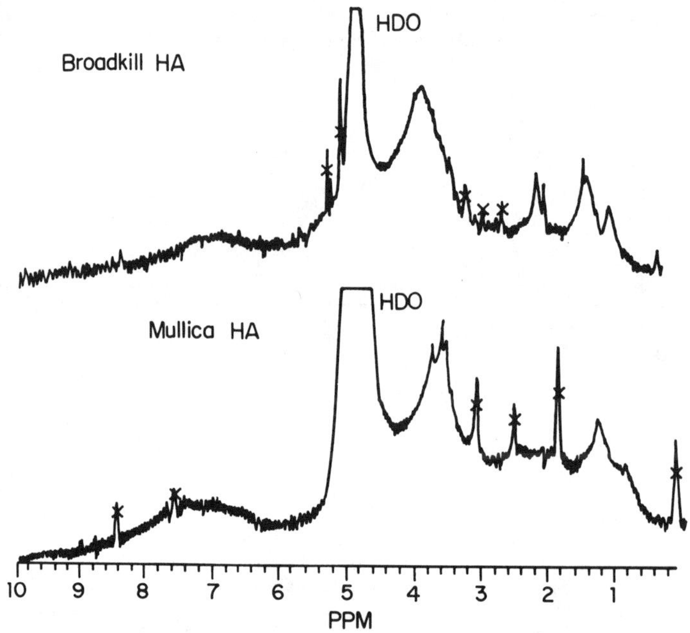

Figure 6. The ^{1}H-NMR spectra of Mullica and Broadkill humic acid. X denote contaminants.

In addition, the proportion of nitrogen in Broadkill humic acid is roughly twice that of the Mullica. It is suspected that the observed nitrogen difference is the result of a higher protein abundance in the Broadkill, but the analytical variability in the measurement of amino acid as a proportion of total nitrogen in the Mullica extracts (Table IV) is too large to allow verification.

Measurements of carbohydrate and protein carbon extracted from humic acid as percentages of the total humic acid carbon are presented in Table IV. Broadkill humic acid has a slightly higher proportion of amino acid and carbohydrate than the Mullica. This is consistent with the slightly greater abundance of aliphatic organic matter in the Broadkill as suggested by the elemental proportions (Table III). Amino acid plus carbohydrate carbon comprises less than 25% of the total humic acid carbon, however, necessitating the use of other analyses to compare the bulk of these substances.

^{1}H-NMR was performed on the humic acid extracts in an effort to obtain a more complete chemical picture. NMR spectra are illustrated in Figure 6. The general features of both substances are similar. Both spectra are characterized by broad peaks between 0.9 to 1.0 ppm, 3.2 to 4.2 ppm and 6 to

8 ppm. The most outstanding feature is the peak between 3.2 and 4.2 ppm. This is an area where hydroxyl protons resonate and is characteristic of carbohydrates. The idea has been suggested that much of the aliphatic portions of both extracts are composed of acid mucopolysaccharides [26]. Such carbohydrate would not be measured by the analytical technique used here to measure humic acid carbohydrate because of the resistance to hydrolysis characteristic of acid mucopolysaccharides.

The major differences between the NMR spectra of the two extracts are (1) a peak at 2ppm in the Broadkill but not observed in the Mullica, (2) the broad resonance between 2 and 3 ppm in the Mullica not evident in the Broadkill, and (3) a large aromatic peak shifted sightly downfield in the Mullica. The Broadkill peak at 2 ppm is characteristic of protons attached to carbons adjacent to carbonyl groups, methyl protons adjacent to phenyl rings, methylene protons adjacent to alkene carbons and the methyl protons of N-acetyl groups associated with aminoglucans. The broad resonance between 2 and 3 ppm in the Mullica could result from benzylic protons attached to a variety of side chains. The larger resonance between 6 and 8 ppm in the Mullica simply implies more aromatic protons, which is consistent with the lower hydrogen saturation previously discussed. The downfield shift suggests the aromatic component of Mullica humic acid has a greater abundance of electron with drawing groups such as carboxyl and carbonyl.

Generally, results of the comparison study suggest that humic acids extracted from the Mullica and Broadkill rivers have basically similar organic compositions, at least at the level of resolution the analytical techniques selected allow. Riverine proportions of carbon, nitrogen, and protein, abundance of amino acid and carbohydrate, and overall ^{1}H-NMR spectral morphology between the extracts are similar. The major differences are a higher abundance of aromatic compounds in the Mullica and a higher proportion of nitrogen in the Broadkill.

The chemical similarity of humic acid from the Broadkill and Mullica waters offers the possibility that their observed estuarine removal may be caused by a common mechanism. Removal experiments suggest salt-induced aggregation is not that mechanism. To further investigate the conclusion, the chemical characteristics of organic matter salted out of the Mullica and Broadkill waters were examined and compared to the humic acid characteristics previously discussed. If humic acid is undergoing aggregation and removal in response to increasing salinity, then salt extracts should have chemical characteristics similar to humic acid extracted from the same waters. The characteristics of the salt extracts (SFR) are summarized in Tables II through VII.

Large chemical differences between the humic acid and SFR extracts from both rivers are evident. Consistent with the results of the mixing experiment, very little SFR could be extracted from Broadkill waters. That which was

Table V. The Percentages of Humic Acid Carbon and Nitrogen Extracted from the SFR and UFR Fractions[a] (Humic acid was extracted from dissolved SFR and UFR solutions by acidification. The results of the extraction are reported in terms of the percentages of carbon and nitrogen extracted from the SFR and UFR fractions, and the percentage of the total riverine concentration of humic acid carbon and nitrogen obtained by extraction from the SFR and UFR fractions.)

River Extract	% HAC[b] in Extract	% HAN[c] in Extract	HAC in Extract as % total	HAN in Extract as % total
Mullica SFR	25.8 ± 4.0	70.6 ± 15.4	18.5 ± 2.2	32.2 ± 4.3
Broadkill SFR	6.9 ± 3.0	102.7 ± 19.3	4.0 ± 2.0	19.1 ± 4.1
Mullica UFR	33.4 ± 2.3	49.3 ± 5.1	108.5 ± 13.8	107.5 ± 22.1
Broadkill UFR	58.8 ± 8.1	85.4 ± 15.9	118.0 ± 50.0	129.0 ± 24.0

[a]± standard deviation.
[b]HAC = humic acid carbon.
[c]HAN = humic acid nitrogen.

Table VI. The Distribution of Amino Acids in the Mullica River and Mullica River Humic Acid and Salt Filter Residue (Concentrations are given in μmol N/l and extracts compared on an equal molar basis.)

Amino Acid	Total Riverine Amino Acid	Humic Acid	Salt Filter Residue
Cysteic Acid[a]	0.22	0.07	0.07
Aspartic Acid	0.36	0.42	0.35
Threonine[a]	0.19	B.D.	B.D.
Serine	0.24	0.28	0.32
Glutamic Acid	0.23	0.25	0.28
Glycine[b]	0.57	0.56	0.11
Alanine[b]	0.30	0.24	B.D.
Valine[b]	0.12	0.14	B.D.
Methionine	B.D.[c]	B.D.	B.D.
Isoleucine	0.08	0.10	0.07
Leucine	0.12	0.14	0.14
Tyrosine	0.04	0.04	0.04
Phenylalanine	0.05	0.07	0.07
Histidine[a]	0.02	0.06	0.11
Lysine	0.09	0.10	0.11
Arginine[a]	0.03	B.D.	B.D.
Proline[a]	0.15	B.D.	B.D.

[a]Amino acids in riverine sample with concentrations different from humic acid and salt filter residue.
[b]Amino acids concentrations different in humic acid than salt filter residue.
[c]B.D. = below detection.

Table VII. Order of Abundance of Amino Acids Making up >60% of the Total

Mullica River Sample	HA	SFR	Rivers from Beck et al. [15]
Glycine	Glycine	Aspartic Acid	Glycine
Aspartic Acid	Aspartic Acid	Serine	Aspartic Acid
Alanine	Serine	Glutamic Acid	Alanine
Serine	Glutamic Acid		Serine
Glutamic Acid			Threonine

extracted in sufficient quantity for measurement bears little chemical resemblance to Broadkill humic acid or Mullica SFR. The rest of the discussion will be directed primarily at Mullica SFR since Broadkill SFR appears so foreign.

SFR was extracted from the Mullica river in much greater abundance than from the Broadkill. Comparison of the proportions of SFR relative to total riverine constituents (Table II) shows that a generally smaller proportion of SFR than humic acid resides in the Mullica waters. Elemental proportions are also quite different (Table III). Mullica SFR is more saturated with respect to hydrogen than the corresponding humic acid. It also contains about three times the nitrogen. In addition, the amino acid and carbohydrate carbon composition of SFR differ from the corresponding humic acid proportions (Table IV). These compositional differences and the observation of a relatively small decrease in the percentage organic carbon compared to the large decrease in other constituents (Table II) suggest that most of the Mullica SFR is composed of substances other than humic acid. This is supported by direct measurements, which indicate that less than 30% of the Mullica SFR can be extracted as humic acid on a carbon basis (Table V).

Even though Mullica SFR and humic acid are not the same, it is evident they overlap. About 71% of the SFR nitrogen can be extracted by acidification (Table V), suggesting most of the SFR nitrogen may be from the humic acid pool. Since the distribution of SFR amino acid is about 70% of the nitrogen (Table IV), it is likely that humic acid nitrogen extracted from Mullica SFR is from protein.

This supposition is supported by the similarity in the amino acid distributions in Mullica humic acid and SFR (Tables VI and VII). Of seventeen amino acids, only glycine, alanine and valine show substantial concentration differences. With the exception of glycine, the order of abundance of amino acids composing >60% of the amino acid concentration in both extracts is the same (Table VII).

To summarize, the riverine organic matter which aggregates in response to increasing salinity is more variable in both composition and quantity than humic acid. The proportion of humic acid found in this organic fraction also appears to be variable, although there is some indication that humic acid and SFR protein are similar. Such variability offers a possible explanation for the observance of large humic acid removals caused by increasing salt concentrations in some waters [6] and none in others [26].

To assess the effect humic acid removal has on the size distribution of organic matter in an estuary, a high-molecular-weight fraction ($>10^5$ molecular weight) of dissolved organic matter was extracted from the Broadkill and Mullica waters. The total amount of humic acid measured in the waters under study (on both a carbon and nitrogen basis) could be extracted from the high-molecular-weight fraction of the respective rivers (Table V). Past research has also identified humic acid as high molecular weight substances [10,11,16,39–42]. On a carbon basis, 30 to 65% of the high-molecular-weight organic matter in these waters is humic acid (Table IV). Therefore, while less than 20% of the estuarine DOC may be lost (it is not clear that any carbon is actually removed), 30 to 60% of the high-molecular-weight organics are removed. Humic acid removal clearly has a substantial effect on the size distribution of the organic matter in estuaries.

CONCLUSIONS

1. Humic acids are removed in all the estuaries studied.
2. Humic acid removal affects 20% or less of the dissolved organic carbon transported through an estuary.
3. The proportion of dissolved organic carbon composed of humic acid carbon is variable with respect to time and geographic location.
4. Removal of humic acid substantially alters the size distribution of organics within an estuary.
5. Salt-induced aggregation is not the mechanism consistently responsible for most humic acid removal.
6. The compositional characteristics of organic matter extracted from the Broadkill and Mullica waters by salt aggregation were very different from those of the humic acids extracted from the same waters, although there was evidence that some Mullica humic acid was included in the Mullica salt extract.
7. Humic acids from watersheds with relatively different drainage basin characteristics are compositionally similar at the resolution of the chemical analyses employed.

ACKNOWLEDGMENTS

I wish to thank Dr. Jonathan H. Sharp for guidance during the completion of this work. I also wish to thank Drs. Thomas Church, Edward Sholkovitz,

George Harvey, Robert Biggs and John Castle for their many helpful ideas and criticisms. A special thanks to Richard Porter for his help with the fieldwork, and Dr. Steven Wofsy, Jane Fox and Deborah Carroll for their editorial comments on the manuscript. Also, thanks to Dr. Rodge Creecly for skillful operation of the NMR instrument.

This research was supported by NSF Grant OCE80-09835 (to J. H. Sharp) and funds from the College of Marine Studies of the University of Delaware. NMR analysis was funded by NIH Grant GM 27616 to the Department of Chemistry, University of Delaware.

REFERENCES

1. Swanson, V. E., and J. G. Palacas. "Humates in Coastal Sands of Northwest Florida," *U.S. Geol. Surv. Bull.* 1214-B:B-1B29 (1965).
2. Kalle, K. "The Problem of Gelbstoff in the Sea," *Oceang. Mar. Biol. Ann. Rev.* 4:91-104 (1966).
3. Sieburth, J. McN., and A. Jensen. "Studies on Algal Substances in the Sea, 1. Gelbstoff (humic material) in Terrestrial and Marine Waters," *J. Exp. Mar. Biol. Ecol.* 2:174-189 (1968).
4. Hair, M. E., and C. R. Bassett. "Dissolved and Particulate Humic Acid in an East Coast Estuary," *Estuarine and Coastal Mar. Sci.* 1:107-111 (1973).
5. Sholkovitz, E. R. "Flocculation of Dissolved Organic and Inorganic Matter During the Mixing of River Water and Seawater," *Geochim. Cosmochim. Acta* 40:831-845 (1976).
6. Eckert, J. M., and E. R. Sholkovitz. "The Flocculation of Iron, Aluminum, and Humates from River Water by Electrolytes," *Geochim. Cosmochim. Acta* 40:847-848 (1976).
7. Sholkovitz, E. R., E. A. Boyle and N. B. Price. "The Removal of Dissolved Humic Acids and Iron During Estuarine Mixing," *Earth and Planetary Sci. Lett.* 40:130-136 (1978).
8. Brown, M. "Transmission Spectroscopy Examination of Natural Waters, Part C," *Estuarine Coastal Mar. Sci.* 5:309-317 (1977).
9. Shapiro, J. "Effect of Yellow Organic Acids on Iron and Other Metals in Water," *J. Am. Water Works Assoc.* 56:1062-1082 (1964).
10. Black, A. P., and R. F. Christman. "Characteristics of Colored Surface Water," *J. Amer. Water Works Assoc.* 55:753-770 (1963).
11. Ghassemi, M., and R. F. Christman. "Properties of the Yellow Organic Acids of Natural Waters," *Limnol. Oceanogr.* 13:583-597 (1968).
12. Lamar, W. L. "Evaluation of Organic Color and Iron in Natural Surface Waters," *U.S. Geol. Surv. Prof. Pap.* 600-D (1968), pp. D24-D29.
13. Martin, D. F., M. T. Doig and R. H. Pierce. "Distribution of Naturally Occurring Chelators (Humic Acids) and Selected Trace Metals in West Coast Florida Streams, 1968-1969," Florida Dept. Natural Resources Prof. Paper Series 12 (1971).
14. Ishirvatari, R. "Chemical Characterization of Fractionated Humic Acids from Lake and Marine Sediments," *Chem. Geol.* 12:113-126 (1973).
15. Beck, K. C., J. H. Reuter and E. M. Purdue. "Organic and Inorganic Geochemistry of Some Coastal Plain Rivers of the Southeastern United States,"*Geochim. Cosmochim. Acta* 38:341-364 (1974).

16. Gjessing, E. T. *Physical and Chemical Characteristics of Aquatic Humus* (Ann Arbor, MI: Ann Arbor Science Publishers, Inc., 1976).
17. Swain, F. M. "Biogeochemistry of Humic Compounds," in *Proceedings of the International Meeting of Humic Substances* (Pudoc, Wageningen: Nieuwersluis, 1972).
18. Stevenson, F. J., and J. H. A. Butler. "Chemistry of Humic Acid and Related Pigments," in *Organic Geochemistry*, G. Eglington and M. T. J. Murphy, Eds. (New York: Springer-Verlag, Inc., 1969), pp. 535-557.
19. Schnitzer, M., and S. U. Khan. *Humic Substance in the Environment.* (New York: Marcel Dekker, Inc., 1972), 327 pp.
20. Theng, B. K. G., J. R. H. Wake and A. M. Posner. "The Humic Acids Extracted by Various Reagents from a Soil," *J. Soil Sci.* 18:349-363 (1976).
21. Levesque, M. "Fluorescence and Gel Filtration of Humic Compounds," *Soil Sci.* 113:346-353 (1972).
22. Christman, R. F., and M. Ghassemi. "Chemical Nature of Organic Color in Water," *J. Am. Water Works Assoc.* 58:723-741 (1966).
23. Ong, H. L., and R. E. Bisque. "Coagulation of Humic Colloids by Metal Ions," *Soil Sci.* 106:220-224 (1968).
24. Coonley, L. S., E. B. Baker and H. D. Holland. "Iron in the Mullica River and in Great Bay, New Jersey," *Chem. Geol.* 7:51-63 (1971).
25. Boyle, E. A., J. M. Edmonds and E. R. Sholkovitz. "The Mechanism of Iron Removal in Estuaries," *Geochim. Cosmochim. Acta* 41:1313-1324 (1977).
26. Fox, L. E. "The Geochemistry of Humic Acid and Iron During Estuarine Mixing," PhD Thesis, University of Delaware, Newark, DE (1981).
27. Sheldon, R. W. "Size Separation of Marine Seston by Membrane and Glass-Fiber Filters," *Limnol. Oceanog.* 17:494-498 (1972).
28. Menzel, D. W., and R. F. Vaccaro. "The Measurement of Dissolved Organic Carbon and Particulate Carbon in Sea-Water," *Limnol. Oceanog.* 3:138-142 (1964).
29. Sharp, J. H. "Improved Analysis for 'Particulate' Organic Carbon and Nitrogen from Seawater," *Limnol. Oceanog.* 19:984-989 (1974).
30. Burney, C. M., and J. McN. Sieburth. "Dissolved Carbohydrates in Seawater, 2. A Spectrophotometric Procedure for Total Carbohydrate Analysis and Polysaccharide Estimation," *Mar. Chem.* 5:15-28 (1977).
31. Johnson, K. M., and J. McN. Sieburth. "Dissolved Carbohydrates in Sea Water, 1. A Precise Spectrophotometric Analysis from Monosaccharides," *Mar. Chem.* 5:1-13 (1977).
32. Solorzano, L., and J. H. Sharp. "Determination of Dissolved Organic Nitrogen in Natural Waters," *Limnol. Oceanog.* 25:751-754 (1980).
33. Strickland, J. D. H., and T. R. Parsons. *A Practical Handbook of Seawater Analysis* (Ottawa, Canada: Fisheries Research Board of Canada, 1972).
34. Boyle, E. A., R. Collier, A. T. Dengler, J. M. Edmond, A. C. Ng and R. F. Stallard. "On the Geochemical Mass Balance in Estuaries," *Geochim. Cosmochim. Acta* 38:1719-1728 (1974).
35. Liss, P. S. In: *Estuarine Chemistry*, J. D. Burton and P. S. Liss, Eds. (1976), pp. 93-127.
36. Officer, C. B. "Discussion of the Behavior of Non-Conservative Dissolved Constituents in Estuaries," *Estuarine and Coastal Mar. Sci.* 9:91-94 (1979).

37. Fox, L. E., and S. C. Wofsy. "Kinetics of Removal of Iron Colloids from Estuaries." (in preparation).
38. Gjessing, E. T. "Reduction of Aquatic Humus in Streams," *Vatten* 1:14-23 (1970).
39. Gjessing, E. T. "Gel- and Ultramembrane Filtration of Aquatic Humus; A Comparison of the Two Methods," *Schw. Z. Hydrol.* 35:286-293 (1973).
40. Gjessing, E. T. "Use of 'Sephadex' Gel for the Estimation of Molecular Weight of Humic Substances in Natural Waters," *Nature* 208:1091-1092 (1965).
41. Schnitzer, M., and S. I. M. Skinner. "Gel Filtration of Fulvic Acid, and Soil Humic Compounds," in *Isotopes and Radiation in Soil Organic Matter Studies* (Vienna, Austria: International Atomic Energy Agency SM-106/2, 1968), pp. 41-55.

CHAPTER 21

AMINO ACID COMPOSITIONS AND ENANTIOMERIC RATIOS OF HUMIC-LIKE SUBSTANCES EXTRACTED FROM FOSSILS AND SEAWATER BY SEP-PAK C_{18} AND AMBERLITE XAD-2 RESINS

Ming-Shan Ho and Jeffrey Bada

Amino Acid Dating Laboratory
Scripps Institution of Oceanography
and Institute of Marine Resources
University of California, San Diego
La Jolla, California

Zuopeng Yan

Institute of Geochemistry
Academia Sinica
Guiyang, Guizhou Province
People's Republic of China

Deming Liu

Institute of Geology and Invertebrate Paleontology
Academia Sinica
Nanjing, Jiangsu Province
People's Republic of China

ABSTRACT

Amino acid-containing components were isolated from a number of natural systems, including fossil bones and shells, and seawater, using XAD-2 and SEP-PAK C_{18} resins. Preliminary results indicate that these methods preferentially isolate peptides or peptide fragments, but not free amino acids. Furthermore, the amino acids in the extracts from fossils appear to be stable with respect to chemical decomposition for periods of several hundred thousand years. The peptide bonds in the XAD-2- and SEP-PAK C_{18}-extractable components in fossils are fairly rapidly hydrolyzed, however. These results were used as a basis for interpreting the results obtained using the XAD-2 and SEP-PAK C_{18} methods to extract amino acid-containing components from seawater. Only surface seawater samples were found to contain significant quantities of SEP-PAK C_{18}-extractable amino acids. This may be because the peptide bonds in the material extractable from surface waters are hydrolyzed at a rate which is rapid in comparison to oceanic mixing times.

INTRODUCTION

Oceanic waters and sediments contain humic-like substances which differ in their general characteristics from humic materials found in terrestrial environments [1–6]. Very little is known about the sources, or mechanisms of formation, of marine humic substances, nor has the stability of this material with respect to chemical diagenesis been ascertained. The function of marine humic material in the overall global carbon balance has not been extensively evaluated. Humic-like materials may constitute the largest fraction of the total dissolved organic carbon present in oceanic waters [7]. Since the total dissolved organic carbon in deep ocean waters has an apparent radiocarbon age of several thousand years [8], humic-like substances in the ocean could be of considerable age [7,9]. The sources of marine humic acid-like material are thought to be either soluble components released from the decomposition of planktonic organisms or the end products of a series of condensation reactions involving simple organic compounds such as amino acids, carbohydrates, fatty acids, etc. [6,7].

One of the components of marine humic materials is amino acids. The geochemical study of amino acids has been an active area of research during the last decade or so, and as a result many amino acid diagenetic pathways have been evaluated, with the most extensively studied reaction probably being racemization [10–13]. By determining the amino acid composition and enantiomeric ratios of amino acids contained in marine humics it may be possible to gain an insight into the diagenesis of marine humic-like substances. One of the problems in using this approach is that amino acids are found to have greatly varying reactivities depending on the form (i.e., free or peptide-bound) in which they exist [11,13]. To obtain information on the rate of diagenesis of amino acids associated with humic-like materials we used the same processing procedures to extract humic-like substances from ocean waters and fossil shells and bones of various ages. By investigating amino acid diagenesis in the material extracted from fossils we hoped to obtain information about the diagenetic pathways for amino acids in materials which have properties somewhat similar to those of humic substances present in ocean waters and sediments.

MATERIALS AND METHODS

Fossil Shell and Bone Samples

Well characterized fossil *Chione* shells collected from various marine terrace localities in San Diego County, California, were used for these investigations. A map showing the various fossil shell localities is shown in Karrow and Bada [14]. The fossil bone samples which were analyzed came from Zhoukoudian, China, the site where the famous Peking Man fossils were found some 50 years ago [15]. Samples from the Upper Pleistocene (upper cave) and Middle Pleistocene (locality 1) deposits were analyzed.

Seawater Samples

Near-shore Pacific Ocean water samples were collected from the Scripps Institution of Oceanography (SIO) pier. Open water Pacific Ocean samples were collected off the coast of Peru; samples from 5 m (15°16.5'S;75°50.5'W) and 4150 m (15°22.0'S;76°02.3'W) were analyzed. After collection of a water sample, about 8 to 10 liters were filtered through glass fiber filter paper (GF/C, pore size 1 to 2μ) which had been combusted at 450 °C for 24 hr. Following the filtration the seawater samples were spiked with a norleucine internal standard, and in the case of the SIO pier samples, processed immediately. The Peru samples were frozen in plastic containers and then shipped to SIO, where they were subsequently analyzed.

Materials

Because of the ubiquitous nature of amino acids it is very important to use only highly purified reagents and ultraclean glassware to minimize contamination. Water was double-glass-distilled and 6 *M* HCl was prepared by two successive azeotropic distillations. Ammonium hydroxide was prepared by dissolving gaseous ammonia in double-distilled water. All organic solvents were redistilled in glass. Sodium chloride used to make blanks was combusted at 450 °C for 24 hr. All glassware was soaked in potassium dichromate/sulfuric acid and exhaustively rinsed with deionized and double-distilled water prior to use.

Amberlite XAD-2 resin (20 to 50 mesh) was obtained from Mallinckrodt and cleaned by refluxing with methanol for 4 hr followed by Soxhlet extraction with methanol for 24 hr. The cleaned XAD-2 resin was then packed in a glass column, washed with 1.5 *N* ammonium hydroxide followed by several column volumes of methanol and finally distilled water. SEP-PAK C_{18} cartridges were obtained from Waters Associates, Inc. (Milford, MA). The cartridges were cleaned and activated by eluting with 20 ml each of 1 *N* HCl, methanol, 1.5 *N* ammonium hydroxide and double-distilled water.

Blank controls for the seawater analyses were carried out using a 35‰ NaCl solution prepared by dissolving NaCl in double-distilled water. The blanks analyzed had the same volume as the seawater samples and were processed in a similar manner. The blank controls for the fossils consisted of simply carrying out the standard processing procedure, with the exception that the fossil sample was omitted.

Analytical Procedures

The general procedures used to extract amino acid containing humic-like materials from fossils and seawater are shown in Figure 1. Following the isolation by either the XAD-2 or the SEP-PAK C_{18} method, the extracts were hydrolyzed in 6 *M* HCl for 24 hr and the amino acid compositions were determined, as were the D/L enantiomeric ratios of several amino acids. Amino acid compositions were determined on a Beckman Spinco

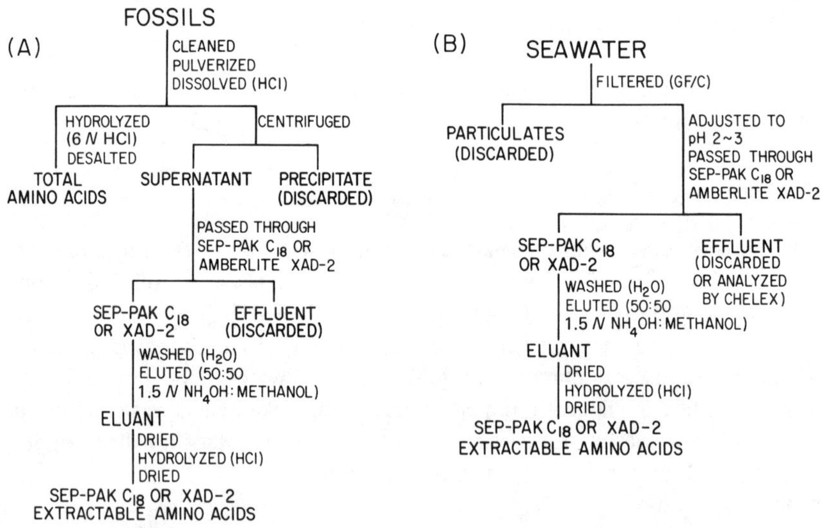

Figure 1. An outline of the procedure used to extract peptide-bound amino acids from fossils (A) and seawater (B) by XAD-2 and SEP-PAK C_{18} resins.

Model 118 automatic amino acid analyzer, interfaced with a Dec PDP-11 computerized data collection and data analysis system. The amino acid enantiomeric ratios were determined using the TPC method [16] with analyses being carried out on a Hewlett-Packard 5710A gas chromatograph using a 10-m glass capillary column.

RESULTS AND DISCUSSION

Fossils

Before interpreting the fossil results, it is important to ascertain which amino acid component in the fossils is being extracted by the XAD-2 and SEP-PAK C_{18} resins. Fossil shells are known to contain free amino acids, HCl-soluble peptides and HCl-insoluble proteinaceous material [10–12,17]. Fossil bones, on the other hand, contain only HCl-soluble peptides and HCl-insoluble proteins with free amino acids being only a minor constituent [18,19]. Evidently because bones are more porous than shells, free amino acids released from protein or peptide hydrolysis rapidly diffuse out of fossil bones and are lost.

To test what components are being extracted by the XAD-2 and SEP-PAK C_{18} methods, we carried out some experiments using norleucine and the tripeptide leucyl-glycyl-glycine. A 1 to 2 *M* HCl, or 35‰ NaCl, solution containing these constituents was carried through the XAD-2 and SEP-PAK C_{18} procedures. Neither norleucine nor the tripeptide were retained to any extent by the two resins. Although tentative, these experiments suggest that the XAD-2 and SEP-PAK C_{18} methods probably extract only HCl-soluble peptides with molecular weights greater than 1000 or so.

The amino acid compositions of some of the XAD-2 and SEP-PAK C_{18} extracts from fossil shells and bones, along with those for modern samples, are given in Figure 2. As can be seen, the amino acid compositions of the modern and fossil shells are similar, with the principal difference being a decrease in the serine and an increase in the alanine contents in the fossil shell extracts. The serine and alanine abundance changes are possibly the result of serine dehydration, which produces alanine [20]. The amino acid composition of the XAD-2 extract from the fossil bone is also similar to that of the modern sample. These results demonstrate that the amino acids in both the XAD-2 and SEP-PAK C_{18} extracts from fossils are stable with respect to decomposition for periods of several hundreds of thousands of years.

Although the amino acid compositions of the XAD-2 and SEP-PAK C_{18} fossil extracts remain fairly constant for many hundreds of thousands of years, there is a large decrease with increasing fossil age in the quantity of

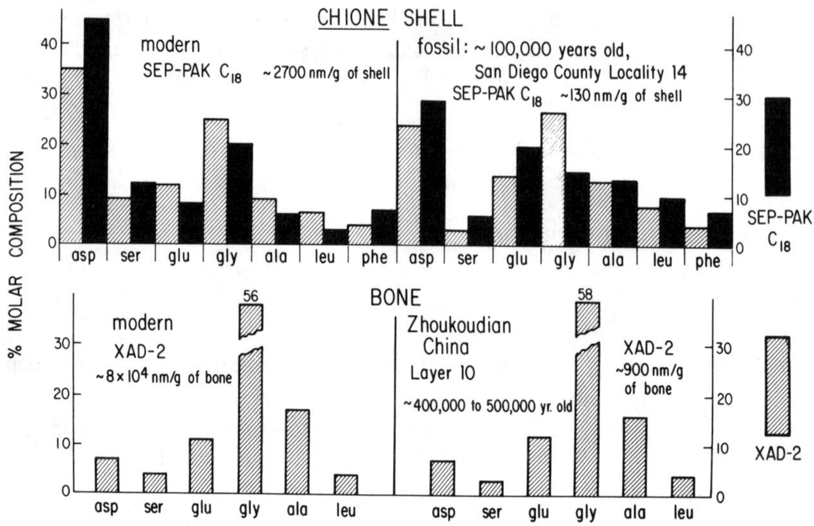

Figure 2. Amino acid compositions of XAD-2 and SEP-PAK C_{18} extracts from modern and fossil *Chione* shells and bones. The blank levels for the two methods were generally negligible in comparison to the amount of amino acids isolated from the fossils. Blank levels are significantly higher for the XAD-2 method, however.

extracted amino acids. For the two fossil samples shown in Figure 2, there is a decrease in the amino acid content extracted by the two methods by a factor of 20 to 100. This observation suggests that the principal diagenetic reaction that has occurred in the component extracted by the XAD-2 and SEP-PAK C_{18} methods is peptide bond hydrolysis. It is well established [10,11,13,17] that one of the principal diagenetic reactions that takes place in fossils is peptide bond hydrolysis. These hydrolysis reactions first generate lower-molecular-weight peptides and eventually free amino acids. Peptide bond hydrolysis may either generate free amino acids and peptide fragments which are not retained by the XAD-2 and SEP-PAK C_{18} resins or generate hydrolysis products which diffuse out of the fossil and are lost. Thus, although the amino acids in the XAD-2 and SEP-PAK C_{18}-extracted components are relatively stable with respect to decomposition, the peptide bonds in the components are relatively unstable and hydrolyze at rates which are rapid in comparison to the age of the fossils.

The amino acid enantiomeric ratios determined for the various extracts from the fossil shells and bones are shown in Tables I and II. Enantiomeric ratios were found to increase with increasing fossil age, indicating that racemization has occurred in the components extracted by the XAD-2 and SEP-PAK

Table I. Amino Acid Enantiomeric Ratios in the Total (T), XAD-2 (X) and SEP-PAK C_{18} (S) Fractions of *Chione* Shells from Marine Terrace Deposits in San Diego County[a]

Locality # and Estimated Ages (yr)	Amino Acid Enantiomeric Ratios (D/L)														
	ASP			ALA			LEU			ALLO/ISO			PHE		
	T	X	S	T	X	S	T	X	S	T	X	S	T	X	S
Modern 0	0.14	0.16			0.05	0.05			0.04	0.02	0.01	0.01		0.07	0.06
#14 ~100,000	0.65	0.46		0.64	0.16	0.20	0.33	0.30	0.12	0.30	0.14	0.05	0.29	0.32	0.15
#130 ~200.000	0.70	0.61	0.50	0.76	0.20	0.25	0.79	0.45	0.19	0.59	~0.3	0.19	0.43	0.51	0.23
#SDSU-2276 ⩾300,000	0.71	0.60		0.83	0.21	0.42	0.76	0.46	0.28	0.66	~0.2	~0.2	0.52	0.56	0.29

[a]A map showing the various localities is given in Karrow and Bada [14].

Table II. Amino Acid Enantiomeric Ratios in the Total (T) and XAD-2 (X) Fractions of Fossil Bones from Zhoukoudian, China (the site where Beijing Man was Discovered)

	Amino Acid Enantiomeric Ratios (D/L)											
	ASP		ALA		GLU		LEU		ALLO/ISO		PHE	
Sample	T	X	T	X	T	X	T	X	T	X	T	X
Modern	0.07	0.16	0.05	0.07	0.05	0.10		0.07				0.06
Zhoukoudian Upper Cave	0.09[a]	0.28	0.03[a]	0.06	0.06[a]	0.06		0.05				0.07
Zhoukoudian Layer 3–4	0.60	0.47	0.55	0.13	0.43	0.18		0.08	0.18			0.17
Zhoukoudian Layer 8–9	0.40	0.54	0.45	0.14	0.30	0.23		0.09				0.17
Zhoukoudian Layer 10	0.53	0.45	0.45	0.17	0.33	0.22		0.10	0.23			0.22

[a]Samples were hydrolyzed 4 hr.

C_{18} resins. The racemization rates (k_i) are relatively slow, however, and suggest that a time period of a million years or more would be required before the amino acids in the XAD-2 and SEP-PAK C_{18} extracts were completely racemized (D/L ~1). The relative racemization rates in the XAD-2 and SEP-PAK C_{18} extracts have the pattern $k_{asp} > k_{ala} \simeq k_{glu} > k_{leu} \simeq k_{iso}$. This is similar to the pattern observed for relative racemization rates of peptide-bound amino acids [10,21]; this result provides additional supporting evidence that the XAD-2 and SEP-PAK C_{18} methods extract peptides from the fossils. The relative racemization rates in the XAD-2 and SEP-PAK C_{18} shell extracts are much different than the relative rates determined for the total amino acid fraction in fossil shells. As has been discussed in detail elsewhere, in shells, the relative racemization rates for the total amino acid fraction are anomalous in comparison with those in other systems; this is probably due to a complex series of hydrolysis and racemization reactions that occur in shells [11,17,21]. Since the XAD-2 and SEP-PAK C_{18} methods extract amino acids which have normal patterns of relative racemization rates, these components may be useful in the future in the racemization dating of fossil shells.

Seawater

The results of the seawater XAD-2 extraction experiments indicated that very little if any of the amino acid material in seawater is extractable using this technique [22]. The level of amino acids isolated by the XAD-2 method for all the samples was very low and close to the levels observed in the 35‰ NaCl blank processed in a similar fashion. The amino acid enantiomeric ratios of the seawater XAD-2 extracts and the amino acids in the blanks were also similar. The amino acids that are detected by the XAD-2 extraction procedure are thus apparently not indigenous to the seawater samples but are rather due to contamination which arises during the XAD-2 extract procedure, possibly from the XAD-2 resin itself.

Figure 3 shows the results of a SEP-PAK C_{18} extraction of various seawater samples. Amino acids were detected in the extracts of surface waters, but appear to be absent or near blank levels in the deep water sample. Comparison of the quantity of SEP-PAK C_{18}-extractable amino acids with the total amino acid content of the various samples determined using the Chelex-100 method [22,23] indicates that the SEP-PAK C_{18} method extracts a large fraction (e.g., >50 to 60%) of the total amino acids present in surface water. However, in deep waters the SEP-PAK C_{18} method only extracts a minor fraction of the total amino acids present.

Based on the results obtained for the various fossil samples, it would appear that the difference in the concentration of SEP-PAK C_{18}-extractable

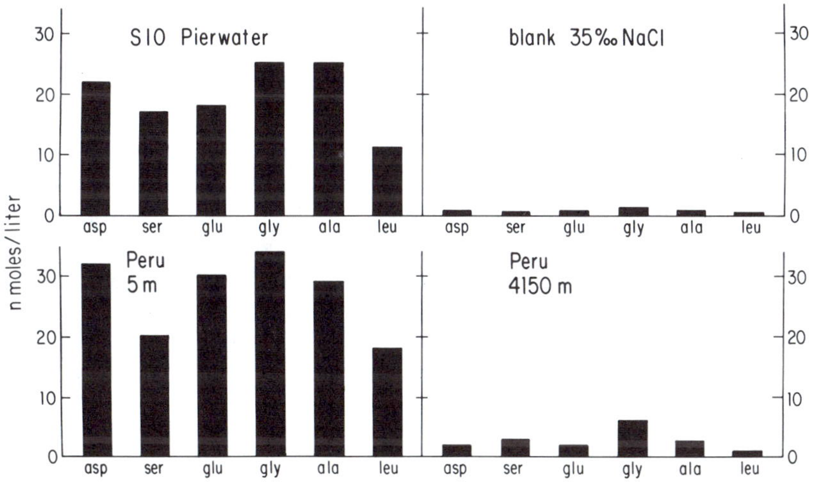

Figure 3. Concentration of amino acid in extracts from various seawater samples isolated by the SEP-PAK C_{18} method. The Peru 5 m sample was collected at 15°16.5'S, 75°50.5'W and the 4150 m sample was from 15°22.0'S, 76°02.3'W. Blank was a 35‰ NaCl solution carried through the same processing procedure as the seawater samples.

amino acids between the surface and deep ocean water samples may be explained by the hydrolysis of peptide bonds in the SEP-PAK C_{18}-extractable component. Peptide-bound amino acids in the SEP-PAK C_{18}-extractable material present in surface waters may be hydrolyzed at a rate which is rapid in comparison to the mixing times in the ocean (~1000 to 2000 years [24]). Peptide hydrolysis may generate fragments which are not retained by the SEP-PAK C_{18} resin. This would account for the observation that in deep waters the bulk of the total amino acids are not extractable by the SEP-PAK C_{18} method. An alternative explanation for the decrease in the SEP-PAK C_{18}-extractable component between surface and deep water is that the material in surface waters is rapidly consumed by organisms. This would be similar to the case for free amino acids [7]. Free amino acids released by planktonic organisms apparently are rapidly consumed in oceanic surface waters, and thus the concentration of free amino acids in the deep ocean is very low. It is not known, however, whether marine organisms can assimilate and metabolize peptide-bound amino acids, although a dipeptide (glycyl-L-phenylalanine) has recently been shown to be taken up by the mussel, *Mytilus edulis* [25].

The D/L enantiomeric ratios of the SEP-PAK C_{18}-extractable amino acids in the surface samples are all very low, indicating that the amino acids in this component are of a recent origin. Enantiomeric ratios in the deep samples were unmeasurable because of the low quantities of amino acids that were extracted.

CONCLUSION

Despite many years of study, very little is known about either the chemical or physical characteristics, or the cycle and fate, of dissolved organic material in the ocean; one major reason is that it is difficult to isolate trace quantities of organics from the much more abundant inorganic constituents of oceanic waters. On the other hand, organic material in fossils of varying ages can be fairly easily isolated in sufficient quantities to carry out detailed investigations of possible diagenetic pathways. The results presented here suggest that insights into the diagenesis of humic substances in seawater can be obtained by investigating the organic material in fossils such as bones and shells.

Our results indicate that the amino acids in humic-like components isolated from fossils are stable with respect to chemical decomposition and racemization for periods of hundreds of thousands of years. The peptide bonds in the material isolated from fossils are fairly rapidly hydrolyzed, however. Applying the same methodology used for fossils to extract humic-like materials from seawater, the results indicates that only oceanic surface waters contain an extractable amino acid-containing component. On the basis of the fossil results, it would appear that one possible explanation for this is that peptide bond hydrolysis of humic-like material in seawater takes place at a rate which is rapid in comparison to oceanic mixing times.

ACKNOWLEDGMENTS

This work was supported by grants from the Research Corporation and the Chemical Oceanography Program of the Office of Naval Research (NOOO14-75-C-1052).

REFERENCES

1. Rashid, M. A., and L. H. King. "Chemical Characteristics of Fractionated Humic Acids Associated with Marine Sediments," *Chem. Geol.* 7:37-43 (1971).
2. Ishiwatari, R. "Molecular Weight Distribution of Humic Acids from Lake and Marine Sediments," *Geochim. J.* 5:121-132 (1971).
3. Nissenbaum, A., and I. R. Kaplan. "Chemical and Isotopic Evidence for the *In Situ* Origin of Marine Humic Substances," *Limnol. Oceanog.* 17:570-582 (1972).
4. Stuermer, D. H., and G. R. Harvey. "Humic Substances from Seawater," *Nature* 250:480-481 (1974).
5. Stuermer, D. H., and G. R. Harvey. "The Isolation of Humic Substances and Alcohol-Soluble Organic Matter from Seawater," *Deep-Sea Res.* 24:303-309 (1977).
6. Gagosian, R. B., and D. H. Stuermer. "The Cycling of Biogenic Compounds and Their Diagenetically Transformed Products in Seawater," *Mar. Chem.* 5:605-632 (1977).
7. Bada, J. L., and C. Lee. "Decomposition and Alteration of Organic Compounds Dissolved in Seawater," *Mar. Chem.* 5:523-534 (1977).
8. Williams, P. M., H. Oeschger and P. Kinney. "Natural Radiocarbon in the Northeast Pacific Ocean," *Nature* 224:256-258 (1969).
9. Bada, J. L., and E. A. Hoopes. "Alanine Enantiomeric Ratio in the Combined Amino Acid Fraction in Seawater," *Nature* 282:822-823 (1979).
10. Bada, J. L., and R. A. Schroeder. "Amino Acid Racemization Reactions and their Geochemical Implications," *Naturwissenschaften* 62:71-79 (1975).
11. Schroeder, R. A., and J. L. Bada. "A Review of the Geochemical Applications of the Amino Acid Racemization Reaction," *Earth Sci. Rev.* 12:347-391 (1976).
12. Masters, P. M., and J. L. Bada. "Amino Acid Racemization Dating of Bone and Shell," *Advances in Chemistry Series, No. 171 Archaeological Chemistry II*, (American Chemical Society) 117-138 (1978).

13. Bada, J. L., and E. H. Man. "Amino Acid Diagenesis in Deep Sea Drilling Project Cores: Kinetics and Mechanisms of some Reactions and their Applications in Geochronology and Paleotemperature and Heat Flow Determinations," *Earth Sci. Rev.* 16: 21-55 (1980).
14. Karrow, P. F., and J. L. Bada. "Amino Acid Racemization Dating of Quaternary Raised Marine Terraces in San Diego County, California," *Geology* 8:200-204 (1980).
15. Chia Laupo, "The Cave Home of Peking Man", Foreign Language Press, Peking, 1975.
16. Hoopes, E. A., E. T. Peltzer and J. L. Bada. "Determination of Amino Acid Enantiomeric Ratios by Gas Liquid Chromatography of the N-Trifluoroacetyl-L-Prolyl-Peptide Methyl Esters," *J. Chromat. Sci.* 16:556-560 (1978).
17. Bada, J. L., and R. A. Schroeder. "Racemization of Isoleucine in Calcareous Marine Sediments: Kinetics and Mechanism," *Earth Planet. Sci. Lett.* 15:1-11 (1972).
18. Bada, J. L. "The Dating of Fossil Bones using the Racemization of Isoleucine," *Earth Planet Sci. Lett.* 15:223-231 (1972).
19. Matsu'ura, S., and N. Ueta. "Fraction Dependent Variation of Aspartic Acid Racemization Age of Fossil Bone," *Nature* 286:883-884 (1980).
20. Bada, J. L., M.-Y. Shou, E. H. Man and R. A. Schroeder. "Decomposition of Hydroxy Amino Acids in Foraminiferal Tests: Kinetics, Mechanism and Geochronological Implications," *Earth Planet. Sci. Lett.* 41:67-76 (1978).
21. Bada, J. L., and M.-Y. Shou. "Kinetics and Mechanism of Amino Acid Racemization in Aqueous Solutions and in Bones," In: *Biogeochemistry of Amico Acids*, P. E. Hare, T. C. Hoering and K. Kings, Eds. (New York, NY: John Wiley and Sons, Inc., 1980), p. 235.
22. Bada, J. L., E. Hoopes and M.-S. Ho. "Combined Amino Acids in Pacific Ocean Waters," *Earth Planet. Sci. Lett.* 58:276-284 (1982).
23. Lee, C., and J. L. Bada. "Amino Acids in Equatorial Pacific Ocean Water," *Earth Planet. Sci. Lett.* 26:61-68 (1975).
24. Craig, H. "The Deep Metabolism: Oxygen Consumption in Abyssal Ocean Water," *J. Geophys. Res.* 6:5078-5086 (1971).
25. Stewart, M. G. "Kinetics of Dipeptide Uptake by the Mussel *Mytilus edulis*," *Comp. Biochem. Physiol.* 69A:311-315 (1981).

CHAPTER 22

ASSOCIATION OF ORGANIC POLLUTANTS WITH HUMIC SUBSTANCES: PARTITIONING EQUILIBRIA AND HYDROLYSIS KINETICS

Edward Michael Perdue

Environmental Sciences/Chemistry Department
Portland State University
Portland, Oregon

ABSTRACT

A kinetic model for the effects of humic substances on the hydrolysis kinetics of hydrophobic organic pollutants is proposed. This model is based on an analogy with the effects of anionic detergent micelles on hydrolysis kinetics. Partitioning equilibria, specific and general acid–base catalysis, and surface micellar catalysis are included in the model. The model is applied to the acid-catalyzed hydrolysis of atrazine and the base-catalyzed hydrolysis of the 1-octyl ester of 2,4-dichlorophenoxyacetic acid.

INTRODUCTION

To describe adequately the mechanisms by which an organic pollutant is dispersed and eventually degraded in the environment, chemical and physical interactions between the pollutant and components of the biosphere must be identified and quantified. Prior research has established that reversible physical adsorption is a fundamentally important interaction, affecting the equilibrium distribution of an organic pollutant between soil, water, biomass and air, as well as the rate at which equilibrium is attained.

The organic carbon component of soils and sediments (primarily humic substances) appears to account for most adsorption of organic pollutants from aqueous solution, whether the mechanism of interaction involves van der Waals interactions, hydrogen bonding or ion exchange [1,2]. In recent years, the use of partitioning equilibrium constants (K_{oc}) to describe the distribution of organic pollutants between aqueous solution and the organic carbon component of soils and sediments has become widespread. The utility of this approach is enhanced by the fact that K_{oc} values can be closely correlated with water solubilities and with octanol–water partition coefficients (K_{ow}), making it possible to estimate K_{oc} values that have not been experimentally determined [3,4]. A more complete discussion of the treatment of partitioning equilibria in the environment was recently published by Karickhoff [5].

The effects of humic substances on pollutant degradation kinetics are much less well understood. One particular degradation process, i.e., hydrolysis, has been the subject of several studies and will be treated in this chapter. Li and Felbeck [6] reported that the hydrolysis of atrazine (2-chloro-4-ethylamino-6-isopropylamino-*s*-triazine) at pH 4.0 was strongly catalyzed by aqueous suspensions of humic acid and by acetic acid solutions. While the latter effect is attributable to *general acid catalysis* of the aqueous-phase hydrolysis reaction (also referred to as *buffer catalysis*), the catalytic effect of humic acid suspensions might also include some contribution from surface-catalyzed hydrolysis of the humic-bound atrazine molecules. In a similar study, Khan [7] has reported that aqueous solutions of fulvic acid also catalyze atrazine hydrolysis. More recently, Perdue and Wolfe [8] observed that the rate of hydrolysis of the 1-octyl ester of 2,4-dichlorophenoxyacetic acid (2,4 DOE) is greatly decreased by humic acid solutions at pH 9 to 10. These observations were consistent with an unreactive humic-bound 2,4 DOE in equilibrium with reactive aqueous-phase 2,4 DOE. Contradictory results were reported by Struif et al. [9], who observed that humic substances accelerated rates of hydrolysis of several alkyl esters of 2,4 D at pH 7.65 in 0.0066 *M* phosphate buffer.

At present, there is no conceptual model for the interaction of organic pollutants with humic substances that can adequately account for the widely variable results described above. A kinetic model based on an analogy with the effects of micellar surfactants on hydrolysis kinetics is proposed in this chapter. This model includes partitioning equilibria, specific and general acid–base catalysis, and surface catalysis. At the very least, the proposed model makes it possible to predict the types of organic pollutants, concentrations of humic substances, etc., for which significant effects on hydrolysis kinetics are most likely to be observed and whether those effects will be positive (rate-enhancing) or negative (rate-retarding).

DEVELOPMENT OF THE CATALYSIS MODEL

Partitioning Equilibria

In principle, the adsorption of a hydrophobic pollutant (P) to a particulate adsorbent should be described by an appropriate adsorption isotherm and the formation of a complex between P and another type of dissolved solute should be described by a simple equilibrium constant. This dichotomy breaks down, however, when P interacts with another substance for which the distinction between dissolved and particulate phases is uncertain. This is the case for humic substances, with "dissolved" humus being operationally defined as that fraction of humus in an aqueous medium that will pass through a 0.45 μm membrane filter. At the very least, "dissolved" humus probably includes fine particulate humus, molecular aggregates that resemble micelles, and even single molecules of colloidal dimensions.

For mathematical convenience, the partitioning of P between aqueous solution and humus will therefore be described by a single distribution coefficient (K_h), with humus concentrations (ρ_h) in terms of organic carbon concentration (g C/g H_2O). The closest analogy in simple systems is probably the micelle, so both "dissolved" and particulate humus will, in effect, be treated as a single micelle-like phase, symbolized by H(m). This model is mathematically expressed in Equations 1 through 3, where P(aq) and P − H(m) are aqueous P and humic-bound P, respectively, and X_{aq} is the fraction of P in the aqueous phase.

$$P(aq) + H(m) \underset{}{\overset{K_h}{\rightleftharpoons}} P - H(m) \tag{1}$$

$$K_h = \frac{[1 - X_{aq}]}{[X_{aq}][\rho_h]} \tag{2}$$

$$X_{aq} = [1/(1 + \rho_h K_h)] \tag{3}$$

This simple partitioning model is of the same form as that used to describe partitioning equilibria in sediment–water suspensions, in which sedimentary organic carbon is regarded as a solvent phase. The K_{oc} values used in those studies are essentially equivalent to K_h [5]. Accordingly, the linear free energy relationships that have been used to estimate K_{oc} values from water solubilities and from octanol–water partition coefficients (K_{ow}) should be useful for estimation of K_h values. Both K_{oc} and K_h values have been assumed to be independent of pH, an assumption that is probably invalid. Unfortunately, insufficient experimental data are currently available to warrant a more complex treatment.

Micellar Catalysis

Having chosen to describe aqueous humus as a micelle-like moiety for purposes of modeling partitioning equilibria, the implications of this model on rates of hydrolytic degradation reactions should be considered. Though the solution chemistry of humic substances is almost certainly more complex than that of simple detergents, the properties of detergent micelles and their effects on kinetics of hydrolysis reactions should serve as useful first approximations in our efforts to describe humus–pollutant interactions.

The subject of micellar catalysis of hydrolysis reactions is discussed by Fendler and Fendler [10]. The effect of a micelle on the rate of a hydrolysis reaction is a function of the distribution of the substrate between the aqueous phase and the micelle, the charge of the micelle, and the type of catalysis (acid or base) that is most important in the aqueous phase. Because humic substances are negatively charged over the environmental pH range, a comparison with anionic micelles such as those of sodium dodecyl sulfate is probably most relevant. In general, anionic micelles accelerate acid-catalyzed hydrolysis reactions and retard base-catalyzed hydrolyses. These effects are attributed to electrostatic stabilization of the transition state for acid catalysis, in which the substrate becomes more positively charged, and destabilization of the transition state for base-catalyzed hydrolysis, in which the substrate becomes more negatively charged. Such predictions are consistent with the effects of humic substances on the acid-catalyzed hydrolysis of atrazine [6,7] and the base-catalyzed hydrolysis of the 1-octyl ester of 2,4 D [8].

In micellar systems, hydrolysis rates are described in terms of the total pollutant concentration $[P_t = P(aq) + P - H(m)]$

$$d[P_t]/dt = -\{k_{aq}[P(aq)] + k_m[P - H(m)]\} \tag{4}$$

where k_{aq} and k_m are pseudo-first-order rate constants for hydrolysis of aqueous P and humic-bound P, respectively. Substitution from Equation 3 yields the pseudo-first-order rate expression

$$d[P_t]/dt = -k_{obs}[P_t] \tag{5}$$

where

$$k_{obs} = \left[\left(\frac{1}{1 + \rho_h K_h}\right)k_{aq} + \left(\frac{\rho_h K_h}{1 + \rho_h K_h}\right)k_m\right] \tag{6}$$

Acid-Base Equilibria in Humic Substances

Because of the potential catalytic effects of Brönsted acids and bases on pollutant hydrolysis reactions (see subsequent discussion), a mathematical model for the distribution of acidic functional groups in humic substances is needed. From an intuitive and a mathematical viewpoint, the most appropriate model is that of Posner [11], who pointed out that pH titration curves of humic substances are consistent with a Gaussian distribution of functional groups around a mean pK value (μ). The distribution can be described by

$$\frac{C_i}{C_T} = \frac{1}{\sigma\sqrt{2\pi}} e^{-1/2\left(\frac{\mu - pK_i}{\sigma}\right)^2} dpK \qquad (7)$$

where C_i/C_T is the mole fraction of ligands in the interval dpK whose acid dissociation constant (K_i) is expressed as a negative logarithm (pK_i)
σ is the standard deviation for the distribution of pK_i values about the mean pK value (μ)

The values of μ and σ that best describe the distribution of functional groups in a humic substance are obtained by curve-fitting titration data. For example, in the titration of humic substances with a strong base such as sodium hydroxide, the average degree of ionization of acidic functional groups ($\bar{\alpha}$) is easily calculated at any point in the titration using Equation 8.

$$\bar{\alpha} = \frac{[H^+] + [Na^+] - [OH^-]}{C_T} \qquad (8)$$

For a complex mixture of *monoprotic* acids, $\bar{\alpha}$ is also given by:

$$\bar{\alpha} = \sum_{i=i}^{n} \left(\frac{K_i}{K_i + [H^+]}\right)\left(\frac{C_i}{C_T}\right) \qquad (9)$$

When Equation 7 and Equation 9 are combined and converted from a discrete model to a continuous model,

$$\bar{\alpha} = \frac{1}{\sigma\sqrt{2\pi}} \int_{-\infty}^{\infty} \left(\frac{1\bar{0}^{pK}}{1\bar{0}^{pK} + [H^+]}\right) e^{-1/2\left(\frac{\mu - pK}{\sigma}\right)^2} dpK \qquad (10)$$

Through nonlinear regression methods, it is possible to determine the set of μ and σ values that generates a set of $\bar{\alpha}$ values from Equation 10 that agrees best with the experimental $\bar{\alpha}$ values (Equation 8). Titration data for aquatic humus from the Satilla River in southeastern Georgia (SRHS) are best described by μ and σ values of 4.00 and 3.94, respectively. The results are presented in Figure 1 in terms of $p\bar{K}$ values, where

$$\bar{K} = [H^+][\bar{\alpha}/(1 - \bar{\alpha})] \quad (11)$$

The titration curves of SRHS are much more accurately described by a bimodal Gaussian distribution, a beta distribution, or by a member of the family of Pearson distributions [12]; however, the simple Gaussian distribution, with only two adjustable parameters, is sufficiently accurate for modeling acid-base catalysis by humic substances.

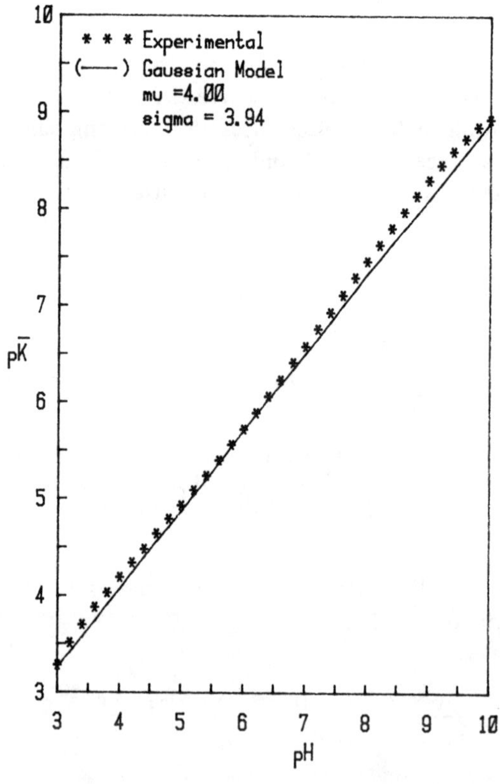

Figure 1. The pH-dependence of $p\bar{K}$ values of aquatic humus predicted from the Gaussian distribution model.

General Acid-Base Catalysis

Many hydrolysis reactions are subject to acid-base catalysis by H_3O^+ and OH^-, with reaction rates being strongly dependent on pH [13]. As a further complication, even at constant pH, other Brönsted acids and bases (e.g., pH buffers) may catalyze hydrolysis reactions, a phenomenon known as *general acid-base catalysis* [14,15]. The term *buffer catalysis* is also commonly used to describe this phenomenon. Because humic substances contain significant concentrations of carboxyl and phenolic hydroxyl groups, the possibility of general acid-base catalysis of aqueous-phase hydrolysis reactions by humic substances must be considered. It should be emphasized that this type of catalysis differs from micellar catalysis, which affects the reactivity of a pollutant only in the micellar phase.

The first major breakthrough in the development of a systematic model of general acid-base catalysis was the realization by Brönsted and Pedersen [16] that, for a wide variety of acid–base-catalyzed reactions, the *kinetic* reactivity of an acidic or basic catalyst is related to the *thermodynamic* acidity or basicity of the catalyst. These empirical relationships are mathematically expressed as the Brönsted Catalysis Laws for acid and base catalysis, and they are the oldest and best established of the so-called linear free energy relationships. Omitting commonly used symmetry corrections, these equations are:

$$k_{HA_i} = G_A(K_i)^{\alpha} \tag{12}$$

$$k_{A_i} = G_B(K_i)^{-\beta} \tag{13}$$

where k_{HA_i} and k_{A_i} are second-order rate constants for acid catalysis by HA_i and base catalysis by A_i
K_i is the acid dissociation constant of HA_i
G_A and G_B are empirical constants for a given substrate
α and β are constants which express the sensitivity of the reacting substrate to the strength of the acidic or basic catalyst

We have very recently utilized the Brönsted equation to develop a rigorous mathematical model for predicting the *maximum* potential significance of general acid–base catalysis in aqueous solutions of simple Brönsted acids and bases such as acetic acid, phosphoric acid, ammonia, etc. [17]. Based on this treatment, it is highly unlikely that buffer catalysis will be significant in the aquatic environment. However, humic substances were not previously included in the model, and their inclusion could possibly increase the potential significance of buffer catalysis.

To examine very briefly the essential features of this model, consider the acid–base-catalyzed hydrolysis of a pollutant in aqueous solution [P(aq)], which obeys pseudo-first-order kinetics at constant pH:

$$d[P(aq)]/dt = -k_{aq}[P(aq)] \tag{14}$$

where

$$k_{aq} = \left(k_{H_2O}[H_2O] + k_{H_3O^+}[H_3O^+] + k_{OH^-}[OH^-]\right) + \sum_{i=i}^{n}\left(k_{HA_i}[HA_i] + k_{A_i}[A_i]\right) \tag{15}$$

and HA_i and A_i are the ith Brönsted acid–base pair in solution. The overall contribution of a particular catalytic species to k_{aq} is expressed as the product of a second-order catalysis rate constant and the concentration of the catalytic species. The first three terms in Equation 15 are properties of the solvent species (H_2O, H_3O^+, OH^-) and can be collectively represented by a pseudo-first-order rate constant (k_w). Their combined contribution is constant at a given pH and is independent of the contributions of other Brönsted acids and bases. The ratio k_{aq}/k_w, which expresses the extent to which the observed rate constant exceeds that value which would be observed in the absence of buffer catalysis, is defined as the general catalysis factor (GCF). Rearranging Equation 15, the GCF is given by:

$$GCF = \frac{k_{aq}}{k_w} = 1 + \frac{\sum_{i=i}^{n}\left(k_{HA_i}[HA_i] + k_{A_i}[A_i]\right)}{k_{H_2O}[H_2O] + k_{H_3O^+}[H_3O^+] + k_{OH^-}[OH^-]} \tag{16}$$

For mathematical consistency with the Brönsted catalysis laws, which separately treat acid and base catalysis, it is useful to assume that the neutral hydrolysis rate constant (k_{H_2O}) is actually a sum of acidic and basic contributions [θk_{H_2O} and $(1 - \theta)k_{H_2O}$, respectively], even though the separate contributions cannot be measured experimentally. When Equation 16 is combined with Brönsted equations (Equations 12 and 13) for all catalysis rate constants, it ultimately can be shown that for a complex mixture of monoprotic catalysts

$$GCF = 1 + C_T\left[\sum_{i=i}^{n}(BCF)_i(C_i/C_T)\right] \tag{17}$$

where C_i is the stoichiometric concentration of the ith monoprotic catalyst
$(BCF)_i$ is defined as the buffer catalysis factor of the ith catalyst

The BCF of the ith catalyst is given by:

$$(\mathrm{BCF})_i = \left[\frac{\left(\frac{\theta[H_3O^+]}{K_i + [H_3O^+]}\right)\left(\frac{K_iC_W}{K_W}\right)^{\alpha} + \left(\frac{(1-\theta)K_i}{K_i + [H_3O^+]}\right)\left(\frac{C_W}{K_i}\right)^{\beta}}{C_W + \theta[H_3O^+]\left(\frac{C_W^2}{K_W}\right)^{\alpha} + \frac{(1-\theta)K_W}{[H_3O^+]}\left(\frac{C_W^2}{K_W}\right)^{\beta}}\right] \tag{18}$$

where C_W and K_W are the molar concentration and ion product of H_2O at 298 K ($55.34\ M$ and 1.00×10^{-14}, respectively)

The entire bracketed term in Equation 17 is the number average buffer catalysis factor for the complex mixture of catalysts ($\overline{\mathrm{BCF}}$) and C_T is the total concentration of monoprotic catalysts in the mixture.

At constant pH, the BCF of the ith catalyst will be maximized when α and β are those of the *hypothetical* substrate with the greatest possible tendency to undergo buffer catalysis by that catalyst at the pH of that particular solution. The α and β values that correspond to this hypothetical substrate are those for which both $\partial(\mathrm{BCF})/\partial\alpha$ and $\partial(\mathrm{BCF})/\partial\beta$ equal zero. It can be shown after extensive algebraic manipulation of the partial derivative expressions that the BCF of the ith catalyst is maximized when $(\alpha + \beta)$ equals one and

$$\alpha = \frac{1}{\log\left(\frac{C_W^2}{K_W}\right)}\left[\log C_W + \mathrm{pH} + \log\left(\frac{\log\frac{K_iC_W}{K_W}}{\log\frac{C_W}{K_i}}\right)\right] \tag{19}$$

These results are applicable to any monoprotic acid at any pH, and they are independent of θ, which will subsequently be assigned a value of 0.5. It can be shown graphically or by simple substitution into Equation 19 that the *optimum* α-value is much more sensitive to pH than to the K_i value of the catalyst. To simplify the subsequent calculations, even though every catalyst in a complex mixture of monoprotic acids has its own unique set of optimum α and β values, the α and β values actually used at a particular pH value will therefore be arbitrarily chosen to be those of the catalyst whose pK_i equals pH. Thus, in the pH 0 to pH 14 range, the optimum α value for a complex mixture of monoprotic acids is approximately given by:

$$\alpha = 0.1543 + 0.0494\ \mathrm{pH} \tag{20}$$

Optimum β values are equal to $(1 - \alpha)$.

To calculate $\overline{BCF}$ (the bracketed term in Equation 17) for a mixture of acids such as those in humic substances, all K_i and C_i values must be known. In the absence of such detailed information, the $\overline{BCF}$ of humic substances cannot be evaluated. It is, however, possible to estimate $\overline{BCF}$ values from a realistic model of the acidic functional groups of humic substances, such as the Gaussian distribution model discussed earlier.

Because the Gaussian distribution model expresses C_i values as an explicit function of K_i and utilizes K_i as the variable of integration (dpK), the previously discussed problems in evaluating Equation 17 for humic substances are effectively eliminated. The $\overline{BCF}$ of SRHS can be calculated at a specified pH using Brönsted α values from Equation 20 and the Gaussian parameters (μ = 4.00, σ = 3.94) that were obtained from the analysis of titration data. With $D = (C_W^2/K_W)^\alpha$, $\overline{BCF}$ is given by:

$$\overline{BCF} = \left(\frac{1}{C_W^\alpha \sigma \sqrt{2\pi}}\right)\left(\frac{D[H_3O^+]}{C_W + D[H_3O^+]}\right) \times \int_{-\infty}^{+\infty} \left(\frac{10^{-\alpha pK}}{[H_3O^+] + 10^{-pK}}\right) e^{-1/2\left(\frac{\mu - pK}{\sigma}\right)^2} dpK \tag{21}$$

In numerical integrations of Equation 21, the actual limits of the integral $[(\mu - 4\sigma)$ and $(\mu + 4\sigma)]$ were chosen to limit computational time while covering most of the area under the Gaussian distribution curve. The results of these calculations are given in Figure 2.

For comparison, Equation 18 and Equation 19 were used to calculate BCF for a simple monoprotic acid whose pK_i is 4.00, the mean value of pK for the aquatic humus sample. The monoprotic acid is clearly a better acid–base catalyst at lower pH values; however, in the pH 6 to pH 9 range that encompasses most natural waters, the humic material is actually a better catalyst than a monoprotic acid with $pK_i = 4.00$.

The maximum potential effect of humic substances on aqueous-phase hydrolysis kinetics is expressed by the GCF, which can be calculated from Equation 17 once $\overline{BCF}$ values are obtained. Before proceeding, however, it is desirable to express the acidic functional group concentration (C_T) in terms of the weight concentration (ρ_h, g C/g H_2O) used earlier in the chapter. For a humic substance that contains 50% (w/w) carbon,

$$C_T = 2A_T\rho_h \tag{22}$$

where A_T is the carboxyl content (mmol/g) of the humic substance

It then follows that

$$GCF = k_{aq}/k_w = 1 + 2A_T(\overline{BCF})\rho_h \tag{23}$$

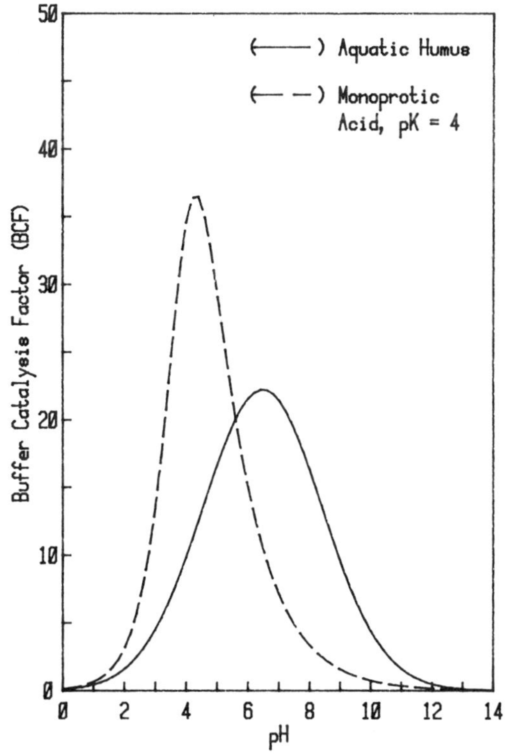

Figure 2. Buffer catalysis factors of aquatic humus and a monoprotic acid with $pK_i = 4.00$.

Overall Catalysis Model

The preceding derivations have provided equations that separately describe partitioning equilibria, micellar catalysis and general acid–base catalysis, the major processes by which humic substances can modify rates of hydrolysis of organic pollutants. These equations (Equations 6 and 23) are easily combined to obtain an overall catalysis model for humic substances.

$$\frac{k_{obs}}{k_w} = \left(\frac{1}{1 + \rho_h K_h}\right) + \left(\frac{2A_T(\overline{BCF})\rho_h}{1 + \rho_h K_h}\right) + \left(\frac{\rho_h K_h}{1 + \rho_h K_h}\right)\left(\frac{k_m}{k_w}\right) \tag{24}$$

The first term on the right side of Equation 24 is the contribution of aqueous-phase hydrolysis catalyzed by H_2O, H_3O^+ and OH^- to k_{obs}/k_w. This term approaches one at low values of ρ_h or K_h. The second term on the right side of Equation 24 is the contribution of general acid–base catalysis by humic substances. At low values of ρ_h, this term increases linearly with ρ_h. It approaches a limiting value of $2A_T(\overline{BCF})/K_h$ at high ρ_h values. The third term in Equation 24 represents micellar catalysis. Its value increases linearly with ρ_h at low ρ_h values and approaches an upper limit of k_m/k_w at high ρ_h values. The overall variation of k_{obs}/k_w with ρ_h is very much dependent on the values of the other variables in Equation 24. Reasonable estimates of practical lower and upper limits for each of these variables are therefore warranted.

Organic carbon concentrations in humus solutions vary from 1 mg/l in dilute natural waters to perhaps 10 g/l in some laboratory experiments. The limits of ρ_h are thus estimated to be 10^{-6} and 10^{-2} g C/g H_2O. By analogy with sediment–water K_{oc} values, it is reasonable to assume that K_h values lie between zero and 10^6 for most aquatic pollutants [5].

The parameter k_m/k_w is the ratio of the pseudo-first-order rate constant for micellar catalysis to that for aqueous catalysis by H_2O, H_3O^+ and OH^-. To estimate reasonable lower and upper limits for this parameter, it will be assumed that the effects of humic substances on hydrolysis kinetics resemble those of anionic micelles not only qualitatively (as discussed earlier) but also quantitatively. Fendler and Fendler [10] have tabulated k_{obs}/k_w values for 39 acid-catalyzed hydrolyses of orthoesters and acetals and 26 base-catalyzed carboxylic ester hydrolyses in anionic surfactant solutions. An average log (k_{obs}/k_w) of $+1.4 \pm 0.6$ is observed for the acid-catalyzed reactions, and an average log (k_{obs}/k_w) of -1.1 ± 0.8 is observed for the base-catalyzed reactions. From Equation 24, it is apparent that the kinetic contribution of micellar catalysis cannot exceed k_{obs}/k_w. When a pollutant is strongly adsorbed, k_{obs}/k_w approaches a limiting value that approximately equals k_m/k_w. While greater or lesser values are certainly not precluded, it therefore seems likely that k_m/k_w values in humus solutions will generally fall between 10^{-2} and 10^{+2}.

Earlier in the chapter, $\overline{BCF}$ values were estimated for an aquatic humus sample (SRHS). If SRHS is representative of other humic substances, $\overline{BCF}$ values should generally fall between zero and 22 (see Figure 2), the maximum $\overline{BCF}$ occurring at a pH of 6.5. Carboxyl contents of humic substances are usually less than 10 mmol/g, and an approximate average value would be 5 mmol/g. This average value will be used in subsequent calculations.

If no partitioning equilibria occur at all, humic substances could affect pollutant hydrolysis kinetics only by general acid–base catalysis. The maximum possible contribution of general acid–base catalysis can be estimated from Equation 24, with $K_h = 0$. At pH 6.5, where a maximum $\overline{BCF}$ of 22 is predicted, and with a humic carbon concentration of 10 g/l ($\rho_h = 0.01$), k_{obs}/k_w is predicted to be 3.2. Thus, even under the most extremely favorable conditions, general acid–base catalysis by humic substances cannot effect more than a modest threefold enhancement in the hydrolysis rate constant of an organic pollutant. It should be reemphasized that this conclusion is independent of the nature of the organic pollutant because $\overline{BCF}$ values are derived for the *hypothetical* substrate with the maximum susceptibility to general acid–base catalysis.

At the opposite extreme, if the pollutant is very extensively bound to humic material, k_{obs}/k_w will approximately equal k_m/k_w. Assuming that the analogy to anionic micelles is reasonable, it follows that partitioning equilibria and micellar catalysis can easily modify k_{obs}/k_w by a factor of 100 or more (the approximate lower and upper limits of k_m/k_w are 0.01 and 100, respectively). The overall effect of humic substances therefore, in most instances, will be almost totally attributable to partitioning equilibria and micellar catalysis, with only a minor effect due to general acid–base catalysis.

The general trends predicted by Equation 24 are presented graphically in Figures 3 through 6 at selected K_h values of zero, 10^2, 10^4 and 10^6. Each figure consists of a family of curves that shows the variation of k_{obs}/k_w with ρ_h at five values of k_m/k_w (0.01, 0.1, 1, 10, 100). In Figure 3, with $K_h = 0$ and no partitioning of the pollutant to the "micellar" humus, the results are independent of k_m/k_w, so only one curve is observed. The increase in k_{obs}/k_w at high ρ_h values is due solely to general acid–base catalysis by the humic substances. Relative to the zero partitioning case (Figure 3), even modest partitioning ($K_h = 10^2$, in Figure 4) leads to noticeable increases in k_{obs}/k_w when $k_m/k_w > 1$ and to minor decreases in k_{obs}/k_w when $k_m/k_w \leqslant 1$. As K_h values are increased to 10^4 (Figure 5) and 10^6 (Figure 6), these trends become even more apparent. In the latter case, k_{obs}/k_w values clearly approach the limiting k_m/k_w values.

APPLICATION OF THE CATALYSIS MODEL

Having examined the general features of the overall catalysis model, it should be validated by application to experimental data. For this purpose, the base-catalyzed hydrolysis of the 1-octyl ester of 2,4-D and the acid-catalyzed hydrolysis of atrazine have been selected. In addition, the predictive capacity of the model is utilized for two other hydrolysis reactions of environmental interest.

Hydrolysis of the Octyl Ester of 2,4-D

Perdue and Wolfe [8] have recently studied the effect of humic acid solutions on the alkaline hydrolysis of the octyl ester of 2,4-D (2,4DOE). In a typical experiment at pH 9.47 and $\rho_h = 0.0005$, a k_{obs}/k_w value of 0.024 was obtained. The carboxyl content (A_T) of the humic acid will be assumed to be approximately 5 mmol/g and its $\overline{BCF}$ will be estimated from the curve

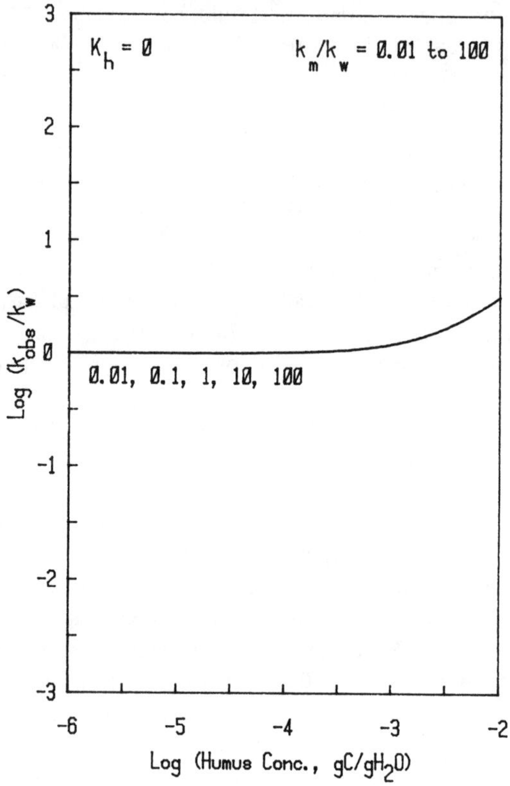

Figure 3. Catalytic effect of humic substances on the hydrolysis of an organic pollutant with $K_h = 0$.

in Figure 2 for aquatic humus (6.8 at pH 9.47). If partitioning equilibria were negligible in this system, general acid–base catalysis by humic acid could potentially result in a k_{obs}/k_w of 1.034. The fact that k_{obs}/k_w is much less than one indicates clearly that partitioning equilibria are quite important and that the humic-bound 2,4DOE is not very reactive. Assuming that k_m/k_w equals zero (the minimum possible value), a minimum K_h value of $10^{4.9}$ is calculated from these data via Equation 25, which is simply a rearrangement of Equation 24.

$$K_h = \frac{\left[1 + 2A_T\overline{BCF}\rho_h - \left(\frac{k_{obs}}{k_w}\right)\right]}{\rho_h\left[\left(\frac{k_{obs}}{k_w}\right) - \left(\frac{k_m}{k_w}\right)\right]} \tag{25}$$

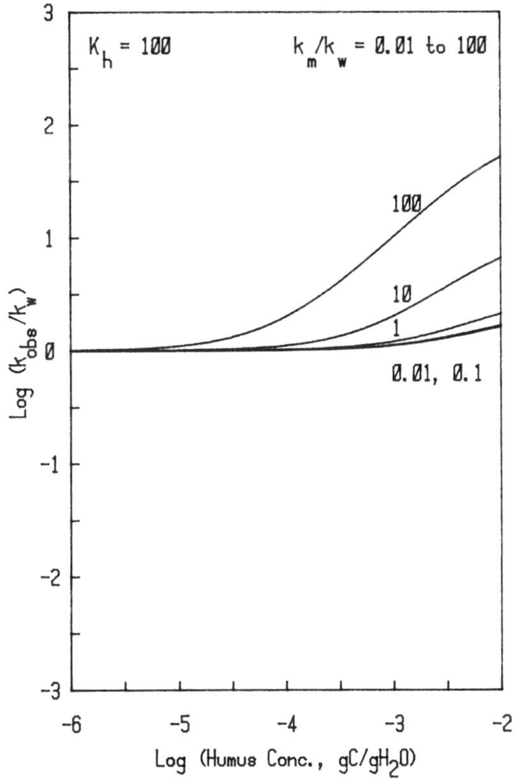

Figure 4. Catalytic effect of humic substances on the hydrolysis of an organic pollutant with K_h = 100.

The minimum K_h value for association of 2,4DOE with humic acid is only slightly smaller than the sediment-water K_{oc} value for 2,4DOE ($10^{5.3}$) [8]. It is apparent from Equation 25 that the value of K_h is related to the assumed value of k_m/k_w, approaching infinity as k_m/k_w approaches k_{obs}/k_w. Because the numerator of Equation 25 is greater than zero in this particular case, the denominator must also be positive. Therefore, it can be stated with certainty that k_m/k_w is less than 0.024. This example points out one problem: unless K_h can be independently determined, a unique value of k_m/k_w cannot be calculated from measurements of k_{obs}/k_w at only one ρ_h value. Even so, the results obtained in this study are obviously in good agreement with predictions based on an analogy to anionic micellar systems, in which base-catalyzed hydrolyses are inhibited. The magnitude of k_m/k_w is close to the practical lower limit of 0.01 (see earlier discussion) and the calculated K_h value is similar to the K_{oc} of 2,4DOE.

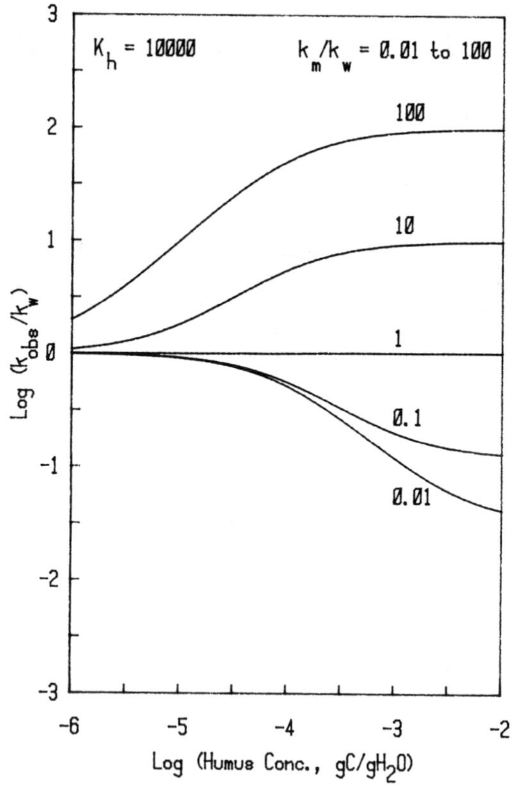

Figure 5. Catalytic effect of humic substances on the hydrolysis of an organic pollutant with K_h = 10,000.

Hydrolysis of Atrazine

Li and Felbeck [6] clearly demonstrated that the atrazine hydrolysis reaction is susceptible to general acid–base catalysis. Their results can be expressed in terms of k_{obs}/k_w through the relationship:

$$k_{obs}/k_w = \frac{t_{1/2}^{(w)}}{t_{1/2}} \tag{26}$$

where $t_{1/2}^{(w)}$ is the hydrolysis half-life in aqueous hydrochloric acid solution at pH 4.0
$t_{1/2}$ is the hydrolysis half-life at the same pH value in an aqueous solution that contains other potential Brönsted acid–base catalysts (e.g., acetic acid or humic acid)

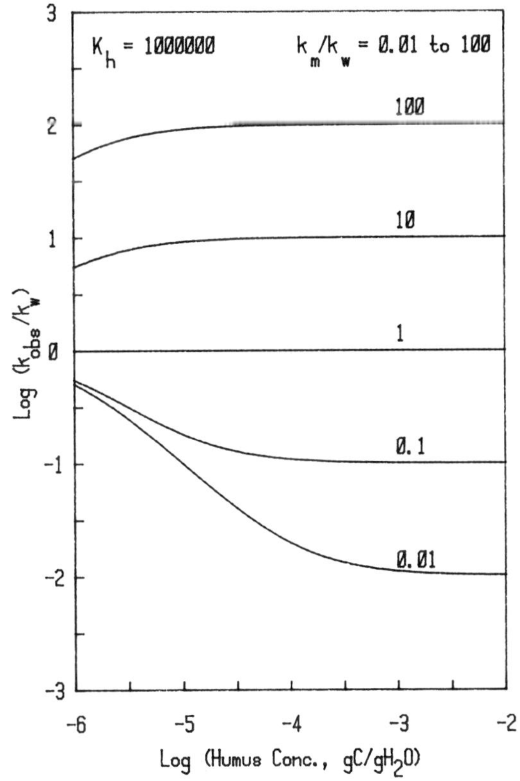

Figure 6. Catalytic effect of humic substances on the hydrolysis of an organic pollutant with $K_h = 1,000,000$.

The experimental k_{obs}/k_w values from Li and Felbeck [6] and the calculated k_{obs}/k_w values from the catalysis model described by Equation 24 are given in Table I. At 25°C, k_{obs}/k_w values of 1.12, 1.77 and 3.44 were obtained at acetic acid concentrations of 0.008 *M*, 0.04 *M*, and 0.08 *M*, respectively. It is interesting to compare these results with the predictions of the general acid–base catalysis model. From Equations 18 and 19, it can be shown that the BCF of acetic acid is 32 at pH 4.0. When this value is used in Equation 17 to calculate k_{obs}/k_w as a function of acetic acid concentration, the predicted *maximum* k_{obs}/k_w values are 1.26, 2.28 and 3.56, respectively. The agreement between experimental results and theoretical predictions is obviously quite good.

Li and Felbeck [6] also reported that humic acid suspensions had a catalytic effect on the atrazine hydrolysis reaction at 40°C. At ρ_h values

Table I. Effect of Acetic Acid and Humic Acid on Atrazine Hydrolysis Rates [6]

Catalyst	Concentration[a]	$(k_{obs}/k_w)_{exp}$	$(k_{obs}/k_w)_{calc}$
Acetic acid	0.008	1.12	1.26
Acetic acid	0.040	1.77	2.28
Acetic acid	0.080	3.44	3.56
Humic acid	0.0025	1.8	1.83
Humic acid	0.0050	2.4	2.45
Humic acid	0.0100	3.0	3.33
Humic acid	0.0200	4.4	4.35

[a]Acetic acid (mol/l); humic acid (g C/g H_2O).

of 0.0025, 0.005, 0.01 and 0.02, k_{obs}/k_w values of approximately 1.8, 2.4, 3.0 and 4.4 were obtained. Not only general acid–base catalysis but also surface micellar catalysis could contribute to k_{obs}/k_w in this system. The maximum contribution of general acid–base catalysis can be estimated from Equation 24 if K_h is assumed to be zero. By using the $\overline{BCF}$ of 10 from Figure 2 (actually calculated for aquatic humus at 25°C) and a carboxyl content of 3 mmol/g (the value reported by Li and Felbeck [6]), the maximum possible k_{obs}/k_w values are predicted to vary from 1.15 to 2.20 with increasing humic acid concentration. In every case, the predictions are too small, implying that partitioning equilibria and micellar catalysis also play a role in the hydrolysis of atrazine. Returning to Equation 24, using a K_h value of 65 and a k_m/k_w value of 6, the predicted values of k_{obs}/k_w are 1.83, 2.45, 3.33 and 4.35, in excellent agreement with experimental observation.

It could be argued that, given the number of adjustable parameters in Equation 24, the observed fit may simply be a mathematical artifact. However, the conceptual model places some constraints on K_h and k_m/k_w, namely, that K_h should approximately equal the K_{oc} of atrazine and that k_m/k_w should be greater than one and probably less than 100 for this acid-catalyzed reaction. Both constraints are met, with $k_m/k_w \sim 6$ and $K_h \sim 10^{1.8}$. The reported K_{oc} of atrazine (at 25°C) is approximately $10^{2.2}$. Therefore, it can be concluded that the conceptual model developed in this chapter efficiently describes the results of Li and Felbeck [6].

General Predictions

The overall catalysis model can also be used in a predictive fashion to estimate the possible range of k_{obs}/k_w for a given pollutant at a specified pH and humus concentration. For example, consider the base-catalyzed hydrolysis of parathion. The catalysis model assumes that k_m/k_w probably lies between 0.01 and one, and the K_{oc} of parathion is $10^{3.7}$. In an organic-rich natural water (25 mg C/l or $\rho_h = 2.5 \times 10^{-5}$) at pH 7, k_{obs}/k_w is predicted to be in the range of 0.89 to 1.00, depending on the value of k_m/k_w. It can be concluded therefore that humic substances in natural waters will not significantly affect the rate of hydrolysis of parathion.

For more strongly associated pollutants, a greater effect on k_{obs}/k_w is predicted. Consider, for example, the base-catalyzed hydrolysis of p,ṕ-DDT, for which $K_{oc} = 10^{5.38}$. Again, k_m/k_w is assumed to be in the 0.01 to 1.0 range. The overall catalysis model predicts that, in the previously described natural water, the k_{obs}/k_w value for p,ṕ-DDT should be in the range of 0.15 to 1.00. The effect of humic substances would clearly be potentially significant in this reaction if the humic-bound DDT is relatively unreactive.

In more extreme environments such as sewage sludge or in the interstitial waters of sediments and soils, where much higher concentrations of humus are possible, the impact of humic substances on pollutant hydrolysis kinetics would naturally be greater. For parathion and p,ṕ-DDT in solutions containing 2.5×10^{-3} g C/g H_2O at pH 7, k_{obs}/k_w values would be in the ranges of 0.12 to 1.04 and 0.01 to 1.00, respectively. At the higher humus concentrations, the hydrolysis reactions of both pollutants would be strongly retarded if the humic-bound pollutant is relatively unreactive.

CONCLUSIONS

While humic substances can modify hydrolysis reaction rates both by general acid–base catalysis and by partitioning equilibria that lead to a humic-bound substrate with a reactivity different from that of the unbound substrate in solution, the latter process will potentially have a much greater effect on hydrolysis kinetics of aquatic pollutants. Using the overall catalysis model described in this chapter, it can be predicted that general acid–base catalysis will be the dominant effect only when K_h values are rather small ($<10^2$). For most common organic pollutants, K_h values are probably greater than 10^2, based on an assumed similarity to K_{oc} values [5].

The two examples considered in this chapter indicate that humic substances may alter the reactivities of bound substrates in much the same manner as do anionic surfactants. Base-catalyzed reactions should be inhibited and acid-catalyzed reactions should be accelerated. It is therefore conceivable that a reaction that is susceptible to both acid and base catalysis in aqueous solution will be retarded by humic substances in base solution and accelerated by humic substances in acid solution.

As more results become available on a greater number and variety of organic compounds, the model presented here can be refined, improving its predictive capabilities. The direct analogy between humic substances and the micelles of anionic surfactants must be corrected for pH effects on the charge of humic substances (decreasing charge with decreasing pH). It is this phenomenon that should cause K_h values to increase with decreasing pH.

As the charge on humic substances decreases, k_m/k_w values should approach one, if the mechanistic rationale for the effects of anionic micelles on hydrolysis kinetics is correct.

REFERENCES

1. Stevenson, F. J. "Organic Matter Reactions Involving Herbicides in Soil," *J. Environ. Qual.* 1:333-343 (1972).
2. Stevenson, F. J. "Role and Function of Humus in Soil with Emphasis on Adsorption of Herbicides and Chelation of Micronutrients," *Bioscience* 22:643-650 (1972).
3. Karickhoff, S. W., D. S. Brown and T. A. Scott. "Sorption of Hydrophobic Pollutants on Natural Sediments," *Water Res.* 13:241-248 (1979).
4. Chiou, C. T., L. J. Peters and V. H. Freed. "A Physical Concept of Soil-Water Equilibria for Nonionic Organic Compounds," *Science* 206: 831-832 (1979).
5. Karickhoff, S. W. "Semi-Empirical Estimation of Sorption of Hydrophobic Pollutants on Natural Sediments and Soils," *Chemosphere* 10:833-846 (1981).
6. Li, G. C., and G. T. Felbeck, Jr. "Atrazine Hydrolysis as Catalyzed by Humic Acids," *Soil Sci.* 114:201-209 (1972).
7. Khan, S. U. "Kinetics of Hydrolysis of Atrazine in Aqueous Fulvic Acid Solution," *Pestic. Sci.* 9:39-43 (1978).
8. Perdue, E. M., and N. L. Wolfe. "Modification of Pollutant Hydrolysis Kinetics in the Presence of Humic Substances," *Environ. Sci. Technol.*, in press.
9. Struif, B., L. Weil and K. E. Quentin. "Verhalten von Herbiziden Phenoxyessigsäuren und ihrer Ester im Gewässer," *Vom Wasser* 45:52-77 (1975).
10. Fendler, J. H., and E. J. Fendler, *Catalysis in Micellar and Macromolecular Systems* (New York: Academic Press, 1975).
11. Posner, A. M. "Titration Curves of Humic Acid," *Intern. Cong. Soil Sci., 8th* (Bucharest, Romania), 161-173 (1964).
12. Parrish, R. S., E. M. Perdue, N. L. Wolfe and J. H. Reuter. "On Estimation of Pearson Mixing Distributions with an Application to Titration," *Technometrics*, submitted for publication.
13. Mabey, W., and T. Mill. "Critical Review of Hydrolysis of Organic Compounds in Water Under Environmental Conditions," *J. Phys. Chem. Ref. Data* 7:383-415 (1978).
14. Bell, R. P. *The Proton in Chemistry*, 2nd ed. (Ithaca, NY: Cornell University Press, 1973).
15. Bell, R. P. In: *Correlation Analysis in Chemistry*, N. B. Chapman and J. Shorter, Eds. (New York: Plenum Press, 1978), p. 55.
16. Brönsted, J. N., and K. Pedersen. "Die Katalytische Zersetzung des Nitramids und ihre Physikalisch-Chemische Bedeutung," *Z. Phys. Chem.* 108:185-235 (1924).
17. Perdue, E. M., and N. L. Wolfe. "Prediction of Buffer Catalysis in Field and Laboratory Studies of Pollutant Hydrolysis Reactions," *Environ. Sci. Technol.*, submitted for publication.

CHAPTER 23

SPECTROFLUORIMETRY OF HUMIC SUBSTANCES FROM ESTUARINE WATERS: PROGRESS OF THE TECHNIQUE

M. Ewald, C. Belin and P. Berger

Groupe d'Océanographie Chimique
Université de Bordeaux I
Talence Cedex, France

H. Etcheber

Institut de Géologie du Bassin d'Aquitaine,
Talence Cedex, France

ABSTRACT

It is shown that humic substances from Gironde estuary can be characterized by spectrofluorimetry in the dissolved state (fulvic acids) and in the particulate state (fulvic and humic acids). Humic substances in the estuary are recognized to be from terrigenous origin when compared to marine autochthonous samples obtained during the Orgon IV cruise.

INTRODUCTION

The situation of humic materials in estuaries has been reviewed by several authors (Reuter and Perdue [1]; Head [2] and references therein). The humics behavior into an estuary is first estimated from dissolved and particulate organic carbon determination. UV-visual light spectrophotometry can also give qualitative and quantitative information [3].

To our knowledge, direct investigation of fulvic acids in the dissolved state has not been performed on estuary waters. Only in marine waters was total determination of humic substances proposed [4,5]. Moreover, the fulvic acids and humic acids extracted from particles swept along the estuary, after filtration and chemical extraction, have not been examined for their absorption and fluorescence properties.

Our purpose is to show the existence of dissolved and particulate fulvic acids in estuarine waters as well as particulate humic acids by way of spectrofluorimetry. This preliminary account of results is a descriptive one. To allow the description of the natural situation, with regard to the spectrofluorimeter capabilities, the study materials were not demetallized. In the following approach metal cations should be taken into account. Especially those which are paramagnetic and are known to quench the fluorescence [6] need attention.

Such information is involved in the understanding of the role of humic substances as an ecological factor in regulation processes of the biocenosis directly or via the pollution occurring from transfer of chemicals by humic substances.

EXPERIMENTAL

Water and suspended matter samples were taken in the middle part of the Gironde (France) estuary during February 1981. pH was 7.6, salinity 2 ‰ and turbidity 1.120 mg/l. Water was filtered on a Whatman GC/C (0.45 to 1 μ) filter, pyrolyzed at 500 °C for 30 min to eliminate organic impurities. The retentate is called particulate organic matter (POM) and the filtrate is called dissolved organic matter (DOM). Filtered water with dissolved material was directly examined by UV-visual light spectrophotometry on Beckman acta MVI with 5-cm-length quartz cells and by spectrofluorimetry in 1 cm × 1 cm quartz cell on a Hitachi-Perkin-Elmer MPF-3 with R 106 photomultiplier. Band pass was 16 nm at excitation and 4 nm at observation. The cells are thermostated at 20°C. Fluorescence spectra were not corrected for the variation of the instrumental response. With regard to the absence of external intercalibration procedure (with other laboratories) to avoid mistakes coming from incorrect calibration, a methodology on this subject will be presented in the remote future. All fluorescence spectra given here are obtained on the same instrument with the same bandwidth conditions. Relative comparisons do not need emission correction for the information given here.

Another correction should be critically examined. First, the solutions should be sufficiently dilute to avoid a screening effect and fluorescence reabsorption. Second, all the examined waters show a significant Tyndall scattering that modifies the blank conditions related to the experiments (Ewald and Jobet-Belin [7], Orgon IV). This effect may be measured by examination of a nonfluorescing polymeric solution [polyvinyl alcohol (PVA), molecular weight 75,000] in water. In these conditions, when the PVA scattering is normalized to that of the real solution, the blank due to polychromatic light coming from stray light through the exitation monochromator [7] was subtracted from the recorded spectrum. This correction was considered to be critical when the filtered water scattering remains high or when the fluorescence intensity is low (i.e., on the same range of intensities as Raman scatter, generally the case for humic acids).

For the particulate organic matter (>0.45 μm), fulvic acids (FA) and humic acids (HA) are extracted by well known chemical extraction methods (Debyser and Gadel [8] and references therein) FA and HA were not demetallized. A blank of all the operations has been done and the background fluorescence obtained at the excitation of 370 nm does not interfere with the fluorescence spectra presented here.

We use as reference materials samples from the Orgon IV cruise on R/V Jean Charcot in November 1978. Cores of marine autochthonous sediments were obtained at a depth of 4750 m deep with a long Reineck corer (KL4). Interstitial waters, superficial waters (5 m over the sediment) and extracted FA and HA from the sediment (10 to 50 cm under the water–sediment interface) were obtained as described earlier [7]. The organic matter sampled here has been shown by other methods to be essentially from autochthonous origin [8,9]. Results obtained on these materials will be considered here as reference material for humic substances of marine origin.

RESULTS

The comparison (Figure 1) of the fluorescence spectrum of filtered water from Gironde estuary with filtered interstitial water (or superficial water) from sediment of Orgon IV–KL4 station shows that FA is the dominant fluorescent organic material in the dissolved state. A significant redshift is obtained with the fluorescence of FA from Gironde water compared to that from Orgon IV interstitial water.

For the particulates it is shown (Figure 2) that from the point of view of the fluorescence spectral information, extracted humics behave respectively like fulvic and humic acids extracted from Orgon IV sediments, but always with a red-shift (only a red broadening of the spectrum for HA).

UV-visual light absorption spectra of these materials (not presented here) show the well known featureless behavior obtained with humic substances. The slope of optical density as a function of the wavelength is higher for FA as compared to HA, as is already known [10].

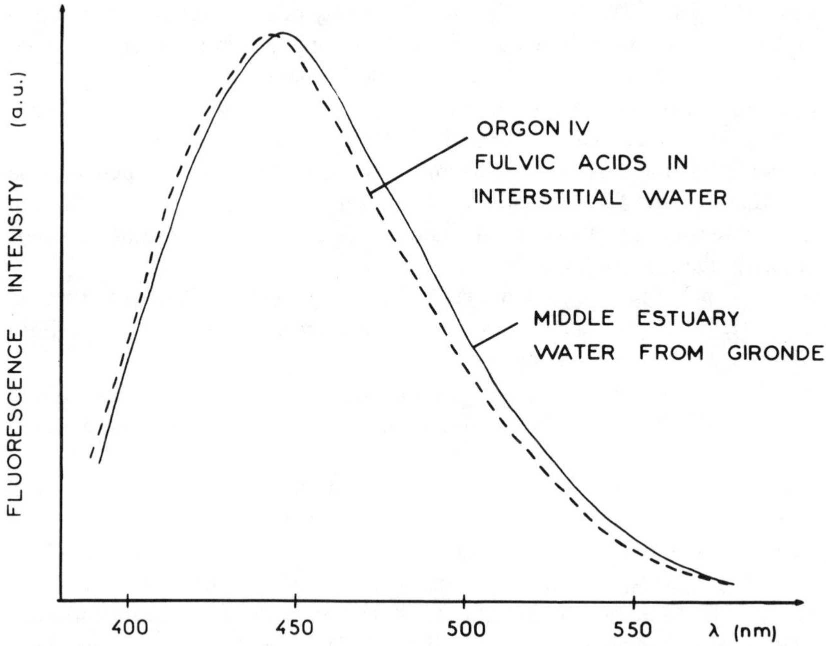

Figure 1. Fluorescence spectra of water from the middle of the Gironde estuary and of interstitial water from deep marine autochthonous sediments (cruise Orgon IV, Ewald and Jobet-Belin [7]). The analysis of the waters was done on the spectrofluorimeter without any chemical process, and with filtration at 0.45 μ. Excitation at 370 nm, bandwidths f_{ex} = 16 nm, f_{ob} = 4 nm. The spectra are corrected for Raman and Tyndall scattering (see the text) and normalized. Fluorescence intensities are in arbitrary units (a.u.) and no relative correspondence in intensities between the two different water spectra can be obtained from these pictures.

DISCUSSION

From the results, which should be considered as a preliminary account for the Gironde estuary, it is shown that Gironde waters contain fluorescent FA in the dissolved state. Dissolved humic concentration should not be as high as that of fulvic acids. In the particulate state FA and HA are found at similar levels of concentration (deduced from organic carbon determination, to be published).

Comparison with Orgon IV results indicates, in all cases, a fluorescence red-shift for Gironde humic materials which should be related to the terrestrial origin of these geopolymers drained by leaching water from soil to river.

It is known that terrestrial humics are derived in a great part from highly aromatic lignin, giving a polymeric structure with aromatic chromophores

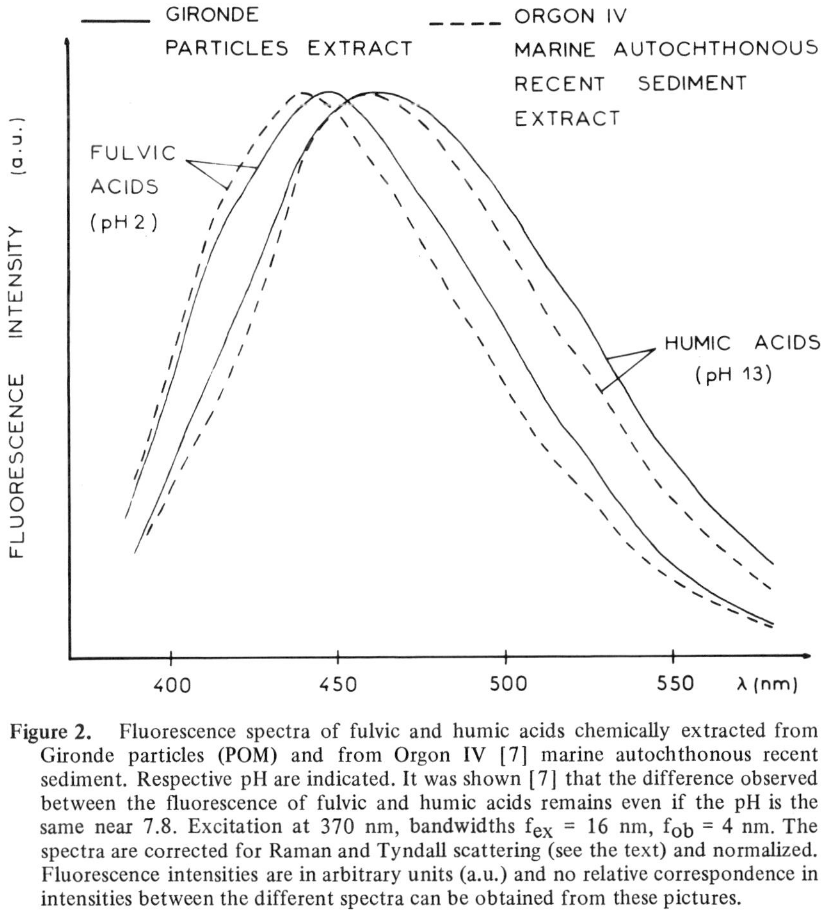

Figure 2. Fluorescence spectra of fulvic and humic acids chemically extracted from Gironde particles (POM) and from Orgon IV [7] marine autochthonous recent sediment. Respective pH are indicated. It was shown [7] that the difference observed between the fluorescence of fulvic and humic acids remains even if the pH is the same near 7.8. Excitation at 370 nm, bandwidths f_{ex} = 16 nm, f_{ob} = 4 nm. The spectra are corrected for Raman and Tyndall scattering (see the text) and normalized. Fluorescence intensities are in arbitrary units (a.u.) and no relative correspondence in intensities between the different spectra can be obtained from these pictures.

containing phenols with carboxylic groups [10], and more conjugated molecules as coumarins [11]. Humics from autochthonous marine origin, on the other hand, cannot generally contain such phenol derivatives, and the fluorescence is blue-shifted [12].

The understanding of the chromophores' role in the fluorescence characteristics of humic substances is the object of increasing interest from the scientific community with regard to the possibility of characterizing and quantifying the concentration of humic substances in water.

Such work is underway in our laboratory in order to present a methodology based on the spectrofluorimetry to characterize the origin and the chemical evolution of humics in water and to attain their specific involvement in the biocenosis and in the chemical pollutant transfer.

ACKNOWLEDGMENTS

We are indebted to Professors J. Joussot-Dubien and M. Vigneaux for continuous interest and support, to Dr. C. Latouche and his group for his collaboration and valuable discussions for the Gironde work.

Funds from CNRS (ATP "Oceanographie Chimique," 1981-1982) and from CNEXO (contract n° 80/2370) are gratefully acknowledged.

REFERENCES

1. Reuter, J. H., and E. M. Perdue. "Importance of Heavy Metal-Organic Matter Interactions in Natural Waters," *Geochim. Cosmochim. Acta* 41:325-334 (1977).
2. Head, P. C. "Organic Processes in Estuaries," in *Estuarine Chemistry*, J. D. Burton and P. S. Liss, Eds. (London: Academic Press Ltd, 1976), pp. 54-85.
3. Krom, M. D., and E. R. Sholkovitz. "Nature and Reaction of Dissolved Organic Matter in the Interstitial Waters of Marine Sediments," *Geochim. Cosmochim. Acta* 41:1565-1573 (1977).
4. Lundquist, K., B. Josefson and G. Nyquist. "Analysis of Lignin Products by Fluorescence Spectroscopy," *Holzforschung* 32:27-32 (1978).
5. Nyquist, G. "Investigation of Some Optical Properties of Seawater with Special Reference to Lignin Sulfonates and Humic Substances," PhD Dissertation, University of Goteborg, Sweden (1979).
6. Saar, R. A., and J. H. Weber. "Comparison of Spectrofluorimetry and Ion-Selective Electrode Potentiometry for Determination of Complexes between Fulvic Acid Heavy-Metal Ions," *Anal. Chem.* 52:2095-2100 (1980).
7. Ewald, M., and C. Jobet-Belin. "Fluorescence d'Acides Fulviques et d'Acides Humiques Extraits d'un Sediment Marin d'Origine Autochtone. Etude Directe de la Fluorescence de l'Eau Sus-Jacente et Interstitielle du Sediment," in *Géochimie Organique des Sediments Marins Profonds*, (Mission Orgon IV, Golfe d'Aden, Mer d'Oman. Edition du CNRS, 1981). pp. 383-403.
8. Debyser, Y., and F. Gadel. "Géochimie des Kérogenes dans les Sediments," in *Géochimie Organique des Sediments Marins Profonds*, (Mission Orgon IV, Golfe d'Aden, Mer d'Oman. Edition du CNRS, 1981), pp. 447-482.
9. Caratini, C., J. Bellet and C. Tissot. "Etude Microscopique de la Matiere Organique: Palynologie et Palynofacies," in *Geochimie Organique des Sediments Marins Profonds*, (Mission Orgon IV, Golfe d'Aden, Mer d'Oman. Edition du CNRS, 1981), pp. 265-307.
10. Schnitzer, M. "Humic Substances: Chemistry and Reactions," in *Soil Organic Matter (Developments in Soil Science 8)*, M. Schnitzer and S. U. Khan, Eds. (New York: Elsevier Scientific Pub. Comp., 1978), pp. 1-58.
11. Larson, R. A., and A. L. Rockwell. "Fluorescence Spectra of Water-Soluble Humic Materials and Some Potential Precursors," *Arch. Hydrobiol.* 89:416-425 (1980).
12. Stuermer, D. H. "The Characterization of Humic Substances in Sea Water," PhD Dissertation, M.I.T. and Woods Hole Oceanographic Institution (1975).

CHAPTER 24

ABIOTIC UPTAKE AND PHOTODEPENDENT RELEASE OF PHOSPHATE FROM HIGH-MOLECULAR-WEIGHT HUMIC–IRON COMPLEXES IN BOG LAKES

David A. Francko

Botany Department
Oklahoma State University
Stillwater, Oklahoma

Robert T. Heath

Biology Department
Kent State University
Kent, Ohio

ABSTRACT

Orthophosphate adsorbed to ferric iron–dissolved humic complexes may be released by a mechanism involving the photoreduction of iron. The majority of soluble phosphorus present in bog water at all times of the year was associated with such photolabile complexes, suggesting that this abiotic phosphorus recycling mechanism may have ecological significance in lakes containing high concentrations of iron and dissolved humic material. Photoreduction of complexed phosphorus–iron–humic associations, with comcomitant release of free orthophosphate, was accomplished in the laboratory using very low dose rates of UV irradiation and was also observed in lake water samples incubated in full sunlight. $^{32}P\text{-}PO_4$ labeling experiments on filtered bog water indicated that sorption of phosphate to higher-molecular-weight dissolved humic material occurs abiotically and that the sorbed $^{32}P\text{-}PO_4$ may be photodependently converted to a low-molecular-weight fraction distinct from orthophosphate.

INTRODUCTION

The functional relationships between iron, phosphate and dissolved humic materials (DHM) in bog lakes under aerobic conditions are complex and not fully understood. In the absence of DHM, ferric iron (Fe^{3+}) and phosphate combine to form insoluble ferric phosphate under oxidizing conditions [1]. In bog lakes, high concentrations of DHM sequester this ferric iron, preventing ferric phosphate precipitation [2–5]. These ferric iron–DHM complexes have been reported to sorb ionic orthophosphate (HPO_4^-) under acidic, aerobic conditions [2,3]. DHM–ferric iron complexes can readily be photoreduced by UV light to form stable ferrous iron–DHM complexes [2,6,8]. The ferrous ion is long-lived in aerobic, acidic bog water [2,3,6–8] provided it is associated with humic materials. However, the ferrous–humic complex may not bind phosphate [3].

These and other investigations on iron–DHM–phosphate complexes in soil and lake water systems [9–12] suggest that photodependent, abiotic processes may affect the release of bound constituents (e.g., phosphates) from DHM–iron complexes. Little is known of the mechanisms or ecological importance of such processes.

Previously we characterized the filterable phosphorus composition of an acid bog lake using two chromatographic systems (Dowex AG1-X8 anion exchange chromatography and Sephadex G-25 gel filtration) [13]. Our results indicated that 60 to 90% of the dissolved phosphorus fraction occurred as a class of high-molecular-weight compounds which cochromatographed with DHM through each chromatographic system. This high-molecular-weight phosphorus fraction (HMWP) did not release detectable quantities of orthophosphate (equivalent to soluble reactive phosphate or "SRP") in the presence of acid molybdate reagents or upon alkaline phosphatase hydrolysis, but released ionic SRP when briefly exposed to low-intensity UV light. The conversion of this "UV-sensitive complexed phosphorus" fraction to a fraction chromatographing as orthophosphate was accompanied by a concomitant decrease in chromatographing DHM, as evidenced by decreases in absorbance at 400 nm. The data imply that UV-sensitive P and cochromatographing DHM were functionally related in this bog system and that this relationship was mediated by the iron oxidation state. Subsequent work suggested that a functional relationship existed between iron–humic material and PO_4 in this bog system [14].

In this work we examined the notion that the release of phosphate bound to DHM was related to photoreduction of bound iron to Fe^{2+} and that uptake and release of phosphate can occur abiotically. Specifically, we report data on seasonal relationships between UV-sensitive complexed phosphorus and cochromatographing DHM in a bog lake; the photodependent release of phosphate from its complexed form by UV light and natural sunlight; kinetic experiments on the involvement of ferric iron photoreduction in phosphate release; and $^{32}P\text{-}PO_4$ sorption and release dynamics in bog lake water samples.

MATERIALS AND METHODS

Study Sites

Two study sites were employed in this investigation. Lake water samples were collected from the 0.1-m strata of Crazy Eddie Bog, a small (0.7 ha), shallow ($\bar{z}$ = 0.7 m) humic bog lake, and from Triangle Bog, a relic bog 2 ha in area ($\bar{z}$ = 6.0 m). Both bodies of water are located in Portage County in northeastern Ohio.

Crazy Eddie Bog is situated in a small old field watershed surrounded by a marsh community dominated by *Sphagnum*. Water inputs to this lake occur largely by surface runoff, resulting in large allochthonous inputs of dissolved humic material, frequently reducing water transparency to <0.5 m. During the ice-free period, pH generally ranged between 5.2 and 5.8. Surface water temperatures in midsummer often approached 30 °C. Other characteristics of this bog and detailed analytical methodology were reported previously [13,14].

Triangle Bog is surrounded by a *Sphagnum* shelf and a *Larex* forest with an ericacious understory. The seasonal pH generally averages 4.5 and the lake frequently supports a late spring phytoplankton bloom dominated by *Gonyostomum semen*.

Analytical Methodology

Orthophosphate (SRP), dissolved complex phosphate (SUP), total iron and ferrous iron were determined colorimetrically [6,8,13–17]. Phosphate and iron concentrations were determined to a precision of ±5% (±SD; N = 2) and ±10% (±SD; N = 3), respectively. Absorbance of light at 400 nm was used to detect DHM in chromatographed and unchromatographed samples adjusted to pH 8.0 immediately before assay [4,5,13,14].

Samples were returned to the laboratory within 30 min of collection and filtered in 25-ml portions through prewashed 0.45-μ Millipore filters using a vacuum differential of 0.5 atm. Filters were analyzed for potential phosphate retention. We detected no filtration losses of SUP, although small (<1.0 μg/l) amounts of SRP were retained on the filters.

Filtered 100-ml portions of lake water were fractionated by ion exchange chromatography on columns of Dowex AG1-X8 anion exchange resin (200 to

400 mesh; Bio-Rad Laboratories) by elution with an increasing gradient of KCl at pH 5.0 [13–15]. Eluted fractions (2.5 ml) were analyzed for SRP, SUP and DHM.

We reported previously that this chromatographic procedure recovers greater than 98% of the SRP and SUP adsorbed to the resin without chemical or physical alteration of the adsorbed species [13,14]. According to this procedure 75 ± 10% (±SD; N = 3) of the DHM adsorbed to the resin could be recovered by elution with the KCl gradient; the remaining DHM apparently bound irreversibly to the top of the resin bed (seen as a discolored band at the top of the resin bed). Based on percentage of recovery of P input, the bound DHM fraction contained little measurable phosphorus. DHM-containing samples (2–5) routinely fractionated late in the ionic gradient, indicating a high affinity for the anionic resin matrix. In chromatographed samples, we reported DHM absorbance (A_{400}nm) as the total eluted absorbance of all DHM-containing fractions [14].

Samples (25 to 100 ml) of filtered lake water, incubated at 22 ± 2 °C, were irradiated for 3 hr with a low intensity of ultraviolet light. An Ultraviolet Products, Inc. Model UVL-22 lamp was positioned 15 cm over the water surface. The effective thickness of irradiated samples ranged from about 1.0 to 2.0 cm. The UV light source delivered a very low dose rate ($\sim 1.4 \times 10^1$ J m^{-2} hr^{-1}) at wavelengths shorter than 400 nm. Samples were irradiated before chromatographic fractionation. Other samples used as controls were incubated in darkness. Alternately, filtered lakewater samples were incubated during midday in full sunlight ($\sim 2.0 \times 10^2$ J m^{-2} hr^{-1} at wavelengths shorter than 400 nm).

UV-sensitive complexed P was determined as the difference in SRP chromatographing at the orthophosphate position in the anionic gradient before and after UV irradiation [13,14]. In unchromatographed samples, UV-sensitive complexed P was determined as the difference in SRP concentration detected before and after irradiation with UV light.

RESULTS AND DISCUSSION

UV-Sensitive P and DHM: Seasonal Correlations

The concentration of DHM varied seasonally in Crazy Eddie Bog. The least amounts were seen in early spring; quantities increased in June, and reached the highest concentrations in late summer and early autumn. Throughout the season the daytime concentrations of UV-sensitive P were linearly related to cochromatographing DHM absorbance eluted in each chromatographed sample (Figure 1). The linear relationship was valid in both relatively unstained (<50 Pt units) and highly stained (>200 Pt units) water samples.

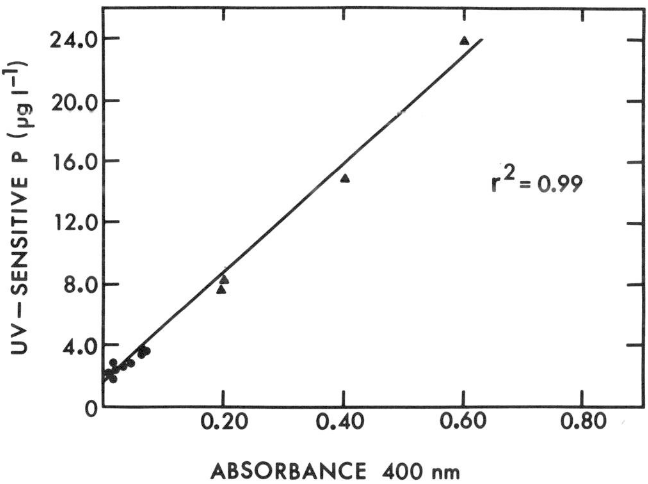

Figure 1. Correlation between total eluted DHM (A_{400} nm in 1-cm cells) and the concentration of UV-sensitive phosphorus in ion exchange chromatographed Crazy Eddie Bog water (y = 35.19X + 1.58). Samples were collected at midday in April and May 1977 (closed circles) when the lake water was only lightly stained and May and June 1976 (triangles) when the lake was heavily stained [14].

These data are consistent with the view that DHM and cochromatographing UV-sensitive complexed P were functionally related and that this complex can be broken by low doses of UV light under laboratory conditions. Consistent with previous work [13], DHM and UV-sensitive P were coeluted as a discrete fraction late in the ionic gradient.

Formation and Photodependent Breakdown of DHM and UV-Sensitive P Under Natural Conditions

These abiotic, photodependent events are of limited interest unless they can be shown to occur under natural conditions. The UV source used in laboratory experiments delivered very low dose rates of UV light, suggesting that UV light occurring in sunlight may alter these DHM-P compounds in situ, resulting in the release of phosphate. Consistent with this view, we observed a diurnal fluctuation in both the concentration of DHM and in the amount of UV-sensitive P (Figure 2). As above, the amounts of DHM and UV-sensitive P were linearly related and fell on the same regression line reported in Figure 1. During daylight hours, both components declined in concentration. Highest concentrations of both DHM and UV-sensitive phosphorus occurred

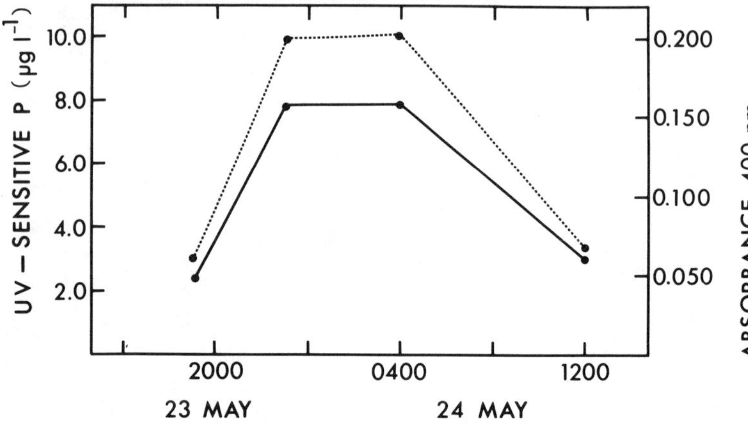

Figure 2. Diurnal changes in total eluted DHM adsorbance (A_{400} nm) (·······) and UV-sensitive phosphorus (——) in ion exchange chromatographed samples (100-ml), 23–24 May 1977. Midday solar irradiation at the surface of the lake averaged 2.1×10^4 J m^{-2} h^{-1} at wavelengths less than 400 nm [14].

during the night. Although such data may be interpreted in several ways, they are consistent with the view that DHM P, which may be formed (abiotically) or synthesized (biotically) at night, is disrupted in a photodependent process in daylight in situ, thereby making phosphate available for microbial assimilation.

Water samples collected during the daylight period were irradiated either with natural sunlight ($\sim 2.0 \times 10^2$ J m^{-2} hr^{-1} at wavelengths <400 nm) or with the artificial UV source. Although samples in natural sunlight received an order of magnitude greater irradiation at wavelengths less than 400 nm and much higher doses of visible light, each light source produced a similar loss rate of DHM (i.e., rate of decrease of absorbance at 400 nm). The photodependent rate of release of SRP from UV-sensitive complexed P compounds was also similar in sunlight and with the artificial UV source (half-lives about 10 min). These data suggest that such photodependent processes may involve only restricted portions of the UV or visible spectrum, and may involve a sequence of events in which light is necessary but is not the ultimate limiting factor.

Similarly, Miles and Brezonik [8] reported that while the destruction of DHM absorbance in the model bog systems they examined was light-dependent, light irradiance did not appear to be the rate-limiting factor in the reaction sequence. The relative importance of visible light (i.e., <620 nm) in these reactions remains unresolved. We have not yet accomplished precise action spectra of either the destruction of DHM absorbance or the release of phosphate from UV-sensitive complexed P. However, the use of broad-band color filters positioned between a source of natural sunlight or the artificial UV source and Crazy Eddie Bog samples indicated that light of a wavelength greater than 620 nm is not involved in these photodependent processes.

Relationship Between UV-Induced Photoreduction of DHM–Iron and Phosphate Release from UV-Sensitive Phosphorus

Because DHM in Crazy Eddie Bog contained iron, and because ferric iron (complexed with DHM) has been implicated in the process whereby phosphate becomes associated with DHM [3], we tested the hypothesis that the photodependent release of SRP from DHM–phosphorus complexes involves the photoreduction of ferric iron. We measured the concentrations of ferrous iron and SRP in a sample of darkly stained (~200 Pt units) filtered bog water during UV-irradiation (sample thicknesses optically thin, i.e., 1.0 to 2.0 cm). Another aliquot of the same filtered bog water sample was incubated simultaneously in darkness. The concentration of reduced iron and SRP increased at the same rate in the irradiated sample (Figure 3). Both reduced iron and SRP reached an upper limit after 60 min. At this time, all of the dissolved phosphorus in the irradiated sample occurred as SRP, and all of the detectable iron was present as Fe^{2+}.

After 90 min the UV light source was turned off and the sample incubated in darkness. Concentrations of both reduced iron and SRP then declined at similar rates. Within 60 min both Fe^{2+} and SRP returned to original values. The amount of detectable Fe^{2+} and SRP in the unirradiated aliquot remained unchanged throughout the experiment. The coincidence of the kinetics of these two processes suggests that the release of SRP from iron-containing DHM was related to the photodependent reduction of ferric iron and that the process was reversible. The data demonstrated that about 25 μg/l of SRP was released upon reduction of about 4.0 mg/l of Fe^{3+}, suggesting that only a portion of ferric iron present in bog water may be associated with phosphate.

UV-induced photoreduction kinetics of DHM-associated ferric iron present in lake water samples were compared with the reduction of a simple organic iron compound (ferric citrate) dissolved in acetate-buffered deionized water (Figure 4). Both DHM and ferric citrate exhibited similar first-order photoreduction kinetics in optically thin (<2.0 cm) samples. In contrast with the behavior of the DHM, ferrous iron reduced by UV irradiation of ferric citrate was not reoxidized, even after incubation in darkness for 3 hr. This suggests that the reoxidation of Fe^{2+} to Fe^{3+} in naturally occurring DHM may have been caused by some moiety in the DHM (e.g., stable free radicals, as

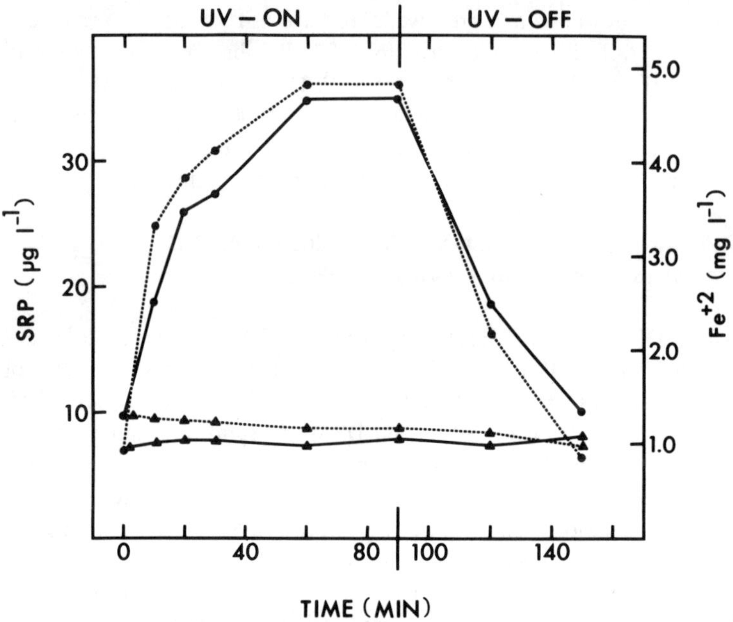

Figure 3. Relationship between ferrous iron (·······) and SRP (——) concentration in filtered, Crazy Eddie Bog water during UV irradiation (•) and in darkness (▲). After 90 min of irradiation, the sample was then incubated in darkness. Total soluble phosphorus and total soluble iron concentration of each lake water sample at t_0 were 36 ± 2 µg/l and 5.0 ± 0.5 mg/l, respectively [14].

suggested by Minderman [18]). Photoreduction kinetics of iron in our bog water samples corresponded well with rates of similar processes reported by Miles and Brezonik in other bog systems [8].

These findings suggest that UV-sensitive high-molecular-weight compounds may be iron-containing humic materials capable of sorbing phosphate to bound Fe^{3+} but releasing sorbed PO_4 on photoreduction of Fe^{3+} to Fe^{2+}. That is, these findings suggest that the uptake and release of phosphate can be accomplished abiotically, release being mediated by availability of UV light and uptake being a function of available Fe^{3+} bound to humic materials.

Abiotic Uptake and Release of ^{32}P-Phosphate by Dissolved Bog Water Compounds

A direct test of the notion that phosphate can be sorbed and released by humic compounds without biotic involvement was performed by radioactive

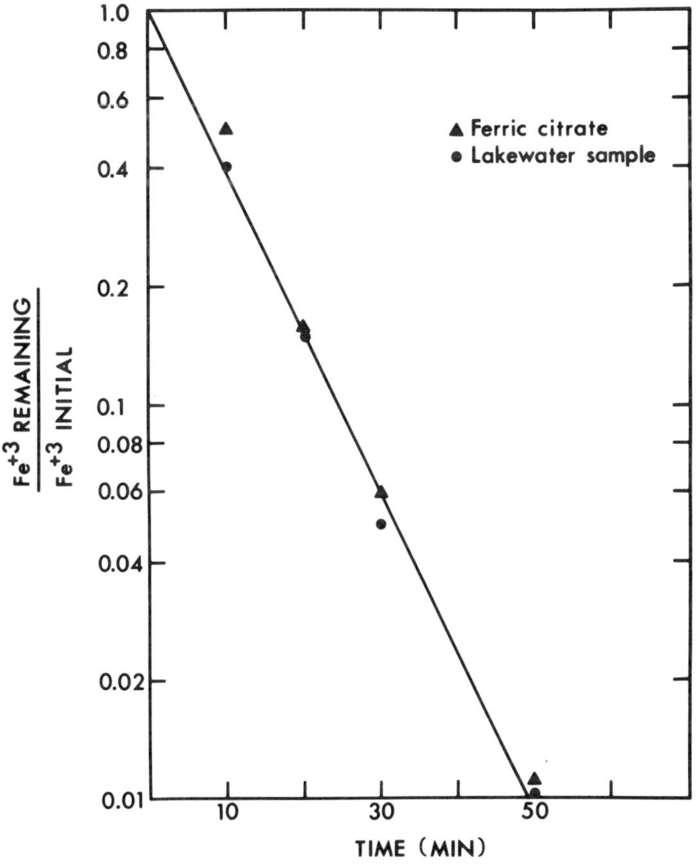

Figure 4. UV-induced photoreduction kinetics of ferric iron in filtered water from Crazy Eddie Bog (pH 5.1) and of ferric citrate. The ferric citrate was dissolved in deionized water, buffered to pH 5.1 with 0.8 *M* potassium acetate to the same salinity (~0.3 meq/l) as the lake water sample. Each sample contained about 5.0 mg/l of ferric iron. Control samples, incubated in darkness, contained no measurable ferrous iron throughout the experiment [14].

labeling of filter-sterilized bog water. Water from Triangle Bog was filtered through a prerinsed Millipore 0.45-μ pore (HAWP) membrane filter into an autoclaved flask. Portions (5 ml) were aseptically transferred to a sterile tube to which was added 50 μCi carrier-free $^{32}P\text{-}PO_4$ in 0.001 *M* potassium acetate at pH 4.5, the pH of the water at the time of collection. Carrier-free $^{32}P\text{-}PO_4$ has a specific activity of 9.13×10^6 Ci/mol, or 1.1×10^{-7} mol/Ci;

thus we added 5.5×10^{-12} mol of $^{32}P\text{-}PO_4$ to 5.0-ml aliquots of membrane-filtered lake water. After 12 hr at 5 °C in the dark, this material was fractionated on Sephadex G-25 in 0.01 *M* potassium acetate and 0.09 *M* KCl, pH 4.5. Following fractionation, 0.2-ml aliquots of 4-ml fractions were dissolved in 2 ml Aquasol and ^{32}P counts per minute detected by liquid scintillation spectrometry. Figure 5 shows that most (>99%) of the ^{32}P chromatographed at the position expected of unbound PO_4 (i.e., at K_{av} = 1.0; where K_{av} equals the elution volume minus void volume divided by the elution volume of very low-molecular-weight compounds minus the void volume). Yet, 34.4 nCi of ^{32}P, corresponding to 3.8×10^{-15} mol of PO_4, chromatographed at the void volume, indicating that it was bound to a compound of molecular weight greater than 3000.

The material chromatographing at the void volume (K_{av} = 0) on G-25 released bound $^{32}P\text{-}PO_4$ during the irradiation with a low dose of UV light. Three fractions containing the majority of the ^{32}P at void volume in Figure 5A were pooled. One portion was held in the dark as another portion was UV-irradiated for 4 hr in a sterile petri plate at 20 °C with the UV source set 1 cm from the surface of the sample. Each portion was rechromatographed on G-25. Blue Dextran (molecular weight > 2×10^6; Pharmacia Fine Chemicals, Inc.) and orthophosphate (1 ppm) were added to each aliquot as internal standards to mark the positions of K_{av} = 0 and K_{av} = 1.0, respectively. Figure 5B indicates that unirradiated material rechromatographed almost exclusively as in the first run. In contrast, 40% of the ^{32}P label was released as a low-molecular-weight form during 4 hr of UV irradiation. The ^{32}P label was released as a compound chromatographing at a position expected of a substance having a molecular weight of several hundred.

Identification of the High-Molecular-Weight Humic Fraction to Which ^{32}P-Phosphate Sorbed

Sephadex G-100 fractionation (Figure 6A) of filtered water from Triangle Bog yielded at least six fractions differing in elution position, absorbance at 400 nm, absorbance at 250 nm, and fluorescence (excitation at 380 to 450 nm, emission at 480 to 550 nm). The general characteristics of each fraction are summarized in Table I.

The ^{32}P-orthophosphate sorbed predominantly to a species of apparent high molecular weight that chromatographed as peak 1. Labeled carrier-free orthophosphate added to filter-sterilized water under sterile conditions sorbed to material originally chromatographing in the void volume of G-25 (Figure 5A). Three fractions containing the majority of the ^{32}P-labeled material excluded from G-25 were then pooled and rechromatographed on G-100 in 0.01 *M* potassium acetate and 0.09 *M* KCl, pH 4.5. Greater than 90% of the ^{32}P eluted in the void volume of G-100 (peak 1; i.e., molecular

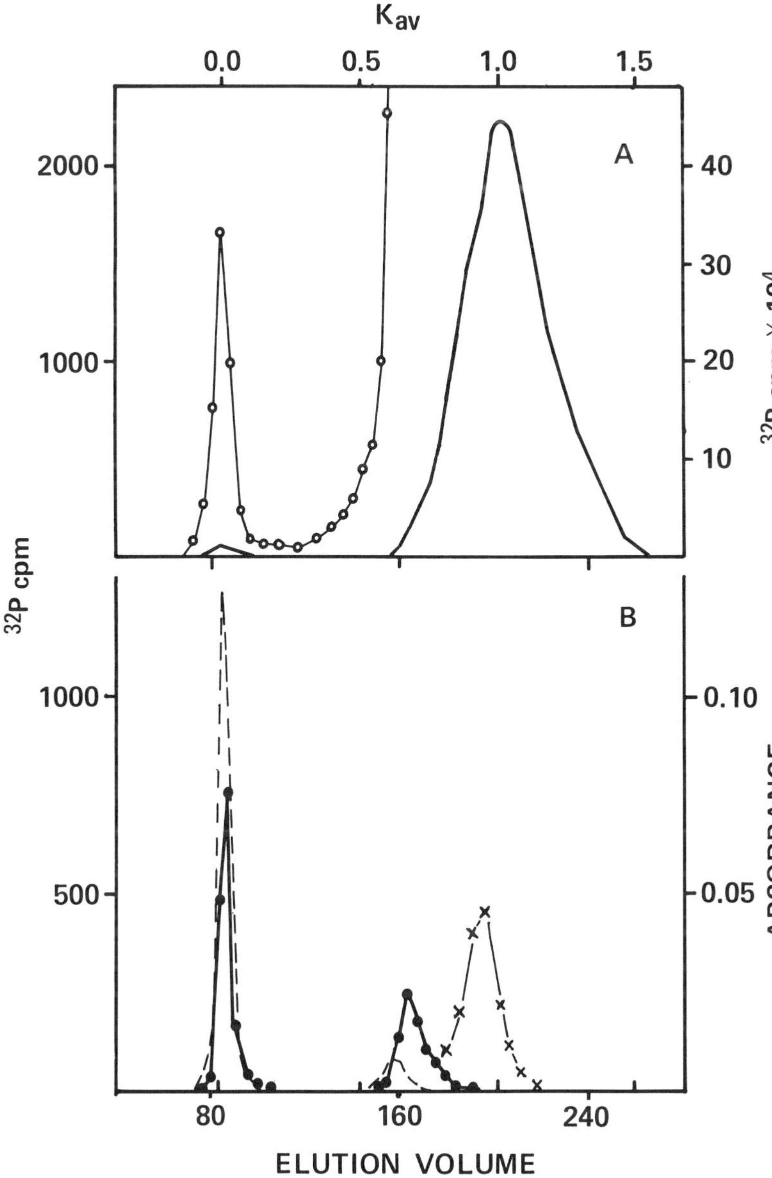

Figure 5. Fractionation of filtered Triangle Bog water on Sephadex G-25 in 0.09 *M* KCl, 0.01 *M* potassium acetate, pH 4.5. (A) ^{32}P counts per minute (cpm) detected by liquid scintillation spectrometry. Solid line (——) indicates ^{32}P cpm × 10^4, open circles (-o-) indicate ^{32}P cpm. Elution position expressed both as elution volume (ml) and relative elution position (K_{av}). (B) ^{32}P cpm of unirradiated (– – –) and irradiated material (-•-). Elution position of orthophosphate internal standard is shown as absorbance of Murphy and Riley colorimetric test read at λ = 885 nm in 1-cm quartz cuvettes.

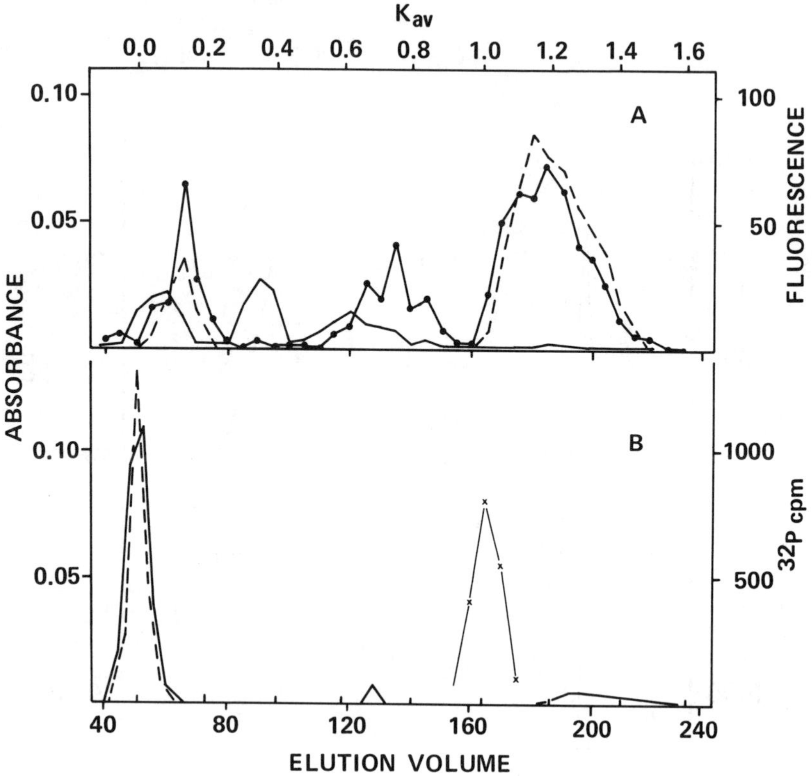

Figure 6. Sephadex G-100 fractionation of filtered Triangle Bog water in 0.09 *M* KCl, 0.01 *M* potassium acetate, pH 4.5. (A) Elution profile of filtered Triangle Bog water analyzed for absorbance at λ = 250 nm (-●-), absorbance at λ = 400 nm (——) and fluorescence of material excited by light λ = 380 to 450 nm and emitting light λ = 480 to 550 nm (- - -). (B) Elution profile of material sorbing ^{32}P-orthophosphate and excluded from G-25 (——). The absorbance of blue dextran (λ -625) (- - -) and of phosphate (Murphy and Riley, λ = 885) (-×-) added as internal standards are shown. Elution position expressed both as elution volume (ml) and relative elution position (K_{av}).

weight > 100,000). Less than 10% of the ^{32}P was associated with the trailing edge of the late-eluting, apparently polydisperse highly fluorescent fraction 6.

CONCLUSIONS

These findings taken together indicate that phosphate can be sorbed to a very high-molecular-weight humic substance. The molecular weight of this

Table I. Characteristics of Filtered Water from Triangle Bog Fractionated on Sephadex G-100

Peak	Elution Position (K_{av})	Absorbance 250 nm	Absorbance 400 nm	Fluorescence 400/500	$^{32}P\text{-}PO_4$ Sorbtion
1	0.00	moderate	moderate	little to none	>90%
2	0.15	high	low	moderate	none
3	0.35	low	high	none	none
4	0.60	moderate	moderate	none	none
5	0.75	high	low	none	none
6	1.0–1.4	high	none	high	<10%

phosphate-sorbing species may be greater than 100,000. Low doses of UV light release sorbed ^{32}P label as a low-molecular-weight species which does not chromatograph in the same position as orthophosphate. Conversely, other experiments indicated that phosphate released from naturally occurring UV-sensitive phosphorus is chemically and chromatographically identical to orthophosphate. The sorption of phosphate may involve an iron (Fe^{3+}) humic complex and release may require photoreduction of iron to Fe^{2+}. The mechanisms of uptake and release of phosphate from these complexes remain unclear, but neither process appears to require the presence of living organisms.

ACKNOWLEDGMENTS

This work has been supported in part by U.S. Environmental Protection Agency grant 10160-HCS-R801936, Office of Water Resources Research grant A-103 MICH, and U.S. Department of Energy grant EY-76-S-02-1599, COO-1599-180, and NSF grant DEB-8012164. We thank Emmy Meijers and Frans Nooy for their technical assistance, and we thank Elizabeth Buchanan and Louise Wagner for their help in preparation of the manuscript.

REFERENCES

1. Wetzel, R. G. *Limnology* (Philadelphia: Saunders, 1975).
2. Koenings, J. P. "In Situ Experiments on the Dissolved and Colloidal State of Iron in an Acid Bog Lake," *Limnol. Oceanog.* 21:674-683 (1976).
3. Koenings, J. P., and F. F. Hooper. "The Influence of Colloidal Organic Matter on Iron-Phosphorus Cycling in an Acid Bog Lake," *Limnol. Oceanog.* 21:684-696 (1976).

4. Shapiro, J. "Effect of Yellow Organic Acids on Iron and Other Metals in Water," *J. Am. Water Works Assoc.* 56:1062-1082 (1964).
5. Shapiro, J. "Yellow Organic Acids of Lakewaters: Differences in Their Composition and Behavior," in *Chemical Environment in the Aquatic Habitat*, H. L. Golterman and R. S. Clymo, Eds. (New York: North-Holland, 1967), p. 202.
6. Stumm, W., and G. F. Lee. "Oxygenation of Ferrous Iron," *Ind. Eng. Chem.* 52:143-146 (1961).
7. Stumm, W., and J. J. Morgan. *Aquatic Chemistry* (New York: John Wiley and Sons, Inc., 1970).
8. Miles, C. J., and P. L. Brezonik. "Oxygen Consumption in Humic-Colored Waters by a Photochemical Ferrous-Ferric Catalytic Cycle," *Environ. Sci. Technol.* 15:1089-1095 (1981).
9. Gjessing, E. T. "The Influence of Ultraviolet Radiation on Aquatic Humus," *Vatten* 2:144-145 (1970).
10. Gjessing, E. T. "Physical and Chemical Characteristics of Aquatic Humus," in *Environmental Biogeochemistry*, Vol. 1, J. O. Nriagu, Ed. (Ann Arbor, MI: Ann Arbor Science Publishers, Inc., 1976), p. 89.
11. Il'in, N. P., and D. S. Orlov. "Photochemical Destruction of Humic Acids," *Soviet Soil Sci.* 13:75-83 (1973).
12. Schnitzer, M. "Some Observations on the Chemistry of Humic Substances," *Agrochimica* 22:216-225 (1978).
13. Francko, D. A., and R. T. Heath. "Functionally Distinct Classes of Complex Phosphorus Compounds in Lakewater," *Limnol. Oceanog.* 24:463-473 (1979).
14. Francko, D. A., and R. T. Heath. "UV-Sensitive Complex Phosphorus: Association with Dissolved Humic Material and Iron in a Bog Lake," *Limnol. Oceanog.* 27:564-569 (1982).
15. Francko, D. A., and R. T. Heath. "Aluminum Sulfate Treatment: Short-Term Effect on Complex Phosphorus Compounds in a Eutrophic Lake," *Hydrobiologia* 78:125-128 (1981).
16. Murphy, J., and J. P. Riley. "Determination of Phosphate in Natural Waters," *Anal. Chim. Acta* 27:31-36 (1962).
17. Grande, J., and J. Beukenkamp. "Gradient Elution Chromotography of Phosphates," *Anal. Chem.* 28:1497-1498 (1956).
18. Minderman, G. "Alternative Approach to the Molecular Structure of Humic Acids: The Spectral Evidence for a Derivative of Humic Acids from Plant-Borne Esters. 2. Infrared and Chemical Analyses," *Neth. J. Agric. Sci.* 27:153-175 (1979).

CHAPTER 25

ENVIRONMENTAL AND HUMAN HEALTH SIGNIFICANCE OF HUMIC MATERIALS: AN OVERVIEW

A. Prakash
Toxic Chemicals Management Centre
Environment Canada
Ottawa, Canada

Donald J. MacGregor
Contaminants Control Branch
Environmental Impact Control Directorate
Environment Canada
Ottawa, Canada

ABSTRACT

Knowledge of the biological effects of humic materials is less developed than knowledge of their chemical characteristics. The most intensively studied beneficial effect is the stimulation of physiological and biochemical processes associated with terrestrial plant metabolism. Lower-molecular-weight humic fractions may be more active in this respect. The ecological implication of various biological effects of humic substances on the productivity and fertility of terrestrial and aquatic ecosystems are reviewed. Due to the special human health implications of the ingestion of drinking water, the beneficial and detrimental effects attributed to aquatic humic substances are also reviewed.

INTRODUCTION

The ubiquity of humic substances in the terrestrial and the aquatic ecosystems is indeed impressive. However, despite their extensive distribution in soils, sediments, lakes, rivers and the sea, and notwithstanding a long history of diligent and systematic research, large gaps remain in our understanding of the origin, nature, fate, chemical structure and functions of humic substances. It has become increasingly clear that humic substances cannot be looked upon as immutable chemical entities but as a dynamic group of heterogeneous compounds which are complex in their structural makeup and prone to change with time and location. Studies on biological effects of humic substances have generally lagged behind progress made in elucidation of their chemical characteristics. This can be attributed partly to lack of enthusiasm on the part of researchers to look deeply into biological significance of humic compounds whose chemical identity and origin remain uncertain, and partly to the fact that much of the humic chemistry of the past has tended to develop as a separate activity without much consideration of biological effects or ecological significance.

Although it is a commonly held view that soil humic substances are very similar to those isolated from water, conclusive comparison studies between soil and aquatic humic substances are very few indeed. The separation of humic substances of both terrestrial and aquatic origin into traditional humic acid and fulvic acid fractions, which is more an analytical classification than a genetic characterization, has also failed to establish them as distinct chemical entities. Comparison of the gross characteristics of humic and fulvic material such as molecular weight distribution, fractionation behavior, elemental composition, absorption spectra, degradation products, etc. has not revealed significant chemical differences between soil and aqueous humic materials [1,2]. However, preliminary applications of proton magnetic resonance spectrometry to terrestrial and aquatic humic material have indicated differences in their aliphatic side-chain content that may be significant [3,4].

From the viewpoint of effects on living systems, however, aquatic humus appears to play an ecologically more significant role than terrestrial humus, perhaps because of its greater relative abundance and bioavailability. A major portion of the total dissolved organic matter in freshwater and seawater is composed of humic or humic-like substances, and despite several studies, the nature and composition of aquatic humus has remained unclear. Though there is little doubt that the primary sources of aquatic humus are soil and marsh leachates, polyphenolic components of planktonic and phytobenthic decomposition products and exudates also contribute significantly to aquatic humus [5]. Aquatic humus of allochthonous as well as autochthonous origin occurs in "dissolved," collodial and organic detrital states. Most of this humus is adsorbed to clay and other mineral particles and remains in the bottom sediments as stable, relatively insoluble organic matter; the remainder stays in "soluble" form in the water column (Figure 1). The characteristic yellowish-brown color seen in lake, river and inshore coastal marine waters is almost totally due to the presence of relatively low-molecular-weight humic and fulvic acid fractions which are biologically active [5].

BIOLOGICAL EFFECTS

The influence of humic substances on biological systems has been considered largely in terms of beneficial effects. The reported beneficial effects have included (a) increases in growth, root initiation, yield, nutrient uptake, chlorophyll synthesis, seed germination, etc., (b) stimulation of the various physiological and biochemical processes associated with cellular metabolism, and (c) attenuation of toxicity of heavy metals and other toxic compounds. Most of these beneficial effects have been explained on the basis of complexation reactions which enhance the availability of nutritionally important trace elements, regenerate and prolong the residence time of essential nutrients in the medium, and modify the fate of toxic compounds. Other responses, such as enhanced permeability of cell membrane, increased metabolism of proteins, RNA and DNA, increased activity of a number of enzymes, respiratory catalysis, etc., have been attributed to the presence of humic compounds and indirectly contribute to some of the above beneficial effects [5].

Humic compounds are also able to invoke inhibitory responses, either due to their specific toxic characteristics or to their presence in high concentrations. For instance, some phenolic constituents of aquatic humus contributed by exudates of brown algae are toxic to other algae, planktonic

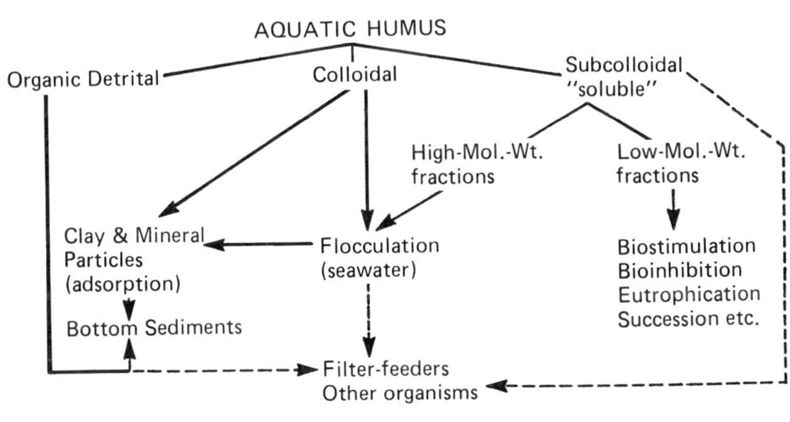

Figure 1. Fate and biological role of humus in the aquatic ecosystem (– – – possible utilization).

larval forms and fouling organisms [6–8]. Excessive complexation resulting from high concentrations of humic substances in the medium is often responsible for either a lack of response or an inhibitory effect on planktonic algae [9].

From the biological standpoint, the metal complexing ability of humic substances is perhaps the most significant factor since it plays a vital role in nutritional physiology of not only unicellular algae but also of higher plants. Sufficient experimental evidence has accumulated in recent years to show that humic substances of lower apparent molecular weights are biologically more active than those of higher apparent molecular weights. It is estimated that between 45 to 85% of organic matter in aquatic humus is composed predominantly of fulvic acid material with a molecular weight less than 700 (Table I). Such low-molecular-weight fractions are able to complex two to six times more metal than other, higher-molecular-weight fractions [10,11]. There is evidence to suggest that metal–organic complexes formed by low-molecular-weight humic substances are capable of entering the plant cell, and that the degree of penetration depends on their molecular size [12,13]. Enhanced solubilization and mobilization of several toxic metals from soils and sediments by humic and fulvic acids is of biological concern because of the possibility of biological accumulation through the food chain.

Certain bacteria, fungi and other microorganisms are reported to be capable of utilizing humic substances as a source of energy [14–16]. Increases in growth and cell yield of a freshwater bacterium (*Pseudomonas* sp) have been attributed to metabolism of fulvic acid by the bacterial population [17]. Humic substances, by virtue of their ability to precipitate alone or together with other organic molecules such as proteins and acidic polysaccharides, may provide additional sources of energy to higher aquatic organisms [9]. Although definitive evidence is lacking, the possibility of inclusion of organic

Table I. Molecular Weight Distribution of Aquatic Humus from Various Sources as Determined by Sephadex Gel-Filtration

Molecular Weight Range	Soil Extract, Aqueous (%)	River Water (%)	Mangrove Leachates (%)	Coastal Marine Water (%)	Seaweed Residues (%)
<700	43.8	63.0	86.4	68.3	22.7
700–1500	56.2	25.8	13.6	23.9	1.5
1500–5000	0.0	11.2	0.0	7.8	8.4
5000–10,000	0.0	0.0	0.0	0.0	8.4
>10,000	0.0	0.0	0.0	0.0	59.0

detrital and particulate humus into the aquatic food chain particularly through filter and detrital feeding forms cannot be ruled out (Figure 1). It is also possible that certain freshwater and marine invertebrates may be able to grow at the expense of dissolved humus [18,19].

Very little attention has been paid to the effects of humic substances on higher animals. However, some of the observations on mammalian toxicology are of interest. Intracardial application of humic acid caused rat mortality, which was ascribed to the possible reaction of humic acids with serum protein [20]. A low antigenic response in blood serum of the rabbit was noticed after intramuscular injection of high-molecular-weight humic material [21]. Significant histological changes in the liver and kidney and less pronounced effects in the spleen and testes have been reported in rats fed a high dose of humic material [22].

ECOLOGICAL IMPLICATIONS

The ecological significance of the various biological effects ennumerated earlier becomes more meaningful when one considers the overall impact or influence of humic substances on the productivity and fertility of terrestrial and aquatic ecosystems. Apart from their role in facilitating the dissolution of most otherwise insoluble metallic salts, humic substances are involved in a variety of reactions in soils, sediments and the water column with major nutrients such as ammonia, nitrates, phosphates and silicates. These interactions result not only in considerably increasing the retention and residence time of the nutrients in the media, but also in enriching and biologically conditioning the media, which have a profound influence on the biological production processes.

Terrestrial Productivity

The role of humic substances in soil fertility and in terrestrial plant growth is generally well established and has been extensively reviewed [23–26]. Increases in growth and yield of plants resulting from humic enrichment have been explained on the basis of direct and indirect effects. Indirect effects involve humic substances as suppliers and regulators of plant nutrients; the processes in this category are concerned mainly with alterations of physical and chemical properties of the soil and generally occur outside the plant. Direct effects, on the other hand, involve entry of the humic substances through the plant roots and participation in the metabolic processes of the plant [1,27]. As a generalization, it can be stated that the productivity of soils in terms of crop yield increases as a function of humus content and that while high-molecular-weight humic material alters the physical properties of soils, the low-molecular-weight substances are involved in chemical reactions in the soil and influence plant metabolic processes.

The most obvious effects of humic substances on terrestrial plants are rapid germination of seeds, initiation and enhancement of root development, promotion of plant growth, and a general increase in yield. These effects have been explained on the basis of stimulation of physiological and biochemical processes associated with cellular metabolism brought about by humic substances. Several studies have suggested that humic substances act as specific sensitizing agents which enhance the permeability of the plant cell membrane, thereby increasing the uptake of nutrients [23,28–30]. The significance of other subtle effects induced in plants by humic substances, such as increased activity of a number of enzymes, altered oxidation–reduction processes, altered protein and carbohydrate metabolism and enhanced chlorophyll synthesis, is not fully understood [1,25,31,32].

From the viewpoint of terrestrial productivity, the role of humic compounds in more complete utilization by plants of nutrients such as nitrogen, phosphorus and potassium is noteworthy [24,30,33]. Likewise, the solubility, mobility, migration, recycling and accumulation of trace metals through complexation properties of humic compounds play significant roles in the mineral nutrition of terrestrial plants [34–36]. The participation of humic compounds in plant respiration either through their acting as respiratory catalysts [31,37], or through some other mechanisms which increase oxygen absorption by plant tissue, is of ecological significance since it enables plants to withstand better the adverse effects of high temperature exposure [38].

Aquatic Productivity

Observations by Pringsheim [39] as long ago as 1912 that the growth of several species of freshwater algae was improved in the presence of aquatic humus served to trigger interest in the ecological significance of humic substances in aquatic productivity. Subsequently, numerous studies on the general and specific role of humic substances, particularly on the growth of planktonic algae in the freshwater environment, found these substances to be stimulatory [40–42]. It has even been suggested that lake eutrophication and nuisance growth of blue-green algae may be linked with the presence of aquatic humus [14,43,44]. Studies on the ecological significance of aquatic humus in marine production processes are relatively recent and have resulted in a reappraisal of earlier concepts on factors which influence the fertility of the sea, particularly in inshore coastal waters [5]. Contemporary views of aquatic productivity favor the "biological conditioning" concept, which implies the presence in the environment of biologically active organic compounds, such as humic substances, which create conditions suitable for the enhanced growth of primary producers.

Detailed studies on the effects of humic compounds derived from diverse sources such as soils, marine sediments, river water, mangrove leachates and decomposed algae on marine phytoplankton have shown that, regardless of the source, humic substances exert a stimulatory effect that is reflected in increased yield, growth rate, radiocarbon uptake, chlorophyll concentration, etc. [9,45,46]. These studies have also indicated that the growth responses of planktonic algae under humic enrichment conditions tended to be concentration-dependent and that low-molecular-weight fractions of humic acid generated higher growth response than higher-molecular-weight fractions. Since such low-molecular-weight fractions are predominant in inshore coastal waters, their ecological influence on phytoplankton productivity and consequently on coastal fertility is clearly implicated. It should also be pointed out that in the inshore coastal waters, the "humus coefficient," i.e., the amount of humic material as a percentage of the total organic material present in the water, is generally higher than in the open ocean. Thus, the differential fertility of coastal and oceanic waters can be rationalized on the basis of relative inputs of autochthonous and allochthonous biologically active compounds including humic substances [5].

Other ecological manifestations of stimulatory effects of humic compounds on planktonic algae are seen in the eutrophication of natural water bodies. Traditionally, eutrophication in the aquatic ecosystem is attributed to an increased supply of inorganic nutrients, resulting in increased or excessive production of algae. However, there appears to be no clearcut relationship between the concentration of inorganic nutrients and the degree of eutrophy. Evidence is mounting that humic compounds, through their complexation action, biologically condition natural waters in favor of high algal production [47,48]. Planktonic algae in nature respond differently to different metal ions present in the medium. Humic substances, by virtue of their complexing property, are expected to exert a species-dependent influence on planktonic algae and may thus play an important ecological role in determining the distribution, abundance and succession of species [9].

Another natural phenomenon that has been indirectly linked with humic enrichment of coastal waters is the development of massive, localized, usually monospecific blooms of marine dinoflagellates. This biological phenomenon, commonly referred to as red tide, occurs worldwide and has resulted in massive mortality of fish and other marine life, and has also in some locations endangered human health. While several factors have been identified which are directly or indirectly implicated in the development of red tides, humic compounds are recognized as one of the significant ecological factors triggering a red tide outbreak [49].

Negative effects of humic substances on aquatic productivity are not well known, except in special ecosystems such as bog lakes where the low level of algal production is attributed to selective absorption of light by the increased amount of yellow humic materials, resulting in reduced photosynthesis. However, there is ample evidence that humic substances, irrespective of their source, inhibit phytoplanktonic growth when present in high concentrations. This growth inhibition appears to be due to increased metal complexation, since excess of a complexing agent in the medium reduces the availability of essential ions to the algae [50,51].

HUMAN HEALTH IMPLICATIONS

Considerable attention has been focussed on colloidal humic substances in public water supplies since the discovery that chlorination of potable water produces chloroform and other trihalomethanes [52,53]. It is now well established that these trihalomethanes are low-yield reaction products of aqueous chlorine with humic substances [54,55], and that modifications to conventional water treatment processes can minimize trihalomethane formation [56,57]. Demonstrations that chloroform is an animal carcinogen, albeit of low potency [58], and concern that the chloroform in drinking water may cause a slightly increased incidence of human cancer [59] have led to the establishment of drinking water standards for trihalomethanes in several countries.

The color caused by colloidal humic substances, above a threshold level, in drinking water is objectionable to most people, and color per se has been included in drinking water standards for many years since experience has shown that this may cause a preference for alternative sources of potable water which, although lower in color, may be unsafe because of the presence of pathogens or colorless toxic substances [60]. Other historical human health concerns relating to humic substances in drinking water have included their possible involvement in the occurrence of endemic goiter in man [61], their role in reducing residual chlorine protection in distribution systems, and their probable presence in the extract employed in the classical carcinogenesis experiments of Heuper and Payne [62].

Humic substances in the human drinking water supply are also involved in a large number of subtler ways in its treatment, distribution and analysis. In certain types of raw water, for example, the presence of humic substances is known to interfere with coagulation and sedimentation in treatment plants [63]. In others, the posttreatment deposition of "humic muds" may occur in water mains [64]. The presence of humic substances in water mains has been linked with the bacteriological formation of red and black discoloration in drinking water and with chemical corrosion in distribution systems [65,66]. Humates are also known interferences in colorimetric, compleximetric and ion-specific electrode determinations of various drinking water parameters [67,68].

The ability of humic materials to complex toxic metals may reduce their immediate toxic effects; on the other hand, it may increase bioaccumulation of such metals, with long-term deleterious effects on human health. The chemical form in which many of these metals are taken up by the biota and their toxicological effects, particularly on humans, are of increasing concern to environmental scientists in view of what has become known of the fate and effect of ionic and methyl mercury. Although considerable attention has been paid by aquatic environmental toxicologists to the comparative toxicities of trace metals and their humate complexes [69], only a few such studies have been carried out with reference to human health. For instance, it has been shown that the acute mammalian toxicities of iron, lead, barium, silver, copper and zinc are substantially enhanced on intravenous injection as their humate complexes; orally ingested lead humate, though, is at least 6% less toxic than lead acetate [70]. No published information is available on the effects of adsorption or complexation of toxic organic substances by fulvic and humic acids on their bioavailability to mammals. However, large increases (50 to 100%) were found in the amounts of ionic material able to permeate the intestine in vitro in the presence of humic acid in experiments with calcium, magnesium, iron, manganese, zinc and sulfate [71].

During the past few years, considerable attention has been given to adsorption, solubilization, mobilization and reactivity of toxic pollutants such as pesticides and herbicides by humic substances [1,72]. Such organic chemicals are able to form stable associations with humic substances and such binding greatly increases the persistence and the possibility of bioaccumulation of these chemicals in the soil and the aquatic environment. Since the adsorption process is reported to be slow, and since the adsorbed organics can be recovered only after chemical derivatization of the humic material [72,73], the possibility exists that the concentrations of "dissolved" organics of this type have been underestimated in waters having an appreciable humic content.

Apart from the adverse human health effects noted above, many beneficial human health effects have also been attributed to humic substances. The consumption of humus-enriched colored water and the use of humic preparations in the treatment of diseases such as rheumatic fever, healing of skin defects, stimulation of wound healing and the prevention of swelling in battle wounds have been documented [22,74,75]. Humic substances in water and sediments have also been employed in hydrotherapy and balneotherapy, which are widely practiced and respected elements of European medicine [76].

ACKNOWLEDGMENTS

Portions of this chapter are based on background documentation for the 1978 Canadian Guidelines for Drinking Water Quality [77]. We wish to express our appreciation to the Department of National Health and Welfare for permission to use this material.

REFERENCES

1. Schnitzer, M., and S. U. Khan. *Humic Substances in the Environment,* (New York: Marcel Dekker Inc., 1972).
2. Rashid, M. A., and A. Prakash. "Chemical Characteristics of Humic Compounds Isolated from Some Decomposed Algae," *J. Fish Res. Bd. Can.* 29:55-60 (1972).
3. Beck, K. C., J. H. Reuter and E. M. Perdue. "Organic and Inorganic Geochemistry of Some Coastal Plain Rivers of the Southeastern United States," *Geochim. Cosmochim. Acta.* 38:341 (1974).
4. Stuermer, D. H., and J. R. Payne, "Investigation of Sea Water and Terrestrial Humic Substances with Carbon-13 and Proton Magnetic Resonance," *Geochim. Cosmochim. Acta.* 40:1109 (1976).
5. Prakash, A. "Terrigenous Organic Matter and Coastal Phytoplankton Fertility," In *Fertility of the Sea.* J. D. Costlow, Ed. (Gordon and Breach, Vol. 2, 1971), pp. 351-368.
6. McLachlan, J., and J. S. Craigie. "Algal Inhibition by Yellow Ultra-Violet Absorbing Substances from *Fucus vesiculosus,*" *Can. J. Bot.* 42:23-33 (1964).
7. Sieburth, J. McN., and J. T. Conover. "*Sargassum* Tannin, an Antibiotic Which Retards Fouling," *Nature*, 208:52-53 (1965).
8. Conover, J. T., and J. McN. Sieburth. "Effect of Tannins Excreted from Phaeophyta on Planktonic Animal Survival in Tide Pools." In *Proc. International Seaweed Symp.* E. C. Young and J. L. McLachan, Eds. (Elmsford, NY: Pergamon Press, Inc., 1966):99-100.
9. Prakash A., A. Jensen and M. A. Rashid. "Humic Substances and Aquatic Productivity," In *Humic Substances, their Structure and Function in the Biosphere.* D. M. Povoledo and H. L. Golterman, Eds. (Wageningen, The Netherlands: PUDOC 1975):259-268.
10. Rashid, M. A. "Role of Humic Acids of Marine Origin and Their Different Molecular Weight Fractions in Complexing Di- and Trivalent Metals." *Soil Sci.* 111:298-306 (1971).
11. Hoffmann, M. R., E. C. Yost, S. J. Eisenreich and W. J. Maier. "Characterization of Soluble and Colloidal Phase Metal Complexes in River Water Ultrafiltration. A Mass Balance Approach." *Environ. Sci. Technol.* 15:655 (1981).
12. Prat, S., M. Smidova and A. L. Cincerova. "Penetration and Effect of Humus Substances (Fractions) on Plant Cells." *International Congr. Biochem.* (5th) Abstr. Commun. 329. Moscow (1961).
13. Aso, S., and I. Sakai. "Studies on the Physiological Effects of Humic Acid. I. Uptake of Humic Acid by Crop Plants and Its Physiological Effects." *Soil Sci. Plant Nutr.* (Tokyo) 9:1 (1963).

14. Shapiro, J. "Chemical and Biological Studies on the Yellow Organic Acids of Lake Water." *Limonol. Oceanog.* 2:161-179 (1957).
15. Mathur, S. P., and E. A. Paul. "Microbial Utilization of Soil Humic Acids," *Can. J. Microbiol.* 13:573-580 (1967).
16. Bhardwaj, K. K. R., and A. C. Gaur. "Growth Promoting Effect of Humic Substances on *Rhizobium trifolia.*" *Indian J. Microbiol.* 12:19-21 (1972).
17. De Haan, H. "Effect of a Fulvic Acid Fraction on the Growth of a *Pseudomonas* from Tjeukemeer (The Netherlands)." *Freshwater Biol.* 4:301-310 (1974).
18. Khailov, K. M., and Z. Z. Finenko. "Organic Macromolecular Compounds Dissolved in Sea Water and Their Inclusion into Food Chains," In *Marine Food Chains* J. M. Steele, Ed. (Oliver and Boyd, 1970), pp 6-18.
19. Khailov, K. M. "Utilization of Dissolved Organic Matter in Sea Water by Echinoderms and Molluscs," *Dokl. Akad. Nauk. SSSR.* 198:443-446 (1971).
20. Klocking, R., H. Friemel and D. Mucke. "Biochemistry of Humic Acids: V. Binding of Humic Acids to Serum Proteins In Vivo (Qualitative and Semiquantitative Electrophoretic Studies)." *Acta Biol. Med. Ger.* 18:9(1967); *Biol. Abstr.* 49. 65800 (1968).
21. Barth, E. F., and N. H. Acheson. "High Molecular Weight Materials in Tap Water." *J. Amer. Water Works Assoc.*, 54:959-964 (1962).
22. Janecek, J., and J. Chalupa. "Biological Effects of Peat Water Humic Acids on Warm-Blooded Organisms." *Arch. Hydrobiol.* 65:515-522 (1969).
23. Khristeva, L. A. "The Participation of Humic Acids and Other Organic Substances in the Nutrition of Higher Plants." *Pochvovedenie* 10:46-59 (1953).
24. Kononova, M. M. *Soil Organic Matter, Its Nature, Its Role in Soil Formulation and in Soil Fertility*, 2nd ed. (New York: Pergamon Press, 1966). p. 544.
25. Khristeva, L. A., K. Solakha, R. Dinkina, V. Kovalenko and A. Gorovaya. "Influence of Physiologically Active Substances in Soil Humus and Fertilizers on Nucleic Acid Conversion, Plant Growth, and Their After Effects on Seed Qualities in Generations." Symposium. *Humus et Planta.* IV: Praha. (1967).
26. Flaig, W. "Uptake of Organic Substances from Soil Organic Matter by Plants and Their Influence on Metabolism," In *Study Week on Organic Matter and Soil Fertility*. Pontificiae Academiae Scientiarvm Scripta Varia 32. (New York: American Elsevier Publishing Co., 1968): 723-770.
27. Flaig, W. "An Introductory Review of Humic Substances: Aspects of Research on Their Genesis, Their Physical and Chemical Properties and Their Effect on Organisms." In *Humic Substances, their Structure and Function in the Biosphere*. D. M. Povoledo and H. L. Golterman. Eds. (Wageningen, The Netherlands: PUDOC 1975):19-42.
28. Prozorovskaya, A. A. "The Effect of Humic Acid and Its Derivatives on the Uptake of Nitrogen, Phosphorus, Potassium and Iron by Plants." Collected Papers. *Organomineral Fertilizers* (Organomineral'nye udobreniya). *Trudy. nach-issles udobr. insektisid. Fungisid.* 127 (1936).
29. Chaminade, R. "Action de L'Acide Humique sur le Développement et la Nutrition Minéral des Végétaux." *Trans. 6th International Cong. Soil. Sci.* Paris. 4:65 (1956).

30. Lee, Y. S., and R. J. Bartlett. "Stimulation of Plant Growth by Humic Substances." *Soil Sci. Soc. Am. J.* 40:876-879 (1976).
31. Biber, V., and K. Magaziner. "The Effect of Humic and Fulvic Acids on the Respiration of Isolated Plant Tissues." *Dokl. Akad. Nauk. SSSR.* 76:609 (1951).
32. Fialova, S. "Influence of Sodium Humate and Nutritive Conditions on the Content of Nucleic Acid, Particularly the Ribosomal Ribonucleic Acid in Wheat Roots." *Biol. Plant.* 11:8-22 (1969).
33. Jelenic, Dj. B., M. Hajdukovic and Z. Aleksic. "The Influence of Humic Substances on Phosphate Utilization from Labelled Superphosphate." *Proc. Symp. The Use of Isotopes in Soil Organic Matter Studies.* (Oxford: Pergamon Press, 1966):85-99.
34. D'yakonova, K. V. "Iron-Humus Complexes and Their Role in Plant Nutrition." *Soviet Soil Sci.* 7:692-698 (1962).
35. Schnitzer, M., and P. A. Paopst. "Effect of a Soil Humic Compound on Root Initiation." *Nature.* 213:598-599 (1967).
36. Rashid, M. A., and J. Leonard. "Modifications in the Solubility and Precipitation Behaviour of Various Metals as a Result of Their Interaction with Sedimentary Humic Acid." *Chem. Geol.* 11:89-97 (1973).
37. Smidova, M. "The Influence of Humic Acid on the Respiration of Plant Roots." *Biol. Plant.* 2:152-164 (1960).
38. Khristeva, L. A., and N. V. Luk'yanenko. "Role of Physiologically Active Substances in the Soil Humic Acids, Bitumens, Vitamin B, C, P-P, A and D in the Life of Plants and Their Replenishment." *Soviet Soil Sci.* 10:1137-1141. (1962).
39. Pringsheim, E. G. "Die Kultur der Algen in Agar." *Beitr. Biol. Pflanz.* 11:305-333 (1912).
40. Nechutova, H., and V. Tichy. "Effect of Humus Substances in the Production and the Quality of the Biomass of *Scenedesmus quadricauda.*" *Arch. Hydrobiol.* Suppl. 39:26-37 (1970).
41. Hoeffner, S. L., and Manahan, S. E. "Influence of Coal Humic Acid on the Growth of *Chlorella vulgaris* Algae." *J. Environ. Sci. Health,* A15:149-161 (1980).
42. Toledo, A. P. P., J. C. Tundisi and V. A. D'aquino. "Humic Acid Influence on the Growth and Copper Tolerance of *Chlorella* sp." *Hydrobiologia.* 71:261-263 (1980).
43. Prat, S. "The Influence of Humus Substances on Algae." *Folia Biol.* 1:321-326 (1955).
44. Lange, W. "Blue-Green Algae and Humic Substances." Paper presented at the 13th Conference of the International Assoc. Great Lakes Res. Buffalo, NY:58-70 (1970).
45. Prakash, A., and M. A. Rashid. "Influence of Humic Substances on the Growth of Marine Phytoplankton: Dinoflagellates." *Limonol. Oceanog.* 13:598-606 (1968).
46. Prakash, A., M. A. Rashid, A. Jensen and D. V. Subba Rao. "Influence of Humic Substances on the Growth of Marine Phytoplankton: Diatoms." *Limnol. Oceanog.* 18:516-524 (1973).
47. Provasoli, L. "Organic Regulation of Phytoplankton Fertility." In *The Sea*, Vol. 2, M. N. Hill, Ed. (London: Interscience, 1963):165-219.
48. Skulburg, O. M. "Algal Cultures as a Means to Assess the Fertilizing

Influence of Pollution." Paper 6. *3rd International Conf. on Water Pollution Res.* Munich. Water Poll. Cont. Fed. Washington, D.C. (1966).
49. Doig, M. T., and D. F. Martin. "The Effect of Naturally Occurring Organic Substances on the Growth of a Red Tide Organism." *Water Res.* 8:601-606 (1974).
50. Provasoli, L., J. J. A. McLaughlin and M. R. Droop. "The Development of Artificial Media for Marine Algae." *Arch. Microbiol.* 25:392-428 (1957).
51. Siegel, A. "Metal-Organic Interactions in the Marine Environment." In *Organic Compounds in Aquatic Environment.* S. D. Faust and J. V. Hunter, Eds. (New York: Marcel Dekker, 1971):265-295.
52. Rook, J. J. "Formation of Haloforms during Chlorination of Natural Waters." *Water Treatment and Examination.* 23, Part 2:234 (1974).
53. Bellar, T. A., J. J. Lichtenberg and R. C. Kroner. "The Occurrence of Organohalides in Chlorinated Drinking Water." *J. Am. Water Works Assoc.* 66:703 (1974).
54. Rook, J. J. "Chlorination Reactions of Fulvic Acids in Natural Waters." *Environ. Sci. Technol.* 11:478 (1977).
55. Stevens, A. A., C. J. Slocum, D. R. Seeger and G. C. Robeck. "Chlorination of Organics in Drinking Water." *J. Am. Water Works Assoc.* 68:615 (1976).
56. Stevens, A. A., and J. M. Symons. "Measurement of Trihalomethane and Precursor Concentration Changes." *J. Am. Water Works Assoc.* 69:546 (1977).
57. Bolton, C. M. "Cincinnati Research in Organics." *J. Am. Water Works Assoc.* 69:405 (1977).
58. Stokinger, H. E. "Toxicology and Drinking Water Contaminants." *J. Am. Water Works Assoc.* 69:399 (1977).
59. Tardiff, R. G. "Health Effects of Organics: Risk and Hazard Assessment of Ingested Chloroform." *J. Am. Water Works Assoc.* 69:658 (1977).
60. "Public Health Service Drinking Water Standards." *Public Health Service Publication No. 956.* U.S. Dept. of Health, Education and Welfare. Rockville, MD (1962), p. 21.
61. Woodward, R. L. "Review of the Relations of Water Quality and Simple Goiter." *J. Am. Water Works Assoc.* 55:887 (1963).
62. Heuper, W. C., and W. W. Payne. "Carcinogenic Effects of Adsorbates of Raw and Finished Water Supplies." *Am. J. Clin. Pathol.* 39:475 (1963).
63. "Research Committee on Color Problems." Report for 1966. *J. Am. Water Works Assoc.* 59:1023 (1967).
64. Gjessing, E. T. "Physical and Chemical Characteristics of Aquatic Humus." (Ann Arbor, MI: Ann Arbor Science, Publishers, Inc., 1976) Ch. 7.
65. Larson, T. E. "Corrosion Phenomenon–Causes and Cures." In *Water Quality and Treatment.* 3rd Ed. (New York: McGraw-Hill Book Company, 1971):311.
66. Van Beneden, G., and E. Leclerc. "Les Matières Humiques, Leur Comportement dans les Eaux ou Général, Leur Role dans la Corrosion des Métaux." *Technol. Water.* (Czech) 8:225 (1964).
67. Menke, C. G., "Removal of Natural Color-Causing Substances in Laboratory Analysis." *J. Am. Water Works Assoc.* 54:303 (1962).

68. "Standard Methods for the Examination of Water and Wastewater." 14th Ed. (*APHA, AWWA, WPCF, Washington, D.C. 1976*):202.
69. Brown, V. M., T. L. Shaw and D. G. Shulen. "Aspects of Water Quality and the Toxicity of Copper to Rainbow Trout." *Water Res.* 9:797-803 (1974).
70. Klocking, R. "Influence of Humic Acids on the Toxicity of Lead." *Proc. Eur. Soc. Toxicol.* 16:258-261 (1975).
71. Visser, S. A. "Some Biological Effects of Humic Acid in the Rat." *Acta Biol. Med.* Ger. 31:569-581 (1973).
72. Khan, S. U., and M. Schnitzer. "The Retention of Hydrophobic Organic Compounds by Humic Acid." *Geochim. Cosmochim. Acta.* 36:745 (1972).
73. Khan, S. U. "Interaction of Humic Acid with Chlorinated Phenoxyacetic and Benzoic Acids." *Environ. Lett.* 4:141 (1973).
74. Hiller, E. "Action of Certain Humic Acids from Bogs in the Treatment of Rheumatic Fever." *Med. Monatsschr.* 6:302 (1952); *Chem. Abstr.* 46:875 (1972).
75. Biber, V. A., and N. S. Bogolyubov. "Biological Activity of Soil and Peat Humic Acid." *Doklady Adad. Nauk SSSR.* 76:313 (1951); *Chem. Abstr.* 45, 3974 (1951).
76. Van Beneden, G. "Les Matières Organiques dans les Eaux et les Agents de Balneotherapie." *Presse Therm. Clim.* 108:195 (1971).
77. Canada Department of National Health and Welfare. *Guidelines for Canadian Drinking Water Quality 1978.* Supporting Documentation, Supply and Services Canada (1979).

CHAPTER 26

EFFECT OF AQUATIC HUMUS ON UPTAKE AND TOXICITY OF CADMIUM TO *Selenastrum capricornutum* Printz

Josef Sedlacek,* Torsten Källqvist and Egil Gjessing

Norwegian Institute for Water Research
Oslo, Norway

ABSTRACT

The influence of aquatic humus on the toxicity and the kinetics of bioaccumulation of cadmium in the green alga *Selenastrum capricornutum* Printz was studied. The results showed that the toxic effect of cadmium on algae was reduced by the addition of humus, and EC_{50} increased from 6 μg/l in growth medium without humus to 150 μg/l at the highest humus concentrations, 8.1 mg C/l. The accumulation studies were carried out with ^{109}Cd- and ^{115m}Cd-labeled solutions of cadmium chloride. It was found that the logarithm of the amount of cadmium accumulated in the algae is proportional to the logarithm of algal biomass. In the concentration range below 5 μg Cd/l, where no toxicity was observed, the presence of humus reduced the cadmium uptake as well as the sorption of cadmium on the walls of the cultivation vessels. After 7 days more than 80% of added ^{109}Cd was taken up by the algae in medium without humus, including the part originally adsorbed on walls. At humus concentrations of 1.6 and 8.1 mg C/l, 40 to 55% and <20%, respectively, of added ^{109}Cd was taken up after 7 days.

At cadmium concentrations above 10 μg/l the accumulated cadmium at a

*Present address: Department of Nuclear Chemistry, Technical University of Prague, Prague, Czechoslovakia.

fixed algal biomass was proportional to the cadmium concentration in solution and inversely proportional to humus concentration.

The results show that natural aquatic humus affects the availability of cadmium to algae. Humus reduces the rate of accumulation and the toxic effect of cadmium due to changes in the chemical speciation of cadmium in solution.

INTRODUCTION

Humic substances represent a major fraction of the organic matter in natural waters. It has been well established that this aquatic humus, apparently in solution, is able to form complexes or chelates with heavy metal ions in natural waters [1,2] and thereby change the state of trace metals in solution [3–7]. This affects their migration in the aquatic environment [8], and their toxicity to living organisms.

Recently, there have been several reports on the influence of inorganic and organic chelators on the uptake, accumulation and toxicity of heavy metals to living organisms. It has been demonstrated [9–11] that chelation of copper reduces copper toxicity to algae by decreasing the activities of free cupric ion. Complexes of copper with natural organic ligands are not toxic to algae [12,13].

The same observations have also been reported for zinc [14,15] and for the toxicity of cadmium to grass shrimps [16].

It should be pointed out, however, that also other parameters like alkalinity, pH, hardness and suspended matter influence metal speciation and toxicity [17–21].

Nearly all the above mentioned studies were carried out with well defined inorganic and organic ligands and other simple conditions, which allowed computation of metal speciation in solution from known stability constants and enabled the direct determination of free ion concentration using ion selective electrodes.

In general relatively little is known about the influence of natural organic substances on the toxicity of heavy metals, particularly cadmium.

As a whole the reports show, first of all, that this is a very complex matter with a number of variables under natural conditions. The bioavailability of cadmium is influenced by other heavy metals in the aquatic environment [22,23], and it is different for various species [24,25] and depends on the chelating agents involved [26–30]. In addition, the chelating agent itself may have a positive [31] or negative biological effect, and also cadmium speciation in natural waters is complicated [7], and most probably different chemical species have different biological effects.

The aim of this chapter is to throw more light on the effect of humus substances on the toxicity of cadmium to green algae *Selenastrum capricornutum* Printz and to provide additional data on the influence of cadmium and humus concentrations on the kinetics of cadmium bioaccumulation by green algae and on cadmium behavior in solution in the presence of aquatic humus.

EXPERIMENTAL

Natural humic water (HW) used in the experiments was collected from the outlet of a small bog area (Hellerudmyra, Baerum, Norway). Water samples collected in 20-liter polythene bottles were stored at 4 °C and filtered through Whatman GF/C glassfiber filters before use. Chemical analyses of water samples taken on January 11, 1980, henceforth designated HW (1), and on June 22, 1979, HW (2), are shown in Table I.

The water samples are characterized by high concentrations of humic substances, as shown by the high values for color and total organic carbon (TOC).

Synthetic water (SW) used in the experiments had similar macrocomposition as the natural water. It was prepared immediately before use by 100X dilution of stock solution in triple-distilled water. The svnthetic water had the following composition:

5.62 mg $NaHCO_3$/l
1.18 mg $MgCO_3$/l
7.17 mg $CaCO_3$/l
10.96 mg $CaCl_2$/l

Cadmium concentration in SW was 0.40 μg/l.

Table I. Chemical Composition of the Humus Water Samples HW (1) and HW (2)

	pH	Electric Cond. (μS/Cm)	Color (mg Pt/l)	TOC (mg/l)	Ca (mg/l)	Mg (mg/l)	K (mg/l)
HW (1)	4.43	30.2	129	16.2	1.25	0.278	0.177
HW (2)	4.83	24.4	260	19.8	1.32	0.271	0.268
	Na (mg/l)	**Cu (μg/l)**	**Cd (μg/l)**	**Al (μg/l)**	**Fe (μg/l)**	**SO_4 (mg/l)**	**Cl (mg/l)**
HW (1)	1.13	16	0.21	350	460	4.9	2.5
HW (2)	1.45	55	<10	450	900	4.5	2.0

Growth medium (GM) used was fully inorganic, without EDTA and trace metal addition, and had the following composition:

476 mg $NaNO_3$/l
59 mg $Ca(NO_3)_2 \cdot 4\ H_2O$/l
31 mg $K_2\ HPO_4$/l
25 mg $MgSO_4 \cdot 7\ H_2O$/l
21.2 mg Na_2CO_3/l

It was sterilized at 120 °C for 20 min and filtered through a Whatman GF/C filter. Cadmium concentration in GM was 0.24 μg/l. In all experiments this medium was diluted ten times with SW or HW + SW. Finally 1 ml/l of a solution containing 483 mg $FeCl_3 \cdot 6\ H_2O$ in 0.1 *M* HCl was added.

Solutions used in experiments were prepared from analytical grade chemicals and triple-distilled water. Their cadmium concentrations were adjusted with stock solution of 1.854 g $CdSO_4$/l in 0.1 *M* HCl. 5×10^{-6} *M* HNO_3, 0.1 *M* NaCl, 0.01 *M* Na_2EDTA, concentrated HNO_3, 10% HNO_3 and acetone were used as washing solutions. Lugol's solution was used for stopping algal growth.

Centrifugation was done at 10,000 rpm using an ordinary laboratory centrifuge. Algal cells were counted on a Coulter Counter model ZB equipped with a Chanellyzer 1000 and a desk computer. Size distributions and particle volumes were calculated.

Radionuclides and Activity Measurement

Radionuclides ^{115m}Cd (CAS. 1) with specific activity 37 MBq/ml (1.7 mg Cd/ml) and ^{109}Cd (CUS. 1) with specific activity 7 MBq/ml (0.156 μg Cd/ml) as $CdCl_2$ in 0.1 *M* MCl were used (supplied by The Radiochemical Centre Ltd., Amersham, England). Detector NaI(Tl) coupled with Scaler-Timer ST-5 (Nuclear Enterprises Ltd., U.K.) was used for activity measurements. Samples for counting were prepared by transferring an aliquot of solution into polyethylene ampoules (~6 ml) or by evaporation of a known volume of solution on a filter paper disk with diameter of 20 mm, which was mounted on Scotch tape fixed in a paper frame. In the same way the membrane filters with separated algae were fixed for counting.

Procedures and Calculations

The accumulation of cadmium was studied at different concentrations of natural aquatic humus. The study was carried out for TOC = 0 mg C/l (SW), TOC = 0.16, 1.62 and 8.1 mg C/l (HW (1) diluted with GM and SW to 1, 10 and 50%, respectively) and TOC = 1.98 mg C/l (10% HW (2)). Two ranges of cadmium concentrations have been tested; low (0.4, 1 and 5 μg Cd/l) and high (17, 56 and 112 μg Cd/l). Experimental solutions with the desired concentration of Cd and humus were prepared from the stock solutions of GM, HW, SW, nonactive Cd and ^{115m}Cd or ^{109}Cd. 160 ml of the solutions were transferred to 250-ml Erlenmeyer flasks and left for three days to equilibrate while being stirred by a magnetic stirring rod.

The equilibrated solutions were inoculated with *Selenastrum capricornutum* Printz from exponentially growing cultures in 10% GM. The cultures were incubated at room temperature on magnetic stirrers illuminated by fluorescent tubes (Osram H). The light intensity was approximately 4000 lux at the culture surface.

Algal wet biomass in 5 ml suspension m_a (μg) was calculated from the cell concentration C_a (cells/l) and average cell volume $\overline{V}_a$ (μm^3/cell) according to Equation 1, assuming that the specific weight of algae = 1 g/cm^3.

$$m_a = 5 \times 10^{-9} \times C_a\overline{V}_a \tag{1}$$

Samples of 5 ml each were taken at intervals for determination of total activity of the solutions (A_s). From 5 ml of the cultures the algae were separated by membrane filtration. The algae were washed by passing 5 ml of 5×10^{-6} *M* HNO_3 through the filter before the filters (θ 25 mm) were used for algal activity measurements (A). The same procedure was performed with a second portion of the same solution containing no algae for determination of the cadmium activity sorbed on the filters (A_F). The activity of cadmium sorbed by algae (A_a) was calculated according to Equation 2.

$$A_a = A - A_F \tag{2}$$

The activities A_s and A_a were expressed as percentages of the initial solution's activity (A_0) and were used for calculation of the concentration factor CF using Equation 3.

$$CF = \frac{A_a/m_a}{(A_s - A_a)/m_s} \tag{3}$$

where m_a = algal mass
m_s = solution mass

Equation 3 can be transformed into Equation 4 using (1) for V_s = 5 ml and assuming a specific density of algae = 1 g/cm^3.

$$CF = \frac{A_a}{A_s - A_a} \times \frac{1}{C_a\overline{V}_a} \tag{4}$$

where C_a = the number of cells per ml
$\overline{V}_a$ = mean volume of the cells

The toxicity tests were performed in 50-ml cultures in 100-ml flat bottom flasks on a shaking table. The light intensity was 6000 lux from Osram H fluorescent tubes, and the temperature was 20 ± 0.5 °C. The growth medium was the same as in the bioaccumulation experiments. Test algae (*Selenastrum capricornutum* Printz) were added from exponentially growing cultures. The test was carried out with different concentrations of Cd in the range 5 to 200 μg/l and with four levels of humus concentration (0, 0.162, 1.62 and 8.1 mg C/l). The test was performed in triplicate.

The cell numbers (n) in the cultures were determined each day for 4 days using a Coulter Counter. The average growth rate (μ) during the period 0 to 3 days from the start was calculated using Equation 5

$$\mu = \frac{\log n_3 - \log n_0}{3 \times \log 2} \tag{5}$$

The growth rates were expressed as percentages of the growth rate in control cultures without cadmium for each concentration of humus. The concentration that reduced the growth rate to 50% of the control was estimated from concentration/response plots.

RESULTS AND DISCUSSION

Uptake at Low Cadmium Concentrations

Figures 1 and 2 show the increase of algal biomass as weight in 5-ml suspension during the accumulation experiments at low cadmium concentrations (0.4 to 5 μg Cd/l). The initial concentration of algae was about 10^7 cells/l. The data for the exponential growth phase were treated by least squares method and the values of slopes μ for Equation 6 are presented in Table II.

$$\log m_a = a + \mu t \tag{6}$$

The data show that the growth rate of the algae was essentially the same at all combinations of cadmium and TOC in the ranges 0.4 to 5 μg Cd/l and 0 to 2 mg C/l. In contrast, the uptake of cadmium by the algae was significantly dependent on the TOC value.

In solutions with initial cadmium concentrations below 5 μg/l it was found that the uptake can be described by Equation 7

$$\log A_a = a + b \log m_a \tag{7}$$

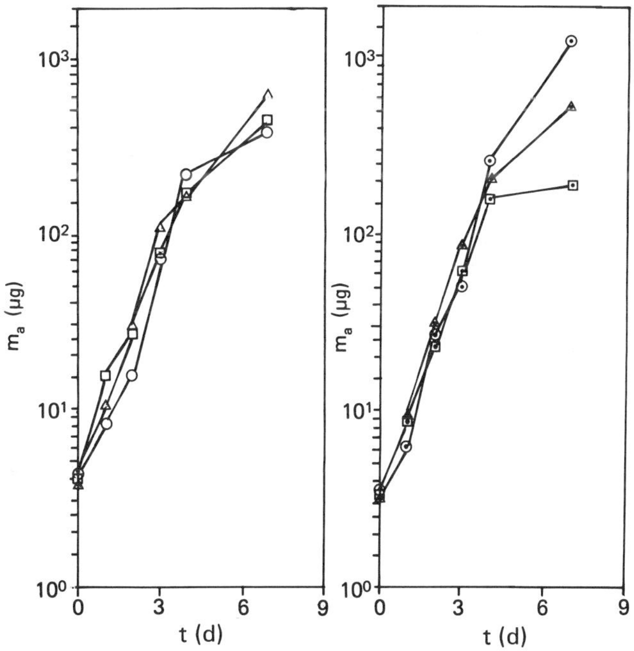

[^{109}Cd, A_0 = 766 cps/5 ml,

~ 0.4 µg Cd/l, TOC: 0 (○), 1.62 (△), 8.1 (□) mg C/l

~ 1 µg Cd/l, TOC: 0 (⊙), 1.62 (▲), 8.1 (⊡) mg C/l.]

Figure 1. Growth of algae (wet weight in 5-ml suspension) in uptake experiments with cadmium concentrations 0.4 and 1 µg/l.

Table II. Influence of Cadmium Concentration and TOC (mg C/l) on Algal Growth Rate (µ)

	µ (hr^{-1})					
C_{Cd} (µg/l)	TOC = 0	TOC = 1.62 (H1)	TOC = 8.1 (H1)	TOC = 0.16 (H1)	TOC = 1.98 (H2)	Figure
0.4	0.4324	0.4286	0.3405			1
1	0.4681	0.4651	0.2990			1
1	0.3382	0.4200		0.5431		2
5	0.4746	0.4557			0.4081	2
$\bar{\mu} \pm S^a$	0.428 ± 0.06	0.442 ± 0.02	0.344 ± 0.06			

[a] $\bar{\mu}$ = arithmetic mean; S = standard deviation.

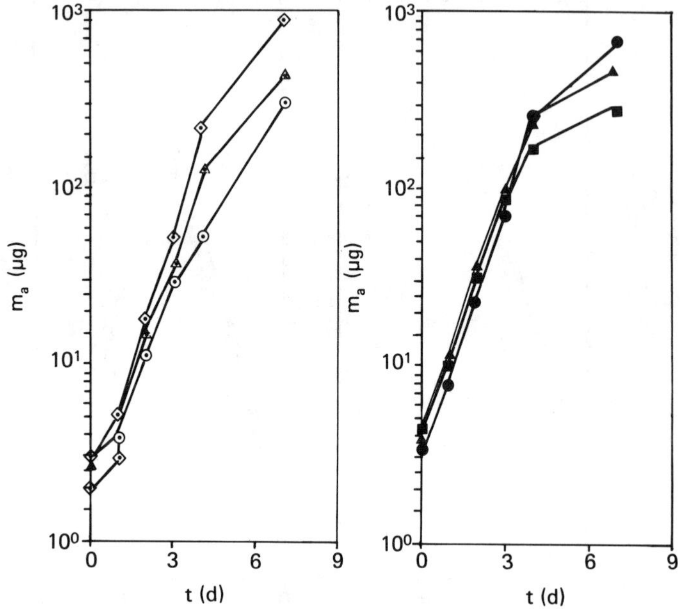

[^{109}Cd, A_0 = 766 cps/5 ml,

~ 1 µg Cd/l TOC: 0 (⊙), 0.16 (◈), 1.62 (▲) mg C/l

~ 5 µg Cd/l TOC: 0 (●), 1.62 (▲), 1.98 (■) mg C/l.]

Figure 2. Growth of algae (wet weight in 5-ml suspension) in uptake experiments with cadmium concentrations 1 and 5 μg/l.

where A_a (cps) is the activity accumulated by algae with wet mass m_a, μg. (Figures 3 and 4)

The values of the constants a and b in Equation 7 are summarized in Table III together with the corresponding correlation coefficient r and number of analyzed values n.

These results show that the values of slopes b are independent of TOC and of C_{Cd}; the average value is $\bar{b} \pm s = 0.816 \pm 0.13$. However, the parameter a depends strongly on TOC, decreasing with the increasing value of TOC. Humus water HW (2) with higher color is more effective in reducing the uptake as compared with HW (1), at approximately the same TOC. In Table III is also shown the solution of Equation 7 for m_a = 100 μg at the different combinations of cadmium and humus concentrations. The activity values

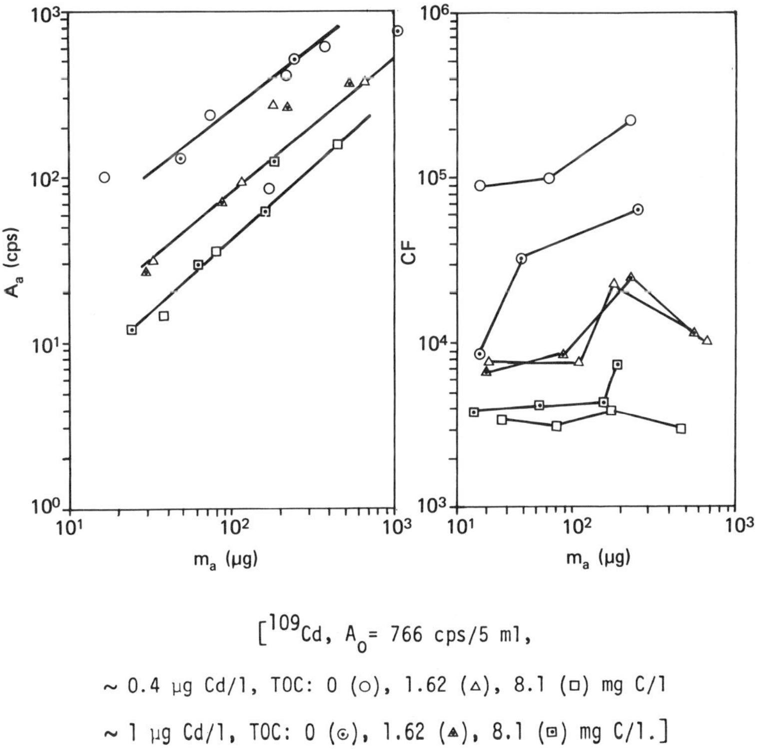

Figure 3. Effect of biomass on activity accumulated by algae (A_a) and concentration factor (CF).

Table III. Influence of TOC on Cadmium Uptake at Low Cadmium Concentrations

TOC (mg C/l)	a	b	r	n	C_{Cd} (μg/l)	A_a/m_a (cps/100 μg)	Figure
0	1.1812	0.6214	0.976	6	0.4 & 1	265	
1.62	0.2068	0.8549	0.998	6	0.4 & 1	82	3
8.1	−0.1566	0.9175	0.977	8	0.4 & 1	32	
0 & 0.162	0.8347	0.7699	0.964	10	1 & 5	236	
1.62	0.5562	0.7466	0.990	8	1 & 5	112	4
1.98	−0.0484	0.9850	0.974	4	5	83	

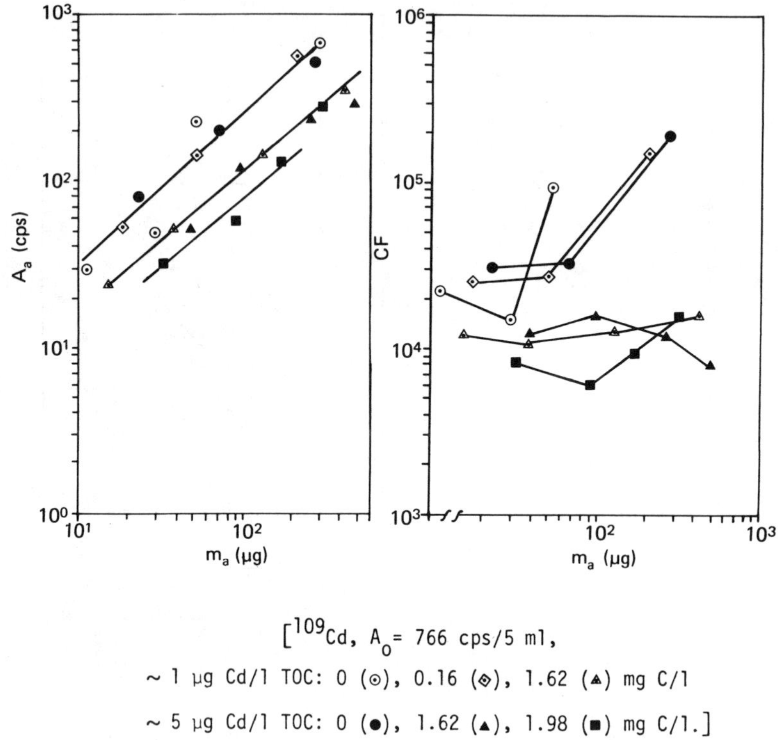

Figure 4. Effect of biomass on activity accumulated by algae (A_a) and concentration factor (CF).

show that the cadmium content of the algae is decreasing with increasing humus concentration at a given cadmium concentration.

The effect of humus on the accumulation of cadmium is also reflected by the concentration factor CF (see Table IV), which is dependent on TOC and is independent of C_{Cd} at constant TOC (see Figures 3 and 4). The concentration factor CF for cadmium decreases in the presence of aquatic humus.

The time-dependence of cadmium accumulation is shown in Figure 5. After 7 days more than 80% of added ^{109}Cd was taken up by the algae in media without humus. At humus concentrations of 1.62 and 8.1 mg C/l, 40 to 50% and <20%, respectively, of added ^{109}Cd was taken up after 7 days. From the changes in total activity in the algal suspension shown in Figure 5 it is seen that cadmium was sorbed on the walls but aquatic humus strongly suppressed the sorption, especially at very low cadmium concentrations. In the case of

Table IV. Influence of TOC (mg C/l) and Cadmium Concentration on Concentration Factor CF

C_{Cd} (μg/l)	$10^{-3} \times$ CF					
	TOC = 0	TOC = 1.62	TOC = 8.1	TOC = 0.162	TOC = 1.98	Figure
0.4	139 ± 75	12.3 ± 7.6	3.4 ± 0.4			3
1	35 ± 28	16 ± 9	5.1 ± 1.8			3
1	43 ± 43	13 ± 2		69 ± 74		4
5	88 ± 98	11.7 ± 3.3			9.8 ± 4	4
$\overline{CF}$ ± S[a]		13.2 ± 2	4.1 ± 1.2			

[a]S = standard deviation.

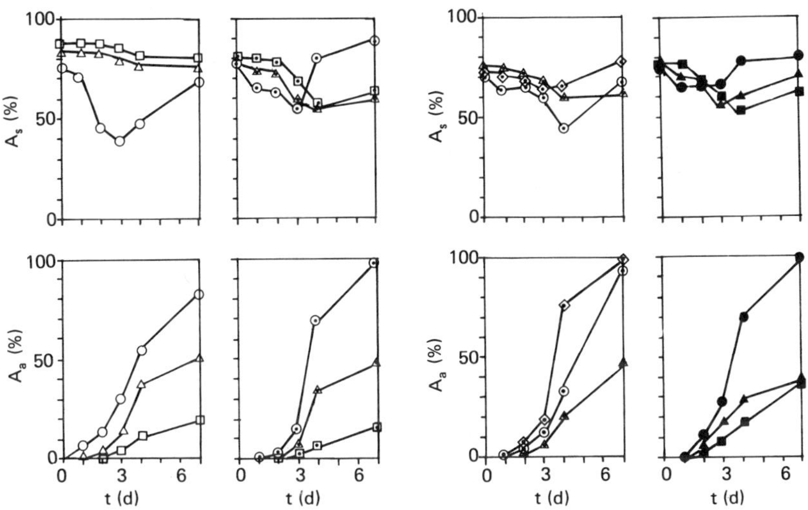

[^{109}Cd, A_0= 766 cps/5 ml,

~ 0.4 µg Cd/l, TOC: 0 (○), 1.62 (△), 8.1 (□) mg C/l

~ 1 µg Cd/l, TOC: 0 (⊙), 0.16 (◈), 1.62 (▲), 8.1 (▣) mg C/l

~ 5 µg Cd/l, TOC: 0 (●), 0.62 (▲), 1.98 (■) mg C/l.]

Figure 5. Variations in total suspension activity (A_s) and algae activity (A_a) in uptake experiments with low cadmium concentrations. A_s as a percentage of initial suspension activity and A_a as a percentage of A_s.

TOC = 0 mg C/l the cadmium initially adsorbed on the walls of the cultivation vessel was also available for accumulation by algae.

It was also found that adsorption of cadmium on membrane filters Millipore MA is lowered in the presence of aquatic humus. The activity of cadmium adsorbed on the filters (A_F) from solutions with cadmium concentration below 5 μg/l, pH 7 to 8 and with volume activity of ^{109}Cd 766 counts per second (cps)/5 ml amounted to 11.9 ± 1.4 cps, 5.3 ± 0.3 cps and 3.8 ± 0.8 cps for TOC 0, 1.62 and 8.1 mg C/l, respectively, when filtration was performed immediately after addition of radiotracer. The corresponding activities adsorbed from the same solutions 3 to 10 days after labeling were 22.5 ± 3.9 cps, 14.1 ± 2.4 cps and 10.5 ± 5.6 cps, respectively.

Beneš and Kopička [32] studied the adsorption behavior of cadmium traces in pure aqueous solutions. They observed that cadmium was strongly adsorbed on cellophane membranes and on the walls of vessels and that the adsorption from aged and fresh solutions was different. This corresponds well with our results found for sorption of cadmium on glass vessels and on membrane filters from more complex solutions.

It has been very well established that adsorption behavior of trace elements depends on their physicochemical state (speciation) [33]. Consequently, it can be concluded that immediately after addition, aquatic humus strongly affects the state of cadmium in solution, and this state changes with the age of solution. There is little doubt that also the bioavailability of cadmium to algae depends on the physicochemical state of cadmium in solution.

Uptake at High Cadmium Concentrations

The biomass production in the experiments at high cadmium concentrations in 5 ml of algae suspension is illustrated in Figures 6, 7 and 8. As can be seen from the figures the growth rates at all humus concentrations decreased with increasing cadmium concentrations. Further the highest growth rates were found at the highest humus concentration. It should be emphasized that the inoculum was lower in this experiment, which may influence the growth rate.

The kinetics of cadmium accumulation are shown in Figures 9, 10 and 11. The accumulation can be described by Equation 8

$$\log m_{Cd} = a + b \log m_a \tag{8}$$

where m_{Cd} (ng) is mass of cadmium accumulated by algae mass m_a (μg)

Values of m_{Cd} were calculated from measured activities of algae and specific activity of ^{115}Cd. The results of linear regression analysis according to Equation 8 are summarized in Table V. The last column in Table V shows the

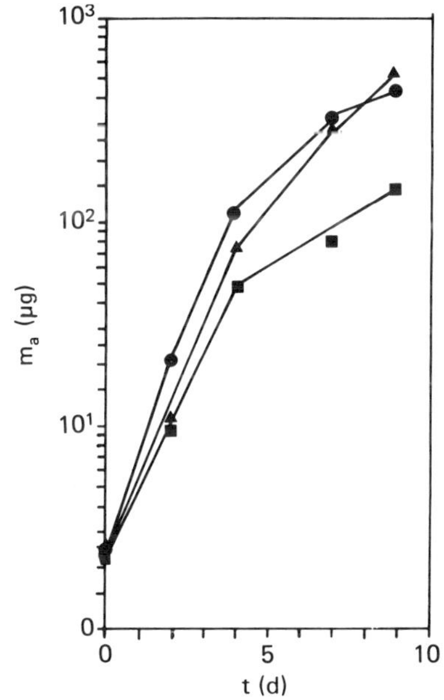

[^{115m}Cd, 8.1 mg TOC/l,
17 (●) 56 (▲), 112 (■) µg Cd/l.]

Figure 8. Growth of algae (wet weight in 5-ml suspension) in uptake experiments at high cadmium concentrations and 8.1 mg TOC/l.

Table VI. Influence of TOC (mg/l) and Cadmium Concentration on Concentration Factor CF

C_{Cd} (µg/l)	10^{-3} × CF		
	TOC = 0	TOC = 1.62	TOC = 8.1
17	18 ± 3	9 ± 2	4 ± 2
56	28 ± 13	21 ± 11	4 ± 2
112	13 ± 3	9 ± 5	3.2 ± 0.7

It was found that also for the higher cadmium concentrations used in this accumulation study the relative sorption of cadmium on membrane filters decreased with increasing cadmium concentration and with the addition of aquatic humus into the solution. Cadmium uptake by algae in both concentration ranges can be described by either Equation 6 or Equation 7. These

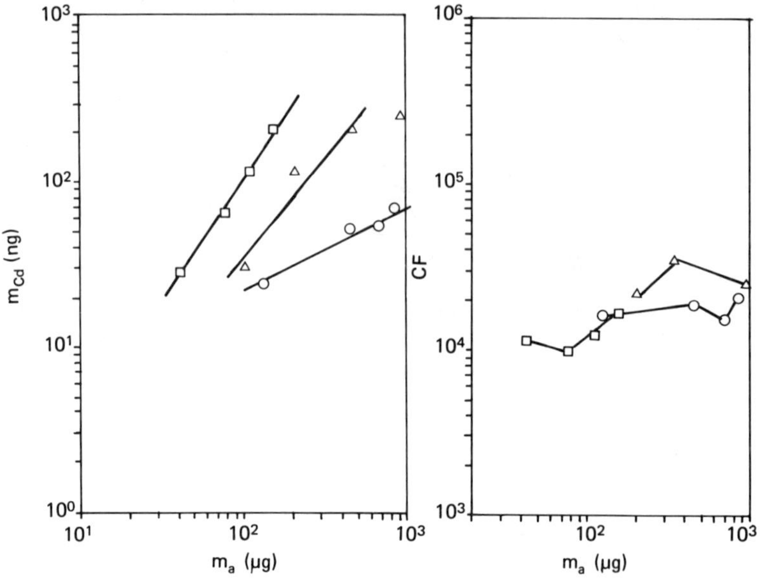

[^{115m}Cd, 0 mg TOC/l,

17 (○), 56 (△), 112 (□) μg Cd/l.]

Figure 9. Effect of biomass on amount of cadmium accumulated by algae (m_{Cd}) and concentration factor (CF) at high cadmium concentrations and 0 mg TOC/l.

equations give the same results, because activity is directly proportional to mass.

Toxicity Tests

The toxic effect of cadmium on *Selenastrum capricornutum* at different concentrations of humus is shown in Figure 12. The growth rate decreased with increasing cadmium concentration at TOC concentrations 0, 0.162 and 1.62 mg/l. At the highest humus concentration (TOC = 8.1 mg/l) 10 and 17 μg Cd/l significantly increased the growth rate as compared to the control cultures and 100 and 200 μg Cd/l caused a reduction of the growth rate. The same observation has been made in a similar investigation [30], but was not made in the uptake experiments. The reason for the apparent stimulation of algal growth by cadmium at high humus concentration is not clear, but may be connected to interactions between cadmium and iron, which is an essential nutrient for algae. In the uptake experiments the solutions were equilibrated

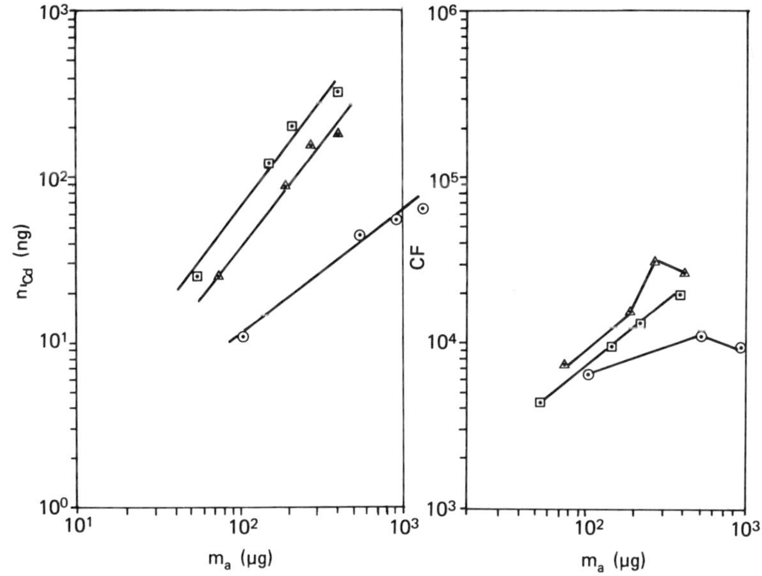

[115mCd, 1.62 mg TOC/l,
17 (⊙), 56(▲), 112(⊡)μg Cd/l.]

Figure 10. Effect of biomass on amount of cadmium accumulated by algae (m_{Cd}) and concentration factor (CF) at high cadmium concentrations and 1.62 mg TOC/l.

for 3 days before algae were added and the speciation of cadmium and iron may have been different from the toxicity tests where equilibration was not performed.

The toxicity tests showed that the toxicity of cadmium was reduced by the addition of humus, as reflected by the EC_{50} values shown in Figure 13. These results are in agreement with the conclusions a, b, c and d mentioned above. The reduction of the toxicity of cadmium to algae in the presence of aquatic humus thus can be explained by the decrease in cadmium uptake. Similar effects, i.e., the reduction of toxicity to algae and the decrease of accumulation by algae in presence of aquatic humus, were found by Sedláček et al. [34] for other heavy metals like barium, zinc and mercury. They also observed that natural aquatic humus and its separate fractions with molecular weight 1000 to 10,000 and 10,000 to 100,000 are more effective in reducing cadmium uptake by algae than fractions with molecular weight below 1000 and above 100,000.

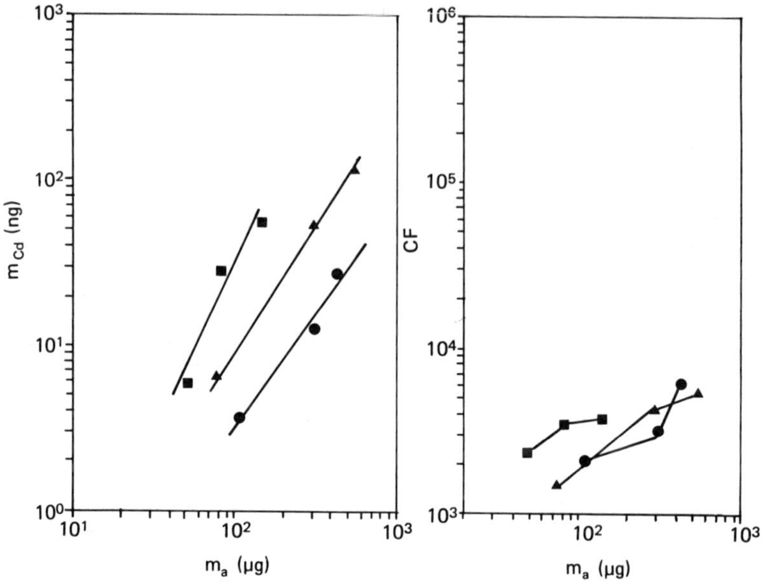

[^{115m}Cd, 8.1 mg TOC/1,

17 (●) 56 (▲), 112 (■) μg Cd/1.]

Figure 11. Effect of biomass on amount of cadmium accumulated by algae (m_{Cd}) and concentration factor (CF) at high cadmium concentrations and 8.1 mg TOC/l.

CONCLUSIONS

1. Aquatic humus significantly reduces cadmium accumulation by algae, which is reflected in a decrease in the concentration factor for cadmium in algae. The logarithm of accumulated cadmium activity or mass in algae is linearly proportional to the logarithm of algal biomass.
2. Reduction of toxicity of cadmium to green algae *Selenastrum capricornutum* Printz by aquatic humus can be explained as a result of changes of cadmium state in solution, which in turn reduces cadmium uptake by algae.
3. Aquatic humus changes the physico-chemical state of cadmium in solution. The presence of humus suppresses the adsorption of cadmium on membrane filters used for algae separation.

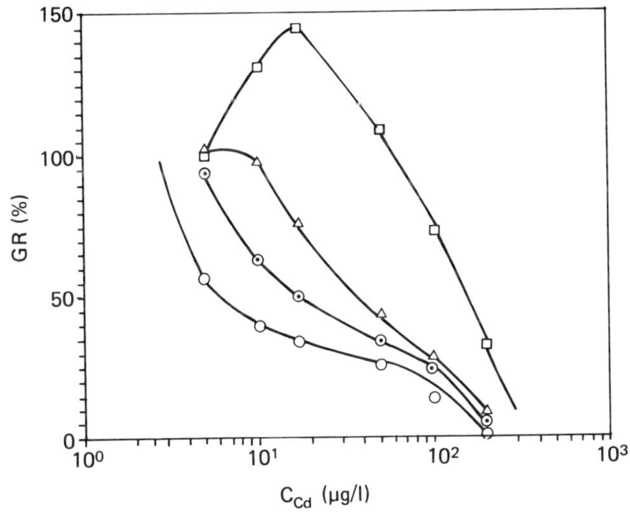

[TOC: 0 (○), 0.162 (⊙) 1.62 (△), 8.1 (□) mg/l.]

Figure 12. Effects of cadmium concentration (C_{Cd}) on growth rate (GR) as a percentage of controls at different concentrations of aquatic humus.

ACKNOWLEDGMENTS

We thank the Royal Norwegian Council for Scientific and Industrial Research for the postdoctorate fellowship grant to Josef Sedláček under which this research was carried out and the Institute for Energy Technology at Kjeller (Norway) for providing facilities for the radiochemical parts of this study.

REFERENCES

1. Mantoura, R. F. C., A. Dickson, and J. P. Riley. "The Complexation of Metals with Humic Materials in Natural Waters," *Estuar. Coast Mar. Sci.* 6:387-408 (1978).
2. Gjessing, E. T. *Physical and Chemical Characteristics of Aquatic Humus.* (Ann Arbor, MI.: Ann Arbor Science Publishers, Inc., 1976), 120 pp.
3. Beneš, P., E. T. Gjessing, and E. Steinnes. "Interaction between Humus and Trace Elements in Fresh Water," *Water Res.* 10:711-716 (1976).

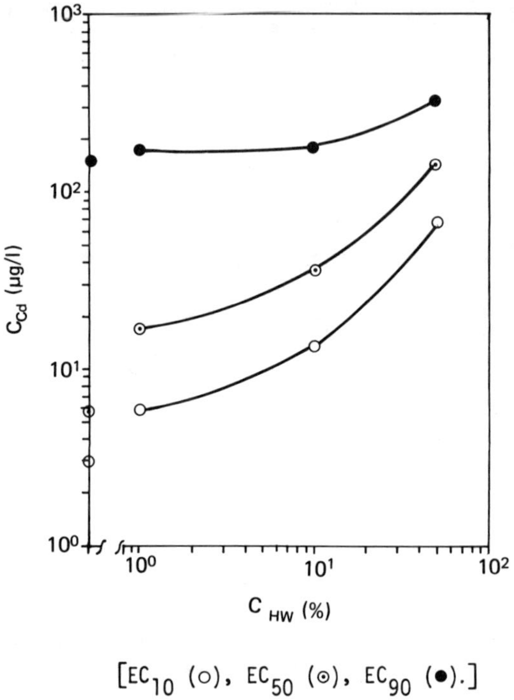

Figure 13. Effect of humus water concentration on EC values for cadmium toxicity to *Selenastrum capricornutum* (100% HW = 16.2 mg C/l).

4. Hart, B. T., and S. H. R. Davies. "Speciation of Fe, Cd, Cu, Pb and Zn in the Yarra River and Estuary, Australia," In: *International Conference on Management and Control of Heavy Metals in the Environment,* (London: CEP Consultants, 1979) pp. 466-471.
5. Steinberg, C. "Species of Dissolved Metals Derived from Oligotrophic Hard Water," *Water Res.* 14:1238-1250 (1980).
6. Florence, T. M., and G. E. Batley. "Determination of the Chemical Forms of Trace Metals in Natural Waters, with Special Reference to Copper, Lead, Cadmium and Zinc," *Tatanta* 24:151-158 (1977).
7. Florence, T. M., and G. E. Batley. "Chemical Speciation in Natural Waters," *CRC Critical Reviews in Anal. Chem.* 9:219-296 (1980).
8. Ling Ong, H., V. E. Swanson, and R. E. Bisque. "Natural Organic Acids as Agents of Chemical Weathering," *U.S. Geol. Survey Prof. Paper* 700 C; C 130-C 137 (1970).
9. Sunda, W., and R. L. Guillard. "The Relationship between Cupric Ion Activity and the Toxicity of Copper to Phytoplankton," *J. Mar. Res.* 34:511-529 (1976).

10. Andrew, R. W., K. E. Biesinger, and G. E. Glass. "Effect of Inorganic Complexing on the Toxicity of Copper to *Daphnia magna*," *Water Res.* 11:309-315 (1977).
11. Anderson, D. M., and F. M. M. Morel. "Copper Sensitivity of *Gonyaulax tamarensis*," *Limnol. Ocenog.* 23:283-295 (1978).
12. Sunda, W. G., and J. A. M. Lewis. "Effect of Complexation by Natural Organic Ligands on the Toxicity of Copper to a Unicellular Alga, *Monochreysis lutheri*," *Limnol. Ocenog.* 23:870-876 (1978).
13. Gächter, R., J. S. Davis, and A. Marés. "Regulation of Copper Availability to Phytoplankton by Macromolecules in Lake Water," *Environ. Sci. Technol.* 12:1416-1421 (1978).
14. Allen, H. E., R. M. Hall, and T. D. Brisbin. "Metal Speciation. Effect on Aquatic Toxicity," *Environ. Sci. Technol.* 14:441-443 (1980).
15. Anderson, M. A., and F. M. M. Morel. "Growth Limitation of a Coastal Diatom by Low Zinc Ionactivity," *Nature* 276:70-71 (1978).
16. Sunda, W. G., D. W. Engel, and R. M. Thuotte. "Effect of Chemical Speciation on Toxicity of Cadmium to Grass Shrimps, *Paleamonetes fugio*: Importance of Free Cadmium Ion," *Environ. Sci. Technol.* 12:409-413 (1978).
17. Chakoumakos, Ch., R. C. Russo, and V. Thurston. "Toxicity of Copper to Cutthroat Trout (*Salmo clarki*) under Different Conditions of Alkalinity, pH and Hardness," *Environ. Sci. Technol.* 13:213-219 (1979).
18. Davies, P. H., J. P. Goettl, Jr., J. R. Sinley, and N. F. Smith. "Acute and Chronic Toxicity of Lead to Rainbow Trout *Salmo gairdneri*, in Hard and Soft Water," *Water Res.* 10:199-206 (1976).
19. Brown, V. M., T. L. Shaw, and D. G. Sharben. "Aspects of Water Quality and the Toxicity of Copper to Rainbow Trout," *Water Res.* 8:797-803 (1974).
20. Nriagu, J. O., Ed. *Cadmium in the Environment. Part 1. Ecological Cycling.* (New York: John Wiley and Sons, Inc., 1980), 682 pp.
21. Klöcking, R. "Influence of Humic Acids on the Toxicity of Lead," *Proceeding of the European Society of Toxicology* 16:258-261 (1975).
22. Ray, S., D. W. McLeese, and D. Pezzack. "Chelation and Interelemental Effects on the Bioaccumulation of Heavy Metals by Marine Invertebrates," In: *Management and Control of Heavy Metals in the Environment* (London: CEP Consultants, Ltd., 1978), pp. 35-38.
23. Gerhards, U., and H. Weller. "The Uptake of Mercury, Cadmium and Nickel by *Chlorella pyrenoidosa*," *Z. Pflanzen-physiolog.* 82:292-300 (1977).
24. Duddridge, J. E., and M. Wainwright. "Heavy Metal Accumulation by Aquatic Fungi and Reduction in Viability of *Gamarus pulex* Fed Cd^{2+} Contaminated Mycelium," *Water Res.* 14:1605-1611 (1980).
25. Cossa, D. "Sorption du Cadmium par une Population de la Diatomée *Phaeodactylum tricornutum* en Culture," *Mar. Biol.* 34:163-167 (1976).
26. Gardiner, J. "The Chemistry of Cadmium in Natural Waters–1. A Study of Cadmium Complex Formation Using Cadmium Specific Ion Electrode," *Water Res.* 8:23-30 (1974).
27. Giesy Jr., J. P., G. J. Leversee, and D. R. Williams. "Effect of Naturally Occurring Aquatic Organic Fractions on Cadmium Toxicity to *Simocephalus serrulatus* (*Daphnidae*) and *Gambusia affinis* (*Poeciliidae*)." *Water Res.* 11:1013-1020 (1977).

28. Podolski, J. E. "Cadmium Bioaccumulation Assays. Their Relationship to Various Ionic Equilibria in Lake Superior Water," *Environ. Sci. Technol.* 13:701-706 (1979).
29. George, S. G., and T. H. Coombs. "The Effect of Chelating Agents on the Uptake and Accumulation of Cadmium by *Mytilus edulis*," *Mar. Biol.* 39:261-268 (1977).
30. Gjessing, E. "The Effect of Aquatic Humus on the Biological Availability of Cadmium," *Arch. Hydrobiol.* 91:144-149 (1981).
31. Prakash, A., and M. A. Rashid. "Influence of Humic Substances on the Growth of Marine Phytoplankton: Dinoflagellates," *Limnol. Oceanog.* 13:598-606 (1968).
32. Beneš, P., and K. Kopička. "The State and Adsorption Behavior of Traces of Cadmium in Aqueous Solutions," *J. Inorg. Nucl. Chem.* 38:2043-2048 (1976).
33. Beneš, P., and V. Majer. *Trace Chemistry of Aqueous Solutions: General Chemistry and Radiochemistry* (Prague, Czechoslovakia: Academia, 1980), 252 pp.
34. Sedláček, J., T. Källqvist, and E. Gjessing. "The Effect of Aquatic Humus on Uptake and Toxicity of Cadmium, Barium, Zinc and Mercury to the Green Algae *Selenastrum capricornutum* Prinz," Norwegian Institute for Water Research, 1. Nat. Report (Oslo, Norway: NTNFs utvalg for drikkevannsforskning, Drikkevannsrapport, 3/81, under preparation).

CHAPTER 27

PRIORITIES IN HUMIC RESEARCH

Russell F. Christman

Department of Environmental Sciences and Engineering
School of Public Health
University of North Carolina at Chapel Hill
Chapel Hill, North Carolina

Egil Gjessing

Norwegian Institute for Water Research
Oslo, Norway

Perhaps chemists lack persistance or maybe the research findings have not been sufficiently encouraging, but whatever the reason, relatively few scientists in the last century have devoted their energies to research on the nature and properties of aquatic and terrestrial humic materials. Those who have struggled valiantly in the last decade have found their patience rewarded by renewed interest in this very ancient subject.

In 1974, J. J. Rook [1] of the Rotterdam Water Works suggested that aquatic humic materials were ubiquitous reaction precursors of halogenated methanes via reaction with chlorine disinfectant. Subsequent analytical surveys in the U.S. [2,3] established that chloroform and its sister trihalomethanes were commonly present in chlorinated drinking waters of the U.S. and The National Cancer Institute reported that $CHCl_3$ is a potential human carcinogen. The number of humic chemistry research reports in the literature has risen dramatically since the recognition of this problem.

The benefits from increased understanding that will accrue from this renewed interest are only partially evident at present. New experimental directions in isolation and purification of natural humic materials, increased identification of structural moieties resulting from controlled degradation (including chlorine degradation) and revised concepts of the mechanisms by which metal ions interact with these complex natural products are among the initial scientific findings.

AQUATIC HUMIC MATERIALS IN THE CARBON CYCLE

Major research questions pertaining to the extent of occurrence, qualitative nature and role of humic materials in natural aquatic systems have been explored in the last decade and are still largely unanswered. During this period factual evidence of the levels of natural organic carbon carried by terrestrial streams has been slowly gathering, but increases in our understanding of the chemical characteristics of these complex natural products have been more modest.

The recent discovery of halogenated organic reaction products in drinking water [1–3] has called attention to the ubiquitous occurrence of complex forms of organic carbon in terrestrial water systems. Knowledge of the annual flux rates of carbon in terrestrial streams to the oceans has undergone significant revision in the last decade (Table I). The increase in inorganic carbon values is probably due to better data from the study of a wider and more representative range of terrestrial streams and not to actual increases in stream carbon content. The quantities of carbon transported annually to the earth's oceans in this manner are large (approximately 1.0 billion tons/yr) compared to annual chemical industry production, but small compared to other carbon transfer routes to the oceans (Table II).

Dissolved organic carbon (DOC) values vary considerably with climatic zone, ranging from 3 mg/l in temperate and arid zones to 6 mg/l in tropical zones and 10 mg/l in subarctic zones. Meybeck [6] has reported a stream volume weighted world average of 5.75 mg DOC/l. Thus, our understanding of natural carbon flux in terrestrial streams has broadened significantly within the last decade. In the 1960s and 1970s research reports noted the presence and significance of complex organic carbon in terrestrial surface waters and groundwaters and today the presence of these organic forms is seen as an important part of a natural global process.

Table I. Estimate of Annual Carbon Flux to Oceans from Terrestrial Streams (10^{12} mol C/yr)

	1973 [4]	1981 [5]
Inorganic carbon (C in HCO_3^-)	37.5	33
Dissolved organic carbon (DOC)	10.7	18
Particulate organic carbon (POC)	5.8	15

Table II. Selected Annual Carbon Fluxes to Oceans (10^{12} mol C/yr) [5]

Route	Annual Flux
Terrestrial streams	66
Ocean photosynthesis	3750
Net dissolution from atmosphere	159

The data will undoubtedly continue to improve, as we still have limited knowledge of the seasonal variations in DOC values among climatic zones. More important to this discussion is the fraction of DOC that is comprised of humic materials. It is attractive to assume that most, if not all, of the DOC in terrestrial streams is natural product organic material. Multiple opportunities exist in the earth's water and exogenic cycles for the transfer of virtually every form of terrestrial carbon to various environmental reservoirs (oceans, atmosphere, sediments, etc.) through dissolution or suspension in water. Qualitatively, the organic carbon pool of terrestrial streams may contain a myriad of structural moieties arising not only from the extraction and/or sorption of components of terrestrial vegetation in all stages of growth and decay, the organic portion of soil and any transformation products formed in subsequent stream transit, but also from the organic structures of soil and aquatic microorganisms and their numerous metabolic and natural degradation products [7,8]. Considering the facts that 90% of the terrestrial biospheric carbon (standing biomass) is tied up in woody/grassy tissue [9] and that microbial communities have had hundreds of millions of years to adapt to the transfer process, it is reasonable to assume that these sources are ubiquitous in nature.

Finding a meaningful factual verification for this attractive assumption is much more difficult. Comparison of $CHCl_3$ yield data on relatively pure aquatic humic and fulvic acid preparations with that of natural waters of known total organic carbon (TOC) may provide one instructive measure. Various yield figures from humic preparations have been published [1, 10–12] and are in the range of 3 to 16 μg $CHCl_3$ (as C)/mg TOC for various chlorine contact times and pH conditions. More recently, Colthurst and Singer [13] reported chloroform formation potential values for humic and fulvic acid material obtained from a North Carolina surface water which averaged 13.5 μg $CHCl_3$ (as C)/mg TOC at pH 7.0. Singer et al. [14] surveyed a variety of North Carolina surface drinking water supplies and reported a correlation between terminal chloroform yields and TOC which shows that 1.0 mg TOC produced 6.46 μg $CHCl_3$ (as C) at pH 6.7.

These studies provide the best available data base for estimating the amount of humic materials in aqueous DOC or TOC since they were performed on waters and humic fractions obtained from the same geographic area and they employed quite similar conditions for $CHCl_3$ generation and measurement. The data suggest that 48% (6.46/13.5) of the organic carbon in North Carolina terrestrial waters was humic and/or fulvic acid material. The limitation in this approach, however, is the assumption that only the humic content of water samples will produce $CHCl_3$ upon exposure to chlorine. Many substances have been shown [15–17] to produce $CHCl_3$ under these conditions, and any nonhumic contributions would make the above estimate unrealistically low.

Aquatic humic materials are strongly adsorbed by certain XAD-8 macroporous resins with excellent elution efficiencies [18]. Thurman and Malcolm [19] have reported recovery factors of 80 to 95% for stream organic carbon bound to XAD-8 resins in their preparative isolation scheme for aquatic humic substances. Thurman's current work on terrestrial stream organic matter at the U.S. Geological Survey provides an interesting and comprehensive perspective on this question. Experience with a variety of samples leads Dr. Thurman to the belief that approximately 10% of the DOC consists of identifiable compounds such as amino acids, carbohydrates, and volatile and fatty acids [20]. The balance consists of carbon in polymeric form (90% of DOC), most of which is humic in nature (50% of DOC), as judged by retention on XAD-8 resins.

It is probably not wise to dwell on the fact that two independent approaches to estimating the humic fraction of DOC agree so closely (48 and 50%), since wide variations probably exist in nature. More importantly, Thurman's observation that additional portions of the DOC are complex and polymeric but nonhumic in terms of XAD-8 retention highlight the fact that the definition of aquatic humic material is entirely operational. Future research will undoubtedly cast more light on important geographic and seasonal variations in both humic and nonhumic organic carbon fluxes. The development of nondestructive separation and concentration methods which account for all dissolved carbon forms is most fundamental to maximizing the value of this future research.

STRUCTURES OF AQUATIC HUMIC MATERIALS

Most of our knowledge of the structural characteristics of aquatic humic materials has come from degradative methods [8,21] which have been patterned after similar work on closely related, natural products such as lignin and soil organic matter [22,23].

Liao et al. [21] recently reported an extensive investigation of the NaOH hydrolysis and $KMnO_4$ oxidation degradation products of the aquatic and fulvic acid fractions of two geographically separated surface waters. The maximum yield of products identified by GC/MS was approximately 12 to 17 wt % of starting material for base hydrolysis and 20 to 25 wt % for alkaline $KMnO_4$ oxidation. Thus, slightly less than one-third of the original material was accounted for among the structural categories indicated in Table III. The dominant oxidative products identified for both humic and fulvic acid fractions were benzenecarboxylic acids containing three or more carboxyl functional groups. In addition, both of these fractions produced relatively high yields of oxalic, malonic and succinic acids. Fulvic acid fractions, however, produced lower oxidative yields of monobasic aliphatic acids and longer-chain (C_8–C_{15}) dibasic acids.

The fact that the majority of starting material was not identified makes it difficult to generalize about the original macromolecular structures of aquatic humic or fulvic material. It can be said, however, that the presence of both aromatic and aliphatic structures is indicated in the data of Liao et al., as is the presence of oxygen in ring-bound aliphatic side-chains (carboxyphenylglyoxylic acids) and in heterocyclic rings (furancarboxylic acids). It is attractive to assume that the predominance of the aromatic polycarboxylic acidity after $KMnO_4$ reaction is due to cleavage of interaromatic carbon to carbon linkages since the base hydrolysis yield of these polyacids is much lower. If this is true, the data of Liao et al. suggest that the aliphatic chains connecting aromatic rings in their aquatic humic and fulvic material are relatively short (C_2–C_4) to account for the predominance of these dibasic acids in product mixtures. Some (less than 3%) of the aromatic acids must be bound originally in ester linkages to account for the base hydrolysis yields and some of the aliphatic and/or aromatic carboxyl groups must be unbound to account for the well documented acidity of aquatic humic material [8,24, 25]. Unfortunately, the present literature does not contain reports of degradative studies of sufficient scope to permit comment on temporal and spatial variations in structural composition.

An inherent weakness in degradative experiments is the opportunity for artifact formation. This issue is explored by Reuter et al. (Chapter 5, this volume). These authors subjected aquatic humic substances from the Satilla River in southeastern Georgia to $KMnO_4$ oxidation under different conditions designed to protect the humic materials during oxidative degradation. Their data suggest that overall yields are dramatically improved if humic substances are methylated prior to the oxidation step, which in turn is conducted in a homogeneous solution of an aprotic solvent (crown ether in CH_2Cl_2). In addition to improved yields, the qualitative composition of the degradation mixture was shifted toward dominance of short-chain dibasic acids (primarily oxalic acid).

It is difficult to compare these findings to those of Liao et al. [21] since different water sources were involved. Furthermore, Liao's quantitative data is based upon use of internal standards, whereas Reuter's is expressed

Table III. Summary Results of $KMnO_4$ Oxidation and NaOH Hydrolysis of Aquatic Humic Acids and Fulvic Acids from Black Lake and Lake Drummond. Yields (mg) Resulted from 1.0 g of Starting Humic Samples [21][a]

Components	Black Lake				Lake Drummond			
	$KMnO_4$ HA	Oxidation FA	NaOH HA	Hydrolysis FA	$KMnO_4$ HA	Oxidation FA	NaOH HA	Hydrolysis FA
Benzenecarboxylic acid methyl esters (29 compounds)	142.71	127.77	5.36	5.56	110.71	104.14	2.96	4.03
Furancarboxylic acid methyl esters (5 compounds)	13.61	9.98	0.14	0.16	21.10	26.64	0.36	0.38
Carboxyphenylglyoxylic acid methyl esters (8 compounds)	4.49	4.70	0.00	0.00	5.77	5.61	0.00	0.00
Aliphatic monobasic acid methyl esters (14 compounds)	4.07	0.82	0.72	2.49	14.84	6.14	0.54	1.99
Aliphatic dibasic acid methyl esters (14 compounds)	47.57	56.48	10.62	4.57	98.06	98.07	11.30	5.66
Aliphatic tribasic acid methyl esters (5 compounds)	0.62	1.47	0.29	0.39	4.94	4.52	0.45	0.51
Sum	213.07	201.22	17.13	13.17	255.42	245.12	15.61	12.57
Identified percentage of total GC peak area	88%	91%	71%	73%	86%	88%	63%	69%

[a]FA = fulvic acid; HA = humic acid. Reprinted with permission of *Environmental Science and Technology*.

only as "relative yield" within each structural category. These caveats notwithstanding, the results of these two investigations amount to the most thorough experimental evidence yet available on the structural composition of aquatic humic substances and establish clear research directions for future work.

REACTIONS OF HUMIC MATERIALS

Two specific reaction properties of aquatic humic materials have been of intensive interest to research workers. The first property, metal ion interactions, has been recognized for many years as being of ecologic significance as well as a means of elucidating reactive functional group properties. The second property, reactivity with dilute aqueous chlorine (HOCl), has been recognized much more recently. Public health concern over the hazards of ingestion of chlorinated humic reaction products has stimulated research in aquatic humic chemistry generally, and has resulted in new insights into the nature of these complex natural products.

Metal–Ion Interactions

Advances in the last few decades in soil and water chemistry, as well as the development of analytical techniques such as anodic stripping voltammetry and atomic absorption spectrophotometry, have extended the sensitivity and accuracy of trace metal data of interest to biologists and chemists. The role of humic materials as a complicating factor in the uptake and utilization of metal ions has been long recognized. With increasing knowledge of the structural features of humic materials and more available trace metal data, recent research has focused on the relation of chemical environmental measurements to the measurements of responses of biological systems.

Early work on metal ion–organic interactions was based on conceptual models which recent research is beginning to question. The new directions depart from the idea that there are equivalent, noninteracting sites for bonding, and suggest the concept that the strength of binding varies with the degree of metal loading on humic materials. This trend has appeared in the literature in the last few years, and is addressed by authors in this volume (Chapters 11, 14, 17 and 22). These authors consider numerical values of binding constants distributed over several orders of magnitude, and their work clearly establishes the need for definition of the nature of this distribution.

Another recent trend in metal ion–organic interaction research has been efforts to obtain chemical measures of "availability" by measuring the dissociation rate of metals bound to humic materials. Biologists and chemists may never agree on the meaning of the word "availability," but present research is at least moving in such a way that an information base is being established that may provide for correlation of rates of specific chemical reactions with rates of growth or the inhibition of growth.

Research in this important area is of course heavily influenced by developments in our fundamental understanding of aquatic and terrestrial humic chemistry. New information on chemical structures and functionalities, especially those of strong binding potential, remains the principal determinant of research directions in this field. Environmental application of these findings also depends on information on how these properties vary spatially and temporally in nature. Humic molecules are polydisperse and molecular size and shape may bear upon the degree of binding of metals, the number of sites available or the ease of metal uptake biologically. In many environments metal–ion concentrations are quite low compared to humic concentrations, and research on relatively high humic-to-metal concentration ratios needs to be done. Chemists must consider the applicability of their data to environmental systems and not assume that measurements made at high metal loading are significant in the actual environment. In addition, bonding at extremely low metal saturation may take place at sites in humic materials that include relatively minor electron donor sites (N, S) rather than at oxygenated sites. Analytical developments in the area of fluorescence, ion-selective electrodes and separation chromatography may yield new methods to detect bonding at extremely low saturation and would be extremely important to this area of inquiry.

Our understanding of the nature of environmental effects associated with metal ion–organic interactions is still primitive. Examples of both positive and negative effects on the uptake rate of metal ions by living organisms can be found in the literature. The so-called "Heidermoore" illness to plants and organisms is thought to result from copper deficiency due to humic interactions. There is also ample evidence that organic ligands including humic substances lower metal ion activity, which inhibits phytoplanktonic growth [26–29]. Metal complexation thus may act in a positive way by decreasing acute toxic effects of heavy metals on aquatic organisms. The same relatively beneficial effect may extend to mammals exposed to heavy metals in the presence of humic materials, although this effect as well as the effects of chronic exposures to metal-bound humics is still uncertain. On the other hand, George and Coombs [30] suggest that accumulation of cadmium by *Mytilus edulis* is facilitated by the presence of humic material. No general conclusions can be made about the effect of metal ion–organic interactions because of the different organisms involved in these various research reports.

Disinfectant Reactions

Since the mid-1970s a variety of research reports [1–3] have established that dilute aqueous chlorine is an effective degrader of humic substances. At first, it was thought that chlorine produced only chloroform and its sister trihalomethanes ($CHCl_2Br$, $CHClBr_2$ and $CHBr_3$ if Br^- is present). Subsequent research has shown that this perception is limited due to an analytical bias for volatile reaction products in purge and trap techniques, and that a variety of materials can be found in humic/chlorine product mixtures [31].

At present only limited generalizations are possible concerning the chemistry of these reactions. First, the yield of $CHCl_3$ on a carbon basis from aquatic humic material is relatively small. Rook [1] and Babcock and Singer [12] have measured yields in the range 3 to 16 μg $CHCl_3$ (as C)/mg TOC, although actual yields vary with the fraction exposed (humic or fulvic acid), contact time and pH. These data suggest that an excess of chlorine produces approximately 1.0 mol $CHCl_3$ per 1000 molecular weight units (or 1.0 mol of $CHCl_3$ for every 42 carbon units, assuming elemental composition of 50% C). Second, all gel filtration size fractions of the aquatic humic substances are capable of producing $CHCl_3$, although some evidence suggests that only the smallest size ranges produce heavily brominated trihalomethanes [32].

Third, it is widely accepted that both the measurable chlorine demand exerted by humic substances exposed to chlorine and the total organic halogen produced by the exposure exceed that which can be accounted for by the production of trihalomethanes [15]. A partial explanation of these observations appears from an examination of the principal classes of nontrihalomethane reaction products that have been identified. These classes include four major classes of carboxylic acids: chlorinated aliphatic acids, unchlorinated aliphatic monocarboxylic acids, unchlorinated aliphatic dicarboxylic acids and unchlorinated aromatic acids. The production of unchlorinated oxidation products accounts for high chlorine demand values relative to measures of halogenated organic products, and establishes the fact that chlorine is an effective oxidative degrader of aquatic humic substances. The tendency for chlorine to produce acidic degradation products from aquatic humic substances is quite similar to the effect of alkaline $KMnO_4$ [21].

Unlike permanganate degradation, chlorine product mixtures are dominated by chlorinated acids, principally dichloroacetic acid, trichloroacetic acid, α,α-dichloropropionic acid, chlorosuccinic acid, dichlorosuccinic and dichloromaleic acid [33–35]. It may be worth noting that the principal chlorine oxidation products are relatively short chain length (C_2–C_4) mono- and dibasic acids, and that this is very similar to the results of $KMnO_4$ degradation studies.

$KMnO_4$ DEGRADATION STUDIES

Two areas are critically important with regard to the development of our knowledge of chlorine humic reactions. The first is the need to understand these reactions mechanistically, which would permit answers to questions such as:

- Are the short-chain chlorinated products independently produced via simultaneous reactions with independent sections of the humic macromolecule?
- Are some of the short-chain chlorinated products the result of chemical instability in aqueous solution of a primary reactions product, i.e., does trichloroacetic acid decompose to produce chloroform?
- What fraction of the total organic halogen (TOX) is accounted for by these identified products?
- What changes occur in TOX measurements with time and how are the product distributions affected?

The second critical area for additional research is the determination of the health effects of these products and product mixtures. Human consumers of chlorinated drinking water ingest many products other than chloroform; yet, the only specified maximum contaminant level in U.S. drinking water at present is for the trihalomethane category, which is based on the effects of chloroform ingestion on laboratory test animals.

REFERENCES

1. Rook, J. J. "Formation of Haloforms During Chlorination of Natural Waters," *Water Treatment Exam.* 23(2):234-243 (1974).
2. Bellar, T. A., J. J. Lichtenberg and R. C. Kroner. "The Occurrence of Organohalides in Chlorinated Drinking Water," *J. Am. Water Works Assoc.* 66(12):703-706 (1974).
3. Symons, J. M., T. A. Bellar, J. K. Carswell, J. DeMarco, K. L. Krapp, G. G. Robeck, D. R. Seeger, C. J. Slocum, B. L. Smith and A. A. Stevens. "National Organic Reconnaissance Survey for Halogenated Organics," *J. Am. Water Works Assoc.* 67(1):634-647 (1975).
4. Garrels, R. M., F. T. MacKenzie and C. Hunt. *Chemical Cycles and the Global Environment* (Los Altos, CA: William Kaufmann, Inc., 1973), Chapter 6.
5. *Carbon Dioxide Effects Research and Assessment Program.* Flux of Organic Carbon by Rivers to the Ocean. Prepared by the Committee on Flux of Organic Carbon to the Ocean (G. E. Likens, Chairman), Division of Biol. Sciences, National Research Council, published by U.S. Dept. of Energy, Office of Energy Research, Washington, DC 20545, Reference: CONF-8009140 UC 11 (April 1981).
6. Meybeck, M. "River Transport of Organic Carbon to the Ocean," in *Carbon Dioxide Effects Research and Assessment Program.* Flux of Organic Carbon by Rivers to the Ocean. Prepared by the Committee on Flux of Organic Carbon to the Ocean (G. E. Likens, Chairman),

Division of Biol. Sciences, National Research Council, published by U.S. Dept. of Energy, Office of Energy Research, Washington, DC 20545, Reference: CONF-8009140 UC 11 (April 1981), p. 219.
7. Steelink, C. "Humates and Other Natural Organic Substances in the Aquatic Environment," *J. Chem. Ed.* 54:599 603 (October 1977).
8. Christman, R. F., and M. Ghassemi. "Chemical Nature of Organic Color in Water," *J. Am. Water Works Assoc.* 58(6):723-741 (1966).
9. Whittaker, R. H., and G. H. Likens. *Carbon and the Biosphere*, U.S. Atomic Energy Commission, AEC Symposium Series, No. 30, p. 281 (August 1973).
10. Rook, J. J. "Haloforms in Drinking Water," *J. Am. Water Works Assoc.* 58(3):168-172 (1976).
11. Stevens, A. A., C. J. Slocum, D. R. Seeger and G. Robeck. "Chlorination of Organics in Drinking Water," *J. Am. Water Works Assoc.* 68:615 (November 1976).
12. Babcock, D. B., and P. C. Singer. "Chlorination and Coagulation of Humic and Fulvic Acids," *J. Am. Water Works Assoc.* 71:149 (March 1979).
13. Colthurst, J. M., and P. C. Singer. "Removing Trihalomethane Precursors by Permanganate Oxidation and Manganese Dioxide Adsorption," *J. Am. Water Works Assoc.* 74:78 (February 1982).
14. Singer, P. C., J. J. Barry, G. M. Panel and A. E. Scrivner. "Trihalomethane Formation in North Carolina Drinking Waters," *J. Am. Water Works Assoc.* 73:392 (August 1981).
15. Morris, J. C., and B. Baum. "Precursors and Mechanisms of Haloform Formation in the Chlorination of Water Supplies," in *Water Chlorination: Environmental Impact and Health Effects, Vol. 2*, R. L. Jolley, H. Gorchev and D. H. Hamilton, Jr., Eds. (Ann Arbor, MI: Ann Arbor Science Publishers, Inc., 1978), pp. 29-48.
16. Hoehn, R. C., C. W. Randall, F. A. Bell, Jr. and P. T. B. Shaffer. "Trihalomethanes and Viruses in a Water Supply," *J. Environ. Eng. Div., ASCE* 103(EES):803-814 (1977).
17. Norwood, D. L., J. D. Johnson and R. F. Christman. "Reactions of Chlorine with Selected Aromatic Models of Aquatic Humic Material," *Environmental Science and Technology* 14:187 (February 1980).
18. Aiken, G. R., E. M. Thurman and R. L. Malcolm. "Comparison of XAD Macroporous Resins for the Concentration of Fulvic Acid from Aqueous Solution," *Anal. Chem.* 51:1799 (September 1979).
19. Thurman, E. M., and R. L. Malcolm. "Preparative Isolation of Aquatic Humic Substances," *Environ. Sci. Tech.* 15:463 (April 1981).
20. Thurman, E. M. Personal communication (March 1982).
21. Liao, W., R. F. Christman, J. D. Johnson, D. S. Millington and J. R. Hass. "Structural Characterization of Aquatic Humic Material," *Environ. Sci. Tech.* 16:403-410 (1982).
22. Schnitzer, M., and S. U. Khan. *Humic Substances in the Environment* (New York: Marcel Dekker, Inc., 1972).
23. Green, G., and C. Steelink. "Structure of Soil Humic Acid," *J. Org. Chem.* 27:170-174 (1962).
24. Shapiro, J. "Chemical and Biological Studies on the Yellow Organic Acids of Lake Water," *Limnol. Oceanog.* 2:161 (1957).
25. Gjessing, E. *Physical and Chemical Characteristics of Aquatic Humus* (Ann Arbor, MI: Ann Arbor Science Publishers, Inc., 1976).

26. Mantoura, R. F. C., A. Dickson and J. P. Riley. "The Complexation of Metals with Humic Materials in Natural Waters," *Estuar. Coast. Mar. Sci.* 6:387-408 (1978).
27. Benes, P., E. T. Gjessing and E. Steinnes. "Interaction between Humus and Trace Elements in Fresh Water," *Water Res.* 10:711-716 (1976).
28. Hart, B. T., and S. H. R. Davies. "Speciation of Fe, Cd, Cu, Pb and Zn in the Yarra River and Estuary, Australia," in *International Conference Management and Control of Heavy Metals in the Environment* (London: CEP Consultants, 1979), pp. 466-471.
29. Gjessing, E. "The Effect of Aquatic Humus on the Biological Availability of Cadmium," *Arch. Hydrobiol.* 91:144-149 (1981).
30. George, S. G., and T. H. Coombs. "The Effect of Chelating Agents on the Uptake and Accumulation of Cadmium by *Mytilus edulis*," *Marine Biol.* 39:261-268 (1977).
31. Christman, R. F., J. D. Johnson, F. K. Pfaender, D. L. Norwood, M. R. Webb, J. R. Hass and M. J. Bobenreith. "Chemical Identification of Aquatic Chlorination Products," in *Water Chlorination: Environmental Impact and Health Effects, Vol. 3*, R. L. Jolley, W. A. Brungs and R. B. Cumming, Eds. (Ann Arbor, MI: Ann Arbor Science Publishers, Inc., 1980).
32. Schnoor, J. L., J. L. Nitzschke, R. D. Lucas and J. N. Veenstra. "Trihalomethane Yields as a Function of Precursor Molecular Weight," *Environ. Sci. Tech.* 13:1134-1138 (1979).
33. Norwood, D. L., J. D. Johnson, R. F. Christman and D. S. Millington. "Chlorination Products from Aquatic Humic Material at Neutral pH," Fourth Water Chlorination Conference, Pacific Grove, CA (1981).
34. Quimly, B. D., M. F. Delaney, P. C. Uden and R. M. Barnes. "Determination of the Aqueous Chrloination Products of Humic Substances by Gas Chromatography with Microwave Emission Detection," *Anal. Chem.* 52:259-263 (1980).
35. Miller, J. W., P. C. Uden and R. M. Barnes. "Determination of Trichloroacetic Acid at the Part-per-Billion Level in Water by Precolumn Trap Enrichment Gas Chromatography with Microwave Plasma Emission Detection," *Anal. Chem.* 54:485-488 (1982).

INDEX